Value Added Products From Bioalgae Based Biorefineries: Opportunities and Challenges

Shailendra Kumar Arya • Madhu Khatri •
Gursharan Singh
Editors

Value Added Products From Bioalgae Based Biorefineries: Opportunities and Challenges

Springer

Editors
Shailendra Kumar Arya
University Institute of Engineering and Technology (UIET), Panjab University
Chandigarh, India

Madhu Khatri
Department of Biotechnology
Panjab University
Chandigarh, India

Gursharan Singh
Lovely Professional University
Phagwara, India

ISBN 978-981-97-1661-6 ISBN 978-981-97-1662-3 (eBook)
https://doi.org/10.1007/978-981-97-1662-3

This Springer imprint is published by the registered company Springer Nature Singapore Pte Ltd.
The registered company address is: 152 Beach Road, #21-01/04 Gateway East, Singapore 189721, Singapore

Foreword

It is with great pleasure and enthusiasm that I present this comprehensive volume on *Value Added Products from Bioalgae Based Biorefineries: Opportunities and Challenges*. This book represents a culmination of extensive research, innovative insights, and collaborative efforts dedicated to exploring the untapped potential of bioalgae in the context of biorefineries. As the editor of this compilation, I am honoured to bring together a diverse group of scholars and experts who have contributed their knowledge and expertise to shed light on the opportunities and challenges associated with this fascinating field.

The world is witnessing a paradigm shift towards sustainable and eco-friendly practices, and bioalgae have emerged as a promising candidate in the quest for renewable resources. Harnessing the power of these microscopic organisms holds the key to addressing crucial global challenges such as climate change, energy security, and resource depletion. This book delves into the multifaceted aspects of bioalgae-based biorefineries, offering readers a comprehensive understanding of the opportunities that lie within this burgeoning domain.

The journey through the pages of this book begins with a fundamental exploration of bioalgae, covering their taxonomy, physiology, and ecological significance. This serves as a crucial foundation for readers to grasp the intricacies of their potential applications in biorefineries. We then progress into the heart of the matter—the value-added products that can be derived from bioalgae through innovative biorefinery processes. From biofuels and bioenergy to high-value chemicals and pharmaceuticals, the chapters in this book provide a detailed examination of the diverse array of products that can be synthesized, paving the way for a sustainable and circular economy.

One of the distinguishing features of this compilation is its focus on the challenges and bottlenecks that researchers and practitioners may encounter in the pursuit of bioalgae-based biorefineries. Acknowledging the obstacles is crucial for devising effective strategies to overcome them. Issues such as scalability, economic viability, and technological constraints are addressed with a pragmatic approach, fostering a holistic understanding of the field.

As deeply immersed in the realms of bioalgae research, I am acutely aware of the collaborative and interdisciplinary nature of this field. The contributors to this volume bring forth a wealth of knowledge from various disciplines, including biology, engineering, chemistry, and environmental science. Their collective expertise contributes to the richness and diversity of perspectives presented in this book.

I would like to express my heartfelt gratitude to all the authors who have dedicated their time and expertise to contribute to this endeavour. Their commitment to advancing the understanding of bioalgae-based biorefineries is commendable, and I am confident that their work will inspire further research and innovation in this field.

In conclusion, *Value Added Products from Bioalgae Based Biorefineries: Opportunities and Challenges* is intended to serve as a valuable resource for researchers, students, and industry professionals seeking to explore the vast potential of bioalgae in the context of sustainable biorefineries. I hope that this compilation sparks new ideas, fosters collaboration, and contributes to the ongoing dialogue surrounding the development of a more sustainable and resilient future.

Sudarshan Sahu

Chandigarh, India

Preface

In the dynamic landscape of sustainable technologies and green innovations, the exploration of bioalgae-based biorefineries represents a significant leap towards an eco-friendlier and resource-efficient future. The book you are about to delve into, *Value Added Products from Bioalgae Based Biorefineries: Opportunities and Challenges*, is a comprehensive and timely contribution to the ever-evolving field of biotechnology.

The utilization of bioalgae, often referred to as the "green gold", has gained momentum as an invaluable resource in the pursuit of renewable energy, sustainable agriculture, and bioproduct development. This volume serves as a beacon, guiding readers through the intricate realms of bioalgae-based biorefineries, offering a nuanced understanding of the opportunities they present and the challenges they pose.

As we stand at the crossroads of environmental stewardship and technological innovation, the authors of this book navigate through the intricacies of harnessing the potential of bioalgae for the production of value-added products. The synergy between the academic rigour and industrial applicability evident in the chapters reflects a collaborative effort to bridge the gap between theory and practice.

The journey begins with an exploration of the foundational principles of bioalgae-based biorefineries, delving into the diversity of algae species, their cultivation techniques, and the underlying biochemical processes. Moving forward, the book unravels the vast array of value-added products that can be derived from algae, spanning biofuels, biofertilizers, biopolymers, and nutraceuticals, among others. Each chapter meticulously dissects the potential applications, elucidating the economic and environmental benefits that arise from the sustainable utilization of bioalgae.

Yet, as with any pioneering field, challenges are inevitable. This book does not shy away from addressing the obstacles and uncertainties that come hand-in-hand with the integration of bioalgae-based biorefineries into mainstream industries. Through insightful analyses and case studies, the authors shed light on the practical hurdles and propose innovative solutions, fostering a constructive dialogue among researchers, policymakers, and industry professionals.

In an era where the urgency to transition towards a bio-based, circular economy is paramount, *Value Added Products from Bioalgae Based Biorefineries: Opportunities and Challenges* emerges as a timely resource. It beckons scientists, engineers, entrepreneurs, and policymakers to engage in a collective effort to harness the full potential of bioalgae, ensuring a sustainable and resilient future for generations to come.

As you embark on this intellectual journey, may the pages of this book inspire new perspectives, fuel innovation, and contribute to the transformative journey towards a greener, more sustainable world.

Chandigarh, India — Shailendra Kumar Arya
Chandigarh, India — Madhu Khatri
Phagwara, Punjab, India — Gursharan Singh

Contents

Editors and Contributors

About the Editors

Shailendra Kumar Arya presently holds the position of Professor at the Department of Biotechnology Engineering within the University Institute of Engineering & Technology, Panjab University, Chandigarh, India. He attained his doctoral degree in Biotechnology from the esteemed Indian Institute of Technology, Banaras Hindu University, Varanasi, in 2010. With an affiliation dating back to 2006, Dr. Arya has been a dedicated member of Panjab University. His primary focus lies in the realm of Enzyme Engineering and Bioremediation, showcasing a profound expertise in enzymes, wastewater treatment, and sustainable development. Dr. Arya boasts an extensive publication record, contributing over 100 research articles to prestigious international journals. Furthermore, he has actively engaged in numerous international and national conferences, where he has been featured as a distinguished speaker. Leading a dynamic research group, Dr. Arya mentors both PhD and Master's students, guiding them in diverse projects encompassing Enzyme Engineering, Wastewater Treatment, Algal Studies, and Bioremediation. The collaborative efforts of his research team contribute significantly to advancing knowledge and innovation in these critical areas.

Madhu Khatri is currently working as Assistant Professor at the Department of Biotechnology, UIET, Panjab University. She received her PhD (Biomedical Engineering and Biotechnology) degree in 2012 from the University of Massachusetts, USA. During her PhD and postdoctoral training, she worked in close collaboration with Harvard School of Public Health and Northeastern University, Boston, USA. She joined Panjab University in 2014. She received the prestigious Wellcome trust/DBT IA fellowship in 2015. Dr. Khatri works in the field of Environmental Nanobiotechnology. She has expertise in the area of nanomaterial synthesis and studying their various application and environmental effects. She has published more than 60 research articles in reputed international journals and participated in various national and international conferences as a speaker. Her research group includes PhD and ME students working on interdisciplinary projects, such as in the areas of nanotoxicology, bioplastics, water purification, biosensors, and bioremediation.

Gursharan Singh is working as an Associate Professor in the Department of Medical & Laboratory Sciences, Lovely Professional University, India. He worked at UIET, Biotechnology Branch, Panjab University, Chandigarh (Sector 25-B), as an Assistant Professor (Adhoc faculty) from February 2013 to June 2017. He received his PhD from Biotechnology Department, Panjab University, Chandigarh (2009, July), and postdoctorate from South Dakota School of Mines & Technology, Rapid City, USA (January 2010 to January 2011). He also did postdoctorate from Agriculture & Agrifood, Canada, Quebec (June 2011 to June 2012). He has published more than 90 articles in scientific journals of repute. Further, his name is present in the list of 2.0% scientists of the world in the area of biotechnology for the last two consecutive years, 2022 and 2023, data released by the Stanford University, USA.

Contributors

Andhere Animish Marine Biotechnology and Bioproducts Laboratory, School of Biosciences and Technology, Vellore Institute of Technology, Vellore, Tamil Nadu, India

Shailendra Kumar Arya University Institute of Engineering and Technology (UIET), Panjab University, Chandigarh, India

Ram Naresh Bharagava Department of Environmental Microbiology (DEM), School of Earth and Environmental Sciences (SEES), Babasaheb Bhimrao Ambedkar University, Lucknow, Uttar Pradesh, India

Nivedita Chaudhary Department of Environmental Science, School of Earth Sciences, Central University of Rajasthan, Ajmer, Rajasthan, India

Brígida Maria Villar da Gama Technology Center, Federal University of Alagoas, Maceió, Alagoas, Brazil

Albanise Enide da Silvaa Technology Center, Federal University of Alagoas, Maceió, Alagoas, Brazil

João Victor Oliveira Nascimento da Silva Technology Center, Federal University of Alagoas, Maceió, Alagoas, Brazil

Carlos Eduardo de Farias Silva Technology Center, Federal University of Alagoas, Maceió, Alagoas, Brazil

M. Devaprakash Department of Molecular Microbiology, School of Biotechnology, Madurai Kamaraj University, Madurai, Tamil Nadu, India

Anuradha Devi Department of Environmental Microbiology (DEM), School of Earth and Environmental Sciences (SEES), Babasaheb Bhimrao Ambedkar University, Lucknow, Uttar Pradesh, India

Luiz Fernando R. Ferreira Postgraduate Program in Genomic Sciences and Biotechnology, Catholic University of Brasília, Brasília, Brazil

João Victor Ferro Technology Center, Federal University of Alagoas, Maceió, Alagoas, Brazil

Rizul Gautam Academy of Scientific and Innovative Research (AcSIR), Ghaziabad, India

Biomedical Applications Group, CSIR Central Scientific Instruments Organisation, Chandigarh, India

M. A. Jayasri Marine Biotechnology and Bioproducts Laboratory, School of Biosciences and Technology, Vellore Institute of Technology, Vellore, Tamil Nadu, India

Rahul Kadam Department of Advanced Energy Engineering, Chosun University, Gwangju, Republic of Korea

Sachin Kajla Environmental Research Group, R&D and Scientific Services Department, Tata Steel Limited, Jamshedpur, Jharkhand, India

Manvinderpreet Kaur Department of Biotechnology Engineering, University Institute of Engineering & Technology, Panjab University, Chandigarh, India

Sandeep Kaur School of Allied Medical Sciences, Lovely Professional University, Phagwara, Punjab, India

Jyoti Kaushal Department of Biotechnology Engineering, University Institute of Engineering & Technology, Panjab University, Chandigarh, India

Madhu Khatri Department of Biotechnology, Panjab University, Chandigarh, India

Chaitanya Kumar Centre for Nano-science and Nano-technology, Panjab University, Chandigarh, India

Priya Kumari Environmental Research Group, R&D and Scientific Services Department, Tata Steel Limited, Jamshedpur, Jharkhand, India

Ritu Kumari Environmental Research Group, R&D and Scientific Services Department, Tata Steel Limited, Jamshedpur, Jharkhand, India

Shiv Kumar Department of Microbiology, Guru Gobind Singh Medical College, Baba Farid University of Health Sciences, Faridkot, Punjab, India

Department of Microbiology, RT-PCR Laboratory, Civil Hospital, Kapurthala, Punjab, India

Vikas Kumar Patel Environmental Research Group, R&D and Scientific Services Department, Tata Steel Limited, Jamshedpur, Jharkhand, India

Plash Kunj Department of Biotechnology Engineering, University Institute of Engineering & Technology, Panjab University, Chandigarh, India

Larissa Rodrigues Macário Technology Center, Federal University of Alagoas, Maceió, Alagoas, Brazil

Iradat Hussain Mafat Department of Chemical Engineering, Indian Institute of Petroleum and Energy, Vishakhapatnam, Andhra Pradesh, India

Josimayra Almeida Medeiros Technology Center, Federal University of Alagoas, Maceió, Alagoas, Brazil

Sikandera I. Mulla Department of Biochemistry, School of Applied Sciences, REVA University, Bangalore, Karnataka, India

Jagriti Nagar Department of Microbiology, Lovely Professional University, Phagwara, India

Sridhar Palla Department of Chemical Engineering, Indian Institute of Petroleum and Energy, Vishakhapatnam, Andhra Pradesh, India

Vaniksha Pal Department of Microbiology, Lovely Professional University, Phagwara, Punjab, India

Jungyu Park Department of Advanced Energy Engineering, Chosun University, Gwangju, Republic of Korea

Shubham Kumar Patial Academy of Scientific and Innovative Research (AcSIR), Ghaziabad, India

Materials Science & Sensor Applications Group, CSIR Central Scientific Instruments Organisation, Chandigarh, India

Shailesh Kumar Patidar Department of Environmental Science, School of Earth Sciences, Central University of Rajasthan, Ajmer, Rajasthan, India

K. Ponmurugan Department of Botany and Microbiology, College of Science, King Saud University, Riyadh, Saudi Arabia

Rohit Rai Faculty of Applied Medical Sciences, Lovely Professional University, Phagwara, Punjab, India

Sandeep Kaur Saggu Department of Biotechnology, Kanya Maha Vidhyalaya, Jalandhar, Punjab, India

Sudarshan Sahu Department of Biotechnology Engineering, University Institute of Engineering & Technology, Panjab University, Chandigarh, India

Gabrielly Karla Silva Santos Technology Center, Federal University of Alagoas, Maceió, Alagoas, Brazil

Christina Saran Department of Environmental Microbiology (DEM), School of Earth and Environmental Sciences (SEES), Babasaheb Bhimrao Ambedkar University, Lucknow, Uttar Pradesh, India

Ganesh Dattatraya Saratale Department of Food Science and Biotechnology, Dongguk University, Seoul, Republic of Korea

Anil Kumar Sarma Chemical Conversion Division, Sardar Swaran Singh National Institute of Bio-Energy (An Autonomous Institute of MNRE, Government of India), Kapurthala, Punjab, India

Monika Sharma Department of Biotechnology Engineering, University Institute of Engineering & Technology, Panjab University, Chandigarh, India

Shalini Sharma Department of Biotechnology Engineering, University Institute of Engineering & Technology, Panjab University, Chandigarh, India

R. Shyam Kumar Department of Biotechnology, Kamaraj College of Engineering and Technology, Virudhunagar, Tamil Nadu, India

Gursharan Singh Lovely Professional University, Phagwara, Punjab, India

Rattandeep Singh School of Bioengineering and Biosciences, Lovely Professional University, Phagwara, Punjab, India

Suman Singh Academy of Scientific and Innovative Research (AcSIR), Ghaziabad, India

Materials Science & Sensor Applications Group, CSIR Central Scientific Instruments Organisation, Chandigarh, India

N. Sivakumar Department of Molecular Microbiology, School of Biotechnology, Madurai Kamaraj University, Madurai, Tamil Nadu, India

Dadi Venkata Surya Department of Chemical Engineering, Pandit Deendayal Energy University, Gandhinagar, Gujarat, India

Naziya Tarannum Department of Environmental Science, School of Earth Sciences, Central University of Rajasthan, Ajmer, Rajasthan, India

R. Thirumalaivasan Department of Molecular Microbiology, School of Biotechnology, Madurai Kamaraj University, Madurai, Tamil Nadu, India

Uplabdhi Tyagi University School of Chemical Technology, Guru Gobind Singh Indraprastha University, Delhi, India

Punam Vishwakarma Faculty of Applied Medical Sciences, Lovely Professional University, Phagwara, Punjab, India

Introduction to Algae-Based Biorefineries in the Context of Sustainable Development

1

Plash Kunj, Sudarshan Sahu, and Shailendra Kumar Arya

Contents

P. Kunj · S. Sahu
Department of Biotechnology Engineering, University Institute of Engineering and Technology, Panjab University, Chandigarh, India

S. K. Arya (✉)
University Institute of Engineering and Technology (UIET), Panjab University, Chandigarh, India
e-mail: skarya@pu.ac.in

S. K. Arya et al. (eds.), *Value Added Products From Bioalgae Based Biorefineries: Opportunities and Challenges*,
https://doi.org/10.1007/978-981-97-1662-3_1

Abstract

Algae has proven to be the potential source of proteins, lipids and carbohydrates. Their unique chemical characteristics and the interaction with the living tissues, i.e. bioactive potential, make it an ideal choice for different industries such as nutraceuticals, food, feed and cosmetics. Also, with the latest trends evolving around the globe for the production of different compounds by green methods, algae prove to be the best fit, and to make this green synthesis sustainable the biomass can be reused making it a biorefinery approach for the production process. Algae-based biorefineries promote not only the sustainable and green methods but also the bioeconomy, i.e. it promotes the use of renewable resources from land and sea. This chapter will be focusing on different algae-based biorefineries, their working, design and need along with their advancement in recent times. Also, in this chapter different species of algae and their useful end-products will be discussed.

Keywords

Algae · Biorefinery · Sustainable development

1.1 Introduction

In 2012, the United Nations suggested 17 sustainable development goals (SDGs) that were adopted in 2015 and are to be accomplished by 2030. Sustainable development goals were the replacement to then existing Millennium development goals (MDGs) which were started in 2000 and were in existence till 2015. MDGs focused on poverty, deadly diseases, hunger, education and other development priorities but it did not focus on the economic, political and environmental issues (United Nations Development Programme n.d.). SDGs, as shown in Fig. 1.1, focuses on clean and safe environment, healthy living, social and economic aspects. The reason behind setting up these goals were increasing population, scarcity of resources, lack of policies and inadequate finances, and the aim is to provide future generation a clean, healthy and risk-free environment (Olabi et al. 2023).

Algae are aquatic organisms having diverse forms, ranging from simple single-celled to multicellular organisms. Algae are photosynthetic organisms that can grow in wide range of environment such as fresh water, terrestrial ecosystem and marine. They can tolerate variable parameters such as temperature, light intensity, pH and salinity. Algae are seen to grow in different environments such as bark of trees, sediments, soil, oceans, sea, ponds, rivers and lakes (Olabi et al. 2023). They are being used as a source of fertilisers, food and medicines from a long time but in recent times due to their high growth rates and their ability to produce large amount of lipids they are being used to produce biofuels (Vieira et al. 2022). Algae contains different organic and chemical substances also known as secondary metabolites or bioactive substances which are produced as an end-product of different metabolic processes which are then converted to biofuels such as biodiesel, bioethanol and

Fig. 1.1 Seventeen sustainability goals set by the United Nations

biomethanol that has proven to be a sustainable approach as this production do not compete with resource (Wang et al. 2023). Most commonly algae can be seen as a coupled approach for the wastewater treatment and biofuel production.

Biorefinery can be defined as a facility that integrates multiple biomass conversion processes leading to the production of bio-products. It is similar to traditional process of production in petroleum refineries but instead of producing crude oil biorefineries use biomass sources such as algae, organic waste, agriculture crops and forestry residue. The main goal of biorefinery is to convert biomass into valuable products and in return not exploiting the fossil fuels (Saral et al. 2021).

Bioeconomy can be described as an economic system that uses biological resources, principles and processes to produce a wide range of goods and services in a sustainable manner. In order to create goods that satisfy human needs, it entails harnessing the power of biological systems and processes as well as living things like plants, animals and microbes. The production of food, forestry, fisheries, agriculture and bio-based materials and energy are all included in the bioeconomy (Pacheco-Torgal 2020). Bioeconomy has some key elements such as renewable resources, sustainable practice, biotechnology and innovation, circular economy, diversification of feedstock and contribution to sustainable development.

1.2 Algae and Its Classification

Algae are broadly classified into two groups, micro- and macroalgae, as illustrated in Fig. 1.2. Microalgae are unicellular, microscopic, photosynthetic organisms that can live in fresh water, soil and saltwater; green, blue-green and golden-brown algae fall into this category. Microalgae are photosynthetic eukaryotes or prokaryotic cyanobacteria, sometimes found as nitrogen fixing organisms. They are capable of

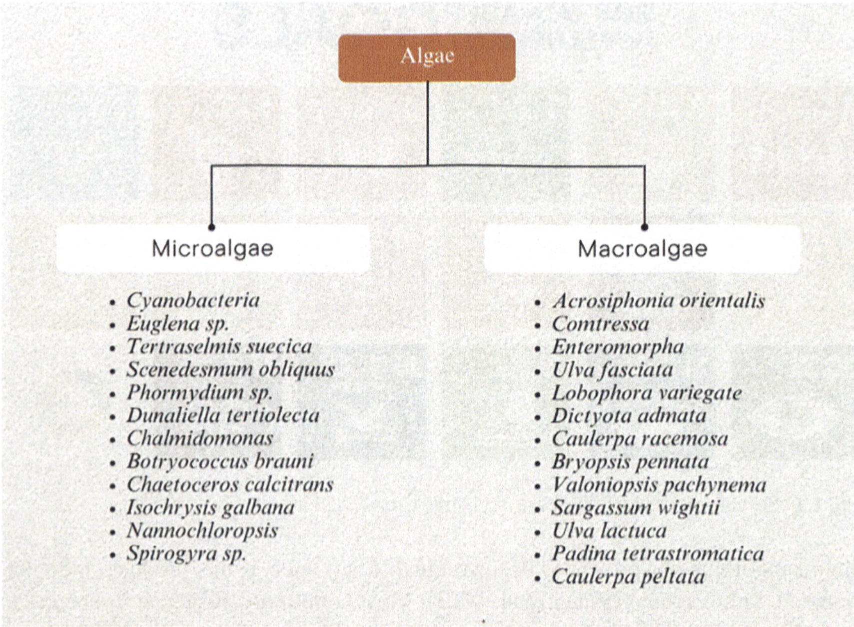

Fig. 1.2 Different types of micro- and macroalgae

carbon fixation and produce large amounts of oxygen. On the other hand, macroalgae are multicellular seaweeds that can be seen in ocean with naked eyes. These macroalgae are capable of filtering and reducing nitrates, nitrites and phosphate level. Brown, green and red algae fall under this category (Unique Benefits of Macroalgae and Microalgae in Agriculture n.d.; Jalilian et al. 2020).

Algae can be classified into eight types depending on their appearance:

1. Bacillariophyta or diatoms
2. Charophyta or stonewort
3. Chlorophyta or green algae
4. Phaeophyta or brown algae
5. Chrysophyta or golden algae
6. Cyanobacteria or blue-green algae
7. Dinophyta or dinoflagellate
8. Rhodophyta or red algae

1.2.1 Bacillariophyta or Diatoms

They are unicellular organisms belonging to the diatom algae. Diatoms appear to be like glass due to their silica wall. Diatoms are essential parts of phytoplankton, which are microscopic organisms that resemble plants and drift in aquatic environments, mostly freshwater and oceans. Some diatoms may be bottom dwellers or adhere to substrates like rocks or other aquatic vegetation, but the majority of diatoms are planktonic, or they float in the water column. Diatoms exhibit their adaptability to various ecological niches by growing on other algae or plants. Bacillariophyta have plastid that is responsible for photosynthesis; this plastid present is brownish in colour due to the presence of chlorophyll a and b along with fucoxanthin, which is a brownish pigment. These algae reproduce asexually through mitosis leading to the two daughter cells, each containing the silica content in the cell wall. In some cases, these algae show symbiotic relationship with protozoa leading to the lack of or modification in their characteristic silica cell wall (University of California Museum of Paleontology n.d.).

1.2.2 Charophyta or Stonewort

They are fresh water algae commonly seen growing in ponds, lakes and slow-flowing rivers. They grow in an anchored form to the thread like substratum rhizoids. These algae consist of shoot extending upwards from the substratum leading to the nods formation which develop into reproductive structure which resembles the structure of mosses. Charophyta are green-coloured algae due to the presence of chlorophyll a and b (Thoughtco.com n.d.).

1.2.3 Chlorophyta or Green Algae

Mostly, Chlorophyta are aquatic but these can be seen growing on tree trunks, snow surface or even in a symbiotic relationship with protozoans, fungi and hydra. Chlorophyta has a variable size range along with the variable colours, mainly green, yellow-green or blackish-green. This variable range of colours is seen due to the presence of carotenoid pigment or higher concentration of chlorophyll (Smithsonian National Museum of Natural History n.d.).

1.2.4 Phaeophyta or Brown Algae

They are entirely marine and usually found in the cold and temperate water shores. They appear to be brown, yellow-brown or beige in colour due to the presence of fucoxanthin (Smithsonian National Museum of Natural History n.d.).

1.2.5 Chrysophyta or Golden Algae

They are unicellular organisms densely found in fresh water and marine environments. They consist of chlorophyll a and c along with a pigment called fucoxanthin which is responsible for golden colour. Brown and golden algae both show similarity in terms of storing food as both store it outside the chloroplast in the form of chrysolaminarin or polysaccharide laminarin. Also, both the algae show structural similarity for the motile cells with unequal flagella (Smithsonian National Museum of Natural History n.d.).

1.2.6 Cyanobacteria or Blue-Green Algae

Although cyanobacteria, due to the absence of membrane-bound nucleus, are classified as bacteria, but due to their appearance they are classified as algae. These algae are capable of changing their appearance depending on their environmental conditions. Cyanobacteria when grown in soil, fresh and salt water with variable temperature range gives blue-green colour. Cyanobacteria can be photosynthetic, nitrogen fixing or even symbionts. They have chlorophyll a which produces free oxygen as a byproduct. Cyanobacteria does not have the organised chloroplast, instead has its own photosynthetic apparatus distributed on the periphery of cytoplasm. These algae have a light-gathering pigment, phycobilin, attached to protein granules that are attached to photosynthetic membrane which is responsible for the exhibiting variety of colours (Hachicha et al. 2022).

1.2.7 Dinophyta or Dinoflagellate

These organisms usually are pelagic unicellular eukaryotic and having two flagella algae, but they can also be benthic. They consist of chlorophyll a and c along with peridinin responsible for both green and brown pigments, respectively. Some of the species falling in this category reproduce asexually and others sexually. These algae have wide range of morphology and size along with the cell coverings called as theca which differentiates them from other algal groups. They show characteristics of both animals and plants as they can show both photosynthetic and non-photosynthetic patterns; they may have cell wall and they are capable of swimming. These algal species turn water reddish-brown leading harm to the sea-life (Smithsonian National Museum of Natural History n.d.).

1.2.8 Rhodophyta or Red Algae

They are mostly marine and multicellular and grow in attachment to the rocks or other algae, but a few species of this algae can grow in fresh water and have unicellular or colonial forms. They lack flagella and have complex structure; usually they have a pit-like connection between cell and their walls having rigid components that are made up of microfibrils and mucilaginous matrix and can be seen in bulk in nature, as shown in Fig. 1.3. They are structurally similar to cyanobacteria having chlorophyll a along with phycobilin (Smithsonian National Museum of Natural History n.d.).

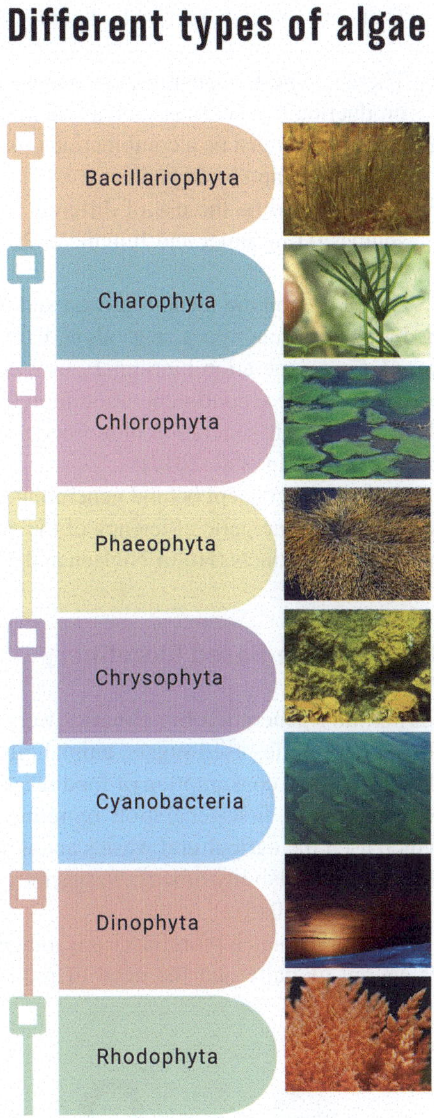

Fig. 1.3 Different types of algae grown in bulk

1.3 Biorefineries

They are defined as a framework where biomass is utilised in an appropriate manner to produce more than one product leading to no harm to the environment and a self-sustainable approach (Narayanan 2024). On a commercial scale, biorefineries in food industry, paper and pulp industry and biofuel industry can be seen. Biorefineries can be used to produce value-added products such as biofuel, bioenergy, biomaterials and biochemicals (Ng et al. 2017).

There are some characteristic features of a biorefinery:

1. There should be generation of energy in gaseous or liquid form along with the production of materials such as chemical, food and feed.
2. Biorefinery must be a combination of several processes such as thermochemical, mechanical and biological.
3. There should be the use of different raw materials from both new and residual sources (Hingsamer and Jungmeier 2019).

Depending on the feedstock used and its generation, biorefineries can be classified into different types, e.g. algae-based, wood-based, palm-based and forest-based. First-generation biorefinery uses edible oil seeds, energy crops, animal fat, food crops etc.; second-generation biorefinery uses lignocellulosic biomass; while third- and fourth-generation biorefineries are algal and other microorganism-based biorefineries (Ng et al. 2017).

The overall concept behind generating biorefinery is to optimise the economical, technical and energetic efficiency of the process for the production of valuable and marketable products (Holm-Nielsen and Ehimen 2014).

1.3.1 Algae-Based Biorefinery

Earlier, first-generation biorefineries were employed for the production of biofuel; these biorefineries used sugars, animal oils, vegetables etc. as feedstocks but due to the competitive consumption of food resources these refineries were criticised. As a replacement to first-generation biorefineries, second generation came into play which used the agricultural wastes and non-edible vegetable oils as feedstocks, but due to low availability of such wastes these biorefineries led to low yield of product (Trivedi et al. 2015).

Various steps in a biorefinery are shown in Fig. 1.4. With the increasing demand of energy globally and the need of reduction on the dependency of fossil fuels,

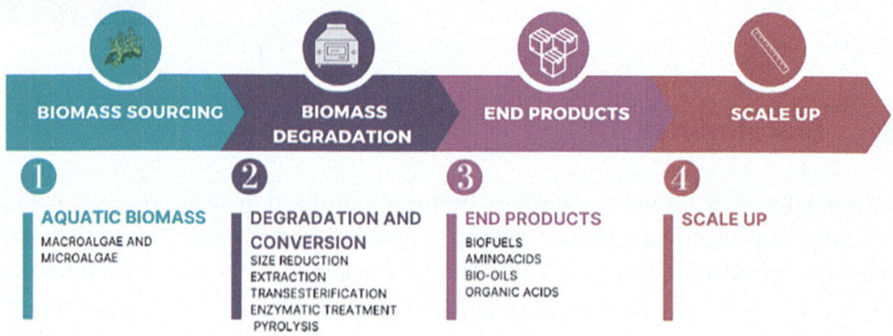

Fig. 1.4 Steps involved in the biorefineries

nowadays algae are being used as feedstocks to develop the biorefineries, which falls in third generation of biorefinery (Holm-Nielsen and Ehimen 2014). Algae helps not only in cost cutting of the production process but also in the reduction of carbon dioxide (CO_2) from the atmosphere along with higher lipid content. Algae can be grown in the wastewater leading to less consumption of fresh water, hence leading to environmentally sustainable feedstock (Trivedi et al. 2015).

1.3.2 Working and Designing of Algae-Based Biorefinery

The entire functioning of a biorefinery can be understood as a three-step process:

1. Algal biomass production and harvesting
2. Transformation processes
3. Downstream processes

As algae are classified into two types depending on their structure, the process of getting an end-product from both the types is slightly different (Andrić et al. 2017).

For microalgae the production process includes cultivation, harvesting, extraction and conversion, whereas for macroalgae the production process includes cultivation, harvesting, extract processing and purification (Seckbach et al. 2010).

To design an algae-based biorefinery, the first step is the selection of algal strain and to find the optimum conditions to grow the strain. At this point, environmental factors should also be kept in the consideration. The selection of strain is done depending on the product needed (Nelson et al. 2023). Next step is to design an appropriate cultivation system for the selected strain. During this cultivation phase, algae is allowed to grow and all the required nutrients are provided to it along with the appropriate conditions of pH, temperature and light supply. Cultivation step is then followed by harvesting. During the harvesting step, different techniques such as filtration, flocculation and centrifugation are used to separate the biomass from the culture. The obtained biomass is then dried using different methods such as sun dry, mechanical, thermal or chemical dewatering (Yaashikaa et al. 2022). The obtained biomass is then grinded well so as to extract out the intracellular compounds from the cells for the extraction of lipids which are basically converted from grinded biomass by various methods such as mechanical pressing, solvent extraction or supercritical fluid extraction. The lipid now extracted is then converted to biofuel using different methods such as transesterification or hydro processing. Algal species and their lipid contents are shown in Table 1.1. After the production of biofuel, there is still a possibility of leftover biomass; this leftover biomass is then used for the extraction of protein by various protein extraction methods such as enzymatic hydrolysis, solubilisation or mechanical separation (Katiyar et al. 2021). Biomass residue is then utilised for the production of biochar, biogas and fertiliser. To find out the environmental impact, the life-cycle assessment (LCA) of entire process is done.

Table 1.1 Various algal species with their lipid content and biomass productivity

Algal species	Lipid productivity (g/L/day)	Lipid content (% dry weight biomass)	Biomass productivity	References
Chlorella vulgaris	0.1837	32.50	0.2341–0.5322	Bošnjaković and Sinaga (2020)
Desmodesmus sp.	0.0198	48.41	–	Bošnjaković and Sinaga (2020)
Isochrysis sp.	0.0378	7.10	0.08	Xiao et al. (2013)
Elliposodium sp.	0.0473	14.00	0.17	Xiao et al. (2013)
Monodus subterraneus	0.0304	16.00	0.19	Xiao et al. (2013)
Chlamydomonas reinhardtii	0.0809	–	3.30	Bošnjaković and Sinaga (2020)
Nannochloropsis oceanica	0.0569	46.14	0.91	Xiao et al. (2013)
Dunaliella salina	0.116	23.10	0.20	Bošnjaković and Sinaga (2020)
Scenedesmus quadricauda	0.1395	38.61	0.36	Bošnjaković and Sinaga (2020)
Oocystis pusilla	0.0494	10.50	0.16	Xiao et al. (2013)

1.3.3 Life-Cycle Assessment of Algae-Based Biorefinery

This analysis is done to get the information about the net energy production and overall sustainability. Life-cycle assessment (LCA) entails assessing the environmental effects related to the biorefinery process's full life cycle, from the extraction of raw materials to the disposal of the finished product. Making educated decisions for sustainable development, streamlining procedures and locating possible environmental hotspots are all made easier with the aid of this assessment (Sander and Murthy 2010).

There are various ways to carry out the LCA of algae-based biorefineries:

1. Cradle-to-Gate LCA: The environmental effects from raw material extraction to the product manufacturing point (gate) are the main focus of this assessment. It offers information on the emissions and resource usage related to the production stage (Silva 2012). This evaluation takes into account the product's complete life cycle, which includes the stages of production, use and end-of-life. It offers a more thorough understanding of how a product affects the environment over the course of its lifetime (Finkbeiner 2014).
2. Socio-Economic LCA: It adds social and economic dimensions to the evaluation, taking into account things like the creation of jobs, health effects on people and financial ramifications (Velasco-Muñoz et al. 2019).
3. Carbon Footprint Assessment: It focuses on the greenhouse gas emissions that are directly related to the biorefinery process, giving information about the products' carbon footprint (Lenzen et al. 2018).

4. Water Footprint Assessment: It examines the amount of water used and any possible environmental effects related to it over the biorefinery's life cycle (Hoekstra and Mekonnen 2012).
5. Cumulative Energy Demand: It helps to identify the energy-intensive stages of the biorefinery life cycle by quantifying the total primary energy demand (Kuo et al. 2021).
6. Techno-Economic Assessment LCA: It examines the biorefinery process's economic viability while taking revenue generation, operating costs and capital costs into account (Davis et al. 2016).
7. Impact Category Assessment: It examines particular environmental impact categories, like toxicity, eutrophication and acidification. This offers a thorough comprehension of the various environmental effects connected to the biorefinery (Huijbregts et al. 2017).

1.3.4 Types of Algae-Based Biorefinery

They can be classified into five types depending on their valuable end-product:

1. Biorefinery for biofuel production
2. Biorefinery for carbohydrate and protein extraction
3. Biorefinery for biochemicals and biomolecules
4. Biorefinery for wastewater treatment
5. Integrated system biorefinery

1.3.5 Application of Algae-Based Biorefinery

1. Production of carbohydrates from algae
2. Production of biodiesel from algae
3. Production of polysaturated fatty acid from algae
4. Production of pigments and proteins from algae
5. Production of carotenoids from algae
6. Production of biogases such as biomethane

1.3.6 Advantages and Disadvantages of Algae-Based Biorefinery

Advantages

1. The main advantage of using algae is that it uses higher amounts of ammonium nitrate and phosphorous compounds leading to simultaneously higher growth rate and reduction of environmental pollution (Alaswad et al. 2015).
2. Algae can be grown in variable environment ranging from non-aerated lands to wastewater, hence leading to no competition for agricultural land (Chisti 2007).

3. Algae-based biorefineries support international efforts to lessen dependency on fossil fuels by advancing the development of sustainable and renewable energy sources (Mata et al. 2010).
4. Algal farming does not directly compete with food production, as some biofuel feedstocks derived from food crops do, which allays worries about food security (Wijffels and Barbosa 2010).
5. Numerous valuable compounds, such as biofuels, high-value chemicals, nutraceuticals and animal feed, can be produced by algae-based biorefineries (Ebhodaghe et al. 2022).

Disadvantages

1. At times, it can be expensive to set up and operate algae-based biorefinery (Davis et al. 2016).
2. It becomes hard to dewater and harvest algae, which overall impacts the entire process (Khoo et al. 2019).
3. The use of genetically modified algae or specific cultivation techniques may be subject to regulatory scrutiny and public perception issues, which could have an impact on the industry's social license (Radakovits et al. 2010).
4. Climate and seasonal variations can have an impact on algae cultivation, which can affect overall reliability and production rates (Singh et al. 2011).
5. Large volumes of algae biomass handling and transportation from cultivation sites to processing facilities can present logistical difficulties and extra expenses (Beal et al. 2012).

1.4 Future Aspects

1. Biofuel Production: Due to their high lipid and carbohydrate content, algae are a valuable source for the production of biofuel. Algal strains and cultivation techniques are being optimised in ongoing research to increase biofuel yields and economic viability and scalability of algae-based biofuels.
2. Pharmaceuticals and Nutraceuticals: Bioactive compounds with potential applications in the pharmaceutical and nutraceutical industries can be found in algae. The utilisation of algae-based systems for the synthesis of medicinal compounds, vaccines and therapeutic proteins is still being investigated.
3. Carbon Capture and Utilisation (CCU): By absorbing CO_2 during photosynthesis, algae can contribute to the process of carbon capture. By incorporating algae-based systems into CCU's industrial processes, the company may be able to reduce greenhouse gas emissions while simultaneously creating useful bio-based products.
4. High-Value Products: Omega-3 fatty acids, pigments, antioxidants and bioactive compounds are just a few examples of the high-value products that can be made from algae. Increasing the variety of valuable products that algae can produce can help algae-based biorefineries remain financially viable.

5. Treatment of Wastewater and Recycling of Nutrients: Algae are able to take up and retain nutrients from wastewater. When algae-based biorefineries are incorporated into wastewater treatment systems, they can remove nutrients and produce biomass for a range of uses.
6. Concepts for Integrated Biorefineries: It is anticipated that algae-based biorefineries will develop into integrated systems that optimise the use of all biomass components. This includes proteins, carbohydrates and other useful substances for a variety of applications, in addition to lipids for biofuels.
7. Scaling Up and Commercialisation: Algae-based biorefineries may be expanded for use in commercial production as technology develops and production methods become more effective. This entails streamlining downstream processing, harvesting techniques and cultivation systems.
8. Genetic Engineering and Selective Breeding: Current research endeavours to enhance algae strains via genetic engineering and selective breeding. One important area of research is customising algae strains for particular uses, like increased productivity or a higher lipid content for the production of biofuel.
9. International Cooperation and Knowledge Sharing: International cooperation among academics, businesses and governments can hasten the development of technologies for algae-based biorefineries. Cooperation and the exchange of knowledge can help solve problems and advance the long-term growth of the algae-based bioeconomy.
10. Supportive Policies and Incentives: Supportive policies and incentives can be very effective in promoting the development of the market for algae-based biorefineries. To promote the adoption of algae-based technologies, governments and businesses can fund research and development (R&D), offer financial incentives and set up regulatory frameworks.

1.5 Conclusion

In conclusion, the implementation of biorefineries based on algae is consistent with the worldwide effort to achieve the Sustainable Development Goals (SDGs) established by the United Nations. These biorefineries provide an adaptable and eco-friendly solution to a number of problems, such as resource scarcity, environmental degradation and population growth. Because of their variety and ability to adapt to various environments, algae have become a promising feedstock for biorefineries. Understanding the potential uses of algae in biorefineries is made easier by their categorisation into micro- and macroalgae. Both macroalgae, such as brown, green and red algae, and microalgae, like *Chlorella* and *Nannochloropsis*, have unique benefits in terms of biomass productivity and lipid content, which are necessary for the production of biofuel. The design of algae-based biorefineries heavily relies on the choice of appropriate algal strains and ideal growing conditions. The idea of biorefineries is a departure from previous iterations of biofuel production, which were criticised for competing with food supplies. This is especially true of those based on algae. Third-generation biorefineries that use algae as their feedstock not only lessen competition but also lower atmospheric carbon dioxide concentrations. Algae-based biorefineries produce biofuels, proteins, biochar, biogas and fertilisers through a three-step process that includes biomass production and harvesting, transformation processes and downstream processes. When assessing the overall sustainability and environmental impact of biorefineries based on algae, life-cycle assessment (LCA) becomes an essential tool. Numerous life-cycle assessments (LCAs), such as cradle-to-gate, socio-economic, carbon footprint, water footprint, cumulative energy demand, techno-economic, and impact category assessments, offer a thorough understanding of the environmental impacts related to the biorefinery process throughout its whole life cycle. Beyond producing biofuel, algae-based biorefineries are used for wastewater treatment, biochemicals, protein and carbohydrate extraction, and biomolecules. High biomass productivity, adaptability in growing conditions, smaller land and water footprint, and the generation of several valuable compounds are the benefits of algae-based biorefineries. Furthermore, the production of food is not directly hampered by the cultivation of algae, which promotes food security. Notwithstanding, it is imperative to tackle certain obstacles and drawbacks, including the requirement for optimal strain selection, cultivation systems and downstream processing. However, the continuous research and development in algae-based biorefineries highlights their potential as a financially feasible and sustainable way to meet global environmental and energy targets. Algae-based biorefineries have the potential to be a major player in creating a future that is healthier, cleaner and more sustainable as technology advances.

References

Alaswad A, Dassisti M, Prescott T, Olabi AG (2015) Technologies and developments of third generation biofuel production. Renew Sust Energ Rev 51:1446–1460. https://doi.org/10.1016/J.RSER.2015.07.058

Andrić I, Pina A, Ferrão P, Fournier J, Lacarrière B, Le Corre O (2017) The 15th international symposium on district heating and cooling assessing the feasibility of using the heat demand-outdoor temperature function for a long-term district heat demand forecast-review under responsibility of the scientific committee of the 15th international symposium on district heating and cooling. https://doi.org/10.1016/j.egypro.2018.07.028

Beal CM, Stillwell AS, King CW, Cohen SM, Berberoglu H, Bhattarai RP, Connelly RL, Webber ME, Hebner RE (2012) Energy return on investment for algal biofuel production coupled with wastewater treatment. Water Environ Res 84(9):692–710. https://doi.org/10.2175/106143012X13378023685718

Bošnjaković M, Sinaga N (2020) The perspective of large-scale production of algae biodiesel. Appl Sci 10(22):1–26. https://doi.org/10.3390/APP10228181

Chisti Y (2007) Biodiesel from microalgae. Biotechnol Adv 25(3):294–306. https://doi.org/10.1016/J.BIOTECHADV.2007.02.001

Davis R, Markham J, Kinchin C, Grundl N, Tan ECD, Humbird D (2016) Process design and economics for the production of algal biomass: algal biomass production in open pond systems and processing through dewatering for downstream conversion. https://doi.org/10.2172/1239893

Ebhodaghe SO, Imanah OE, Ndibe H (2022) Biofuels from microalgae biomass: a review of conversion processes and procedures. Arab J Chem 15(2):103591. https://doi.org/10.1016/J.ARABJC.2021.103591

Finkbeiner M (2014) The international standards as the constitution of life cycle assessment: the ISO 14040 series and its offspring. Springer, Cham, pp 85–106. https://doi.org/10.1007/978-94-017-8697-3_3

Hachicha R, Elleuch F, Hlima HB, Dubessay P, de Baynast H, Delattre C, Pierre G, Hachicha R, Abdelkafi S, Michaud P, Fendri I (2022) Biomolecules from microalgae and cyanobacteria: applications and market survey. Appl Sci 12(4):1924. https://doi.org/10.3390/APP12041924

Hingsamer M, Jungmeier G (2019) Biorefineries. In: The role of bioenergy in the emerging bioeconomy: resources, technologies, sustainability and policy. Elsevier, Amsterdam, pp 179–222. https://doi.org/10.1016/B978-0-12-813056-8.00005-4

Hoekstra AY, Mekonnen MM (2012) The water footprint of humanity. Proc Natl Acad Sci USA 109(9):3232–3237. https://doi.org/10.1073/PNAS.1109936109/-/DCSUPPLEMENTAL

Holm-Nielsen JB, Ehimen EA (2014) Biorefinery plant design, engineering and process optimisation. In: Advances in biorefineries: biomass and waste supply chain exploitation. Elsevier, Amsterdam, pp 89–111. https://doi.org/10.1533/9780857097385.1.89

Huijbregts MAJ, Steinmann ZJN, Elshout PMF, Stam G, Verones F, Vieira M, Zijp M, Hollander A, van Zelm R (2017) ReCiPe2016: a harmonised life cycle impact assessment method at midpoint and endpoint level. Int J Life Cycle Assess 22(2):138–147. https://doi.org/10.1007/S11367-016-1246-Y

Jalilian N, Najafpour GD, Khajouei M (2020) Macro and micro algae in pollution control and biofuel production – a review. ChemBioEng Rev 7(1):18–33. https://doi.org/10.1002/CBEN.201900014

Katiyar R, Banerjee S, Arora A (2021) Recent advances in the integrated biorefinery concept for the valorization of algal biomass through sustainable routes. Biofuels Bioprod Biorefin 15(3):879–898. https://doi.org/10.1002/BBB.2187

Khoo CG, Dasan YK, Lam MK, Lee KT (2019) Algae biorefinery: review on a broad spectrum of downstream processes and products. Bioresour Technol 292:121964. https://doi.org/10.1016/J.BIORTECH.2019.121964

Kuo CM, Sun YL, Lin CH, Lin CH, Wu HT, Lin CS (2021) Cultivation and biorefinery of microalgae (Chlorella sp.) for producing biofuels and other byproducts: a review. Sustainability 13(23):13480. https://doi.org/10.3390/SU132313480

Lenzen M, Sun YY, Faturay F, Ting YP, Geschke A, Malik A (2018) The carbon footprint of global tourism. Nat Clim Chang 8(6):522–528. https://doi.org/10.1038/s41558-018-0141-x

Mata TM, Martins AA, Caetano NS (2010) Microalgae for biodiesel production and other applications: a review. Renew Sust Energ Rev 14(1):217–232. https://doi.org/10.1016/J.RSER.2009.07.020

Narayanan M (2024) Promising biorefinery products from marine macro and microalgal biomass: a review. Renew Sust Energ Rev 190:114081. https://doi.org/10.1016/J.RSER.2023.114081

Nelson RS, Knoshaug EP, Spiller R, Nagle N, VanWychen S, Wiatrowski M, Davis R, Pienkos PT, Kruger JS (2023) Muconic acid production from algae hydrolysate as a high-value co-product of an algae biorefinery. Algal Res 75:103300. https://doi.org/10.1016/J.ALGAL.2023.103300

Ng DKS, Ng KS, Ng RTL (2017) Integrated biorefineries. In: Encyclopedia of sustainable technologies. Elsevier, Amsterdam, pp 299–314. https://doi.org/10.1016/B978-0-12-409548-9.10138-1

Olabi AG, Shehata N, Sayed ET, Rodriguez C, Anyanwu RC, Russell C, Abdelkareem MA (2023) Role of microalgae in achieving sustainable development goals and circular economy. Sci Total Environ 854:158689. https://doi.org/10.1016/J.SCITOTENV.2022.158689

Pacheco-Torgal F (2020) Introduction to biobased materials and biotechnologies for eco-efficient construction. In: Bio-based materials and biotechnologies for eco-efficient construction. Elsevier, Amsterdam, pp 1–16. https://doi.org/10.1016/B978-0-12-819481-2.00001-5

Radakovits R, Jinkerson RE, Darzins A, Posewitz MC (2010) Genetic engineering of algae for enhanced biofuel production. Eukaryot Cell 9(4):486. https://doi.org/10.1128/EC.00364-09

Sander K, Murthy GS (2010) Life cycle analysis of algae biodiesel. Int J Life Cycle Assess 15(7):704–714. https://doi.org/10.1007/S11367-010-0194-1/FIGURES/5

Saral JS, Ajmal RS, Ranganathan P (2021) Bioeconomy of hydrocarbon biorefinery processes. In: Hydrocarbon biorefinery: sustainable processing of biomass for hydrocarbon biofuels. Elsevier, Amsterdam, pp 355–385. https://doi.org/10.1016/B978-0-12-823306-1.00011-X

Seckbach J, Einav R, Israel A (eds) (2010) Seaweeds and their role in globally changing environments. Springer, Cham, p 15. https://doi.org/10.1007/978-90-481-8569-6

Silva N (2012) Life cycle assessment in product innovation. In: Life cycle assessment handbook: a guide for environmentally sustainable products. Wiley, Hoboken, pp 329–343. https://doi.org/10.1002/9781118528372.ch15

Singh A, Nigam PS, Murphy JD (2011) Renewable fuels from algae: an answer to debatable land based fuels. Bioresour Technol 102(1):10–16. https://doi.org/10.1016/J.BIORTECH.2010.06.032

Smithsonian National Museum of Natural History (n.d.) Algae classification. https://naturalhistory.si.edu/research/botany/research/algae/algae-classification. Accessed 28 December 2023

Thoughtco.com (n.d.) 7 major types of algae. https://www.thoughtco.com/major-types-of-algae-373409. Accessed 3 December 2023

Trivedi J, Aila M, Bangwal DP, Kaul S, Garg MO (2015) Algae based biorefinery—how to make sense? Renew Sust Energ Rev 47:295–307. https://doi.org/10.1016/J.RSER.2015.03.052

Unique Benefits of Macroalgae & Microalgae in Agriculture (n.d.). https://phycoterra.com/blog/benefits-of-macroalgae-and-microalgae-in-agriculture/. Accessed 29 December 2023

United Nations Development Programme (n.d.) Background on the goals. https://www.undp.org/sdg-accelerator/background-goals. Accessed 30 December 2023

University of California Museum of Paleontology (n.d.) Introduction to the Bacillariophyta. https://ucmp.berkeley.edu/chromista/bacillariophyta.html. Accessed 29 December 2023

Velasco-Muñoz JF, Aznar-Sánchez JA, Batlles-dela Fuente A, Fidelibus MD (2019) Sustainable irrigation in agriculture: an analysis of global research. Water 11(9):1758. https://doi.org/10.3390/W11091758

Vieira VV, Cadoret JP, Acien FG, Benemann J (2022) Clarification of most relevant concepts related to the microalgae production sector. Processes 10(1):175. https://doi.org/10.3390/PR10010175

Wang X, Zhang Y, Xia C, Alqahtani A, Sharma A, Pugazhendhi A (2023) A review on optimistic biorefinery products: biofuel and bioproducts from algae biomass. Fuel 338:127378. https://doi.org/10.1016/J.FUEL.2022.127378

Wijffels RH, Barbosa MJ (2010) An outlook on microalgal biofuels. Science 329(5993):796–799. https://doi.org/10.1126/SCIENCE.1189003

Xiao M, Shin HJ, Dong Q (2013) Advances in cultivation and processing techniques for microalgal biodiesel: a review. Korean J Chem Eng 30(12):2119–2126. https://doi.org/10.1007/S11814-013-0161-1

Yaashikaa PR, Keerthana Devi M, Senthil Kumar P (2022) Algal biofuels: technological perspective on cultivation, fuel extraction and engineering genetic pathway for enhancing productivity. Fuel 320:123814. https://doi.org/10.1016/J.FUEL.2022.123814

Algae as a Promising Feedstock for Biorefineries

2

Rahul Kadam and Jungyu Park

Contents

Abstract

In recent years, the focus of research has shifted to third-generation fuels, particularly algae. This is in response to challenges such as the energy crisis, global warming, climate change, and increased socio-economic awareness of the impact of fossil-based fuels. Algae biomass, rich in biological elements like lipids, proteins, and carbohydrates, emerges as a valuable feedstock for producing energy-dense products such as bio-oil, biochar, and syngas. However, the development of algal biofuels faces a significant dilemma due to high production costs, primarily associated with cultivation expenses. Previous literature has highlighted the economic impracticality of relying solely on the "only fuel" option, given the

R. Kadam · J. Park (✉)
Department of Advanced Energy Engineering, Chosun University, Gwangju, Republic of Korea
e-mail: jp@chosun.ac.kr

S. K. Arya et al. (eds.), *Value Added Products From Bioalgae Based Biorefineries: Opportunities and Challenges*,
https://doi.org/10.1007/978-981-97-1662-3_2

substantial production costs and associated expenditures. Consequently, extracting additional value-added products from algal biomass becomes imperative to justify the substantial investment in algae production and enhance its economic viability. In this context, the algae biorefinery approach emerges as a promising solution, allowing for the simultaneous production of multiple products alongside fuels in a single operation. Algae biorefineries can serve as a platform for a range of products, including various fuels such as biodiesel, biomethane, syngas, and byproducts used in the food, pharmaceutical, and cosmetics industries. The success of a biorefinery depends on creating seamless connections between the various input and output product streams. Overall, algae biorefineries open up new opportunities for the utilization of algae biomass in a multiproduct system.

Keywords

Algae biomass · Algal biorefinery · Bioremediation · Sustainability · Bioeconomy

Abbreviations

ASTM	American Society for Testing and Materials
BOD	Biochemical oxygen demand
C/N ratio	Carbon to nitrogen ratio
COD	Chemical oxygen demand
FAO	Food and Agriculture Organization
FDA	Food and Drug Administration
GHG	Greenhouse gas
HEFA	Hydrotreated fatty acid and esters
HHV	Higher heating value
HPH	High-pressure homogenization
HTL	Hydrothermal liquefaction
IEA	International Energy Agency
MAAs	Mycosporine-like amino acids
NOx	Nitrogen oxides
PAHs	Polycyclic aromatic hydrocarbons
PEF	Pure electric field pretreatment
PHA	Polyhydroxyalkanoates
PHB	Polyhydroxybutyrate
PTA	Phosphotransacetylase
PUFA	Poly-unsaturated fatty acid
SOx	Sulfur oxides

2.1 Introduction

The world's heavy reliance on limited fossil fuel resources to meet its ongoing energy needs poses a significant and urgent threat to the environment. The overuse of fossil fuels has led to global warming, which is driven by the massive release of greenhouse gases (GHGs) produced by these fuels. This crisis manifests as a significant challenge for humanity, impacting both energy security and sustainability on a global scale. Climate change, which is now indisputable, has manifested in the form of heatwaves, melting glaciers, rising sea levels, intensified storms and flooding, and shifting rainfall patterns (Pires 2017). Consequently, climate change has emerged as a focal point for researchers and policy decisions worldwide. To mitigate climate change, the global economy must shift from its current linear, fossil-fuel-centric model to a circular bioeconomy. In this context, there is growing interest in the exploration of second-generation biofuels derived from lignocellulosic biomass, such as agricultural residues. However, it is worth noting that the utilization of this biomass requires intensive and costly pretreatment due to its complicated structure (Limayem and Ricke 2012). Recently, the focus has shifted to algae as a potential third-generation feedstock, aligning with the principles of the circular economy.

Algae, highly adaptable multicellular photosynthetic organisms, flourish across diverse habitats, spanning lakes, rivers, oceans, and reservoirs. They are typically categorized as Rhodophyta (red algae), Phaeophyta (brown algae), and Chlorophyta (green algae) (Salami et al. 2021). Among these groups, green algae, thriving primarily through photosynthesis, dominate aquatic ecosystems. Their appearance varies greatly, including single celled and multicellular forms, with sizes ranging from 0.2 μm to about 65 cm across approximately 50,000 different algal species (Christaki et al. 2011). Algae are photosynthetic organisms with nuclei that can turn CO_2 and water into important O_2 and organic matter, providing key sources of carbon and energy. They are broadly divided into macroalgae and microalgae, and together they play important roles as natural reserves for simple life forms and primary resources for humanity. While terrestrial plants manage a modest 3–6% carbon fixation rate, algae significantly contribute to almost 40% of carbon fixation (Yamasaki 2003). Furthermore, algae demonstrate exceptional resource efficiency. Producing 1 kg/ha of algae requires only 0.0029 mg/ha of nitrogen and 0.002 mg/ha of phosphorous, in stark contrast to the resource demands of lignocellulosic biomass. Beyond these benefits, algae boast diverse biochemical compositions, including proteins, lipids, and carbohydrates, yielding an array of products (Barsanti and Gualtieri 2022). As a third-generation biofuel, algae offer several distinct benefits, including rapid growth, superior CO_2 fixation capabilities, and the ability to thrive without the need for arable land.

Recently, a lot of research has focused on finding out how algae can be used to make useful things like food, fuel, and other valuable products in a sustainable way. Algal biomass yields specific products of interest, notably oils, antioxidants, carotenoids, and other metabolites, which find applications in therapeutic and dietary contexts (Sathasivam et al. 2019). While algae are a good source of protein, amino acids, and fatty acids that humans can consume, growing algae for commercial use requires a lot of attention. While considering the limitations of arable land availability, the cultivation of algae in aquatic environments emerges as a promising means to meet global energy demands. Recent advancements in designing photobioreactors, efficient harvesting methods, and integrated approaches to algae farming have greatly increased the production of biofuels from algae (Brennan and Owende 2010). Furthermore, it is worth noting that high-value commodities such as proteins, vitamins, and lipids can be extracted from algal biomass prior to its utilization in biofuel production. This holistic approach to processing algae, which involves various biochemical steps before the biofuel conversion, is commonly referred to as an "algal biorefinery."

The concept of an algae biorefinery is not new and has been extensively discussed. It involves transforming algae biomass sustainably to produce biofuels and high-value products (Trivedi et al. 2015). Currently, the economic viability of biofuel production from algae is hindered by the high capital cost per unit of fuel. To make algae biorefineries economically feasible, generating high value-added products is crucial. The goal of an algae biorefinery is to optimize resource use, maximize profitability, and minimize waste. Integrating biofuel production from algae with biochemical processing is key to implementing algae biorefineries. This approach has the potential to be economically viable, reduce waste, and make better use of resources.

2.2 Algae Biomass

Worldwide, the algae market is expected to grow significantly, with a projected increase of US$ 4,148,000 based on the 2019–2025 market analysis report, driven by a growth rate of 6.6%. Notably, various algal species possess carbon-concentrating mechanisms, rendering them intricate yet valuable feedstock for bioenergy production. Recent findings highlight the algae's remarkable capacity to absorb sulfur and nitrogen oxides (SO_x and NO_x) emissions, subsequently transforming them into valuable resources (Singh and Thakur 2015). These unique characteristics have garnered significant attention, positioning algae as a novel and promising biomass source for biofuel production.

Algae biomass is typically categorized into two groups—macroalgae and microalgae—based on their morphology, size, and aquatic habitat colonization (Barsanti and Gualtieri 2022). They possess the remarkable ability to catalyze the conversion of CO_2 into a diverse range of products. Algae's primary biochemical constituents consist of proteins, carbohydrates, and lipids. Table 2.1 presents the proximate and elemental composition of various micro and macroalgal species, while Table 2.2 offers insights into their biochemical composition (Becker 2007; Sahu et al. 2013; Suganya et al. 2016).

Table 2.1 Proximate and elemental composition of various micro and macro algal species

Algae	Proximate analysis (wt.%)				Elemental analysis (wt.%)					HHV (MJ kg^{-1})	References
	MC	Ash	VM	FC	C	H	O	N	S		
Microalgae											
Chlorella vulgaris	4.4	15.9	67.2	12.4	44.8	6.8	40.4	7.0	1.0	–	López-González et al. (2014)
Chlorella pyrenoidosa	10	9	64	17	46.8	6.8	25.8	0.8	8.4	21.5	Duan et al. (2015)
Chlorella sorokiniana	3.8	7.9	73.2	15.1	45	7.6	35.5	3.8	–	2.4	Chen et al. (2014)
Chlamydomonas sp.	3.5	5.2	75.5	15.6	40.3	7.3	44.5	2.6	–	17.4	Chen et al. (2014)
Nannochloropsis gaditana	28.5	7.3	–	–	55.9	7.8	23.3	2.8	2.8	27	López-González et al. (2014)
Pavlova sp.	10.1	27.2	59.9	12.8	33.9	4.2	31.1	3.5	–	12.9	Aysu et al. (2018)
Macroalgae											
Macrocystis pyrifera	8	26	42	–	41	6.1	–	3.1	–	14	Ross et al. (2009)
Laminaria digitata	5.3	26.5	61.3	–	33.5	4.4	–	0.8	0.7	11.3	Yanik et al. (2013)
Sargassum natans	10.4	29	48.8	11.6	25.9	5.5	24.1	3.5	1.2	8.6	Wang et al. (2013)
Laminaria hyperborea	12	17	54	–	50	7.6	–	Yanik et al. (2013)	–	16	Ross et al. (2009)
Enteromorpha clathrata	10.1	21.1	57.9	10.7	32.6	4.9	24.6	4.4	2	12	Wang et al. (2013)
Fucus serratus	4.6	21.4	65.4	–	38.8	4.9	–	1.2	0.8	13.5	Yanik et al. (2013)

HHV: higher heating value, *MC: moisture content*, *VM: volatile matter*, *FC: fixed carbon*, *C:* carbon, *H:* hydrogen, *O:* oxygen, *N:* nitrogen, *S:* sulfur

Table 2.2 General biochemical composition of different types of algae

Algae	Carbohydrates	Protein	Lipids
Microalgae			
Chlorella vulgaris	12–17	51–58	14–22
Chlorella pyrenoidosa	26	57	2
Anabaena cylindrica	25–30	43–56	4–7
Aphanizomenon flos-aquae	23	62	3
Euglena gracilis	14–18	39–61	14–20
Scenedesmus obliquus	10–17	50–56	12–14
Dunaliella salina	32	57	6
Spirulina maxima	13–16	60–71	6–7
Macroalgae			
Codium tomentosum	29.25	5.06	7.15
Hypnea valentiae	11.8–23.6	11.8–13.0	9.6–11.6
Halimeda macroloba	5.40	32.63	9.89
Ulva lactuca	11.6–13.2	11.4–12.6	9.6–11.4
Caulerpa racemosa	33.75	8.78	10.63
Chaetomorpha aerea	31.50	10.13	8.50
Microdictyon agardhianum	27	20.93	9.40
Valoniopsis pachynema	31.50	8.78	9.09

2.2.1 Macroalgae

Macroalgae, commonly known as seaweed, are multicellular marine algae typically found near the seabed. They play a significant role in preventing eutrophication and pollution (Notoya 2010). In their natural habitat, they thrive on rocky substrates, exhibiting rapid growth rates of approximately 2–4 m per year that can extend up to 60 m (McHugh 2003). Many important factors affect pigment growth and composition of macroalgae, such as light, temperature, salinity, nutrients, and pollution. Among these factors, light is crucial as it categorizes macroalgae into brown, red, and green algae (Chan et al. 2006). Regarding chemical composition, macroalgae, except for green algae, are characterized by low levels of oils, starch, and lignin but contain significant amounts of water and minerals, including alkali metals. They also tend to have lower protein and lipid content but higher carbohydrate levels (Sambusiti et al. 2015). On the other hand, the carbohydrates in green, red, and brown algae vary in quality and quantity depending on the species. Green algae primarily consist of manna, ulvan, starch, and cellulose, with starch accounting for approximately 1–4%. Ulvan, a distinctive feature of green alga, is composed of D-glucuronic acid, D-xylose, L-rhamnose, and sulfate (Dhargalkar and Pereira 2005). Brown algae's major polysaccharides include alginate, laminarin, fucoidan, cellulose, and mannitol (Andrade et al. 2004). Lastly, red algae are composed of carrageenan (75%) and agar (52%), which are their primary saccharides (Whyte and Englar 1981).

In addition to their natural growth in aquatic environments, macroalgae are actively cultivated in open systems, both offshore and near shores, for large-scale production. Although they are grown globally, their primary markets are found in Asian countries such as South Korea, Japan, Malaysia, and China. China and Indonesia take the lead as major producers, yielding a combined production of approximately 21–23 million tons, while South Korea and the Philippines contribute over 1 million ton (Lee et al. 2020). Currently, the majority of macroalgae, about 94%, is cultivated through aquaculture, with the remaining 6% harvested from natural sources (Tiwari and Troy 2015). Four main techniques are employed for microalgae cultivation today, which include floating rafts, tension-lag platforms, fixed off-bottom long lines, and rock-based farming (Lee et al. 2020). These methods hold an advantage due to their independence from arable land, a valuable resource. Floating raft farming employs rafts or wooden boats to suspend macroalgae seedlings on the sea surface. The tension-lag platform method, although less common due to its cost, involves vertical floating platforms. In the fixed off-bottom long line approach, macroalgae are grown on ropes ranging from 10 to 20 m in length, anchored to the seabed. Rock-based farming entails attaching macroalgae seedlings to rocky substances until they reach full growth. Following growth, macroalgae are harvested manually or mechanically, depending on the scale of production, and subjected to post-harvest processing. Upon reaching full growth, macroalgae are harvested either manually or with mechanical aids like boats and dredgers. In comparison, harvesting macroalgae is a more straightforward and direct process, facilitated by their larger overall structure (Sudhakar et al. 2018). In post-harvest processing, macroalgae undergo various pretreatments, depending on the scale of production. Large-scale production often employs formaldehyde preservation techniques, while small-scale operations opt for natural sun drying (Sudhakar et al. 2018). Natural sun drying involves exposing macroalgae to the sun for less than 24 h, reducing moisture content from 90 to 10%, which saves money compared to the technical drying process. It offers a promising avenue for sustainable resource utilization, offering innovative techniques and global market potential to address economic and environmental challenges.

2.2.2 Microalgae

Microalgae, also known as microphytes, are simple unicellular microorganisms with a size range spanning from 1 to 400 μm. They consist of approximately 20,000 distinct species and are typically found in marine and freshwater habitats (Andrade et al. 2018). Their size varies significantly depending on the specific species. Remarkably, microalgae lack roots, stems, and leaves yet possess the capacity for photosynthesis, generating O_2 and utilizing CO_2 to fuel their growth. Regarding chemical composition, microalgae primarily consist of proteins, carbohydrates, and lipids. They are also accompanied by a plethora of compounds such as pigments, fatty acids, vitamins, and more. The specific composition varies depending on the species, and cultivation methods. Temperature and light levels significantly impact microalgae, prompting the accumulation of carbohydrates within them (Markou et al. 2012). It is essential to emphasize that the pivotal consideration for microalgae lies in their energy production capabilities. Additionally, various cultivation methods for microalgae hold tremendous value. The annual production of microalgae has consistently risen over the years, reaching an impressive 20,000 tons annually, showcasing the substantial potential they offer (Lee et al. 2020). However, despite their potential, the challenge of expensive cultivation and harvesting processes remains a considerable hurdle, with costs that exceed those of macroalgae by US $185 per ton (Bae et al. 2011). Commonly used cultivation methods for macroalgae include phototrophic, heterotrophic, mixotrophic, and photoheterotrophic approaches. These methods differ primarily in their energy supply, carbon source, and reactor type. Phototrophic cultivation in open ponds or photobioreactor systems stands out as the most widely employed and cost-effective method (Milano et al. 2016). This stands in contrast to heterotrophic cultivation, involving conventional fermenters, where both energy and carbon sources are derived from organic carbon. Mixotrophic cultivation combines both phototrophic and heterotrophic methods in a closed bioreactor system. Unlikely other mechanisms, photoheterotrophic cultivation involves utilizing light as an energy source, with organic components serving as the primary carbon sources. The techniques like heterotrophic, mixotrophic, and photoheterotrophic cultivation are infrequently used due to their high equipment costs and elevated contamination risk (Lee et al. 2020).

The harvesting of microalgae remains a significant concern in microalgae production, accounting for approximately 20–30% of the overall biofuel production costs. Over the year, numerous harvesting techniques have been developed. However, a common challenge persists: the high energy consumption associated with microalgae harvesting. This is primarily due to their dispersed suspension nature, which necessitates surface dewatering. Over time, several pretreatment methods have been explored, including physical, chemical, and electrical-assisted techniques (Abdullah et al. 2019; Chen et al. 2011; Goh et al. 2019). Physical harvesting methods consist of sedimentation, centrifugation, flotation, filtration, and screening. Sedimentation relies on gravitational forces, making it a relatively cost-effective and energy-efficient option. Centrifugation, while straightforward, is considered more costly compared to alternative methods. Flotation, a form of dispersion, utilizes air bubbles to generate uplift forces for particle separation. Filtration and screening involve the passage of microalgae through screens that retain the microalgae while allowing unwanted materials to pass through. Despite its cost, centrifugation remains the most employed method due to its simplicity and efficiency. Chemical harvesting methods involve separation through flocculation with auto-flocculation, bioflocculation, and chemical coagulation. These chemical techniques are well established and suitable for large-scale applications (Chen et al. 2011). Emerging techniques like electrophoresis, electro-flocculation, and electro-flotation explore the separation of negatively charged microalgae based on their charge differences (Goh et al. 2019). The selection of a harvesting method should be driven by the desired end products. For lower value-added end products like biofuels, cost-effective methods are recommended, while higher value-added products can justify the use of more expensive techniques.

2.3 Algae Biorefinery: A Sustainable Approach

The term "biorefinery" is like the refining of crude oil in the petroleum industry, where various fuels are produced as primary products alongside additional chemical byproducts. Similarly, the concept of biorefining is applicable to biomass, involving the processing of algae to yield biofuels and other value-added products. This approach has been employed for years, aligning with the International Energy Agency's (IEA) Task 42 (2009) vision, which emphasizes sustainable biomass processing to generate a range of marketable products and energy (Sonnenberg et al. 2007). The central objective of algae-based biorefineries is to optimize resource utilization, maximize profit, and minimize waste production. Figure 2.1 depicts the schematic flowsheet of sustainable and efficient integrated algal biorefinery, which yields a variety of products.

Algae biomass undergoes bioconversion into biofuels and other value-added products through three primary routes: physiochemical, thermochemical, and biochemical conversion. The physiochemical pathway employs techniques like mechanical, microwave, ultrasound, and shockwave pretreatment for cell wall disruption. Mechanical pretreatment involves utilizing a bead and ball mill to disrupt the cell wall, while ultrasound induces structural modifications to starch granules,

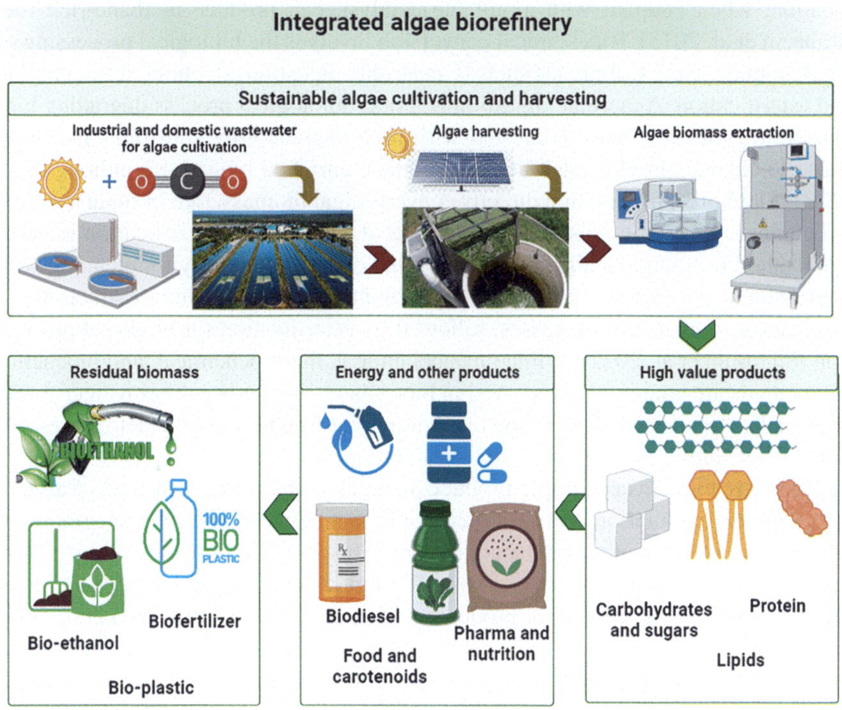

Fig. 2.1 Schematic flow sheet of sustainable, efficient, and integrated algal biorefinery

enhancing enzymatic hydrolysis for bioethanol production. Microwave pretreatment is particularly effective, achieving over 80% cell lysis (Mohan et al. 2020). Shockwave pretreatment subjects algae to high-voltage electric shock waves, disrupting cell structure (Lee et al. 2022). Newly developed techniques like pure electric field pretreatment (PEF) and high-pressure homogenization (HPH) have also demonstrated significant cell disruption efficiency. In the thermochemical conversion route, biomass undergoes thermal breakdown and chemical reformation to yield fuel products through pyrolysis, gasification, hydrothermal liquefaction (HTL), and direct combustion. The success of thermochemical conversion is contingent on algae water content. Methods are broadly classified into dry processes (e.g., torrefaction, pyrolysis, gasification) and wet processes (e.g., hydrothermal liquefaction, hydrothermal gasification, hydrothermal carbonization). Pyrolysis, a process of thermal degradation in the absence of oxygen, generates biofuels with a medium-low calorific value. Gasification converts carbonaceous materials into syngas (Chandra et al. 2019), while torrefaction, performed in inert atmospheres, facilitates biomass dehydration to a solid carbonaceous product (Mohan et al. 2020). The wet process, such as hydrothermal liquefaction (HTL), requires a temperature of 250–450 °C and a pressure of 100–350 bar for depolymerization as well as re-polymerization (Yin et al. 2010). This process yields bio-oil enriched with furan, phenol, acid, and ester derivatives, with high compositions of aliphatic, phenolic, alcoholic, carboxylic, and hydroxyl groups (Kumar et al. 2018). Hydrothermal gasification, when coupled with a suitable catalyst, can produce methane-rich fuel (Raheem et al. 2015). Biochemical conversion involves the biological processing of algal biomass using techniques such as anaerobic digestion, alcoholic fermentation, and esterification. Anaerobic digestion is a microbiological process degrading biomass into biofuel, often co-digested with high-degradability substrates like food waste to enhance biodegradability and produce enriched biogas (Khanthong et al. 2023). Alcoholic fermentation directly converts algal biomass, rich in sugars, starch, or cellulose, into ethanol using yeast (Ding et al. 2020). Transesterification is a crucial method for reducing algal oil viscosity and enhancing fluidity for blending with conventional petroleum fuels. Dried algal biomass undergoes lipid extraction via mechanical or chemical processes, followed by esterification for biodiesel production (Sharmila et al. 2022). While physiochemical, thermochemical, and biochemical methods are commonly employed in biorefinery, each process has its merits and drawbacks. A careful consideration of environmental and economic factors is essential to determining the sustainability of each process.

Choosing to recover multiple products in an algae biorefinery instead of aiming for a singular output represents a strategic approach. This involves the integration of one or more waste management technologies. This strategy aims to maximize profitability while minimizing environmental implications. Algae-based biorefinery presents a promising avenue for producing clean energy, reducing GHG emissions, and meeting carbon neutrality targets (Olabi et al. 2023). In the past decade, algal research has been dedicated to exploring the integration of various waste management techniques into algae biorefineries. Various approaches have been explored, including the cultivation of algae in wastewater to convert atmospheric CO_2 into algae production, a potential candidate for biorefinery (Kashyap et al. 2023). The first approach involves implementing numerous cascading biorefinery strategies, categorized based on the value of the main target compound. The first category focuses on producing low-value compounds like energy while valorizing the remaining biomass (Maurya et al. 2016). The second approach integrates the use of medium-value bulk products (lipids, proteins, and carbohydrates) alongside the valorization of residual biomass (Wijffels et al. 2010). The third approach prioritizes the production of high-value products coupled with the utilization of residual biomass—an economically feasible biorefinery approach (Bastiaens et al. 2017). Similarly, cultivating algae in wastewater proves to be a cost-efficient source of organic matter, manure, and biofuel production. The organic carbon present in wastewater enhances the intracellular constituents of algal cells by redirecting carbon flux, resulting in improved intracellular energy reserves (Katiyar et al. 2018). Emphasizing the generation of multiple products through integrated approaches could significantly contribute to the development of a sustainable algal biorefinery.

2.4 Algal Biomass Product Spectrum

Presently, macroalgae and microalgae have found extensive applications as feedstock for generating a diverse range of products, primarily classified into two categories: energy products and non-energy products. Within the domain of energy products, algae produce various biofuels such as biodiesel, biohydrogen, biogas, and bioethanol. In contrast, non-energy products such as carbohydrates, proteins, pigments, and biomaterials are also produced from algae.

Table 2.3 presents some of the promising algae species that are currently being used to produce energy as well as non-energy products (Salami et al. 2021; Uma et al. 2023) and Fig. 2.2 shows prominent algal-derived energy as well as non-energy products from an integrated algal biorefinery.

2.4.1 Energy (Primary) Products

Algae are a prolific source of major energy products, primarily in the form of biofuels consisting of biodiesel, bioethanol, biogas, and biohydrogen.

2.4.1.1 Biodiesel

Biodiesel, also known as methyl or ethyl esters of fatty acids, is typically produced through a process called transesterification of triglycerides. Biodiesel shares resemblances with petroleum fuels, boasting a higher heating value (HHV) ranging from 39 to 41 MJ/kg, along with key attributes such as flash point and cetane number (Demirbas 2009). Biodiesel can be derived from various substrates, including vegetable oils and fats. Biodiesel production from algae offers several distinct advantages, including the absence of the need for arable land and minimal water requirements. Biodiesel obtained from algae possesses the environmentally friendly

Table 2.3 Some prominent algal species used to produce both energy and non-energy products

Sr. No	Algae species	Application
1	*Chlorella vulgaris*	Biofuels
2	*Chlorella sorokiniana DSR*	Biofuels and bioremediation
3	*Schizochytrium* sp.	Biofuels and supplements
4	*Neochloris oleoabundans*	CO_2 biofixation and biodiesel
5	*Ulva sp.*	Human health foods and animal feed
6	*Undaria pinnatifida*	Vitamins, minerals, and edible fiber
7	*Rosenvingea intricata*	Fertilizer
8	*Chlorella variabilis*	Biodiesel
9	*Nannochloropsis* sp.	Biofuels and aquafeeds
10	*Dunaliella* sp. *ABRIINW-11*	Pharmaceutical and aquaculture
11	*Sargassum wightii*	Fertilizer
12	*Anabaena azollae*	Fertilizer
13	*Scenedesmus obliquus*	Biodiesel
14	*Tetraselmis* spp.	Biomethane
15	*Cyanobacteria* sp.	Bio-oil
16	*Scenedesmus dimorphus*	Bioethanol
17	*Botryococcus braunii*	Hydrocarbons: pigments
18	*Dunaliella salina*	Pigments (beta-carotene; bixin)

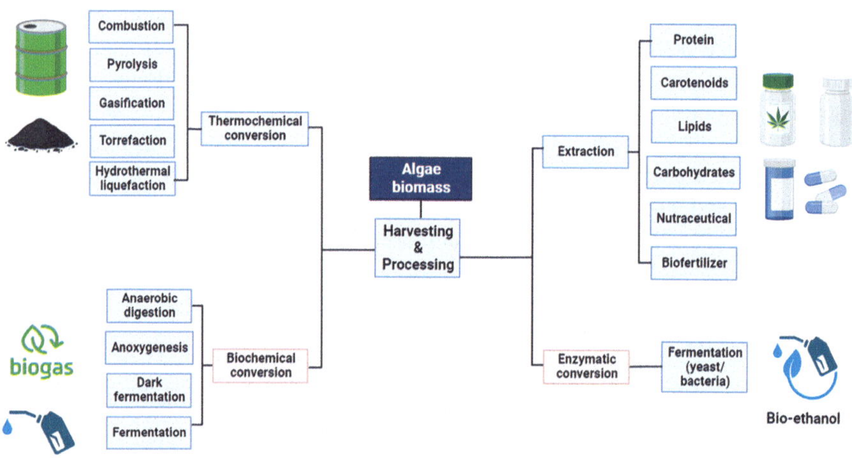

Fig. 2.2 Prominent algal-derived energy as well as non-energy products

characteristic of being naturally carbon neutral, as it assimilates CO_2 during growth and releases it during consumption (Najafi et al. 2011). Transesterification stands out as the most widely employed method for biodiesel production, effectively converting raw and viscous algal lipids into lower molecular weight fatty acid alkyl esters (Rawat et al. 2011). Figure 2.3 illustrates a schematic representation of the biodiesel production process from algae. A prime advantage of biodiesel usage lies

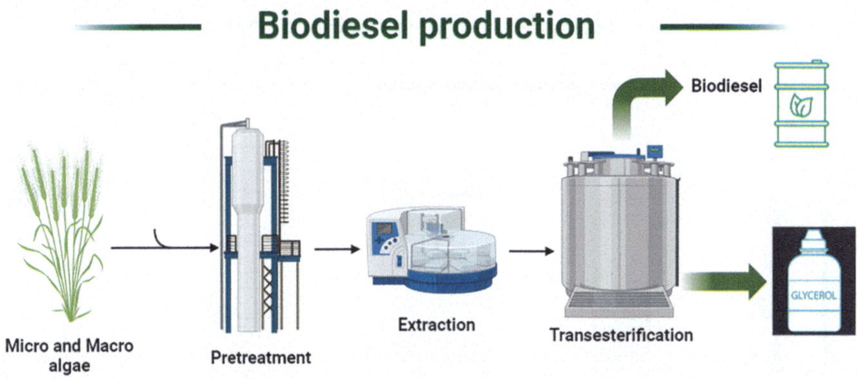

Fig. 2.3 Schematic representation of biodiesel production from algae

in its seamless compatibility with existing engine designs and infrastructure, requiring no substantial modifications.

Furthermore, biodiesel obtained from algae exhibits favorable combustion profiles and generates fewer emissions comparable to petroleum-based fuels (Liu et al. 2013; Ullah et al. 2014). The lipid saturation index, a crucial determinant of biodiesel stability and performance, meets satisfactory levels for algal biodiesel. It also aligns with International Standards for Vehicles (EN14214). A typical property comparison between petroleum fuels and algal biodiesel reveals that biodiesel exhibits lower viscosity and higher density in comparison to petroleum fuels (Talebi et al. 2013). Besides producing algal biodiesel, the process also generates glycerol. This glycerol can be a valuable carbon source, and it has the potential to be transformed into essential metabolic products like organic acids, single-cell oil, and mannitol. This transformation can be achieved by using eukaryotic microorganisms such as yeast and fungi.

2.4.1.2 Bioethanol

Algae can also be used to produce bioethanol because they have a lot of cellulosic material, which gives plenty of carbohydrates for bioethanol production.

Bioethanol can be produced through sugar fermentation or a combination of hydrolysis and fermentation to convert cellulosic biomass. Figure 2.4 delineates the bioethanol production process from algae. In the case of algae, particularly microalgae, they are composed of over 50% starch and lack lignin, rendering them an attractive feedstock for bioethanol production (Chandra et al. 2019). Prior to bioethanol production, careful strain selection and growth considerations are paramount when utilizing algae as a feedstock. This can be achieved through two approaches: utilizing macroalgae from their natural habitat or cultivating microalgae under favorable conditions. With macroalgae, bioethanol can be produced by converting polysaccharides into simple sugars through fermentation. Ethanol production has been reported for various species of macroalgae, i.e., green, red, and brown algae. However, microalgae hold an economic advantage over conventional

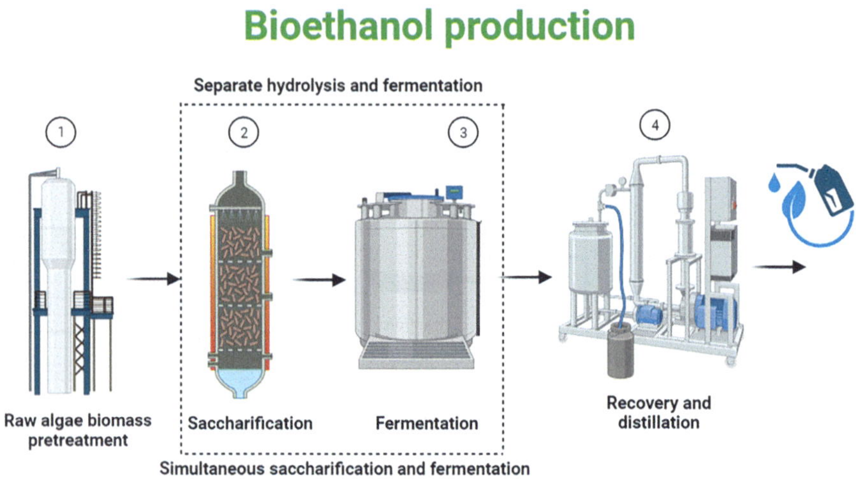

Fig. 2.4 Simplified illustration of bioethanol production from algae

feedstocks due to their rapid growth and reliable supply (Harun et al. 2010). Bioethanol production from macroalgae typically yields 0.10 g from 1 g of dried macroalgae, which is three times lower than the yield from microalgae. Furthermore, the CO_2 generated during the ethanol fermentation process can be recycled for microalgal cultivation. Microalgae species such as *Chlorococcum*, *Chlamydomonas*, and *Chlorella* exhibit better conversion rates compared to others. Brown algae, in particular, stand out due to their high carbohydrate content and faster growth rates achievable with current farming technologies (Harun et al. 2010; Nguyen et al. 2009). Beyond bioethanol, macroalgae can also yield biobutanol through a process known as acetone-butanol fermentation, utilizing species like *Clostridium* sp. (Daroch et al. 2013). The production of bioethanol from algae offers significant potential when considered within an integrated process for additional resource recovery.

2.4.1.3 Biomethane

Biomethane, derived from the anaerobic digestion of organic matter, serves as an economically viable energy source derived from biomass. It usually consists of CH_4 (55–75%) and CO_2 (25–45%). While algae holds great promise as a feedstock for the anaerobic digestion process, the inherent resistance of algal cells can pose challenges to the overall efficiency of the process (Chandra et al. 2019). Additionally, algal biomass's lower carbon to nitrogen (C/N) ratio can impact the methanogens within the reactor. Despite its considerable potential, biogas production from algae often faces feasibility issues due to the high costs associated with cultivation and processing (Trivedi et al. 2015). Enhancing biomethane yield from algal biomass can be achieved through various pretreatment methods, including enzymatic and thermal approaches.

Moreover, algal biomass can be co-digested with other feedstocks, such as crude glycerin and other materials. Figure 2.5 provides a schematic visualization of the process for biomethane production from algae. Biomethane production becomes more economically advantageous when the lipid content in the biomass is below 40%, as lipids can be extracted prior to subjecting the algal biomass to anaerobic digestion. An alternative approach to generating biomethane from algal biomass involves gasification, a process that partially oxidizes the biomass at high temperatures ranging from 800 to 1000 °C. This process produces a gas known as syngas, comprising gases like CH_4, H_2, CO_2, and N_2, which can be utilized as fuel in diesel or gas turbines (Clark and Deswarte 2014; Searchinger et al. 2008).

2.4.1.4 Biohydrogen

Biohydrogen stands as a distinctive biofuel that can be derived from algal biomass, because of the presence of unique photosynthetic pigments that convert solar radiation into chemical energy. In particular, microalgae and cyanobacteria excel at generating biohydrogen through processes such as biophotolysis or by catabolizing their endogenous substrates (Chandra et al. 2015). Cyanobacteria employ three distinct pathways for biohydrogen production: hydrogenases, fermentative, and nitrogen fixation pathways (Mandotra et al. 2021). Figure 2.6 illustrates the comprehensive process of biohydrogen production from algae. In the light-dependent pathway, biophotolysis takes place within the membrane-bound photosynthetic system when exposed to light.

This process can be achieved through either direct or indirect means. The fermentative pathway involves the transfer of electrons from the light-dependent photosynthetic carbon storage, i.e., glycogen, to hydrogenases. The third pathway

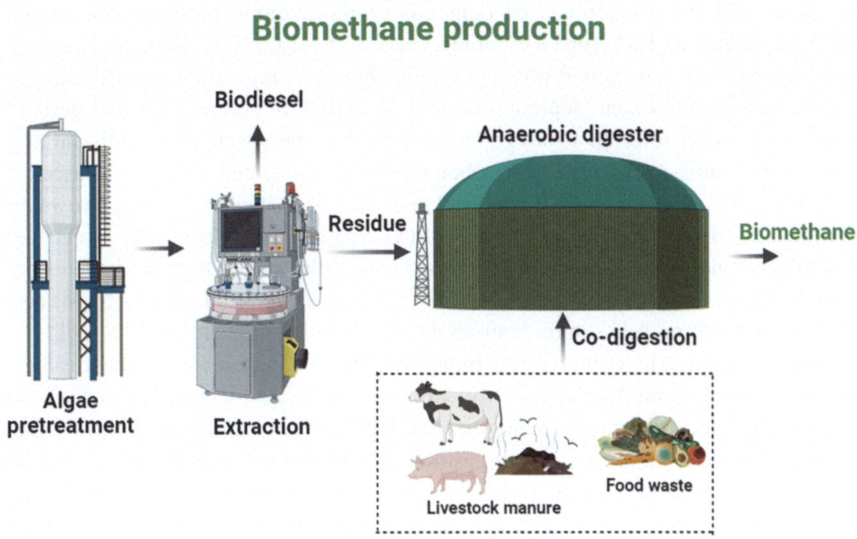

Fig. 2.5 Schematic visualization of biomethane production from algae

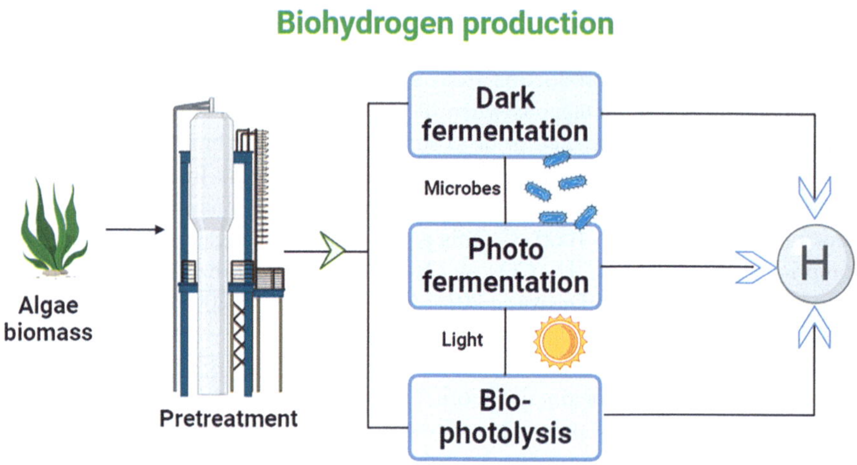

Fig. 2.6 Schematic illustration of algal biohydrogen production process

employs the nitrogenase enzyme, effectively segregating oxygenic photosynthesis from the nitrogen fixation reaction (Kumar et al. 2010; Mandotra et al. 2021). From an economic perspective, achieving a cost range of US$ 10–20 per GJ for biohydrogen production presents notable challenges due to technical and engineering complexities. Nevertheless, the potential for biohydrogen generation from algal biomass represents a significant opportunity within the realm of biofuels.

2.4.1.5 Biojet Fuel

Algae also offers a promising avenue to produce jet fuel, a renewable solution that can help meet the escalating fuel demands of the aviation industry. Traditional fossil-fuel-based jet fuel emits a combination of CO_2, H_2O, NOx, SOx, and particulates, which have detrimental effects on atmospheric composition, contributing to climate change and ozone depletion (Lee et al. 2010). In contrast, jet fuel derived from algae, often referred to as green jet fuel, has the potential to substantially reduce GHG emissions by 60–80% when replacing traditional fossil-fuel-based jet fuel. The conversion of algal oil into jet fuel can be achieved through two main methods. Figure 2.7 depicts the processes involved in the production of biojet fuel from algae. One widely recognized method is the hydrotreated fatty acid and esters (HEFA) process, which is American Society for Testing and Materials (ASTM) certified. This resulting fuel can be seamlessly blended with conventional jet fuel.

Another approach to producing biojet fuel from algae involves the Fischer-Tropsch process, a method commonly employed to derive high-quality fuels from natural gas, coal, or biomass (Sandquist and Guell 2012). Additionally, liquid fuel can be extracted from algae through gasification, involving the synthesis of gas followed by the conversion into liquid hydrocarbons (Sandquist and Guell 2012). However, it is imperative to prioritize safety in aviation, and this has led to limited research in the field of algae-based jet fuel due to stringent requirements that must

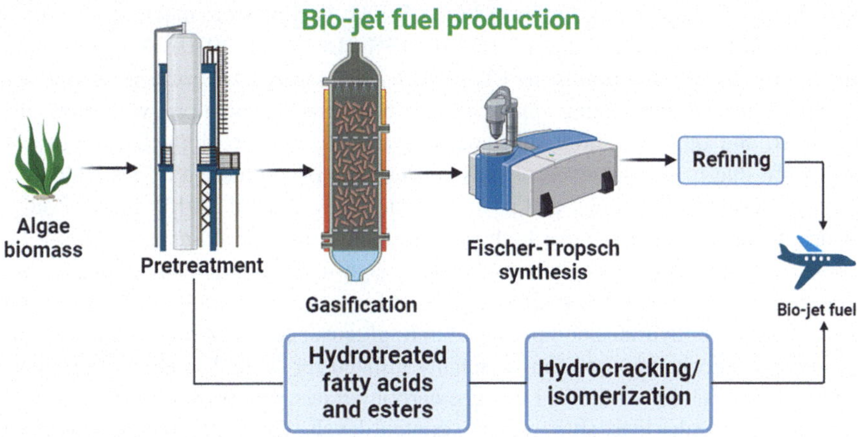

Fig. 2.7 Schematic representation of producing biojet fuel from algae

be met. Although a complete shift to 100% algae-based fuel may not be feasible for the aviation industry, yet there is potential to blend biojet fuel with conventional jet fuel at a 50:50 ratio. This blending approach has become a focal point for researchers in the biojet fuel industry.

2.4.2 Non-Energy (Secondary) Byproducts

The algae can biosynthesize, process, and discharge a variety of diverse and complementary metabolites and over 15,000 biomaterials, which have immense importance in nourishment, pharmaceuticals, the cosmetic industry, and other applications. A wide range of byproducts from algae have already been identified based on their presence in algae and their anticipated market. Algae cultivated for human and animal consumption typically boast a rich nutritional profile, including high protein content (60–70%), omega-3 fatty acids, polysaccharides, minerals, and vitamins (A, B, C, and E), along with enzymes, hormones, and pigments (de Jesus Raposo et al. 2013).

2.4.2.1 Nutritional Supplement

Proteins, essential macromolecules in living organisms, consist of amino acid monomers polymerized into chains. They play a crucial role in maintaining normal functionalities and are vital substrates for plants and animals, contributing to the production of hormones and enzymes. Algae-based proteins, a crucial component in human nutrition, play a pivotal role in averting malnutrition. Algal proteins find diverse applications, including animal feed, industrial enzymes, and bioplastics. Recognizing their significance, the Food and Agriculture Organization (FAO) has emphasized the importance of algal proteins over conventional sources, positioning them as highly competitive in the food market. Microalgae exhibit significantly higher protein content, ranging from 14 to 65.20%, in contrast to macroalgae, which typically fall within the range of 1.12–14.13% (Lee et al. 2020). A widely used algae for protein extraction is the *Cyanobacterium* species *Arthrospira*, commonly known as *Spirulina*, boasting approximately 60% raw protein along with minerals, vitamins, and other essential substances. *Spirulina*, with its high digestibility of over 86% due to the presence of polysaccharides, is commercially available as a dietary supplement in tablet and powder forms (Trivedi et al. 2015). Several other algae species reported to have high protein content include *Anabaena*, *Chlorella*, *Dunaliella*, and *Euglena*. Carbohydrates, a significant component of algae, are stored as reserve materials, such as starch, or form the main component of the cell wall. Algal carbohydrates comprising starch, glucose, cellulose/hemicellulose, and various polysaccharides are conventionally utilized for biofuel production, primarily bioethanol (John et al. 2011). Algae carbohydrates, comprised of sugar monomers or polysaccharides, show varying content levels. Generally, macroalgae have higher carbohydrate contents (24.55–51.4%) than microalgae (3.28–30.21%), primarily for extracellular protection and cell adhesion (Lee et al. 2020). However, studies have indicated a competition between lipids and starch due to the glycerol-3-phosphate produced during glycolysis (Trivedi et al. 2015). Various strategies, such as irradiance, CO_2 supplementation, and pH–temperature variation, are employed to enhance the carbohydrate content of algae. Polysaccharides derived from algae represent high-value products, particularly in the pharmaceutical industry, serving as antioxidants, antitumor, anticoagulant, anti-inflammatory, and antiviral agents. Algal-based foods are considered excellent sources of folate, vitamin B-12, and niacin (Uma et al. 2023). *Spirulina* and *Chlorella*, a specific microalgal species, stand out as excellent sources of vitamins B-1, B-2, B-5, and B-12, and vitamin A. Algae-based vitamins offer proven health benefits, with vitamin A boosting immune function and B-3 regulating cholesterol metabolism (Erdman Jr et al. 2012). It is essential to note that obtaining proteins, vitamins, and carbohydrates from algae requires deeper insights into extraction and utilization.

2.4.2.2 Pharmaceutical and Cosmetics Industry

In addition to their role in nutrition, algae offer numerous applications beneficial to humanity, particularly in the pharmaceutical and cosmetic industries. Algae are rich in various pigments, depending on their chemical composition. Chlorophyll, with its antioxidant properties, is noteworthy for relieving constipation, regulating digestion, and aiding in the recovery of damaged liver cells. Certain microalgal species, like *Arthrospira* and *Chlorella*, have gained significant traction in the skincare industry. Some cosmetic professions have gone as far as establishing their own systems for microalgal production. These microalgae extracts find application in a diverse range of skincare products. They contribute to anti-aging creams, refreshing formulations, and regenerative care items, and act as anti-irritants in peeling products. Additionally, these extracts are utilized in formulations for sun protection and hair care (McHugh 2003). Alongside, algae produce pigments like phycobiliproteins, enhancing the efficiency of light energy utilization and acting as a shield against solar radiation by producing carotenoids. They are commercially available in forms like beta-carotene, lutein, lycopene, astaxanthin, zeaxanthin, annatto, and canthaxanthin. Another significant pigment obtained from algae is astaxanthin, a powerful ketocarotenoid known for its antioxidant properties, which provide protection against chronic inflammatory diseases such as those of the eyes and skin, cancer, and liver-related problems (Trivedi et al. 2015). Other carotenoids like lutein, zeaxanthin, and canthaxanthin find applications in chicken skin coloration and pharmaceuticals. Algae have also been explored as a source of fatty acids, particularly long-chain poly-unsaturated fatty acids (PUFAs), including omega-3-fatty acids (Mehta et al. 2018). The increased demand for PUFAs, driven by growing health awareness, positions algae as an alternative source to traditional fish oils. Long-chain PUFAs offer significant medical benefits, supporting body development, mental health, and preventing conditions such as heart attacks, hypertension, depression, rheumatoid arthritis, asthma, Crohn's disease, and ulcers (Adarme-Vega et al. 2012). Similarly, phytosterols, primarily derived from vegetable oil, nuts, seeds, and whole grains, have gained prominence for their health-promoting activities. With increasing demand, algae serve as a source of phytosterols for the food and pharmaceutical industries. Algae can produce a wide range of phytosterols depending on their taxonomic affiliation, including brassicasterol, sitosterol, and stigmasterol. Phytosterols have various health impacts, such as immunomodulatory effects, hypercholesterolemic properties, cancer prevention, and antidiabetic properties.

2.4.2.3 Industrial Products

Algal biomass can be refined to yield valuable bioproducts, including biopolymers, bioplastics, biolubricants, dyes, and commodity chemicals like aldehydes and acids. Some algae, particularly macroalgae, exhibit gel-forming and water-dissolving characteristics, making them useful in various industrial applications. Polyhydroxyalkanoates (PHA), a type of bioplastic, can be produced from algae through bioplastic synthesis, which involves the regulation of enzyme expression and the C/N ratio. This process requires control over two enzymes, polyhydroxybutyrate (PHB) synthase and phosphotransacetylase (PTA). However, the cost of PHA derived from algae is relatively high due to the increased C/N ratio and energy costs (Chandra et al. 2019). Carrageenan, extracted from algae, constitutes a water-soluble group of polysaccharides extensively employed as emulsifiers and stabilizers in various food products, including chocolate milk, ice cream, milk puddings, jellies, jams, meat products, and pet feed products. Algae-derived fibers, crucial as raw materials, find applications in military gear such as fireproof clothing. Additionally, certain algae species, such as *Cyanobacteria* and marine species, provide ultraviolet (UV) protection. Due to their potent UV absorption and photoreactive properties, mycosporine-like amino acids (MAAs) and scytonemin can serve as active ingredients in cosmetics and pharmaceuticals (Chandra et al. 2019). Furthermore, algae contribute to the production of hydrocolloids, alginates, agars, agaroses, and phenolic compounds, expanding their industrial applications. This diverse range of bioproducts highlights the versatility and potential of algae in various sectors.

2.4.2.4 Fertilizer and Wastewater Treatment

Considering the detrimental environmental effects of extensive chemical fertilizer use, which counters the sustainable agriculture goals and contribute to problems such as soil and water pollution, biofertilizers derived from algae emerge as a promising alternative. Algae are known to release plant growth regulators, including auxin and cytokinin, as well as essential macro and microelements, amino acids, and vitamins. These compounds enhance cellular metabolism and contribute to increased crop yield (Mazur et al. 2001). Seaweed-based fertilizers are widely acknowledged for their efficacy as biological fertilizers. With notable nitrogen, phosphorous, and potassium content, they play a crucial role in enhancing soil fertility. The fiber content in algae plays a crucial role in soil texture softening and moisture retention. Marine algae such as *Enteromorpha flexuosa*, *Ascophyllum nodosum*, *Laminaria digitata*, and *Fucus vesiculosus* exhibit potential for sustainable biofertilizer production, although this is currently considered a secondary goal. The thermochemical conversion process known as pyrolysis, which transforms biomass into bio-oil, syngas, and biochar, provides another avenue for generating biofertilizer from algae (Lehmann 2006). Furthermore, the utilization of biofertilizers obtained from algae not only supplies nutrients to the soil but also minimizes environmental impact while enhancing soil biodiversity.

Moreover, algae have emerged as significant players in the wastewater treatment sector. Their capability to capture plant nutrients, heavy metals, pesticides, and toxic compounds makes them valuable in this context. Algae's photosynthetic activity generates O_2, promoting aerobic bacterial degradation of organic compounds, and, in turn, provides CO_2, facilitating photosynthesis and the uptake of inorganic nutrients (Arnold 2013). Algae can be employed for various wastewater treatment purposes, including the removal of coliform bacteria, reducing chemical (COD) and biochemical oxygen demand (BOD), and removing N and P (Trivedi et al. 2015). Particularly, green algae such as *Euglena*, *Chlamydomonas*, and *Chlorella*, when introduced into large and shallow wastewater treatment tanks, offer a rapid and effective means of treating wastewater. These algae consume nitrates and phosphates, releasing oxygen through photosynthesis, thereby enhancing the mobility of aerobic bacteria to decompose wastewater raw materials. Additionally, algal species like *Hydrodictyon* sp., *Oedogonium* sp., and *Rhizoclonium* sp. can be utilized for the bioremediation of heavy metals from wastewater (Saunders et al. 2012). Given their demonstrated efficacy in wastewater treatment, the widespread adoption of algae technology in this domain is anticipated in the future.

2.5 Environmental Implications of Algae Biorefinery

The holistic approach of algae biorefinery to produce diverse products contributes to the overall environmental sustainability and economic viability of the process. The attainment of carbon neutrality in algae biorefinery becomes feasible when algae growth is sustained through wastewater and flue gas treatment. This not only purifies the environment but also yields enriched algae biomass that can undergo further processing for the generation of high-value products. Algae play a crucial role in mitigating anthropogenic emissions by serving as a sink for atmospheric CO_2. Their ability to accumulate inorganic carbon in their cytoplasm enables them to fix both CO_2 and industrial flue gases (Pires et al. 2012). Algae offer a versatile capability to sequester CO_2 from three distinct origins: atmospheric CO_2, emissions from power plants and industrial activities, and CO_2 derived from soluble carbonates. The absorption of atmospheric CO_2 constitutes a fundamental mechanism, serving as a carbon sink and relying on the transfer of mass from air to algae within their aquatic growth environments during the process of photosynthesis. Utilizing flue gas emissions from industrial processes, such as those from fuel-fired power plants, as a CO_2 source for microalgal growth holds significant promise. This approach has the potential to effectively reduce CO_2 levels and represents a highly encouraging alternative to current strategies for mitigating GHG emissions. Numerous studies have highlighted the tolerance and growth capacity of various algal species at elevated CO_2 levels. For instance, *Chlorococcum littorale*, a marine algae species, exhibits tolerance and growth at concentrations as high as 40% CO_2. Similarly, *Scenedesmus obliquus* and *Spirulina* sp. have demonstrated exceptional efficacy in fixing CO_2 (Trivedi et al. 2015). The microalga species *Selenastrum* sp. distinguishes itself with the unique capability to utilize carbonate salt and CO_2 as carbon sources. This is complemented by its high extracellular carbonic anhydrase activity, which facilitates efficient CO_2 assimilation (Huertas et al. 2000). While the current challenges, such as the high production cost of biodiesel, pose hurdles to its commercialization, the utilization of microalgae as a biofuel resource emerges as a potential solution. This approach not only addresses the hurdles but also contributes to CO_2 biomitigation by algae.

Likewise, algae cultivation utilizing wastewater emerges as a promising strategy both economically and environmentally, for achieving algae growth while addressing environmental concerns. This approach presents a unique opportunity for wastewater treatment and nutrient supplementation for algae growth, utilizing nutrient-rich wastewater. The advantages include the recycling of nutrients into algae biomass, reduced treatment costs, and the production of enriched biomass. Furthermore, it mitigates the impact of sewage effluent and industrial nitrogenous waste, contributing to biodiversity. The removal of nitrogen and carbon from wastewater helps to prevent the eutrophication of water bodies (Suganya et al. 2016). Previous studies have reported successful wastewater treatment using algae. For example, Aslan and Kapdan (2006) achieved average nitrogen and phosphorus removal efficiencies of 72% and 28%, respectively, using *Chlorella vulgaris*. Martınez et al. (2000) demonstrated significant phosphorus and nitrogen removal from urban wastewater using *S. obliquus*. Hodaifa et al. (2008) observed a 67.4% decrease in BOD_5 levels through the cultivation of *S. obliquus* in 25% diluted industrial wastewater derived from olive oil extraction. Algae such as *Chlorella* sp., *Scenedesmus* sp., *Spirulina* sp., *Nannochloropsis*, *Botryococcus braunii*, and *Phormidium bohneri* have been extensively employed to remove contaminants from wastewater due to their distinct molecular mechanisms for distinguishing between essential and harmful substances (Suganya et al. 2016). Algae can be employed for the biodegradation of hazardous or toxic compounds, generating the necessary oxygen for bacteria to facilitate the biodegradation of pollutants like polycyclic aromatic hydrocarbons (PAHs), phenolics, and organic solvents. The amount of oxygen produced through photosynthesis in microalgal cultivation diminishes or eliminates the necessity for external mechanical aeration. Muñoz and Guieysse (2006) demonstrated that *Spirulina* sp. functions as a biosorbent, effectively absorbing heavy metals such as Cr^{3+}, Cd^{2+}, and Cu^{2+}. The integrated approach to algal biorefinery holds immense potential as the global community strives toward achieving sustainability goals.

2.6 Conclusion and Future Perspective

The vast potential of algae biomass lies in its ability to generate a diverse array of products owing to its varied biochemical composition. Consequently, relying solely on the production of biofuels is not economically feasible, and the viability of alternative options hinges on economic considerations. Biorefineries emerge as a valuable approach for maximizing the value of algal biomass by targeting various compounds for different markets. A successful algae biorefinery implementation can yield not only biofuels but also a range of non-energy products. However, challenges persist, particularly in integrating diverse biomass conversion technologies within a comprehensive algae biorefinery operation. A crucial initial step involves establishing technologies for the commercialization of both fuel and non-fuel-based products. The extraction of products, whether energy or non-energy, is heavily reliant on the content of lipids, carbohydrates, and proteins, which varies based on the nature of algae, specific strains, and growth conditions. Therefore, making intelligent choices regarding algal strains and growth conditions is essential to obtaining the desired products. To achieve a successful algae biorefinery, considerations must extend to the marketing potential of the product and its integration with various industries. In instances where algae find application in the food industry, strict compliance with regulations set by the Food and Drug Administration (FDA) is imperative to ensure safety for human consumption. Efficient biomass generation is a prerequisite for building a successful biorefinery. Hence, capturing CO_2 from wastewater through algae provides a sustainable solution to mitigate emissions of this gas into the atmosphere, simultaneously leading to reduced operational costs. Nevertheless, research studies are essential to evaluate integration strategies for optimizing algal growth and achieving desired pollutant concentrations in the final effluent. The successful large-scale implementation of bioremediation of wastewater and CO_2 biofixation by algae could significantly elevate interest in current bio-based industries. Combining algal farming, novel byproduct biosynthesis, and biofuel production using a biorefinery strategy is anticipated to enhance the overall cost-effectiveness of biofuels derived from algae technology. This dual-benefit approach not only serves as a biomass source for biofuel production but also minimizes environmental impact. However, it is crucial to persist in investing and exploring the potential of algae biorefineries, propelling further research and development in this promising field. This will pave the way for innovative solutions to environmental and energy challenges, shaping a more sustainable future for generations to come.

References

Abdullah B, Muhammad SAF, Shokravi Z, Ismail S, Kassim KA, Mahmood AN, Aziz MMA (2019) Fourth generation biofuel: a review on risks and mitigation strategies. Renew Sust Energ Rev 107:37–50

Adarme-Vega TC, Lim DK, Timmins M, Vernen F, Li Y, Schenk PM (2012) Microalgal biofactories: a promising approach towards sustainable omega-3 fatty acid production. Microb Cell Factories 11(1):1–10

Andrade LR, Salgado LT, Farina M, Pereira MS, Mourao PA, Amado Filho GM (2004) Ultrastructure of acidic polysaccharides from the cell walls of brown algae. J Struct Biol 145(3):216–225

Andrade L, Batista F, Lira T, Barrozo M, Vieira L (2018) Characterization and product formation during the catalytic and non-catalytic pyrolysis of the green microalgae Chlamydomonas reinhardtii. Renew Energy 119:731–740

Arnold M (2013) Sustainable algal biomass products by cultivation in waste water flows. Phytochem Rev 22:969–991

Aslan S, Kapdan IK (2006) Batch kinetics of nitrogen and phosphorus removal from synthetic wastewater by algae. Ecol Eng 28(1):64–70

Aysu T, Fermoso J, Sanna A (2018) Ceria on alumina support for catalytic pyrolysis of Pavlova sp. microalgae to high-quality bio-oils. J Energy Chem 27(3):874–882

Bae YJ, Ryu C, Jeon J-K, Park J, Suh DJ, Suh Y-W, Chang D, Park Y-K (2011) The characteristics of bio-oil produced from the pyrolysis of three marine macroalgae. Bioresour Technol 102(3):3512–3520

Barsanti L, Gualtieri P (2022) Algae: anatomy, biochemistry, and biotechnology. CRC Press, Boca Raton

Bastiaens L, Van Roy S, Thomassen G, Elst K (2017) Biorefinery of algae: technical and economic considerations. In: Microalgae-based biofuels and bioproducts. Elsevier, Amsterdam, pp 327–345

Becker EW (2007) Micro-algae as a source of protein. Biotechnol Adv 25(2):207–210

Brennan L, Owende P (2010) Biofuels from microalgae—a review of technologies for production, processing, and extractions of biofuels and co-products. Renew Sust Energ Rev 14(2):557–577

Chan C-X, Ho C-L, Phang S-M (2006) Trends in seaweed research. Trends Plant Sci 11(4):165–166

Chandra R, Arora S, Rohit M, Mohan SV (2015) Lipid metabolism in response to individual short chain fatty acids during mixotrophic mode of microalgal cultivation: influence on biodiesel saturation and protein profile. Bioresour Technol 188:169–176

Chandra R, Iqbal HM, Vishal G, Lee H-S, Nagra S (2019) Algal biorefinery: a sustainable approach to valorize algal-based biomass towards multiple product recovery. Bioresour Technol 278:346–359

Chen C-Y, Yeh K-L, Aisyah R, Lee D-J, Chang J-S (2011) Cultivation, photobioreactor design and harvesting of microalgae for biodiesel production: a critical review. Bioresour Technol 102(1):71–81

Chen W-H, Huang M-Y, Chang J-S, Chen C-Y (2014) Thermal decomposition dynamics and severity of microalgae residues in torrefaction. Bioresour Technol 169:258–264

Christaki E, Florou-Paneri P, Bonos E (2011) Microalgae: a novel ingredient in nutrition. Int J Food Sci Nutr 62(8):794–799

Clark JH, Deswarte F (2014) Introduction to chemicals from biomass. Wiley, Hoboken

Daroch M, Geng S, Wang G (2013) Recent advances in liquid biofuel production from algal feedstocks. Appl Energy 102:1371–1381

de Jesus Raposo MF, de Morais RMSC, de Morais AMMB (2013) Health applications of bioactive compounds from marine microalgae. Life Sci 93(15):479–486
Demirbas A (2009) Progress and recent trends in biodiesel fuels. Energy Convers Manag 50(1):14–34
Dhargalkar V, Pereira N (2005) Seaweed: promising plant of the millennium. Sci Cult 71(3-4):60–66
Ding L, Cheng J, Lin R, Deng C, Zhou J, Murphy JD (2020) Improving biohydrogen and biomethane co-production via two-stage dark fermentation and anaerobic digestion of the pretreated seaweed Laminaria digitata. J Clean Prod 251:119666
Duan P, Jin B, Xu Y, Wang F (2015) Co-pyrolysis of microalgae and waste rubber tire in supercritical ethanol. Chem Eng J 269:262–271
Erdman JW Jr, Macdonald IA, Zeisel SH (2012) Present knowledge in nutrition. Wiley, Hoboken
Goh BHH, Ong HC, Cheah MY, Chen W-H, Yu KL, Mahlia TMI (2019) Sustainability of direct biodiesel synthesis from microalgae biomass: a critical review. Renew Sust Energ Rev 107:59–74
Harun R, Danquah MK, Forde GM (2010) Microalgal biomass as a fermentation feedstock for bioethanol production. J Chem Technol Biotechnol 85(2):199–203
Hodaifa G, Martínez ME, Sánchez S (2008) Use of industrial wastewater from olive-oil extraction for biomass production of Scenedesmus obliquus. Bioresour Technol 99(5):1111–1117
Huertas IE, Colman B, Espie GS, Lubian LM (2000) Active transport of CO_2 by three species of marine microalgae. J Phycol 36(2):314–320
John RP, Anisha G, Nampoothiri KM, Pandey A (2011) Micro and macroalgal biomass: a renewable source for bioethanol. Bioresour Technol 102(1):186–193
Kashyap M, Chakraborty S, Kumari A, Rai A, Varjani S, Vinayak V (2023) Strategies and challenges to enhance commercial viability of algal biorefineries for biofuel production. Bioresour Technol 387:129551
Katiyar R, Bharti RK, Gurjar BR, Kumar A, Biswas S, Pruthi V (2018) Utilization of de-oiled algal biomass for enhancing vehicular quality biodiesel production from Chlorella sp. in mixotrophic cultivation systems. Renew Energy 122:80–88
Khanthong K, Kadam R, Kim T, Park J (2023) Synergetic effects of anaerobic co-digestion of food waste and algae on biogas production. Bioresour Technol 382:129208
Kumar K, Mella-Herrera RA, Golden JW (2010) Cyanobacterial heterocysts. Cold Spring Harb Perspect Biol 2(4):a000315
Kumar PK, Krishna SV, Verma K, Pooja K, Bhagawan D, Srilatha K, Himabindu V (2018) Bio oil production from microalgae via hydrothermal liquefaction technology under subcritical water conditions. J Microbiol Methods 153:108–117
Lee DS, Pitari G, Grewe V, Gierens K, Penner JE, Petzold A, Prather M, Schumann U, Bais A, Berntsen T (2010) Transport impacts on atmosphere and climate: aviation. Atmos Environ 44(37):4678–4734
Lee XJ, Ong HC, Gan YY, Chen W-H, Mahlia TMI (2020) State of art review on conventional and advanced pyrolysis of macroalgae and microalgae for biochar, bio-oil and bio-syngas production. Energy Convers Manag 210:112707
Lee J, Jang H, Kang S, Kim K, Park J (2022) Shockwave pre-treatment enhances the physicochemical availability and anaerobic mono-and co-digestion of highly concentrated algae. J Environ Chem Eng 10(6):108993
Lehmann J (2006) Black is the new green. Nature 442:624–626
Limayem A, Ricke SC (2012) Lignocellulosic biomass for bioethanol production: current perspectives, potential issues and future prospects. Prog Energy Combust Sci 38(4):449–467
Liu X, Saydah B, Eranki P, Colosi LM, Mitchell BG, Rhodes J, Clarens AF (2013) Pilot-scale data provide enhanced estimates of the life cycle energy and emissions profile of algae biofuels produced via hydrothermal liquefaction. Bioresour Technol 148:163–171
López-González D, Fernandez-Lopez M, Valverde J, Sanchez-Silva L (2014) Pyrolysis of three different types of microalgae: kinetic and evolved gas analysis. Energy 73:33–43
Mandotra S, Sharma C, Srivastava N, Ahluwalia A, Ramteke P (2021) Current prospects and future developments in algal bio-hydrogen production: a review. Biomass Convers Biorefinery 13:1–18

Markou G, Angelidaki I, Georgakakis D (2012) Microalgal carbohydrates: an overview of the factors influencing carbohydrates production, and of main bioconversion technologies for production of biofuels. Appl Microbiol Biotechnol 96:631–645

Martınez M, Sánchez S, Jimenez J, El Yousfi F, Munoz L (2000) Nitrogen and phosphorus removal from urban wastewater by the microalga Scenedesmus obliquus. Bioresour Technol 73(3):263–272

Maurya R, Paliwal C, Ghosh T, Pancha I, Chokshi K, Mitra M, Ghosh A, Mishra S (2016) Applications of de-oiled microalgal biomass towards development of sustainable biorefinery. Bioresour Technol 214:787–796

Mazur H, Konop A, Synak R (2001) Indole-3-acetic acid in the culture medium of two axenic green microalgae. J Appl Phycol 13:35–42

McHugh DJ (2003) A guide to the seaweed industry. FAO Fish Tech Pap 441:105

Mehta P, Singh D, Saxena R, Rani R, Gupta RP, Puri SK, Mathur AS (2018) High-value coproducts from algae—an innovational way to deal with advance algal industry. In: Waste to wealth. Springer, Cham, pp 343–363

Milano J, Ong HC, Masjuki H, Chong W, Lam MK, Loh PK, Vellayan V (2016) Microalgae biofuels as an alternative to fossil fuel for power generation. Renew Sust Energ Rev 58:180–197

Mohan SV, Hemalatha M, Chakraborty D, Chatterjee S, Ranadheer P, Kona R (2020) Algal biorefinery models with self-sustainable closed loop approach: trends and prospective for blue-bioeconomy. Bioresour Technol 295:122128

Muñoz R, Guieysse B (2006) Algal–bacterial processes for the treatment of hazardous contaminants: a review. Water Res 40(15):2799–2815

Najafi G, Ghobadian B, Yusaf TF (2011) Algae as a sustainable energy source for biofuel production in Iran: a case study. Renew Sust Energ Rev 15(8):3870–3876

Nguyen MT, Choi SP, Lee J, Lee JH, Sim SJ (2009) Hydrothermal acid pretreatment of Chlamydomonas reinhardtii biomass for ethanol production. J Microbiol Biotechnol 19(2):161–166

Notoya M (2010) Production of biofuel by macroalgae with preservation of marine resources and environment. In: Seaweeds and their role in globally changing environments. Springer, Cham, pp 217–228

Olabi A, Shehata N, Sayed ET, Rodriguez C, Anyanwu RC, Russell C, Abdelkareem MA (2023) Role of microalgae in achieving sustainable development goals and circular economy. Sci Total Environ 854:158689

Pires JC (2017) COP21: the algae opportunity? Renew Sust Energ Rev 79:867–877

Pires J, Alvim-Ferraz M, Martins F, Simões M (2012) Carbon dioxide capture from flue gases using microalgae: engineering aspects and biorefinery concept. Renew Sust Energ Rev 16(5):3043–3053

Raheem A, Azlina WW, Yap YT, Danquah MK, Harun R (2015) Thermochemical conversion of microalgal biomass for biofuel production. Renew Sust Energ Rev 49:990–999

Rawat I, Kumar RR, Mutanda T, Bux F (2011) Dual role of microalgae: phycoremediation of domestic wastewater and biomass production for sustainable biofuels production. Appl Energy 88(10):3411–3424

Ross A, Anastasakis K, Kubacki M, Jones J (2009) Investigation of the pyrolysis behaviour of brown algae before and after pre-treatment using PY-GC/MS and TGA. J Anal Appl Pyrolysis 85(1-2):3–10

Sahu A, Pancha I, Jain D, Paliwal C, Ghosh T, Patidar S, Bhattacharya S, Mishra S (2013) Fatty acids as biomarkers of microalgae. Phytochemistry 89:53–58

Salami R, Kordi M, Bolouri P, Delangiz N, Asgari Lajayer B (2021) Algae-based biorefinery as a sustainable renewable resource. Circ Econ Sustain 1:1349–1365

Sambusiti C, Bellucci M, Zabaniotou A, Beneduce L, Monlau F (2015) Algae as promising feedstocks for fermentative biohydrogen production according to a biorefinery approach: a comprehensive review. Renew Sust Energ Rev 44:20–36

Sandquist J, Guell BM (2012) Overview of biofuels for aviation. Chem Eng Trans 29:1147–1152

Sathasivam R, Radhakrishnan R, Hashem A, Abd Allah EF (2019) Microalgae metabolites: a rich source for food and medicine. Saudi J Biol Sci 26(4):709–722
Saunders RJ, Paul NA, Hu Y, de Nys R (2012) Sustainable sources of biomass for bioremediation of heavy metals in waste water derived from coal-fired power generation. PLoS One 7(5):e36470
Searchinger T, Heimlich R, Houghton RA, Dong F, Elobeid A, Fabiosa J, Tokgoz S, Hayes D, Yu T-H (2008) Use of US croplands for biofuels increases greenhouse gases through emissions from land-use change. Science 319(5867):1238–1240
Sharmila VG, Banu JR, Kumar MD, Kumar SA, Kumar G (2022) Algal biorefinery towards decarbonization: economic and environmental consideration. Bioresour Technol 364:128103
Singh J, Thakur IS (2015) Evaluation of cyanobacterial endolith Leptolyngbya sp. ISTCY101, for integrated wastewater treatment and biodiesel production: a toxicological perspective. Algal Res 11:294–303
Sonnenberg A, Baars J, Hendrickx P (2007) IEA bioenergy task 42 biorefinery. Wageningen University and Research Centre, Wageningen
Sudhakar K, Mamat R, Samykano M, Azmi W, Ishak W, Yusaf T (2018) An overview of marine macroalgae as bioresource. Renew Sust Energ Rev 91:165–179
Suganya T, Varman M, Masjuki H, Renganathan S (2016) Macroalgae and microalgae as a potential source for commercial applications along with biofuels production: a biorefinery approach. Renew Sust Energ Rev 55:909–941
Talebi AF, Mohtashami SK, Tabatabaei M, Tohidfar M, Bagheri A, Zeinalabedini M, Mirzaei HH, Mirzajanzadeh M, Shafaroudi SM, Bakhtiari S (2013) Fatty acids profiling: a selective criterion for screening microalgae strains for biodiesel production. Algal Res 2(3):258–267
Tiwari BK, Troy DJ (2015) Seaweed sustainability–food and nonfood applications. In: Seaweed sustainability. Elsevier, pp 1–6
Trivedi J, Aila M, Bangwal D, Kaul S, Garg M (2015) Algae based biorefinery—how to make sense? Renew Sust Energ Rev 47:295–307
Ullah K, Ahmad M, Sharma VK, Lu P, Harvey A, Zafar M, Sultana S, Anyanwu C (2014) Algal biomass as a global source of transport fuels: overview and development perspectives. Prog Nat Sci Mater Int 24(4):329–339
Uma V, Usmani Z, Sharma M, Diwan D, Sharma M, Guo M, Tuohy MG, Makatsoris C, Zhao X, Thakur VK (2023) Valorisation of algal biomass to value-added metabolites: emerging trends and opportunities. Phytochem Rev 22(4):1015–1040
Wang S, Wang Q, Jiang X, Han X, Ji H (2013) Compositional analysis of bio-oil derived from pyrolysis of seaweed. Energy Convers Manag 68:273–280
Whyte JN, Englar JR (1981) The agar component of the red seaweed Gelidium purpurascens. Phytochemistry 20(2):237–240
Wijffels RH, Barbosa MJ, Eppink MH (2010) Microalgae for the production of bulk chemicals and biofuels. Biofuels Bioprod Biorefin 4(3):287–295
Yamasaki A (2003) An overview of CO_2 mitigation options for global warming—emphasizing CO_2 sequestration options. J Chem Eng Jpn 36(4):361–375
Yanik J, Stahl R, Troeger N, Sinag A (2013) Pyrolysis of algal biomass. J Anal Appl Pyrolysis 103:134–141
Yin S, Dolan R, Harris M, Tan Z (2010) Subcritical hydrothermal liquefaction of cattle manure to bio-oil: effects of conversion parameters on bio-oil yield and characterization of bio-oil. Bioresour Technol 101(10):3657–3664

Cultivation of Algae: Techniques and Challenges

3

Chaitanya Kumar, Monika Sharma, Manvinderpreet Kaur, Shailendra Kumar Arya, and Madhu Khatri

Contents

C. Kumar
Centre for Nano-Science and Nano-Technology, Panjab University, Chandigarh, India

M. Sharma · M. Kaur · S. K. Arya · M. Khatri (✉)
University Institute of Engineering Technology, Panjab University, Chandigarh, India

S. K. Arya et al. (eds.), *Value Added Products From Bioalgae Based Biorefineries: Opportunities and Challenges*,
https://doi.org/10.1007/978-981-97-1662-3_3

Abstract

Algal farming, especially microalgae cultivation, has great promise for several uses, such as the generation of biofuel, food additives, nutraceuticals and wastewater cleanup. However, there are a number of obstacles to successful large-scale cultivation, including cost-effectiveness, growth parameter optimisation and efficient technology. The main methods used in algal production are explored in this chapter, along with their benefits and drawbacks. These methods include heterotrophic fermentation, closed photobioreactors and open pond systems. It goes on to discuss the main obstacles to wider adoption, including harvesting, light management, nutritional control and contamination. In order to overcome these obstacles and realise the full potential of algae as a resource-rich and sustainable solution for a wide range of global concerns, the chapter concludes by highlighting the necessity of technical developments and creative methods.

Keywords

Microalgae cultivation · Biofuel · Food supplements · Wastewater remediation

Abbreviations

ATP	Adenosine triphosphate
C	Carbon
Ca	Calcium
Ca^{2+}	Calcium ions
Co	Cobalt
CO_2	Carbon dioxide
Cu	Copper
Fe	Iron
K	Potassium
K^+	Potassium ions
Mg	Magnesium
Mg^{2+}	Magnesium ions
Mn	Manganese
Mo	Molybdenum
N	Nitrogen
NH_4	Ammonium
NO_2	Nitrite
NO_3	Nitrate
P	Phosphorus
PBRs	Photobioreactors
PO_4^{-3}	Phosphate
S	Sulphur
SO_4^{-2}	Sulphate
Zn	Zinc

3.1 Introduction

There is a long history of algae farming since many societies have long valued these photosynthetic creatures (Fabris et al. 2020). Modern applications for algae farming have replaced more conventional ones throughout time, because of technological breakthroughs and increased public awareness of the advantages algae provide for the environment. Due to the great efficiency in absorbing solar energy and their capacity to synthesise organic molecules through photosynthesis, algae are seen as a possible source for biofuels (Shuba and Kifle 2018). Additionally, they are necessary for nutrient cycling and bioremediation, which efficiently remove pollutants and nutrients from wastewater and promote environmental sustainability (Pande et al. 2020). Algal products have potential health advantages and nutritional value and are utilised in the culinary, cosmetic and pharmaceutical sectors (Wells et al. 2017). They are rich in critical nutrients, including macro and trace elements, vitamins, proteins and bioactive chemicals. As the main primary producers in aquatic environments and the source of almost half of the oxygen produced worldwide, algae play a vital and important role in the ecosystem (Raven and Geider 2003). The issue of algae cultivation is gaining more attention due to its relevance in several fields (Kumar et al. 2020). The search for sustainable and alternative energy sources is one of the main forces. Because algae can be transformed into biofuels and are high in lipids, this presents a viable option for lowering reliance on fossil fuels and lessening the effects on the environments. In addition to producing biofuel, algae farming is essential for treating wastewater (Kube et al. 2018). As they have capacity to absorb elements like phosphorus (P) and nitrogen (N), algae may help purify bodies of water. Algae's ability to function as an environmentally friendly wastewater treatment agent and a renewable energy source highlights how important they are in the fight against today's environmental problems (Ahmad et al. 2022), as shown in Fig. 3.1. Furthermore, algae farming expands into the food and nutrition sectors (Matos 2017). Some algae species have high protein, vitamin and essential fatty acid content, which makes them useful as functional food components and nutritional supplements. The possibility of algae as a nutrient-dense, sustainable food source is growing in importance as the world's population rises.

Diverse photosynthetic creatures, algae may be found in a wide range of habitats, from freshwater to marine settings (Neto and Pinto 2018). They include a vast variety of organisms, ranging from massive seaweeds to tiny phytoplankton. For growth techniques to be efficient, it is necessary to comprehend the diversity found within the kingdom of algae. Algae are distinguished by their capacity to use photosynthesis, a process essential to the Earth's carbon (C) cycle, to transform sunlight into energy (Zeng et al. 2021). Due to this special aptitude, algae are positioned to play a major role in both the creation of oxygen and the sequestration of carbon dioxide (CO_2), underscoring their ecological significance. It is essential to comprehend the unique traits of many algae species in order to customise growth methods for maximum biomass production and targeted biochemical results (Veerabadhran et al. 2021). Algae are categorised using a taxonomy that comprises several groupings including diatoms, brown, red and green algae (Bellinger and Sigee 2015). Different

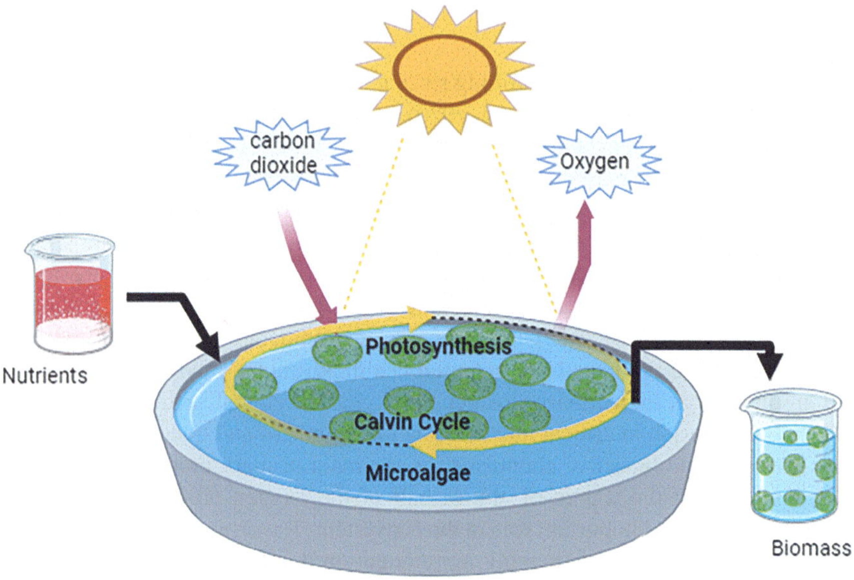

Fig. 3.1 Cultivation of algae

groups will have different traits, and these distinctions may influence how each group cultivates. For instance, because of their high lipid content, green algae, which are frequently found in freshwater habitats, are frequently the subject of biofuel production attempts (Rastogi et al. 2018). Algae have a wide range of uses outside of usual applications (Aditya et al. 2016). Algae are used in the pharmaceutical business to synthesise several chemicals having medicinal qualities in addition to producing biofuel. The cosmetics sector also takes use of algae's potential to produce skincare products that are high in antioxidants and other healthy ingredients (Arora et al. 2021). Algae are also an important source of sustainable feed for aquaculture (Nagappan et al. 2021). The aquaculture industry's overall sustainability is enhanced by the use of algae as a nutrient-rich, renewable substitute for conventional fishmeal in fish and shrimp farms. In this chapter we will discuss the cultivation techniques and the challenges that are generally encountered during the algal cultivation techniques.

3.2 Cultivation Techniques

3.2.1 Open Pond System

As open pond systems, as depicted in Fig. 3.2, are managed and reasonably priced, it is a good way to grow algae. These systems are made up of sizable outdoor ponds that get direct sunlight, allowing algae to thrive and multiply through photosynthesis (Gupta et al. 2015). The technique has been widely used in many different

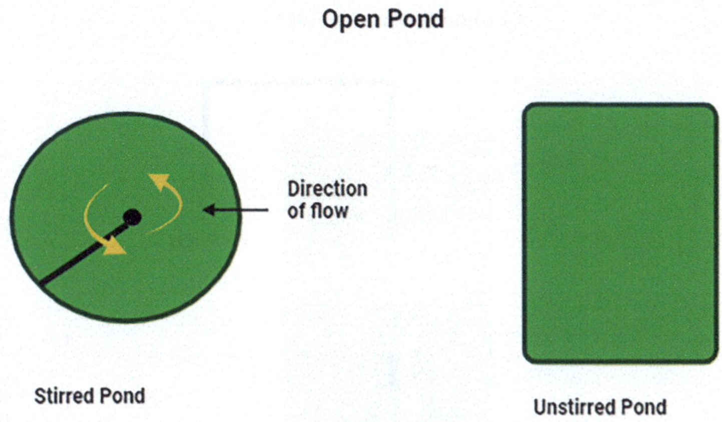

Fig. 3.2 Open pond for algae cultivation

applications, such as the creation of nutritional supplements and biofuel. They are usually shallow, so they receive maximum sunshine exposure. Although the ponds' shapes might vary, they are often circular or rectangular (Dharmaprabhakaran et al. 2020). The layout makes it easier to grow a wide variety of algae species, each of which has unique environmental needs. The fact that open pond systems rely on direct sunshine is one of their main advantages (Raeisossadati et al. 2019). As photosynthetic creatures, algae use a process called photosynthesis to transform sunlight into energy. The technique is more affordable since it uses less artificial lighting due to its dependence on solar energy. Simple devices like paddlewheels are frequently used in ponds to move the water around and keep it from stagnating (Yen et al. 2019). This circulation keeps the environment well-mixed for the best development conditions while guaranteeing that the algae receive carbon dioxide and nutrients. They are so scalable, hence expanding the cultivation area is a simple process. This scalability is essential for adjusting to changing production needs, which qualifies the approach for use in both industrial and research settings. Open pond systems are susceptible to pollution since they are open to the outside world (Bwire et al. 2020). The weather has an impact on the productivity of open pond systems (Quiroz et al. 2021). Variations in temperature and precipitation, and harsh weather can all have an effect on the pace of development of algae and the total amount of biomass produced. Water evaporation can occur in open ponds, particularly in areas with high temperatures (Kleiman et al. 2021). This may cause the remaining water's nutrient concentration to rise, which might have an impact on the growth of algae.

3.2.2 Closed Photobioreactors (PBRs)

Closed photobioreactors or PBRs, in Fig. 3.3, are an advanced, well-managed algae cultivation technology (Chang et al. 2017). PBRs are enclosed settings that provide

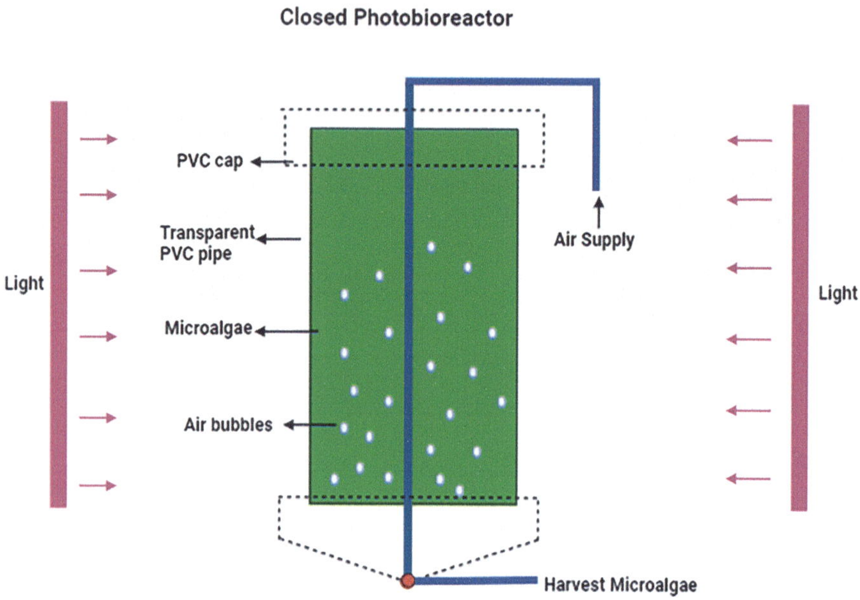

Fig. 3.3 Closed photobioreactor for algae cultivation

precise control over a range of characteristics, such as light, temperature and fertiliser delivery, in contrast to open pond systems. With its benefits in improved growing conditions and contaminant control, this technique is commonly used in research and industrial settings. Usually made of transparent materials like plastic or glass, closed PBRs provide enclosed systems that offer a regulated environment for the growing of algae (Koller 2015). The closed construction allows for exact control of cultivation conditions while reducing the possibility of external contamination. PBRs provide exact control over how much light is exposed (Lau et al. 2022). The best possible light quality and intensity for photosynthesis may be achieved by placing artificial lighting devices, such Light Emitting Diodes, at strategic locations. This ability to regulate light levels helps algae develop in a predictable and repeatable manner. The growth of thermophilic or mesophilic algae species is supported by stable culture conditions that are created by keeping the PBR at a constant temperature (Jankowska et al. 2017). In closed PBRs, the delivery of nutrients may be carefully controlled (Zhang et al. 2020). Algae get the ideal ratio of nitrogen, phosphate (PO_4^{-3}) and other vital nutrients, thanks to this careful nutrient control. Systems for recycling nutrients can be included to reduce their negative effects on the environment. Closed systems are appropriate for both large-scale industrial applications and small-scale research because they are frequently more scalable than open pond systems (Borowitzka and Vonshak 2017). It may need a larger initial investment than open pond systems, which might be prohibitive, particularly for smaller-scale enterprises (Enzmann et al. 2019). The energy required to maintain regulated conditions and the requirement for sterilisation operations to avoid contamination may result in higher operating costs for closed systems. Closed photobioreactors can be quite intricate technically, requiring knowledge of their design, use and upkeep. Operators may find this intricacy challenging, particularly if they lack specialised skills (Chang et al. 2017). It can be difficult to provide homogeneous light dispersion inside the reactor, particularly in large-scale systems. Variations in the proliferation of microorganisms might result from uneven light exposure.

3.2.3 Raceway Ponds

Raceway ponds shown in Fig. 3.4 are used to cultivate microorganisms, mainly cyanobacteria and microalgae. Usually shallow, round or oval basins, these ponds use paddlewheels or other mechanical devices to move the water around (Shen 2021). Raceway ponds are a popular and economical option for large-scale algal growing, particularly for uses including the generation of high-value chemicals and biofuel and wastewater treatment. The layout makes it easier for water to flow, which promotes effective mixing and exposes microbes to light (Rogers et al. 2014). To generate flow in the pond, paddlewheels and other mechanical devices are employed. This circulation aids in the growth of microorganisms, guarantees the uniform distribution of nutrients and light, and prevents sedimentation. Sunlight from the sun is necessary for photosynthesis in raceway ponds (Ritchie and Larkum 2012). This lowers the energy expenses related to artificial illumination, increasing the system's overall cost-effectiveness. Raceway ponds can be utilised for a number of purposes, such as wastewater treatment, high-value chemical synthesis, biofuel generation and nutraceutical manufacturing. These systems' adaptability to various purposes originates from their flexibility. Raceway ponds' open design facilitates

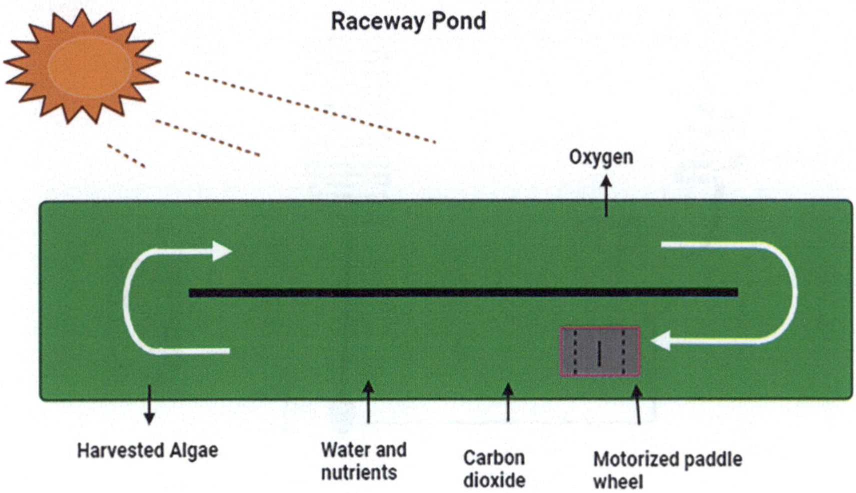

Fig. 3.4 Raceway pond cultivation for algae

natural gas exchange with the environment, which encourages the interchange of carbon dioxide and oxygen and may be advantageous for the growth of microorganisms (Kumar et al. 2015). Raceway ponds are dependent on sunlight, which comes from the weather. Seasonal fluctuations in sunshine or cloudy days might have an effect on productivity (White and Ryan 2015). Raceway ponds need a large amount of land, and the site may be constrained by factors like climate and available area. It is more vulnerable to outside contaminants like insects, competing microbes or pollution. Controlling contamination can be difficult (McBride et al. 2017).

3.2.4 Tubular Photobioreactors

Closed systems called tubular photobioreactors (PBRs), as shown in Fig. 3.5, are used to cultivate microorganisms, mainly cyanobacteria and microalgae (Chanquia et al. 2022). These reactors are made up of long, transparent tubes that are used to circulate a culture media that contains microorganisms. The purpose of tubular photobioreactors (PBRs) is to efficiently absorb solar radiation while maintaining regulated growth conditions. Transparent tubes are stacked in a variety of patterns, including coils, spirals, and horizontal or vertical loops, to create tubular PBRs. The tube's layout optimises the amount of sunlight that microbes are exposed to. As they grow under carefully regulated circumstances and efficiently collect light, tubular PBRs are renowned for their great production. This qualifies them for uses like the

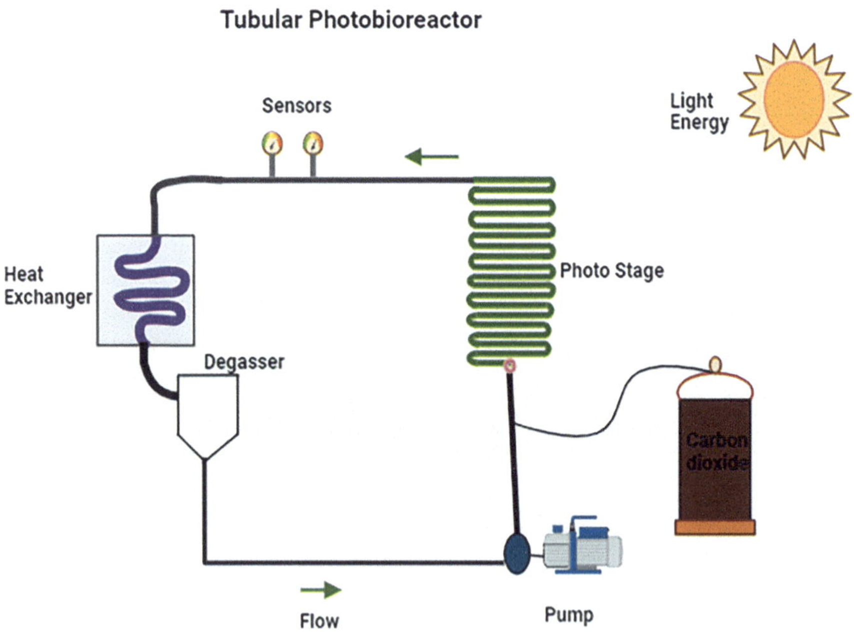

Fig. 3.5 Tubular photobioreactor algae cultivation

synthesis of high-value chemicals and the manufacture of biofuel (Xiaogang et al. 2020). Tubular PBRs are closed systems; they provide greater contamination control than open pond systems. The possibility of contamination from outside sources is reduced by the closed architecture.

They are made up with a comparatively tiny footprint, and are appropriate for spaces that are tight on space (Al-Dailami et al. 2022). This is advantageous in places where there is a shortage of land. The design and operation of tubular PBRs can be technically challenging, requiring knowledge in bioprocess optimisation and engineering (Chanquia et al. 2022). It can be difficult to provide consistent light dispersion throughout the tubes, particularly in large-scale systems. Changes in the exposure to light can affect the development of microorganisms. Because of the possibility of biofilm growth, fouling and contamination over time, cleaning and maintaining the tubes can be difficult (Torzillo and Chini Zittelli 2015). When building a tubular PBR, the initial capital expenses may be more than those of an open pond system. They could, nonetheless, be more affordable than certain closed systems.

3.2.5 Hybrid Systems

When discussing hybrid systems in the context of algal cultivation, it usually refers to the combining of several techniques or technologies to maximise the general productivity and efficiency of microbe culture (Xiong et al. 2021). Often designed to maximise the benefits of both open pond and closed photobioreactor systems while minimising their drawbacks, these hybrid systems incorporate elements from both types of systems, as depicted in Fig. 3.6.

3.2.5.1 Open Pond with Closed Photobioreactors

This hybrid method combines closed photobioreactors with open pond systems (Yun et al. 2018). To guarantee the purity of the culture and avoid contamination, the first microbe cultivation can be done in closed systems. The biomass may be moved to open ponds for further development in the presence of natural sunlight once it reaches a particular density.

3.2.5.2 Two-Stage Systems

In two-stage systems, several cultivating methodologies are used one after the other (Narala et al. 2016). For instance, the first phases of culture may take place in closed photobioreactors, and then the material would be transferred to open pond systems for further growth. This strategy benefits from the cost-effectiveness of open pond systems for later stages and enables more control over early growing conditions.

3.2.5.3 Indoor and Outdoor Cultivation

Certain hybrid methods blend outdoor growing in open ponds with indoor cultivation (e.g. in closed fermenters or photobioreactors) (Apel and Weuster-Botz 2015).

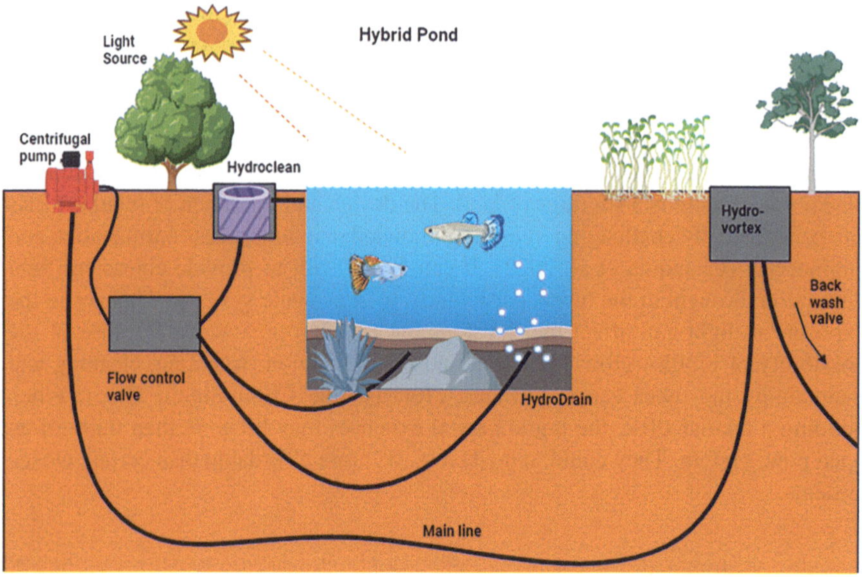

Fig. 3.6 Hybrid pond for algae cultivation

Early growth may be cultivated under controlled conditions indoors, and for larger-scale, more economical production, the biomass can be moved to open ponds.

3.2.5.4 Pond–Photobioreactor Integration

Pond and photobioreactor properties can be combined in hybrid systems (Liu et al. 2019). For instance, closed tubes or panels can be added to open pond systems to create controlled environments for particular farming phases. This allows for the financial benefits of open systems to be reaped while maintaining a high degree of productivity.

The goal of hybrid systems is to combine the benefits of many growing techniques, maximising output by using affordable open systems for large-scale growth and supplying controlled environments when necessary (Wicker et al. 2023). In hybrid setups, closed systems aid in preventing contamination in the first phases of cultivation, guaranteeing a clean culture that can be moved to open systems for additional growth (Ahmad et al. 2021).

In terms of accommodating particular needs and maximising resource usage in accordance with the traits of the microorganisms being grown, hybrid systems are flexible (Xiong et al. 2021). By fusing several technologies, hybrid systems can offer a more resilient and flexible production system by reducing the hazards connected with depending just on one cultivation strategy. It can be difficult to design, install and run hybrid systems; it takes knowledge of both closed and open cultivation techniques (Adesanya et al. 2014). The costs of putting hybrid systems into place might differ; thus it is important to do a thorough economic study to assess the viability and affordability of the hybrid strategy. In a hybrid setup it is difficult to

Table 3.1 Various algae cultivation techniques

Cultivation methods	Definition	Pros	Cons
Open ponds	Ponds that are shallow and exposed to natural sunlight	Low initial costs, scalability, and natural sunshine	The possibility of contamination, as well as the reliance on the weather
Photobioreactors that are closed	Closed systems with strict environmental controls	Contamination control and precise variable control	Capital costs are higher and energy consumption is high
Raceway ponds	Channels that are open and oval in shape, with paddlewheel circulation	Cost-effective, scalable, and natural sunlight utilisation	There is a possibility of contamination, and there is little environmental control
Photobioreactors in tubular form	Transparent, closed tubes for controlled algae cultivation	Productivity is high, and pollution is kept to a minimum	Capital expenditures and technical complexity
Systems with hybrid components	Optimisation through the integration of open and closed systems	Productivity and flexibility have been increased	Complexity, with the possibility of increased costs
Systems with two stages	The sequential administration of various cultivation technologies	Controlled growth stages, and adaptability	Technical complexities, as well as possible inefficiency
Cultivation both indoors and outside	Controlled indoor settings are combined with outside exposure	Adaptability and year-round operation	Capital expenditures and energy costs
Integration of a pond with a photobioreactor	Closed systems are used in open pond situations	Increased productivity and adaptability	Technical difficulties and potential complexities

smoothly integrate several cultivation methods, especially when it comes to control systems, monitoring and fertiliser supply (Miškovičová and Masojídek 2023). All pros and cons are presented in Table 3.1.

3.3 Nutrient Management

3.3.1 Nutrient Requirements

Carbon (C): Obtained from carbon dioxide (CO_2) during photosynthesis, carbon is a fundamental building block for organic compounds.

Nitrogen (N): Nitrogen is crucial for the synthesis of amino acids, proteins and chlorophyll. Common nitrogen sources include nitrate (NO_3), nitrite (NO_2) and ammonium (NH_4).

Phosphorus (P): Phosphorus is essential for the formation of nucleic acids (deoxyribonucleic acid [DNA] and ribonucleic acid [RNA]), adenosine triphosphate (ATP) and phospholipids. Phosphate (PO_4^{-3}) is a common phosphorus source.

Sulphur (S): Sulphur is a component of amino acids and proteins. Sulphate (SO_4^{-2}) is a common sulphur source for algae.

Potassium (K): Potassium is involved in enzyme activation, osmoregulation and photosynthesis. Potassium ions (K^+) are essential for cellular processes.

Calcium (Ca): Calcium is important for cell wall structure and stability. Calcium ions (Ca^{2+}) may be present in the culture medium.

Magnesium (Mg): Magnesium is a central component of chlorophyll and is essential for photosynthesis. Magnesium ions (Mg^{2+}) are often added to the culture medium.

Trace Elements (Micronutrients): Algae require trace elements in small amounts for various cellular functions. Common trace elements include iron (Fe), manganese (Mn), zinc (Zn), copper (Cu), molybdenum (Mo) and cobalt (Co).

Vitamins: Some algae may require vitamins for growth. B-vitamins, such as biotin and thiamine, are examples of vitamins that may be essential for certain algae.

3.4 Harvesting and Extraction

3.4.1 Harvesting Techniques

3.4.1.1 Centrifugation

A crucial separation method used in many biological processes, including algal culture, is centrifugation (Najjar and Abu-Shamleh 2020). Using this technique, a sample, typically an algal culture, is put inside a rotor, which is a spinning chamber. The denser components, such as the algae cells, travel outwards and form a compact pellet as the rotor spins at high speeds, creating centrifugal force, while the less dense liquid component, called the supernatant, stays closer to the centre.

The supernatant can then be properly decanted or aspirated to effectively separate the collected algal biomass. Centrifugation is a very efficient method for concentrating microorganisms, which makes it easier to extract important substances like lipids or proteins (Martínez et al. 2020). Applications for the scalable approach range from large-scale industrial operations to laboratory-scale studies. Critical elements affecting separation effectiveness and cell viability are centrifugation parameters, such as rotating speed and duration.

Centrifugation is a quick and effective method of collecting biomass, but it has certain disadvantages as well. These include the possibility of cell damage from strong pressures, high energy consumption, and the requirement for specialised equipment, all of which can raise operating costs (Milledge and Heaven 2013). Notwithstanding these drawbacks, centrifugation's adaptability and broad range of applications make it a vital instrument in the fields of biotechnology and algae culture.

3.4.1.2 Filtration

In algal cultivation, filtration is a harvesting process used to extract algal biomass from the liquid culture media (Singh and Patidar 2018). Using a porous medium, like a filter, this technique holds the algal cells in place while letting the liquid (supernatant) flow through. The properties of the algae species being grown determine the selection of filter material and pore size. Usually, the procedure is passing the algal culture through the filter, whereupon the algal cells collect and create a concentrated biomass on the filter surface. Filtration can be used at several operational sizes, ranging from large-scale industrial operations to laboratory studies.

Filtration has the benefits of simplicity, economy and scalability (Yin et al. 2020). It is very helpful for handling huge amounts of culture and uses less energy than certain other harvesting techniques. Filtration is also compatible with continuous processing, which makes it a good fit for several industrial uses. The possibility of filter clogging might provide difficulties, particularly in high-density cultures or ones that contain extracellular polymers. To guarantee peak performance, filtering settings must be optimised and maintained on a regular basis. With its benefits in simplicity and scalability, filtration is a flexible and effective method of harvesting algal biomass; nevertheless, its successful application requires careful consideration of the features of the filter and any clogging problems.

3.4.1.3 Flocculation

In algal culture, flocculation is a common harvesting technique that improves algal cell aggregation and sedimentation (Matter et al. 2019). The procedure is introducing flocculating agents, typically polymers, to the algal growth. Larger flocs are formed as a result of these chemicals' interactions with the algal cells, which neutralise charges or create bridges between cells. Flocculation facilitates the quick settling of algal biomass, which makes harvesting easier. The algal flocs settle more quickly under the effect of gravity once the flocculants are added. The biomass and the cleared liquid medium, referred to as the supernatant, may be readily separated once the biomass has settled.

Microalgae may be concentrated using this approach, which increases the effectiveness of later stages in the harvesting process. It is especially helpful for large-scale operations where efficient and quick biomass recovery is crucial (Li et al. 2020). Compared to certain other ways of harvesting, flocculation uses less energy and works with a variety of algae species. Notwithstanding its benefits, there are several drawbacks that should be considered, such as determining the ideal flocculant dose and if further dewatering procedures are necessary. The flocculation method for algal harvesting must be implemented carefully, taking into account both the flocculant employed and the properties of the algal strain.

3.5 Extraction Methods

3.5.1 Solvent Extraction

An essential step in the development of algae is solvent extraction, especially when trying to extract useful substances like pigments and lipids (Li et al. 2020). First, the biomass from the algae is harvested, dried and then ground to provide more surface area. Lipids and other target substances are then dissolved using solvents like ethanol or hexane. Following extraction, the solvent-rich phase is separated, and the extract is concentrated and the solvent is recovered via evaporation. For the synthesis of biodiesel from lipid-rich extracts, additional processing techniques like transesterification are frequently used. When choosing a solvent, safety, environmental effect and extraction efficiency must all be taken into account. Selectivity problems, scalability limitations and the process's economic feasibility are among the challenges (Zapata-Boada et al. 2022). In quest of possible benefits, researchers investigate alternate methods such as ultrasonic-assisted extraction and supercritical fluid extraction. Systems for recovering solvents can cut expenses and lessen their negative effects on the environment. To extract useful components from algae for use in nutraceuticals and biofuels, the whole solvent extraction process is essential.

3.5.2 Mechanical Methods

Physical pressures are used in mechanical techniques of algae extraction to break down cell structures and liberate intracellular chemicals. One method is bead beating, which involves agitating algal cells with beads to break down their cell walls and promote chemical release. Algae are exposed to extreme pressure during high-pressure homogenisation, which disrupts their cells. For smaller-scale extractions, mechanical shear, which involves pressures like mixing or cutting, is employed. High-frequency sound waves are used in ultrasound-assisted extraction to break down cell walls and improve drug release. Ball milling helps in the extraction process by using the mechanical energy of balls to break apart cells. Lipids may be extracted from algae efficiently using mechanical processes in order to produce biodiesel (Bharte and Desai 2021). These methods, which are adaptable and scalable, are essential to sectors looking to extract sustainable substances from algae for wastewater treatment, nutraceuticals and biofuel. To provide a process that is both economically and sustainably feasible, optimisation is necessary to strike a balance between cell viability and extraction efficiency.

3.6 Challenges in Algae Cultivation

3.6.1 Contamination

In nature, interactions between microalgae and other creatures like bacteria, fungus, viruses and others are common (Yao et al. 2019). Grazers, bacteria, fungus, photosynthetic organisms and viruses are the categories of microorganisms that are frequently regarded as contaminants in microalgae farming systems (Molina-Grima et al. 2022). Because the total culture conditions are difficult to manage due to seasonal influences, cultivation in an open pond is limited to microalgae that can survive even in harsh environmental conditions (Jerney and Spilling 2020). One significant disadvantage impacting the open pond is the greater danger of contamination from air. The main source of contamination in open pond cultivation is biological contamination generated by a variety of bacteria (McBride et al. 2014). The large contact between open pond liquid culture and air promotes CO_2 input, releasing excess dissolved O_2 and providing a good habitat for diverse biotic pollutants.

Tubular photobioreactors (PBRs) are the most popular in commercial microalgae growing (Jerney and Spilling 2020). Closed system PBRs are less impacted by the external environment than open pond. As a result, controlling parameters impacting culture like temperature, light, pH and nutrients is easy. However, the closed PBR system is more expensive to build than the open pond, and commercialising large-scale microalgae culture is challenging. Furthermore, the influence on microalgae growth can be minimised by properly removing O_2 emitted by microalgae.

3.7 Sources of Contamination

3.7.1 Zooplankton

Zooplankton are heterotrophic creatures found in lakes, rivers and swamps, and are abundant in ponds with a high organic matter content (Montemezzani et al. 2015). Zooplankton consists of small (20 m) unicellular (protozoa) organisms as well as bigger multicellular rotifers and micro-crustaceans (>200 m), including copepods and cladocerans (up to 5 mm). Zooplankton serve as the primary trophic link between bacteria, algae and higher eaters.

Various aquatic species, such as amoebas, ciliates, rotifers and various zooplankton, feed on microalgae. As a result, grazing by these zooplankton predators must be monitored and avoided. Amoebas, ciliates, rotifers, and zooplankton are examples of aquatic animals that are important because they feed on microalgae and affect algal cultures. Different chemicals are used in algal farms to control zooplankton grazers. For example, quinine sulphate was found to have a 24-hour Liquid Crystal 100 of 12–14 mg L^{-1}, but *Dunaliella* was found to have a 72-hour Electrochemical 50% of 14.5 ± 0.6 mg L^{-1}. It is crucial to keep an eye on these zooplankton predators and steer clear of their grazing in order to preserve the equilibrium in algal cultures and avoid negative impacts on biomass buildup and lipid synthesis (Moreno-Garrido and

Canavate 2001). However, early detection of grazers is now possible to prevent their negative impacts on algal crops. For example, FlowCAM (Fluid Imaging Technologies LLC, Maine) is a combination of a flow cytometer and a camera that uses continuous digital imaging to characterise particles in a fluid sample to prevent their detrimental effects on algal cells (Wang et al. 2017).

3.7.2 Bacteria

Microalgal product may be utilised as an energy source by bacterial species and bacteria can generate and supply needed components for microalgal growth; the two organisms live together (Han et al. 2016). However, when there is insufficient nutrient input in the culture medium, microalgae and bacteria fight for nutrients, and microalgal growth may be restricted (Lam et al. 2018). Furthermore, bacterial contamination can be lethal for microalgae mass production when bacterial species have an algicidal action.

Bacteria, as one of the biotic pollutants having an algicidal action, destroy microalgae in two ways. Bacterial contamination is anticipated to have an impact on algal biomass accumulation, lipid content and nutrient clearance (Ma et al. 2014). Bacteria have been found to both inhibit and boost algal growth. Some bacteria have the potential to harm algae by generating enzymes that break down the algal cell wall and cause immediate lysis of algal cells via direct contact.

Algae producers may employ closed photobioreactors that reduce the amount of contaminating organisms in order to prevent bacterial contamination (Fulbright et al. 2018). Bacteria are known to penetrate and cohabitate in closed systems, but little is known about these communities.

3.7.3 Fungi

The symbiotic relationship between fungi and algae can result in antagonism (Cooley et al. 2011). Fungi feed on microalgae and fight with them for resources (Lam et al. 2018). Non-pathogenic fungi have little impact, but pathogenic fungus must be controlled because fungal zoospores travel across dense algal populations, contaminating host algae by attaching to and infecting them. Although fungi are parasitic on microalgae, it may grow in the absence of a host if organic supplies are abundant in the culture. As a result, organic source control is essential. Complete elimination of unwanted biological parasites on a large scale is neither cost-effective nor feasible (Cooper and Smith 2015). Fungal infection happens in a series of critical steps (Laezza et al. 2022). The fungus attack happens within a few days, and the cell walls of the microalgae are damaged. This step is most likely mediated by specific enzymes. Pathogens then settle, encyst and germinate within the host. Microalgae typically lose their green colour and turn brown. Young microalgae are not infected throughout their motile period, but they are targeted after they reach maturity.

In short, it is critical to carefully manage the cleaning and sanitisation steps in order to avoid major contaminations by pathogenic fungi, especially in closed systems (Molina-Grima et al. 2022). Inadequate facility sterilisation, and regulation of inocula, culture media, water supply and aeration gases can all lead to contamination.

3.8 Light Intensity and Quality

Microalgal growth is influenced by various environmental conditions such as light, pH, salinity, temperature and availability of inorganic carbon (Shu et al. 2012). One of the main elements that regulates the physiological processes in microalgae is light. Algal cell growth, metabolites, photosynthesis, morphogenesis and development may be impacted by light wavelengths in addition to intensity. Light is a necessary component of photosynthetic activity and autotrophic development (Singh and Singh 2015). Chlorophyll a and b, the main pigments in algae that harvest light, and are sensitive to blue and red light. Microalgae's metabolic makeup changes in response to variations in light levels (Krzemińska et al. 2014). The rate at which photosynthesis occurs is directly influenced by light intensity, and this relationship is recognised to be represented by a logarithmic curve (Williams and Laurens 2010). At low light levels, the rate of photosynthesis increases almost linearly with increasing lighting. Still, additional sunlight does not boost the rate of photosynthesis in the region of saturating light intensities (Maltsev et al. 2021). A decrease in the effectiveness and rate of photosynthesis, or photoinhibition, can result from microalgae plants exposed to intense light. This damage might be caused by photooxidative stress on the photosynthetic apparatus.

3.8.1 Optimising Light Conditions

- **Light intensity:** Light has a quantitative quality called intensity (Maltsev et al. 2021). When algae cells are stressed with different light intensities and spectra, remarkable changes in the biomass and biomolecule composition of microalgae have been recorded (Elisabeth et al. 2021). The total lipid content of the microalgae *Pavlova lutheri* has been found to decrease as light intensity increases. Published data indicate that the range of 26–400 μmol photons m^{-2} s^{-1} represents the ideal values of light intensity at which the maximum growth rate is recorded for various taxonomic groups and species of algae. Algal cultures require a light source, which could be daylight (Belcher and Swale 1982). The optimal location for the culture vessels is near a north-looking window, but if a window facing another direction must be used, care must be given. Make sure that no direct sunlight ever hits the tubes.
- **Selecting the appropriate light source**: The type of light source also plays a vital role in the growth of algae (Satthong et al. 2019). Coloured LED lights are the best to use because of their monochrome and small peak width (Li et al. 2021). Nevertheless, since they do not provide light in a broad spectrum, it is better to use them in combination with other light colours.

3.8.2 Temperature Control

One of the primary variables influencing the metabolic makeup, nutrient absorption, growth rate and cell size of microalgae is temperature (Randhir et al. 2020). Temperature swings between 10 and 45 °C occur in temperate climate microalgal cultures, particularly in outdoor PBRs, temperatures that most commercialised algae species cannot sustain (Ras et al. 2013). Indeed, most microalgae can undertake photosynthesis at temperatures ranging from 15 to 30 °C, with optimal growth temperatures ranging from 20 to 25°C (García-González et al. 2005). Optimal growth temperatures allow the cell to conduct photosynthesis without affecting its fundamental biochemical and/or physiological states (Shekh et al. 2022). Mesophilic species prefer temperatures ranging from 20 to 25 °C, while thermophiles (*Chaetoceros*, *Anacystis nidulans*) and psychrophilic strains (*Asterionella formosa*) can survive temperatures as low as 17 °C. Temperature control in open pond systems exposed to the elements is difficult and cannot be simply modified to meet the needs. However, some control is achievable by adjusting the mixing speed with paddlewheels and the culture volume in open ponds. When relative air humidity is low, higher mixing can speed up evaporative cooling, and culture volumes can be raised to make them less susceptible to day–night temperature variations. Through engineering solutions, optimum temperatures and high biomass productivities may be maintained in closed PBRs. When opposed to open pond systems, closed PBRs can reach higher cell densities that are free of contamination for longer periods of time.

The main effect of temperature on photosynthesis is a decrease in the activity of ribulose-1,5-bisphosphate (Rubisco), a dual-functioning enzyme (Salvucci and Crafts-Brandner 2004). Depending on the proportional levels of O_2 and CO_2 in the chloroplasts, it can serve as an oxygenase or a carboxylase. The activity of Rubisco enzyme in CO_2 fixation increases with increasing temperature until a specific point and then decreases. As a result of its effect on ribulose's affinity for CO_2, temperature is a limiting factor for algal growth rate and biomass production (Khan et al. 2018). Temperature can also be utilised as a stress treatment to induce the creation of beneficial metabolites.

3.8.3 Managing Temperature Fluctuations

One particular energy input of importance is related to the need for temperature management, as it has been proven to have a significant impact on algal development (Juneja et al. 2013). Most large-scale algal bioreactors and open ponds or raceways are designed to operate out in the open, where the culture may be subjected to large fluctuations in internal temperature as a result of ambient air temperature, evaporation and internal heating of the culture due to radiant thermal energy absorption (Santos-Ballardo et al. 2016). While various approaches to bioreactor or pond design may alleviate concerns associated with temperature changes in the culture, the importance of these design parameters must be examined for each strain. Closed photobioreactors have more regulated conditions with less risk of unwanted invasions by parasites and zooplankton grazers, but they can suffer from excessive internal temperatures and dense cultures that lead to self-shading (Nalley et al. 2018). In open pond, paddlewheels are commonly used to maintain stable temperature, nutrient concentration and CO_2 delivery throughout the open pond by increasing mass transfer (Costa et al. 2019).

3.9 Conclusion

In conclusion, growing algae offers a viable solution to a number of environmental and financial issues. But in order to reach its full potential, it must overcome the operational and technological obstacles that are preventing its widespread use at the moment. It is imperative that research and development efforts continue to be directed on improving light utilisation, minimising energy usage, establishing cost-effective harvesting procedures and optimising farming systems. Through promoting cooperation among scientists, engineers and politicians, we can unleash algae's enormous potential and prepare the ground for a time when this adaptable and sustainable resource powers society. Unlocking a range of solutions, from food security and environmental cleanup to biofuel generation and carbon sequestration, depends on the successful development of robust and scalable algae growth systems. We can harness the potential of algae to create a more sustainable and greener future if we embrace the difficulties and invest in creative solutions.

References

Adesanya VO, Cadena E, Scott SA, Smith AG (2014) Life cycle assessment on microalgal biodiesel production using a hybrid cultivation system. Bioresour Technol 163:343–355

Aditya T, Bitu G, Mercy Eleanor G (2016) The role of algae in pharmaceutical development. J Pharm Nanotechnol 4:82–89

Ahmad I, Abdullah N, Koji I, Yuzir A, Muhammad SE (2021) Evolution of photobioreactors: a review based on microalgal perspective. IOP Conf Ser Mater Sci Eng 1142:012004

Ahmad A, Banat F, Alsafar H, Hasan SW (2022) Algae biotechnology for industrial wastewater treatment, bioenergy production, and high-value bioproducts. Sci Total Environ 806:150585

Al-Dailami A, Koji I, Ahmad I, Goto M (2022) Potential of photobioreactors (PBRs) in cultivation of microalgae. J Adv Res Appl Sci Eng Technol 27(1):32–44

Apel AC, Weuster-Botz D (2015) Engineering solutions for open microalgae mass cultivation and realistic indoor simulation of outdoor environments. Bioprocess Biosyst Eng 38:995–1008

Arora K, Kumar P, Bose D, Li X, Kulshrestha S (2021) Potential applications of algae in biochemical and bioenergy sector. 3 Biotech 11(6):296

Belcher H, Swale E (1982) Culturing algae. A guide for schools and colleges. Institute of Terrestrial Ecology, Cambridge

Bellinger EG, Sigee DC (2015) Freshwater algae: identification, enumeration and use as bioindicators. Wiley, Hoboken

Bharte S, Desai K (2021) Techniques for harvesting, cell disruption and lipid extraction of microalgae for biofuel production. Biofuels 12(3):285–305

Borowitzka MA, Vonshak A (2017) Scaling up microalgal cultures to commercial scale. Eur J Phycol 52(4):407–418
Bwire G, Sack DA, Kagirita A, Obala T, Debes AK, Ram M, Komakech H, George CM, Orach CG (2020) The quality of drinking and domestic water from the surface water sources (lakes, rivers, irrigation canals and ponds) and springs in cholera prone communities of Uganda: an analysis of vital physicochemical parameters. BMC Public Health 20(1):1–18
Chang J-S, Show P-L, Ling T-C, Chen C-Y, Ho S-H, Tan C-H, Nagarajan D, Phong W-N (2017) Photobioreactors. In: Current developments in biotechnology and bioengineering. Elsevier, Amsterdam, pp 313–352
Chanquia SN, Vernet G, Kara S (2022) Photobioreactors for cultivation and synthesis: specifications, challenges, and perspectives. Eng Life Sci 22(12):712–724
Cooley DR, Mullins RF, Bradley PM, Wilce RT (2011) Culture of the upper littoral zone marine alga Pseudendoclonium submarinum induces pathogenic interaction with the fungus Cladosporium cladosporioides. Phycologia 50(5):541–547
Cooper MB, Smith AG (2015) Exploring mutualistic interactions between microalgae and bacteria in the omics age. Curr Opin Plant Biol 26:147–153
Costa JAV, Freitas BCB, Santos TD, Mitchell BG, Morais MG (2019) Open pond systems for microalgal culture. In: Biofuels from algae. Elsevier, Amsterdam, pp 199–223
Dharmaprabhakaran T, Karthikeyan S, Periyasamy M, Mahendran G (2020) Algal biodiesel-promising source to power CI engines. Mater Today Proc 33:2870–2873
Elisabeth B, Rayen F, Behnam T (2021) Microalgae culture quality indicators: a review. Crit Rev Biotechnol 41(4):457–473
Enzmann F, Stöckl M, Zeng AP, Holtmann D (2019) Same but different–scale up and numbering up in electrobiotechnology and photobiotechnology. Eng Life Sci 19(2):121–132
Fabris M, Abbriano RM, Pernice M, Sutherland DL, Commault AS, Hall CC, Labeeuw L, McCauley JI, Kuzhiuparambil U, Ray P (2020) Emerging technologies in algal biotechnology: toward the establishment of a sustainable, algae-based bioeconomy. Front Plant Sci 11:279
Fulbright SP, Robbins-Pianka A, Berg-Lyons D, Knight R, Reardon KF, Chisholm ST (2018) Bacterial community changes in an industrial algae production system. Algal Res 31:147–156
García-González M, Moreno J, Manzano JC, Florencio FJ, Guerrero MG (2005) Production of Dunaliella salina biomass rich in 9-cis-β-carotene and lutein in a closed tubular photobioreactor. J Biotechnol 115(1):81–90
Gupta PL, Lee S-M, Choi H-J (2015) A mini review: photobioreactors for large scale algal cultivation. World J Microbiol Biotechnol 31:1409–1417
Han J, Zhang L, Wang S, Yang G, Zhao L, Pan K (2016) Co-culturing bacteria and microalgae in organic carbon containing medium. J Biol Res 23(1):1–9
Jankowska E, Sahu AK, Oleskowicz-Popiel P (2017) Biogas from microalgae: review on microalgae's cultivation, harvesting and pretreatment for anaerobic digestion. Renew Sust Energ Rev 75:692–709
Jerney J, Spilling K (2020) Large scale cultivation of microalgae: open and closed systems. Biofuels Algae Methods Protocols 1980:1–8
Juneja A, Ceballos RM, Murthy GS (2013) Effects of environmental factors and nutrient availability on the biochemical composition of algae for biofuels production: a review. Energies 6(9):4607–4638
Khan MI, Shin JH, Kim JD (2018) The promising future of microalgae: current status, challenges, and optimization of a sustainable and renewable industry for biofuels, feed, and other products. Microb Cell Factories 17(1):1–21
Kleiman RM, Characklis GW, Kern JD, Gerlach R (2021) Characterizing weather-related biophysical and financial risks in algal biofuel production. Appl Energy 294:116960
Koller M (2015) Design of closed photobioreactors for algal cultivation. In: Algal biorefineries: volume 2: products and refinery design. Springer, Cham, pp 133–186
Krzemińska I, Pawlik-Skowrońska B, Trzcińska M, Tys J (2014) Influence of photoperiods on the growth rate and biomass productivity of green microalgae. Bioprocess Biosyst Eng 37:735–741

Kube M, Jefferson B, Fan L, Roddick F (2018) The impact of wastewater characteristics, algal species selection and immobilisation on simultaneous nitrogen and phosphorus removal. Algal Res 31:478–488

Kumar K, Mishra SK, Shrivastav A, Park MS, Yang J-W (2015) Recent trends in the mass cultivation of algae in raceway ponds. Renew Sust Energ Rev 51:875–885

Kumar M, Sun Y, Rathour R, Pandey A, Thakur IS, Tsang DC (2020) Algae as potential feedstock for the production of biofuels and value-added products: Opportunities and challenges. Sci Total Environ 716:137116

Laezza C, Salbitani G, Carfagna S (2022) Fungal contamination in microalgal cultivation: biological and biotechnological aspects of fungi-microalgae interaction. J Fungi 8(10):1099

Lam TP, Lee T-M, Chen C-Y, Chang J-S (2018) Strategies to control biological contaminants during microalgal cultivation in open ponds. Bioresour Technol 252:180–187

Lau CC, Teh KY, Afifudeen CW, Yee W, Aziz A, Cha TS (2022) Bright as day and dark as night: light-dependant energy for lipid biosynthesis and production in microalgae. World J Microbiol Biotechnol 38(4):70

Li S, Hu T, Xu Y, Wang J, Chu R, Yin Z, Mo F, Zhu L (2020) A review on flocculation as an efficient method to harvest energy microalgae: mechanisms, performances, influencing factors and perspectives. Renew Sust Energ Rev 131:110005

Li X, Manuel J, Slavens S, Crunkleton DW, Johannes TW (2021) Interactive effects of light quality and culturing temperature on algal cell size, biomass doubling time, protein content, and carbohydrate content. Appl Microbiol Biotechnol 105:587–597

Liu W, Chen Y, Wang J, Liu T (2019) Biomass productivity of Scenedesmus dimorphus (Chlorophyceae) was improved by using an open pond–photobioreactor hybrid system. Eur J Phycol 54(2):127–134

Ma X, Zhou W, Fu Z, Cheng Y, Min M, Liu Y, Zhang Y, Chen P, Ruan R (2014) Effect of wastewater-borne bacteria on algal growth and nutrients removal in wastewater-based algae cultivation system. Bioresour Technol 167:8–13

Maltsev Y, Maltseva K, Kulikovskiy M, Maltseva S (2021) Influence of light conditions on microalgae growth and content of lipids, carotenoids, and fatty acid composition. Biology 10(10):1060

Martínez JM, Delso C, Álvarez I, Raso J (2020) Pulsed electric field-assisted extraction of valuable compounds from microorganisms. Compr Rev Food Sci Food Saf 19(2):530–552

Matos ÂP (2017) The impact of microalgae in food science and technology. J Am Oil Chem Soc 94:1333–1350

Matter IA, Bui VKH, Jung M, Seo JY, Kim Y-E, Lee Y-C, Oh Y-K (2019) Flocculation harvesting techniques for microalgae: a review. Appl Sci 9(15):3069

McBride RC, Lopez S, Meenach C, Burnett M, Lee PA, Nohilly F, Behnke C (2014) Contamination management in low cost open algae ponds for biofuels production. Ind Biotechnol 10(3):221–227

McBride RC, Smith VH, Carney LT, Lane TW (2017) Crop protection in open ponds. In: Slocombe SP, Benemann JR (eds) Microalgal production for biomass and high-value products. CRC Press, Boca Raton, pp 165–182

Milledge JJ, Heaven S (2013) A review of the harvesting of micro-algae for biofuel production. Rev Environ Sci Biotechnol 12:165–178

Miškovičová V, Masojídek J (2023) Transdisciplinary collaboration in architecture: integrating microalgae biotechnologies for human and non-human perspectives. Arch Pap Fac Archit Des 28(2):10–20

Molina-Grima E, García-Camacho F, Acién-Fernández FG, Sánchez-Mirón A, Plouviez M, Shene C, Chisti Y (2022) Pathogens and predators impacting commercial production of microalgae and cyanobacteria. Biotechnol Adv 55:107884

Montemezzani V, Duggan IC, Hogg ID, Craggs RJ (2015) A review of potential methods for zooplankton control in wastewater treatment high rate algal ponds and algal production raceways. Algal Res 11:211–226

Moreno-Garrido I, Canavate J (2001) Assessing chemical compounds for controlling predator ciliates in outdoor mass cultures of the green algae Dunaliella salina. Aquac Eng 24(2):107–114

Nagappan S, Das P, AbdulQuadir M, Thaher M, Khan S, Mahata C, Al-Jabri H, Vatland AK, Kumar G (2021) Potential of microalgae as a sustainable feed ingredient for aquaculture. J Biotechnol 341:1–20

Najjar YS, Abu-Shamleh A (2020) Harvesting of microalgae by centrifugation for biodiesel production: a review. Algal Res 51:102046

Nalley J, O'Donnell D, Litchman E (2018) Temperature effects on growth rates and fatty acid content in freshwater algae and cyanobacteria. Algal Res 35:500–507

Narala RR, Garg S, Sharma KK, Thomas-Hall SR, Deme M, Li Y, Schenk PM (2016) Comparison of microalgae cultivation in photobioreactor, open raceway pond, and a two-stage hybrid system. Front Energy Res 4:29

Neto AI, Pinto IS (2018) Introduction to the marine algae. In: Marine macro-and microalgae: an overview, vol 1. Routledge, London

Pande V, Pandey SC, Sati D, Pande V, Samant M (2020) Bioremediation: an emerging effective approach towards environment restoration. Environ Sustain 3:91–103

Quiroz D, Greene JM, McGowen J, Quinn JC (2021) Geographical assessment of open pond algal productivity and evaporation losses across the United States. Algal Res 60:102483

Raeisossadati M, Moheimani NR, Parlevliet D (2019) Luminescent solar concentrator panels for increasing the efficiency of mass microalgal production. Renew Sust Energ Rev 101:47–59

Randhir A, Laird DW, Maker G, Trengove R, Moheimani NR (2020) Microalgae: a potential sustainable commercial source of sterols. Algal Res 46:101772

Ras M, Steyer J-P, Bernard O (2013) Temperature effect on microalgae: a crucial factor for outdoor production. Rev Environ Sci Biotechnol 12(2):153–164

Rastogi RP, Pandey A, Larroche C, Madamwar D (2018) Algal green energy–R&D and technological perspectives for biodiesel production. Renew Sust Energ Rev 82:2946–2969

Raven JA, Geider RJ (2003) Adaptation, acclimation and regulation in algal photosynthesis. In: Photosynthesis in algae. Springer, Cham, pp 385–412

Ritchie R, Larkum A (2012) Modelling photosynthesis in shallow algal production ponds. Photosynthetica 50:481–500

Rogers JN, Rosenberg JN, Guzman BJ, Oh VH, Mimbela LE, Ghassemi A, Betenbaugh MJ, Oyler GA, Donohue MD (2014) A critical analysis of paddlewheel-driven raceway ponds for algal biofuel production at commercial scales. Algal Res 4:76–88

Salvucci ME, Crafts-Brandner SJ (2004) Relationship between the heat tolerance of photosynthesis and the thermal stability of Rubisco activase in plants from contrasting thermal environments. Plant Physiol 134(4):1460–1470

Santos-Ballardo DU, Rendón-Unceta MDC, Rossi S, Vázquez-Gómez R, Hernández-Verdugo S, Valdez-Ortiz A (2016) Effects of outdoor cultures on the growth and lipid production of Phaeodactylum tricornutum using closed photobioreactors. World J Microbiol Biotechnol 32:1–10

Satthong S, Saego K, Kitrungloadjanaporn P, Nuttavut N, Amornsamankul S, Triampo W (2019) Modeling the effects of light sources on the growth of algae. Adv Differ Eq 2019:1–6

Shekh A, Sharma A, Schenk PM, Kumar G, Mudliar S (2022) Microalgae cultivation: photobioreactors, CO_2 utilization, and value-added products of industrial importance. J Chem Technol Biotechnol 97(5):1064–1085

Shen C (2021) Investigation of vertical mixing in raceway pond systems using computational fluid dynamics. Colorado State University, Fort Collins

Shu CH, Tsai CC, Liao WH, Chen KY, Huang HC (2012) Effects of light quality on the accumulation of oil in a mixed culture of Chlorella sp. and Saccharomyces cerevisiae. J Chem Technol Biotechnol 87(5):601–607

Shuba ES, Kifle D (2018) Microalgae to biofuels: 'promising' alternative and renewable energy, review. Renew Sust Energ Rev 81:743–755

Singh G, Patidar S (2018) Microalgae harvesting techniques: a review. J Environ Manag 217:499–508

Singh S, Singh P (2015) Effect of temperature and light on the growth of algae species: a review. Renew Sust Energ Rev 50:431–444

Torzillo G, Chini Zittelli G (2015) Tubular photobioreactors. In: Algal biorefineries: volume 2: products and refinery design. Springer, Cham, pp 187–212
Veerabadhran M, Natesan S, MubarakAli D, Xu S, Yang F (2021) Using different cultivation strategies and methods for the production of microalgal biomass as a raw material for the generation of bioproducts. Chemosphere 285:131436
Wang Y, Castillo-Keller M, Eustance E, Sommerfeld M (2017) Early detection and quantification of zooplankton grazers in algal cultures by FlowCAM. Algal Res 21:98–102
Wells ML, Potin P, Craigie JS, Raven JA, Merchant SS, Helliwell KE, Smith AG, Camire ME, Brawley SH (2017) Algae as nutritional and functional food sources: revisiting our understanding. J Appl Phycol 29:949–982
White RL, Ryan RA (2015) Long-term cultivation of algae in open-raceway ponds: lessons from the field. Ind Biotechnol 11(4):213–220
Wicker RJ, Kwon E, Khan E, Kumar V, Bhatnagar A (2023) The potential of mixed-species biofilms to address remaining challenges for economically-feasible microalgal biorefineries: a review. Chem Eng J 451:138481
Williams PJ, Laurens LM (2010) Microalgae as biodiesel & biomass feedstocks: review & analysis of the biochemistry, energetics & economics. Energy Environ Sci 3(5):554–590
Xiaogang H, Jalalah M, Jingyuan W, Zheng Y, Li X, Salama E-S (2020) Microalgal growth coupled with wastewater treatment in open and closed systems for advanced biofuel generation. Biomass Convers Biorefinery 2020:1–20
Xiong Q, Hu L-X, Liu Y-S, Zhao J-L, He L-Y, Ying G-G (2021) Microalgae-based technology for antibiotics removal: from mechanisms to application of innovational hybrid systems. Environ Int 155:106594
Yao S, Lyu S, An Y, Lu J, Gjermansen C, Schramm A (2019) Microalgae–bacteria symbiosis in microalgal growth and biofuel production: a review. J Appl Microbiol 126(2):359–368
Yen H-W, Hu I-C, Chen C-Y, Nagarajan D, Chang J-S (2019) Design of photobioreactors for algal cultivation. In: Biofuels from algae. Elsevier, Amsterdam, pp 225–256
Yin Z, Zhu L, Li S, Hu T, Chu R, Mo F, Hu D, Liu C, Li B (2020) A comprehensive review on cultivation and harvesting of microalgae for biodiesel production: environmental pollution control and future directions. Bioresour Technol 301:122804
Yun J-H, Cho D-H, Lee S, Heo J, Tran Q-G, Chang YK, Kim H-S (2018) Hybrid operation of photobioreactor and wastewater-fed open raceway ponds enhances the dominance of target algal species and algal biomass production. Algal Res 29:319–329
Zapata-Boada S, Gonzalez-Miquel M, Jobson M, Cuellar-Franca R (2022) Solvent screening methodology considering techno-economic and environmental sustainability criteria for algae lipid extraction. In: Computer aided chemical engineering, vol 49. Elsevier, Amsterdam, pp 1039–1044
Zeng J, Wang Z, Chen G (2021) Biological characteristics of energy conversion in carbon fixation by microalgae. Renew Sust Energ Rev 152:111661
Zhang X, Davidson E, Zou T, Lassaletta L, Quan Z, Li T, Zhang W (2020) Quantifying nutrient budgets for sustainable nutrient management. Glob Biogeochem Cycles 34(3):e2018GB006060

Algae Biomass: Importance, Harvesting Techniques, Extraction Methods, and Associated Challenges

4

Rizul Gautam, Shubham Kumar Patial, and Suman Singh

Contents

R. Gautam (✉)
Academy of Scientific and Innovative Research (AcSIR), Ghaziabad, India

Biomedical Applications Group, CSIR Central Scientific Instruments Organisation, Chandigarh, India
e-mail: rizul.csio19a@acsir.res.in

S. K. Patial · S. Singh
Academy of Scientific and Innovative Research (AcSIR), Ghaziabad, India

Materials Science and Sensor Applications Group, CSIR Central Scientific Instruments Organisation, Chandigarh, India

S. K. Arya et al. (eds.), *Value Added Products From Bioalgae Based Biorefineries: Opportunities and Challenges*,
https://doi.org/10.1007/978-981-97-1662-3_4

Abstract

The utilization of algae biomass has emerged as a pivotal aspect in the realms of sustainable development and renewable resources. This chapter delves into the multifaceted landscape of algae biomass, elucidating its paramount importance, various harvesting techniques, extraction methods, and the inherent challenges associated with these processes. Meticulously exploring both traditional methods and recent research, the chapter provides a comprehensive overview of the advancements in algae biomass utilization. In this chapter, an in-depth analysis of harvesting techniques unfolds, ranging from traditional methods to cutting-edge technologies that are deeply elaborated. The advantages and disadvantages of each technique are scrutinized, shedding light on the diverse considerations that impact their efficacy. Moving on to extraction methods, the chapter delineates the extraction processes for lipids, proteins, pigments, and other bioactive compounds. Recent innovations in extraction technologies are thoroughly explored, showcasing the strides made in enhancing efficiency and sustainability. Noteworthy research findings are incorporated to underscore the ongoing efforts in optimizing extraction processes for maximum yield and minimum environmental impact. As the discourse progresses, the challenges associated with algae biomass extraction and harvesting are dissected, encompassing economic constraints, scalability issues, and environmental concerns. The interconnected nature of these challenges emphasizes the need for a holistic approach to address them effectively.

Keywords

Algae biomass · Harvesting · Extraction · Microalgae · Biotechnology · Bioenergy

Abbreviations

AEAE	Aqueous enzymatic-assisted extraction
DAF	Dissolved air flotation
DiAF	Dispersed air flotation
DOF	Dissolved ozone flotation
EAE	Enzyme-assisted extraction
ECF	Electro-coagulation flotation
MAE	Microwave-assisted extraction
MF	Microfiltration
PAN	Polyacrylonitrile
PES	Polyether sulfone
PTFE	Polytetrafluoroethylene
PVDF	Polyvinylidene fluoride
SE	Soxhlet extraction
SFE	Supercritical fluid extraction
TAGs	Triacylglycerols

UAE	Ultrasound-assisted extraction
UF	Ultrafiltration

4.1 Introduction

Algae, a diverse group of photosynthetic organisms found in various aquatic environments, play an integral role in ecosystems worldwide. These organisms, ranging from microscopic unicellular species to larger multicellular forms like seaweeds, are vital for the planet's health. Algae are primary producers that contribute significantly to oxygen production through photosynthesis, generating a substantial portion of the Earth's atmospheric oxygen and aiding in carbon dioxide (CO_2) absorption, thereby mitigating climate change. They form the base of aquatic food chains, providing nutrition for various organisms, and contribute to nutrient cycling in aquatic ecosystems (Faruk et al. 2023).

Beyond their ecological significance, algae hold immense importance in various industrial sectors due to their rich biochemical composition. Algae serve as a promising resource for biofuel production, as they contain high lipid content suitable for biodiesel extraction. Moreover, certain algae species are sources of bioactive compounds such as omega-3 fatty acids, antioxidants, pigments like astaxanthin, and vitamins, offering potential applications in nutraceuticals and pharmaceuticals. Algae-based systems are employed in wastewater treatment, acting as a natural filtration system to remove pollutants and excess nutrients from water bodies. Some species of algae are consumed directly as food due to their nutritional value, while others are utilized as additives in food and feed industries (Khan et al. 2018).

The harvesting of algae involves various methods such as filtration, centrifugation, flocculation, and settling to separate algae from the culture medium or natural habitat. Once harvested, the extraction of valuable compounds from algae is performed using different techniques. Solvent extraction involves using solvents like hexane or ethanol to extract lipids, pigments, and other hydrophobic compounds from algae. Supercritical fluid extraction (SFE) utilizes controlled pressure and temperature using supercritical fluids such as CO_2 to obtain extracts, especially lipids. Microwave-assisted extraction (MAE) and ultrasound-assisted extraction (UAE) apply microwave energy and ultrasonic waves, respectively, to disrupt algal cells and enhance the efficiency of compound extraction (Ananthi et al. 2021; Quitério et al. 2022).

Algae represent a valuable and versatile resource with ecological, industrial, and commercial significance. Their diverse applications in various sectors, coupled with innovative harvesting and extraction methods, pave the way for further exploration and utilization of these organisms in sustainable and eco-friendly practices across industries.

4.1.1 Importance of Algae Biomass

Algae biomass plays a pivotal role in various sectors owing to its wide-ranging applications and ecological significance. Its importance stems from its multifaceted nature: algae are pivotal in the global carbon cycle, contributing significantly to oxygen production and serving as efficient carbon dioxide sinks, thus mitigating climate change. Moreover, their capacity for nutrient uptake aids in cleaning polluted water bodies, providing a natural solution to water purification and environmental remediation. Additionally, algae hold immense promise as a renewable energy source, being a primary feedstock for the production of biofuels such as biodiesel and bioethanol. Furthermore, certain algal species boast high nutritional value, being rich in proteins, essential fatty acids, vitamins, and minerals, making them valuable as dietary supplements in both human nutrition and animal feed. Industries also benefit from algal biomass, utilizing it in pharmaceuticals, cosmetics, food production, and agriculture. Algae serve as a platform for innovative research, contributing to advancements in biotechnology, genetic engineering, and biorefineries. Overall, the versatility and sustainability of algae biomass underscore its importance as a critical resource in various fields, offering potential solutions to environmental, energy, and nutritional challenges while fostering advancements in research and industry (Mengstie et al. 2023).

4.1.2 Significance of Harvesting and Extraction

Harvesting and extraction techniques are pivotal stages in the utilization of algae biomass, serving as crucial determinants of its economic viability and industrial applicability. Algae, being tiny and dispersed in natural environments, necessitate effective harvesting methods to concentrate and separate them from their growth medium. The significance of efficient harvesting techniques lies in their ability to attain high algae concentrations while minimizing energy consumption and cost. Various methods such as centrifugation, filtration, flotation, and sedimentation are employed based on the algae species, desired product, and scale of operation. Post harvest, extraction processes become vital to isolate and recover valuable compounds present in algae biomass. Extraction methods aim to access lipids, carbohydrates, proteins, pigments, and bioactive compounds contained within the algae cells. The choice of extraction technique directly impacts the purity, yield, and quality of the extracted compounds. Techniques like solvent extraction, supercritical fluid extraction (SFE), microwave-assisted extraction (MAE), ultrasound-assisted extraction (UAE), enzyme-assisted extraction (EAE), and osmotic pressure methods are employed to extract targeted components. Optimization of these processes is essential to enhance extraction efficiency, reduce environmental impact, and ensure economic viability in utilizing algae biomass for various applications, ranging from biofuels to pharmaceuticals and nutraceuticals. The detailed descriptions of the harvesting and extraction techniques are covered in the sections below and shown in Fig. 4.1 (Michalak and Chojnacka 2014; Singh and Patidar 2018).

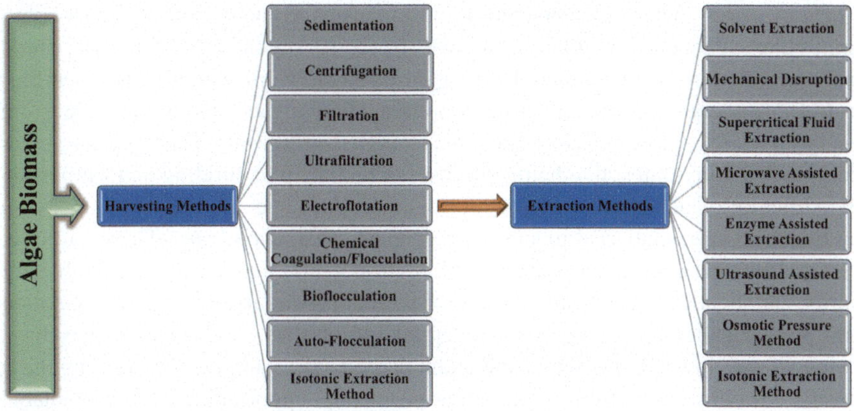

Fig. 4.1 Diagram showing different harvesting and extraction techniques for algae biomass

4.2 Algae Biomass Composition

Algae are incredibly diverse organisms that offer a wide range of compounds with significant potential for various applications, including nutraceuticals and biofuels. Here is a detailed breakdown of their biomass composition and their potential in the context of lipids, proteins, carbohydrates, pigments, secondary metabolites, nutraceuticals, and biofuels.

4.2.1 Lipids, Proteins, and Carbohydrates

Lipids (Fats/Oils) Certain species of microalgae can accumulate high amounts of lipids like glycerolipids, hydrocarbons, waxes, and sterols, which can range from 1 to 70% or more of their dry weight. Among these, glycerolipids are the most abundant and well-described of microalgae lipid classes (Morales et al. 2021). All these lipids are often rich in valuable fatty acids such as omega-3 e.g., Eicosapentaenoic acid (EPA) and Docosahexaenoic acid (DHA) and omega-6. Microalgae like *Chlorella*, *Nannochloropsis*, and certain species of diatoms and green algae are known for their lipid content. Lipids derived from algae have gained attention for biodiesel production due to their high productivity and potential sustainability compared to traditional plant-based oils (Viegas et al. 2015).

Proteins Algae contain significant amounts of proteins, ranging from 10 to 50% or more of their dry weight. These proteins are considered high-quality proteins, containing essential amino acids necessary for human and animal nutrition. *Spirulina*, *Chlorella*, and some species of marine microalgae are known for their protein content. Algae-derived proteins are being explored for use in functional foods, animal feeds, and dietary supplements (Bleakley and Hayes 2017).

Carbohydrates Algae, a diverse group of organisms, contain a variety of carbohydrates like sugars (such as glucose, sucrose, fructose), starches, and cellulose. These carbohydrates make up around 10–50% of an algae's dry weight. They serve as immediate energy sources and structural elements in algae cells. Some algae store energy in starch, while cellulose provides strength to cell walls. There are also other complex carbohydrates, like hemicelluloses, which help in building and supporting the structure of algae cells. Moreover, algae produce specific carbohydrates with unique functions, such as alginates, carrageenans, agar, and ulvans. These carbohydrates differ between algae species and environments, making them valuable for various applications. They can be used for biofuel production through fermentation and as raw materials for making biodegradable plastics. Additionally, they have potential uses in food, medicine, and technology. Understanding and utilizing these diverse carbohydrates from algae offer promising opportunities for sustainable innovations across multiple industries (Ebhodaghe et al. 2022).

4.2.2 Pigments and Secondary Metabolites

Pigments Algae possess a diverse array of pigments including chlorophylls (green pigments), carotenoids (yellow, orange, or red pigments), and phycobiliproteins (red or blue pigments) that enable them to capture light energy, crucial for photosynthesis. These pigments, like chlorophylls, carotenes, xanthophylls, and phycobiliproteins, absorb light across the visible spectrum. Chlorophyll-a, found in all algal species, efficiently harnesses light energy from the visible region of sunlight. It consists of a porphyrin ring comprising four pyrrole molecules and Mg^{2+} atoms at its center. Carotenoids, including α-carotene and β-carotene, are hydrocarbons or oxygen-containing compounds termed oxycarotenes or xanthophylls. Xanthophylls, like lutein, astaxanthin, canthaxanthin, and fucoxanthin, are more polar and soluble in polar solutions compared to carotenes. These pigments are prominent in various algae species like *Scenedesmus*, *Haematococcus lacustris*, *Spirulina platensis, Dunaliella salina, Chlorella,* and *Botryococcus braunii*. They play essential roles in capturing light for photosynthesis and have diverse applications in algae-based industries. Additionally, these pigments have applications in food colorants, cosmetics, and pharmaceuticals. For instance, astaxanthin (a carotenoid) derived from algae has antioxidant properties and is used in nutraceuticals and aquaculture feed (Zhang et al. 2016).

Secondary Metabolites Algae produce an array of secondary metabolites like antioxidants, polysaccharides, phenolic compounds, and bioactive molecules. These compounds possess potential health benefits, such as antioxidant, anti-inflammatory, and antimicrobial properties. Some algae-derived compounds are being researched for their therapeutic potential in treating various diseases. Algal secondary metabolites, comprising alkaloids, flavonoids, glycosides, terpenoids, and phenazines, are extensively employed in diverse health sectors. Despite their comprehensive documentation, a comprehensive assessment of the biological functions and activities of these metabolites remains imperative. Ongoing research endeavors focus on investigating the therapeutic potential of select algae-derived compounds in combating various diseases. The multifaceted nature of microalgal secondary metabolites promises versatile applications across sectors such as food, feed, aquaculture, biomedicine, veterinary medicine, cosmetics, and health-related industries, thereby driving the comprehensive exploration and exploitation of algae as a potent source of biologically impactful compounds (Neeti et al. 2023).

4.2.3 Nutraceutical and Biofuel Potential

Nutraceutical Potential Algae-derived compounds, including proteins, lipids, pigments, and bioactive molecules, offer potential health benefits. *Spirulina* and *Chlorella*, for instance, are used as dietary supplements due to their rich nutrient content, including proteins, vitamins, minerals, and antioxidants. Compounds like phycocyanin (from *Spirulina*) and fucoxanthin (from brown algae) are studied for their health-promoting properties.

Biofuel Potential Algae have gained significant attention as a potential source of biofuels due to their high growth rates and lipid content. Lipids extracted from algae can be converted into biodiesel, and carbohydrates can be fermented to produce bioethanol or other biofuels. Algae have various commercial applications, including producing nutritional supplements, sewage treatment, and serving as coloring agents. One of the most promising uses of algae is as a renewable source for biofuels. The vegetable oil extracted from algae can be used directly as a raw material for biodiesel production or further refined into different biofuels like renewable diesel and jet fuel. These algae-based biofuels can directly substitute conventional petroleum fuels in engines without any modifications. Additionally, carbohydrates derived from algae, such as sugars, can undergo fermentation to produce ethanol, butanol, and other products like plastics and biochemicals. Furthermore, the biomass obtained from algae can be utilized for pyrolysis oil or generating combined heat and power. Overall, algae-derived biofuels meet the same specifications as traditional petroleum fuels, making them viable alternatives in various industries while offering sustainability benefits (Viegas et al. 2015).

4.3 Algae Harvesting Techniques

Algae harvesting constitutes a pivotal stage in the extraction of microalgae from their cultivation environments, relying on specific characteristics like size, density, and final product specifications. An array of techniques, encompassing mechanical, chemical, biological, and electrical methods, either independently or in combination, are deployed. These methods include coagulation/flocculation, flotation, electrical processes, filtration, and centrifugation (Singh and Patidar 2018). The effectiveness of the harvesting process is contingent on its cost-effectiveness, energy efficiency, and adaptability to various microalgal strains, influencing downstream processing considerations (Danquah et al. 2009). Method selection is guided by end-product requirements, considering factors such as moisture content, salt concentrations, and potential cell damage. Ideally, the process should be non-toxic to prevent contamination of the microalgal biomass and facilitate the recycling of the cultivation medium. Mechanical methods, often coupled with preceding chemical or biological coagulation/flocculation, emerge as the most reliable for harvesting microalgal biomass (Molina Grima et al. 2003). Combining techniques, such as flocculation–sedimentation with centrifugation, proves to be cost-effective, while emerging biological approaches hold promise for further reductions in operational costs.

Sedimentation techniques entail allowing harvested microalgal biomass to settle in a quiescent environment, utilizing gravitational forces for separation from the liquid medium. Gravity sedimentation, acknowledged for its cost-effectiveness and energy efficiency, is commonly employed, especially in large-scale cultivation for low-value products like biofuels (Molina Grima et al. 2003). Often, a coagulation/flocculation step precedes gravity sedimentation to expedite the process, with optimal results achieved using lamella-type separators and sedimentation tanks. **Centrifugation**, an advanced method replacing gravity sedimentation, employs centrifugal force for high-efficiency separation, albeit with high energy demands. Various centrifuge options exist, each with its pros and cons, influencing harvesting efficiency based on microalgal species and intended biomass value (Pahl et al. 2013).

In the realm of **filtration**, also known as the membrane process, specialized membranes are employed to selectively separate algae from the surrounding water. Recognized for its environmentally friendly and cost-effective nature, the membrane process plays a pivotal role in algae cultivation for diverse applications, including biofuel and bioproduct development (Wu et al. 2012). Different filtration forms, such as *microfiltration* (MF), *macrofiltration*, *ultrafiltration* (UF), and others, offer varied advantages and applications based on solvent/solute properties, hydrodynamic conditions, and membrane characteristics. Membrane materials, including polyvinylidene fluoride (PVDF), polyacrylonitrile (PAN), polyether sulfone (PES), and polytetrafluoroethylene (PTFE), are commonly used (Dassey and Theegala 2014; Drexler and Yeh 2014). While ultrafiltration (UF) presents a potential substitute for microalgae recovery, challenges like high flux requirements and elevated operating costs limit its widespread adoption (Mo et al. 2015; Rawat et al. 2011). High-gradient magnetic filtration is an effective method that can be applied efficiently. Utilizing biosorption of submicron-sized magnetic particles holds significant promise for microalgal harvesting, with reported efficiencies exceeding 95% (Liu et al. 2013). These particles must be cost-effective, reusable, chemically stable, and possess high adsorption capacity to ensure optimal performance.

Flotation, a gravity separation process utilizing air or gas bubbles, proves to be a rapid and effective method, especially for low-density microalgal species, surpassing sedimentation in certain aspects. The success of flotation relies on factors like particle size, collision likelihood, and adhesion, often involving flocculants, coagulation, and surfactants. Various flotation methods include dissolved air flotation (DAF) and dispersed air flotation (DiAF). DAF is a technique that employs minute bubbles, typically ranging from 10 to 100 mm in size. These bubbles are generated by dissolving air in water under exceptionally high pressure, as outlined by Edzwald and Chen et al. (2011) and Edzwald (1993). The presence of these bubbles compels suspended algae cells to rise to the water's surface, facilitating their removal through skimming. It is important to note that DAF is recognized as an energy-intensive process due to the high pressure required. However, it is crucial to maintain appropriately sized bubbles, as oversized bubbles have the potential to disrupt the formed floc, as highlighted by Park et al. (2011). On the other hand, DiAF bubbles ranging in size from 700 to 1500 mm are generated through the continuous passage of air. This is achieved either by passing air through a porous material, such as diffusers or spargers, or by utilizing a high-speed mechanical agitator. Although this system is characterized by lower energy consumption, it demands expensive equipment and involves a high-pressure drop to generate the required bubbles (Chen et al. 2011). **Electro-flotation** is a technique that involves the flotation of microalgae cells to the water surface, accompanied by the generation of fine hydrogen bubbles through electrolysis. The evolution of hydrogen at the cathode serves to separate the microalgal biomass from the growth medium. The hydrogen bubbles adhere to the microalgal flocs, inducing their ascent to the surface. Noteworthy advantages of this method include the absence of the need for chemicals and its non-species-specific nature. However, drawbacks such as cathode fouling and high power requirements represent significant challenges (Chen et al. 2011).

In contemporary times, the technique of flotation with ozone, known as dissolved ozone flotation (DOF), has garnered increased attention of researchers, despite being perceived as a costly approach for water and wastewater treatment. Unlike conventional methods using atmospheric air, DOF employs ozone gas to generate charged bubbles. Ozone, recognized for its potent oxidizing properties, acts on soluble organic compounds, and the charged bubbles contribute to the separation of microalgae. The synergistic interaction of these two processes holds the potential for enhanced treatment effects. Moreover, biopolymers released during cell lysis serve as coagulants, further aiding in effective separation. Successful applications of Ozoflotation include the removal of *Chlorella vulgaris* and *Scenedesmus obliquus*, with impressive turbidity removal rates of 98% and 95%, respectively (Cheng et al. 2010, 2011).

The exploration of **electrical methods** for microalgal harvesting reveals their versatility across species, providing an environmentally friendly alternative without the need for chemical additives (Zenouzi et al. 2013). By leveraging the negative charge of microalgal cells, applied electrical fields facilitate their separation through electrophoresis and electro-flocculation. Electro-flotation, akin to dissolved air flotation (DAF), stands out for generating hydrogen bubbles through water electrolysis, offering a potential cost-effective alternative (Show and Lee 2013). Electro-coagulation flotation (ECF), an amalgamation of electro-coagulation with electro-flotation, exhibits advantages such as non-species specificity, easy control, minimal chemical usage, improved results, and low power consumption compared to centrifugation. However, challenges include the need for electrode replacement, increased suspension temperature, pH alterations, and residual metals in the algal concentrate (Vandamme et al. 2011).

Chemical coagulation/flocculation plays a crucial role in economically optimizing microalgal harvesting, especially for large volumes of microalgal cultures. This step concentrates the suspension, reducing energy demand significantly. Coagulation involves pH adjustment or electrolyte addition, while flocculation includes the addition of cationic polymers to the broth (Vandamme et al. 2012). This harvesting step involves concentrating the suspension by a factor of 20–100. By doing so, it augments the effective particle size before the dewatering process, thereby substantially diminishing the energy required (Ndikubwimana et al. 2016). **Bioflocculation**, where microorganisms produce bioflocculants to flocculate algae, involves bioflocculants like Extracellular polymeric substances (EPS) from bacteria, microalgae, and fungi. Although this method has applications in water and wastewater treatment, its economic feasibility is challenged by the cost of adding EPS in crude or purified form (Chen et al. 2015; Shen et al. 2009). Certain microalgae exhibit **autoflocculation**, a natural process responding to environmental stresses, enabling rapid and inexpensive biomass recovery without supplementary chemicals. Researchers have observed promising results with autoflocculation induced at elevated pH and through electro-flocculation technology. The latter ensures non-species-specific, easily controlled algae harvesting with low chemical usage and enhanced results. However, challenges include electrode replacement needs, increased suspension temperature, pH changes, and residual metals in the algal concentrate (Vandamme et al. 2011).

The harvesting of microalgae is a multifaceted process crucial for separating these microscopic organisms from their cultivation environments. The selection of a suitable harvesting method is influenced by factors such as cost-effectiveness, energy efficiency, adaptability to different microalgal strains, and downstream processing considerations. Mechanical methods, particularly when combined with coagulation/flocculation or other techniques, emerge as reliable options for harvesting microalgal biomass. Sedimentation techniques, utilizing gravity or advanced centrifugation, offer cost-effective and energy-efficient solutions, especially for large-scale cultivation. Filtration, employing specialized membranes, is recognized for its environmental friendliness and cost-effectiveness in diverse applications, although challenges like high operating costs limit the widespread adoption of ultrafiltration.

Flotation, a gravity separation process, proves effective for low-density microalgal species, with various methods presenting different trade-offs between energy consumption and equipment costs. Electrical methods, such as electro-flotation and electro-coagulation-flotation, provide environmentally friendly alternatives without the need for chemical additives, albeit with specific challenges like electrode replacement and residual metals. Chemical coagulation/flocculation, including bioflocculation and autoflocculation, plays a crucial role in economically optimizing microalgal harvesting, with considerations for the cost of additives and the feasibility of natural flocculation processes. Table 4.1 shows the different algae and their harvesting techniques.

The continuous exploration of diverse harvesting methods reflects the dynamic nature of microalgal cultivation, aiming to meet the growing demand for applications such as biofuels and bioproducts. Future advancements in harvesting technologies are likely to focus on further enhancing cost-effectiveness.

Table 4.1 Harvesting techniques for different microalgae species with their advantages and disadvantages

Sr. No.	Microalgae species	Techniques	Advantages	Disadvantages	References
1.	*Dunaliella salina*	Filtration	• Low energy consumption • No chemical required • Cost-effective • High efficiency	• Slow process • Not suitable for large algae • Regular cleaning of membranes	Zenouzi et al. (2013)
2.	*Chlorella vulgaris*	Electrolytic flotation	• Less energy required • Chemical-free • Suitable for large scale • Less time required	• High operating cost • Not feasible for marine microalgae	Xu et al. (2018)
3.	*Chlorella sp.*	Chemical flocculation	• Suitable for large scale • Time saving • Less energy consumption • Simple and feasible method	• Use of toxic chemicals • Highly pH dependent • Efficiency depends upon coagulants	Kim et al. (2017)
4.	*Desmodesmus brasiliensis*	Microbial bioflocculation	• Cost-effective • Biodegradable process	• Contamination to microalgae • Culture medium is non-recyclable	Ndikubwimana et al. (2016)
5.	*Scenedesmus* sp.	Centrifugation	• Easy and fast process • High recovery rates	• Costly method • Require heavy duty machinery • Cell damage possible	Oliveira et al. (2018)
6.	*Chlorella zofingiensis*	Autoflocculation	• Cost-effective and non-toxic process • Recycling of culture medium is possible	• Microbial contamination • Cellular composition may change	Zhang et al. (2016)

4.4 Challenges and Considerations in Harvesting

Harvesting algae for various applications poses unique challenges and considerations crucial for successful implementation. Achieving selectivity and specificity is vital due to the diverse aquatic environment, requiring efficient separation of algae from microorganisms. Techniques like centrifugation, filtration, and flocculation address selectivity challenges with distinct advantages (Christenson and Sims 2011). Energy consumption is a significant concern impacting economic viability and environmental sustainability. Traditional methods demand substantial energy inputs, prompting research into low-energy harvesting technologies and the integration of renewable sources for greener practices (Wiley et al. 2009). Cost-effectiveness is pivotal for algae-based applications, influencing biofuel production and wastewater treatment. Advances in harvesting technologies focus on scalable, economically viable solutions for large-scale projects (Pahl et al. 2013). Minimizing the environmental footprint is critical, emphasizing eco-friendly agents, water recycling, and environmentally conscious harvesting equipment. Scale-up challenges add complexity, requiring solutions for process uniformity, equipment adaptation, and optimized conditions (Pragya et al. 2013).

Thus, addressing selectivity, energy consumption, cost-effectiveness, environmental impact, and scale-up challenges is essential for developing sustainable and economically viable algae harvesting technologies that are aligned with green practices and environmental stewardship. Ongoing research is crucial for a greener and more conscious future.

4.5 Algae Biomass Extraction Methods

Algal biomass can undergo diverse processing techniques aimed at extracting various compounds, with mechanical-based treatments being particularly noteworthy (Cherubini et al. 2009). When microalgae cultures reach a growth plateau, biomass concentration occurs, enabling extraction from either dry or wet biomass. Dewatering via centrifugation followed by cell disruption aims to rupture cellular walls, facilitating the release of intracellular microalgae components not externally secreted. Techniques involve disruption, breakage, or disintegration methods (Dong et al. 2016). Conversely, obtaining dry biomass involves thermal drying post dewatering, typically yielding a paste-like biomass with a dry weight exceeding 85% (Xu et al. 2011).

Pretreating microalgae biomass stands as a critical step in recovering lipids and other valuable products, enhancing extraction yields significantly (Karemore and Sen 2016). Microalgae cell walls comprise fibrillar matrices and crystalline polymers, with variations across species, necessitating tailored pretreatment methods (Kumari et al. 2018). This specific pretreatment strategy addresses the diverse compositions of microalgae cell walls, optimizing subsequent extraction processes for enhanced recovery of targeted valuable compounds.

Highlighting more about algae biomass extraction methods, it varies depending on the specific type of algae and the desired compounds to be extracted. Following are the description of common extraction methods used in algae biomass processing:

4.5.1 Traditional Practices

Traditional practices for algae biomass extraction methods relied on simplistic yet functional techniques. Predating modern scientific methodologies, these traditional approaches centered on basic extraction methods to obtain valuable components from algae. Among the prevalent practices was sun drying, where harvested algae biomass was spread out to naturally dry under the sun's heat, particularly common in coastal areas with ample sunlight (Chacón-Lee and González-Mariño 2010). Boiling or heating algae in water or other solvents constituted another method to extract compounds, allowing their collection from the resulting solution. Manual grinding or crushing of algae biomass using rudimentary tools like stones or pestles helped break cell walls and release intracellular content for extraction. Filtration and sedimentation techniques were employed thereafter to separate extracted compounds from residual biomass (Rhea et al. 2017). Some traditions involved using vegetable oils as solvents for extracting lipids or hydrophobic compounds from algae. Algal biomass was soaked or mixed with oils, and the oils were then separated to retrieve the extracted compounds. Additionally, traditional maceration practices included soaking algae in solvents like water, alcohol, or oil for an extended duration to extract specific compounds through soaking, followed by separation. While these methods were simplistic and lacked the precision and efficiency of modern scientific techniques, they laid the groundwork for contemporary extraction methods by leveraging indigenous knowledge, forming a basis for the evolution of current algae biomass extraction practices.

4.5.2 Pretreatment Practice

Pretreatment methods play a pivotal role in the extraction of compounds from algae biomass, representing a critical initial step that significantly influences the efficiency and effectiveness of subsequent extraction processes. This process involves various methodologies like mechanical–physical, chemical, and enzymatic treatments, each designed to disrupt the structural integrity of the cell wall and facilitate the release of target compounds.

Mechanical–Physical Methods These techniques rely on physical forces to disrupt cell walls. Grinding, homogenization, ultrasonication, or bead beating is utilized to mechanically break the cellular structure, aiding the release of intracellular components. These methods facilitate efficient extraction but may demand substantial energy input.

Neto et al. (2013) conducted an evaluation using various mechanical disruption approaches, such as bead milling, ultrasonication, and high-pressure homogenization, aimed at enhancing lipid extraction from *Thalassiosira weissflogii* microalgae. Their study meticulously scrutinized the impact of these methods on lipid yields, offering valuable insights into the efficacy of mechanical disruption for extracting lipids from algae.

Chemical Methods These techniques involve the utilization of chemical agents such as acids (e.g., sulfuric acid, hydrochloric acid [HCl]), alkalis (e.g., sodium hydroxide), enzymes, or surfactants to break down the rigid cell walls of algae. The process begins by subjecting the algae biomass to controlled treatment with the selected chemical agent, adjusting parameters like concentration, temperature, and duration. Acid treatments typically hydrolyze cellulosic material, while alkali treatments modify cell wall structures through saponification or polymer solubilization. This pretreatment weakens the cell walls, rendering them more porous and accessible, thereby improving subsequent extraction processes. Chemical pretreatment has been shown to significantly enhance the release of lipids, proteins, carbohydrates, pigments, and other desired compounds from algae biomass. For instance, studies highlighted in the literature demonstrated notable increases in lipid extraction efficiency from microalgae through acid pretreatment. However, careful optimization of pretreatment conditions is crucial to avoid degradation of target compounds and ensure process safety. Additionally, these methods exhibit potential for scaling up in industrial settings, offering promise in optimizing extraction processes for various valuable compounds from algae biomass, thereby contributing to the advancement of algae-based industries (Zhou et al. 2022).

Enzymatic Methods Enzymatic treatment, especially using enzymes like cellulase, plays a crucial role in breaking down the complex structure of algae biomass, particularly cellulosic chains, to release valuable sugars efficiently. This approach is highly selective and effective, as it specifically targets the linkages between sugars without creating inhibitors that could hinder subsequent fermentation processes. Despite its effectiveness, the use of purified enzymes for biofuel production through enzymatic treatment is often hindered by its high cost, posing an economic challenge. To address this, alternative strategies have been explored, such as cultivating microbes capable of secreting cellulose-degrading enzymes, although limited data on this method are available. Another idea involves using immobilized enzymes, but this technique can compromise hydrolytic activity due to inadequate contact between the enzymes and the algae biomass. Researchers have also investigated a two-stage process combining acidic hydrolysis with enzymatic treatment. This approach involves pretreating the algae with acid before enzymatic treatment, which significantly boosts the conversion rate of algal biomass compared to using solely acidic or enzymatic processes. However, this method presents economic challenges, requiring an additional step and increased capital investment due to the need for acid neutralization before enzymatic treatment. Despite these hurdles, enzymatic treatment remains crucial for the efficient breakdown of complex algal structures, offering a potential pathway for extracting valuable sugars from algae biomass. Ultimately, these pretreatment strategies play a crucial role in priming algae biomass for subsequent extraction processes, significantly enhancing extraction efficiency, and maximizing the yield of valuable compounds derived from the algae biomass (Gui et al. 2021; Kim et al. 2011; Wang et al. 2011).

4.5.3 Advances in Extraction Techniques

4.5.3.1 Solvent Extraction

Solvent extraction serves as a prevalent technique for extracting valuable compounds from algae biomass, offering distinct advantages over traditional methods like chemical and enzymatic pretreatment. This method involves immersing the algae biomass in solvents to dissolve specific compounds, followed by separation to isolate the dissolved compounds. Solvent extraction demonstrates remarkable versatility, capable of extracting a wide array of compounds, including lipids, pigments, proteins, and carbohydrates, from algae biomass with high selectivity. Compared to enzymatic pretreatment methods, solvent extraction can be less dependent on specific optimal conditions required for enzymatic activity, simplifying the extraction process. This method is further divided into three categories as:

- *Folch Method:* The method involves utilizing a chloroform–methanol mixture (in a 2:1 ratio by volume) to extract lipids from endogenous cell sources. In this method, homogenized cells undergo equilibration with saline solution, after which the resulting mixture separates into two layers, with lipids predominantly settling in the upper phase. This technique stands as one of the earliest endeavors in lipid extraction, serving as a foundational model for subsequent extraction protocols with refinements. Even today, modified iterations of this method persist in lipid estimation from algal sources using spectrophotometric analysis. Its primary advantage lies in the rapid and facile processing of a large quantity of samples. However, compared to more contemporary procedures, it exhibits lower sensitivity (Folch et al. 1957).
- *Soxhlet Method:* Soxhlet extraction (SE) represents a method employed to transfer partially soluble constituents from a solid sample into a liquid phase (solvent) using a Soxhlet extractor. This technique commonly utilizes non-polar solvents such as hexane to obtain neutral lipids from the sample. The process involves placing the solid sample within a filter paper thimble installed in the main chamber of the Soxhlet apparatus. Subsequently, the solvent is heated to reflux and cycles into the main chamber, facilitating the recovery of less soluble compounds by the solvent. It is noted that as the polarity of the extraction solvent increases, there is a higher yield in the extraction of microalgae due to the retrieval of complex lipids and pigments. This factor is critical, considering that total lipid extracts utilizing polar solvents are intricate, encompassing various metabolites apart from lipids. Parameters influencing Soxhlet extraction encompass solvent

selection, sample particle size, and extraction duration. However, Soxhlet extraction typically operates on a laboratory scale, necessitates substantial solvent consumption, and involves extended extraction periods (Baumgardt et al. 2015; Sharif et al. 2014).

- *Bligh and Dyer Technique:* The Bligh and Dyer method, a well-established technique akin to the Folch method, facilitates the concurrent extraction of lipids and the partitioning of proteins within the interface of two liquid phases. This method stands out due to variations in solvent compositions and ratios, involving the extraction of lipids from a homogenized cell suspension using a 1:2 (v/v) chloroform/methanol mixture. The resultant lipid-rich chloroform phase is isolated and subjected to further processing, a widely adopted gravimetric procedure for estimating lipids, employed not only in small-scale laboratory settings but also extensively in pilot and large-scale extraction operations (Ranjith Kumar et al. 2015). To refine and improve the Bligh and Dyer method, several modifications have been introduced. For instance, incorporating 1M NaCl instead of water in the process prevents the binding of denatured lipids to acidic lipids, contributing to enhanced recovery and preservation of lipids. Moreover, the integration of 0.2M phosphoric acid or HCl into the salt solution has been shown to improve the efficiency of lipid recovery while reducing separation times (Jensen 2008). Additionally, the supplementation of the water phase with 0.5% acetic acid (v/v) has demonstrated increased recovery rates for acidic phospholipids (Weerheim et al. 2002).

Overall, organic solvent extraction methods such as the Bligh and Dyer technique remain pivotal for extracting valuable products from microalgae. The selection of solvents depends on the polarity of the compounds of interest. Notably, non-polar solvents are preferred for extracting non-polar molecules like triacylglycerols (TAGs), which are crucial for biodiesel production (Ranjith Kumar et al. 2015). These refinement and adaptation methods underscore the ongoing efforts to optimize solvent-based extraction techniques, aiming to enhance lipid recovery efficiency and expand their applicability across various industrial scales in the realm of microalgae-based product recovery.

4.5.3.2 Mechanical Disruption

Mechanical disruption, also known as mechanical cell disruption or mechanical extraction, is a method utilized to rupture the rigid cell walls of algae to release intracellular compounds. This technique aims to facilitate the extraction of valuable components such as lipids, proteins, carbohydrates, pigments, and other bioactive compounds from algae biomass. There are several mechanical methods employed in this process, each with its own principles and equipment:

- *Bead Milling:* Bead milling, a mechanical method of disrupting microalgae cells, involves grinding and agitation using solid glass beads, typically ranging from 0.3 to 0.5 mm in size, made of zirconia-silica or zirconium oxide. The efficiency of this process relies on factors such as biomass concentration, flow rate, agitation type and speed, and temperature (Ghasemi Naghdi et al. 2016). There are two primary approaches to bead milling: shaking vessels and agitated beads. Shaking vessels involve vibrating platforms that move the beads and microalgae cells, causing collisions that disrupt the cells. Studies report varied lipid recovery rates; for instance, Ryckebosch et al. achieved 40% lipid recovery from *Phaeodactylum tricornutum*, while Zheng et al. obtained 11% lipid recovery from *Chlorella vulgaris* (Catarina et al. 2021; Zheng et al. 2012). On the other hand, agitated beads involve a rotating agitator inside the culture vessel, generating heat that aids in the disruption process. Lee et al. demonstrated oil yields between 7.9 and 8.1 g/L from *Botryococcus* sp., *Chlorella vulgaris*, and *Scenedesmus* sp. cultures using this method (Lee et al. 2010). Bead beating is another mechanical cell disruption technique where high-speed spinning of a biomass slurry with fine beads causes cell damage. It offers efficient disruption without dewatering requirements, reducing processing costs. Utilizing beads of optimal diameter (0.5 mm) made of materials like zirconia-silica or zirconium oxide enhances disruption rates and extraction efficiency for microalgal cells (Hopkins 1991). Both methods, bead milling and bead beating, present effective mechanical approaches for disrupting microalgae cells, impacting lipid recovery and extraction efficiencies.

- *Expeller Press or Pressing:* The use of mechanical presses, such as the expeller press, stands as a traditional yet effective method for extracting value-added products, including lipids, proteins, and pigments, from various sources, including algal biomass (Demirbaş 2008). This method involves subjecting dried algal biomass to high mechanical pressure to crush and rupture cells, thereby facilitating oil extraction from the biomass. Variations in pressure force, algal strain, and press configurations significantly impact extraction efficiency (Ramesh 2013). However, the process necessitates low-moisture content in the input materials, demanding energy-intensive drying processes that can contribute up to 30% of the production costs. Although the technique yields oil recovery rates typically ranging between 70 and 75%, challenges like decreased lipid recovery with excessive pressure, increased heat generation, and difficulties due to the rigid cell wall structure of algae hinder optimal oil extraction. To enhance oil recovery, mechanical crushing is sometimes combined with chemical methods, but this can escalate processing times and expenses (Ramesh 2013). Moreover, the presence of pigments along with oil poses challenges, necessitating additional processes like pigment removal via solvent extraction or activated carbon adsorption, adding to overall costs. Additionally, mechanical press methods demand high-cost maintenance and skilled labor, and exhibit lower efficiency compared to alternative extraction approaches (Ranjith Kumar et al. 2015).

Although the above-mentioned methods aim to rupture the rigid cell walls of algae, releasing intracellular constituents like lipids, proteins, pigments, and carbohydrates, these methods typically demand higher energy inputs compared to chemical or enzymatic approaches. Furthermore, the heat generated during mechanical disruption processes can potentially damage sensitive compounds, necessitating the integration of cooling systems during extraction. The inclusion of a cooling system adds to the energy and equipment costs, contributing to the overall operational expenses of the extraction process. Despite these challenges, advancements in mechanical disruption methods continue to be explored and optimized to enhance extraction efficiency while addressing concerns regarding energy consumption and product integrity in extracting heat-sensitive compounds from algae biomass (Lee et al. 2012).

4.5.3.3 Supercritical Fluid Extraction (SFE)

Supercritical fluid extraction (SFE) involves utilizing solvents at temperatures and pressures beyond their critical points, resulting in fluid with properties similar to both gases and liquids. The density, viscosity, and diffusivity of the supercritical fluid differ from traditional liquids, offering enhanced transport properties. CO_2 is frequently used in SFE due to its "green" nature, safety, and ease of separation from extracts without leaving any residues. However, its low polarity restricts extraction of polar substances, but this limitation can be overcome by adding polar modifiers or co-solvents. Despite this drawback, SFE with CO_2 presents an opportunity for diverse applications in various sectors such as agriculture, dietary supplements, and cosmetics, providing extracts useful for products like fertilizers, dietary supplements, and cosmetic ingredients. This method uses supercritical fluids (like CO_2) under high pressure and temperature to extract compounds from algae. SFE is environmentally friendly and can selectively extract lipids and other compounds without leaving behind solvent residues (Ibáñez et al. 2012).

4.5.3.4 Microwave-Assisted Extraction

Microwave-assisted extraction (MAE) stands out as a rapid and efficient technique utilizing microwave energy to extract a diverse array of valuable compounds from algae biomass. This innovative method relies on microwaves interacting with dielectric polar substances, primarily water, within the cellular structure, generating heat through rapid oscillating electric fields. This process induces inter- and intramolecular movements, leading to the formation of water vapor and an increase in pressure that ruptures cell walls, facilitating the release of intracellular contents (Ghasemi Naghdi et al. 2016). The application of microwaves in pretreating microalgae results in multiple microfissures within cell walls, enhancing the recovery of bio-oil. Notably, this extraction method disrupts hydrogen bonds and promotes the migration of dissolved ions, facilitating enhanced solvent penetration into the matrix and the subsequent extraction of active compounds (Ranjith Kumar et al. 2015). Recognized for its speed, simplicity, safety, and cost-effectiveness, MAE eliminates the need for prior sample dewatering, making it particularly advantageous for lipid extraction. Its economic feasibility extends to a wide range of bioactive compounds, including polysaccharides (e.g., fucoidan), iodine, bromine, pigments (e.g., fucoxanthin), docosahexaenoic acid, phenols, phytosterols, and phytol, extracted from various algae species (literature examples listed). Compared to supercritical fluid extraction (SFE), MAE has been identified as a more economically viable option (Kadam et al. 2013). The process parameters crucial for effective MAE encompass extraction time, temperature, dielectric properties of the mixture, solid/liquid ratio, and the type and concentration of the solvent. While MAE demonstrates promise in biodiesel production through transesterification due to its short reaction time, low operational cost, and efficient extraction of algal oils, the major drawback lies in the high maintenance cost when scaling up to commercial settings. Overall, microwave-assisted extraction emerges as a versatile and effective method for obtaining a diverse range of valuable compounds from algae biomass. Its speed, efficiency, and cost-effectiveness position MAE as an attractive extraction technique in the field of biotechnology, finding applications in various industries, including pharmaceuticals, nutraceuticals, cosmetics, and food additives.

4.5.3.5 Enzyme-Assisted Extraction (EAE)

Enzymatic extraction methods offer a targeted approach employing specific enzymes to disrupt the rigid cell walls of algae and release lipid bodies, thus separating the lipid fraction from the lipid/protein matrix (Soto et al. 2017). This enzymatic lysis serves as an alternative to mechanical cell disruption, with enzymes tailored to the microalgae species, notably cellulase and lipase, to break down polysaccharides like cellulose and hemicellulose in algal cell walls and access lipids contained within a sac surrounded by phospholipids (Kumari et al. 2018). A significant advantage of EAE is the utilization of widely used food-grade enzymes like neutral and alkaline proteases, α-amylase, cellulase, and pepsin, which contribute to gentle yet efficient cell disruption (Liang et al. 2012). Among enzymatic extraction techniques, aqueous enzymatic-assisted extraction (AEAE) stands out for its high selectivity, operating under mild conditions such as neutral pH and moderate temperatures (ranging from 25 to 37 °C), and does not necessitate intensive drying steps (Soto et al. 2017). For instance, a study by Chen et al. demonstrated an enhanced method involving enzymatic lysis coupled with thermal treatment for lipid extraction from *Nannochloropsis oceanica*, optimizing extraction parameters at 37 °C, pH 5.0, 1.3% cellulase, liquid/solid ratio of 15 mL/g, and a 5-h incubation period, resulting in lipid yields of up to 28.8% (Chen et al. 2016).

The enzymatic extraction process typically involves several steps, including biomass harvesting, conditioning, enzyme addition, incubation to disrupt algal cell walls, solvent addition if necessary, centrifugation, and lipid fraction recovery (Soto et al. 2017). Moreover, enzymatic digestion can serve as a means for saccharification of carbohydrate biomass, aiding in bioethanol production after the lipid removal stage. This approach exhibits promise in efficiently extracting lipids from microalgae while preserving the integrity of valuable compounds and facilitating downstream applications in biofuel production and other biotechnological endeavors.

4.5.3.6 Ultrasound-Assisted Extraction (UAE)

Ultrasound-assisted extraction (UAE) is a method that employs sound waves to induce pressure variations and acoustic cavitations within the medium. These acoustic cavitations, formed due to the oscillation and collapse of microbubbles generated by the sound waves, transform sound energy into mechanical energy, causing disruption of the cell and its walls. This disruption leads to a reduction in particle size, facilitating better contact between the solvent and extracted compounds (Routray and Orsat 2017). UAE can be utilized as a pretreatment method, either in an ultrasonic bath or using an ultrasound probe instrument. This extraction technique operates with simplicity and efficiency, ensuring faster extraction times, high reproducibility, and minimal energy consumption compared to standard methods. Moreover, it works at lower temperatures, reducing the risk of thermal degradation of biomolecules. Ultrasound functions via two primary mechanisms: cavitation and acoustic streaming. Cavitation involves the creation of microbubbles due to ultrasound application, which generates pressure on the cells, causing them to rupture. Acoustic streaming, on the other hand, facilitates the mixing of the algal culture, enhancing the extraction process. Ultrasound-assisted extraction methods typically use two types of sonicators: horn and bath. Horn sonicators utilize piezoelectric generators that vibrate at specific amplitudes, whereas sonicator baths employ transducers placed at the bottom of the reactor to create ultrasonic waves. The choice between the two types depends on the operation mode—either batch or continuous. This process offers distinct advantages such as operating at lower temperatures in comparison to methods like microwave reactors or autoclaves. It eliminates the need for additional costly chemicals or beads used in other extraction methods. However, prolonged ultrasound exposure may generate free radicals, potentially impacting the quality of the extracted oil. In summary, ultrasound-assisted lipid extraction provides an efficient, cost-effective, and environmentally friendly method for extracting lipids from microalgae, while also addressing various challenges associated with conventional extraction techniques (Chemat et al. 2011; Hosikian et al. 2010; Lee et al. 2012).

4.5.3.7 Osmotic Pressure Method

A novel and eco-friendly approach employing osmotic pressure emerges as a cost-effective alternative among various extraction methods (Adam et al. 2012). The method manipulates algal cell walls by swiftly altering the salt concentration in the surrounding aqueous medium, disrupting the osmotic balance within the cells. Osmotic stress, occurring in two forms—hyper-osmotic and hypo-osmotic—impacts the cells differently. Hyper-osmotic stress, caused by higher salt concentrations outside the cells, leads to cell shrinkage and envelope damage due to fluid loss. Conversely, hypo-osmotic stress arises from lower external salt concentrations, resulting in cell swelling or bursting as fluid rushes in to balance osmotic pressure. Hypo-osmotic shock, commonly used for extracting intracellular substances, has shown promise in lipid extraction from various microalgal species like *Chlamydomonas reinhardtii*, *Botryococcus* sp., *Chlorella vulgaris*, and *Scenedesmus* sp. (Lee et al. 2010; Yoo et al. 2012). Researchers suggest that the osmotic pressure method appears to be a straightforward, efficient, and uncomplicated technique for extracting lipids from microalgae. Nonetheless, further exploration is necessary to evaluate lipid extraction using osmotic pressure across diverse algal species, including marine strains, and to assess its viability at pilot- and production-scale levels (Kim et al. 2012).

4.5.3.8 Isotonic Extraction Method

The utilization of ionic liquids for algal lipid extraction represents an innovative and emerging pretreatment technology. Extensive research by various investigators has focused on employing ionic liquids for microbial extraction, showcasing its bio-friendly nature despite the complexity of the procedures (Li et al. 2010; Ninomiya et al. 2012). This concept revolves around replacing hazardous organic solvents with ionic liquids, often referred to as environmentally benign designer solvents. Ionic liquids, non-aqueous salt solutions, maintain a liquid state at moderate temperatures between 0 and 140 °C, comprising an asymmetric organic cation and either an inorganic or organic anion. These unique solutions offer tailored solvent properties, including specific polarity, hydrophobicity, conductivity, and solubility, customizable to particular extraction requirements. However, the application of eco-friendly ionic liquid extraction in microalgal species, such as *Chlorella vulgaris*, remains relatively scarce (Kim et al. 2012). The economic and technical viability of this method is yet to be fully explored, preventing a conclusive judgment regarding its superiority over other algal oil extraction techniques. Despite this, the method as presented in Table 4.2 holds promise as a potentially eco-friendly substitute for traditional organic solvents.

Common challenges across these techniques include achieving high extraction efficiency without compromising compound quality, ensuring scalability, reducing environmental impact, minimizing energy consumption, and addressing regulatory and safety concerns associated with extraction solvents or methods. Each method has its advantages and limitations, requiring continuous research and innovation to overcome challenges and improve the efficiency and sustainability of extracting valuable compounds from algae biomass.

4.6 Challenges and Considerations in Extraction

Extraction of valuable compounds from algae biomass is a complex process that involves various challenges and considerations:

- Cell Wall Complexity: Algae possess diverse and complex cell wall structures that differ across species. The composition and rigidity of these cell walls can make it difficult to access and extract intracellular compounds efficiently.
- Selectivity of Extraction: Different compounds within algae biomass may have varying polarities and solubilities, making it challenging to selectively extract specific target compounds without contaminating the extract with unwanted substances.
- Scalability: Transitioning from laboratory-scale extraction to commercial-scale operations poses challenges in maintaining efficiency, reproducibility, and cost-effectiveness while dealing with larger volumes of biomass.
- Energy Consumption: Some extraction methods require high energy inputs, such as high temperatures or pressures, which can increase operational costs and environmental impact.

Table 4.2 Different extraction techniques along with technique involved and major challenges

Sr. No.	Extraction method	Technique	Challenges
1.	Solvent extraction	• Involves organic solvents or hydrocarbons • Dissolves and extracts compounds from algae biomass	• Solvent toxicity, safety concern, environmental impact, and compound extraction selectivity • Requires solvent removal from final product
2.	Microwave-assisted extraction	• Utilizes microwave energy to heat intracellular liquids • Causes cell rupture and compound release	• Potential for thermal degradation of sensitive compounds • Uniformity in extraction efficiency across various algal species • High equipment costs
3.	Ultrasound-assisted extraction	• Employs ultrasound waves for cell disruption • Enhances compound extraction	• Equipment cost and potential degradation of compounds due to prolonged sonication • Optimizing parameters for consistent and efficient extraction
4.	Enzyme-assisted extraction	• Utilizes enzymes to break down cell walls and release intracellular compounds	• Identifying suitable enzymes and optimizing conditions for different algal species • Ensuring cost-effectiveness at scale
5.	Osmotic pressure extraction	• Uses changes in salt concentrations to disrupt cell walls and extract compounds	• Precise control of osmotic conditions • Potential cell damage due to excessive stress • Scalability for large-scale applications
6.	Ionic liquid extraction	• Utilizes ionic liquids as alternative solvents for extracting compounds	• Limited research on algal species • Economic viability • Technical feasibility at scale • Ensuring the environmental sustainability of ionic liquids

- Solvent Selection: The choice of solvent for extraction is critical. Common solvents might pose environmental and health risks. Identifying eco-friendly and safe solvents that efficiently extract the desired compounds is a challenge.
- Environmental Impact: Ensuring that the extraction process is environmentally sustainable, produces minimal waste, and does not harm the ecosystem from which algae are sourced is important.
- Compound Stability: Certain compounds extracted from algae biomass might be sensitive to heat, light, or other extraction conditions, leading to degradation and affecting the quality of the final product.

- Regulatory Compliance: Meeting regulatory standards regarding the use of solvents, extraction methods, and final product safety is crucial, especially for applications in food, pharmaceuticals, or cosmetics.
- Economic Viability: Balancing extraction efficiency, cost-effectiveness, and quality of the extracted compounds to ensure that the process is economically feasible, especially for large-scale production, is vital.
- Process Optimization: Continuous research and development are necessary to optimize extraction methods, parameters, and technologies for enhanced efficiency, reduced extraction time, and improved yield of target compounds.

Addressing these challenges requires interdisciplinary approaches involving expertise in biology, chemistry, engineering, and environmental sciences. Advances in extraction technologies, such as novel solvents, innovative extraction methods (like microwave or ultrasound-assisted extraction), and genetic engineering to alter algae compositions, may help overcome some of these challenges in the future (Figueroa-Torres and Theodoropoulos 2023; Milledge 2011; Sarwer et al. 2022).

4.7 Conclusion

In the culmination of this comprehensive exploration into algae biomass, it becomes evident that its utilization stands at the nexus of innovation and sustainability. The significance of algae biomass extends beyond conventional applications, with its role expanding into diverse domains such as biofuel production, bioremediation, and the extraction of bioactive compounds. The detailed examination of harvesting techniques unraveled a spectrum of methodologies, each with its own set of advantages and disadvantages. Traditional methods, though time-tested, face challenges in scalability, while modern techniques, leveraging advancements in technology, present promising solutions yet demand careful consideration of their environmental impact. The nuanced understanding of these techniques is imperative for devising sustainable strategies tailored to specific contexts. Similarly, the exploration of extraction methods uncovered a landscape of possibilities, from lipid extraction for biofuel production to the extraction of valuable pigments and proteins. Recent research and technological innovations have significantly enhanced the efficiency of these processes, paving the way for more economically viable and environmentally friendly practices. However, the journey toward algae biomass utilization is not without hurdles. Economic constraints, scalability issues, and environmental concerns form a triad of challenges that necessitate a concerted effort from researchers, industries, and policymakers. Overcoming these challenges demands a multidisciplinary approach that integrates technological innovation, economic viability, and environmental stewardship. In light of the current advancements and challenges, the future of algae biomass holds promise. Ongoing research initiatives, coupled with the increasing recognition of the importance of sustainable resources, position algae biomass as a key player in the transition toward a greener and more sustainable future. As we chart the course forward, it is imperative to foster collaboration, embrace innovation, and engage in a continuous dialogue to unlock the full potential of algae biomass in addressing the global challenges of energy security and environmental sustainability. This chapter serves as a roadmap, guiding future endeavors and inspiring a collective commitment to harness the untapped potential of algae biomass for the benefit of society and the planet.

Acknowledgments Authors acknowledge AcSIR and CSIR-Central Scientific Instruments Organisation, Chandigarh, India, for providing infrastructure support. SKP acknowledges CSIR for CSIR-NET JRF fellowship.

References

Adam F, Abert-Vian M, Peltier G, Chemat F (2012) "Solvent-free" ultrasound-assisted extraction of lipids from fresh microalgae cells: a green, clean and scalable process. Bioresour Technol 114:457–465. https://doi.org/10.1016/j.biortech.2012.02.096

Ananthi V, Balaji P, Sindhu R, Kim S-H, Pugazhendhi A, Arun A (2021) A critical review on different harvesting techniques for algal based biodiesel production. Sci Total Environ 780:146467. https://doi.org/10.1016/j.scitotenv.2021.146467

Baumgardt F, Zandona Filho A, Brandalize M, Costa DC, Antoniosi Filho N, Abreu P, Corazza M, Pereira Ramos L (2015) Lipid content and fatty acid profile of Nannochloropsis oculata before and after extraction with conventional solvents and/or compressed fluids. J Supercrit Fluids 108:3. https://doi.org/10.1016/j.supflu.2015.11.003

Bleakley S, Hayes M (2017) Algal proteins: extraction, application, and challenges concerning production. Foods 6(5):33. https://doi.org/10.3390/foods6050033

Catarina S, Furigo A, Oliveira J, Derner R, Feller R, José V, Oliveira D, Leonardo R, Rorig Elisa C, Moecke H, Mazzutti S (2021) Microalgae biomass as a source of natural compounds: chemical characterization and new approaches for lipid extraction and culture harvesting

Chacón-Lee TL, González-Mariño GE (2010) Microalgae for "healthy" foods—possibilities and challenges. Compr Rev Food Sci Food Saf 9(6):655–675. https://doi.org/10.1111/j.1541-4337.2010.00132.x

Chemat F, Zill EH, Khan MK (2011) Applications of ultrasound in food technology: processing, preservation and extraction. Ultrason Sonochem 18(4):813–835. https://doi.org/10.1016/j.ultsonch.2010.11.023

Chen C-Y, Yeh K-L, Aisyah R, Lee D-J, Chang J-S (2011) Cultivation, photobioreactor design and harvesting of microalgae for biodiesel production: a critical review. Bioresour Technol 102(1):71–81. https://doi.org/10.1016/j.biortech.2010.06.159

Chen C-L, Chang J-S, Lee D-J (2015) Dewatering and drying methods for microalgae. Dry Technol 33(4):443–454. https://doi.org/10.1080/07373937.2014.997881

Chen L, Li R, Ren X, Liu T (2016) Improved aqueous extraction of microalgal lipid by combined enzymatic and thermal lysis from wet biomass of Nannochloropsis oceanica. Bioresour Technol 214:138–143. https://doi.org/10.1016/j.biortech.2016.04.031

Cheng Y-L, Juang Y-C, Liao G-Y, Ho S-H, Yeh K-L, Chen C-Y, Chang J-S, Liu J-C, Lee D-J (2010) Dispersed ozone flotation of Chlorella vulgaris. Bioresour Technol 101(23):9092–9096. https://doi.org/10.1016/j.biortech.2010.07.016

Cheng Y-L, Juang Y-C, Liao G-Y, Tsai P-W, Ho S-H, Yeh K-L, Chen C-Y, Chang J-S, Liu J-C, Chen W-M, Lee D-J (2011) Harvesting of Scenedesmus obliquus FSP-3 using dispersed ozone flotation. Bioresour Technol 102(1):82–87. https://doi.org/10.1016/j.biortech.2010.04.083

Cherubini F, Jungmeier G, Wellisch M, Willke T, Skiadas I, Van Ree R, de Jong E (2009) Toward a common classification approach for biorefinery systems. Biofuels Bioprod Biorefin 3(5):534–546. https://doi.org/10.1002/bbb.172

Christenson L, Sims R (2011) Production and harvesting of microalgae for wastewater treatment, biofuels, and bioproducts. Biotechnol Adv 29(6):686–702. https://doi.org/10.1016/j.biotechadv.2011.05.015

Danquah MK, Gladman B, Moheimani N, Forde GM (2009) Microalgal growth characteristics and subsequent influence on dewatering efficiency. Chem Eng J 151(1):73–78. https://doi.org/10.1016/j.cej.2009.01.047

Dassey AJ, Theegala CS (2014) Reducing electrocoagulation harvesting costs for practical microalgal biodiesel production. Environ Technol 35(6):691–697. https://doi.org/10.1080/09593330.2013.842602

Demirbaş A (2008) Production of biodiesel from algae oils. Energy Sources 31(2):163–168. https://doi.org/10.1080/15567030701521775

Dong T, Knoshaug EP, Davis R, Laurens LML, Van Wychen S, Pienkos PT, Nagle N (2016) Combined algal processing: a novel integrated biorefinery process to produce algal biofuels and bioproducts. Algal Res 19:316–323. https://doi.org/10.1016/j.algal.2015.12.021

Drexler ILC, Yeh DH (2014) Membrane applications for microalgae cultivation and harvesting: a review. Rev Environ Sci Biotechnol 13(4):487–504. https://doi.org/10.1007/s11157-014-9350-6

Ebhodaghe SO, Imanah OE, Ndibe H (2022) Biofuels from microalgae biomass: a review of conversion processes and procedures. Arab J Chem 15(2):103591. https://doi.org/10.1016/j.arabjc.2021.103591

Edzwald JK (1993) Algae, bubbles, coagulants, and dissolved air flotation. Water Sci Technol 27(10):67–81. https://doi.org/10.2166/wst.1993.0207

Faruk U, Nadaf R, Toragall M, Nadaf U, Shivasharana C (2023) The role of algae in sustainable environment: a review. J Algal Biomass Utln 11:28–34

Figueroa-Torres GM, Theodoropoulos C (2023) Techno-economic analysis of a microalgae-based biorefinery network for biofuels and value-added products. Bioresour Technol Rep 23:101524. https://doi.org/10.1016/j.biteb.2023.101524

Folch J, Lees M, Sloane Stanley GH (1957) A simple method for the isolation and purification of total lipides from animal tissues. J Biol Chem 226(1):497–509

Ghasemi Naghdi F, González González LM, Chan W, Schenk PM (2016) Progress on lipid extraction from wet algal biomass for biodiesel production. Microb Biotechnol 9(6):718–726. https://doi.org/10.1111/1751-7915.12360

Gui J, Chen S, Luo G, Wu Z, Fan Y, Yao L, Xu H (2021) Nutrient deficiency and an algicidal bacterium improved the lipid profiles of a novel promising oleaginous dinoflagellate, *Prorocentrum donghaiense*, for biodiesel production. Appl Environ Microbiol 87(19):e01159. https://doi.org/10.1128/AEM.01159-21

Hopkins TR (1991) Physical and chemical cell disruption for the recovery of intracellular proteins. Bioprocess Technol 12:57–83

Hosikian A, Lim S, Halim R, Danquah MK (2010) Chlorophyll extraction from microalgae: a review on the process engineering aspects. Int J Chem Eng 2010:391632. https://doi.org/10.1155/2010/391632

Ibáñez E, Herrero M, Mendiola JA, Castro-Puyana M (2012) Extraction and characterization of bioactive compounds with health benefits from marine resources: macro and micro algae, cyanobacteria, and invertebrates. In: Marine bioactive compounds. Springer, Cham

Jensen SK (2008) Improved bligh and dyer extraction procedure. Lipid Technol 20(12):280–281. https://doi.org/10.1002/lite.200800074

Kadam SU, Tiwari BK, O'Donnell CP (2013) Application of novel extraction technologies for bioactives from marine algae. J Agric Food Chem 61(20):4667–4675. https://doi.org/10.1021/jf400819p

Karemore A, Sen R (2016) Downstream processing of microalgal feedstock for lipid and carbohydrate in a biorefinery concept: a holistic approach for biofuel applications. RSC Adv 6(35):29486–29496. https://doi.org/10.1039/C6RA01477A

Khan MI, Shin JH, Kim JD (2018) The promising future of microalgae: current status, challenges, and optimization of a sustainable and renewable industry for biofuels, feed, and other products. Microb Cell Factories 17(1):36. https://doi.org/10.1186/s12934-018-0879-x

Kim NJ, Li H, Jung K, Chang HN, Lee PC (2011) Ethanol production from marine algal hydrolysates using Escherichia coli KO11. Bioresour Technol 102(16):7466–7469. https://doi.org/10.1016/j.biortech.2011.04.071

Kim YH, Choi YK, Park J, Lee S, Yang YH, Kim HJ, Park TJ, Hwan Kim Y, Lee SH (2012) Ionic liquid-mediated extraction of lipids from algal biomass. Bioresour Technol 109:312–315. https://doi.org/10.1016/j.biortech.2011.04.064

Kim DY, Lee K, Lee J, Lee YH, Han JI, Park JY, Oh YK (2017) Acidified-flocculation process for harvesting of microalgae: coagulant reutilization and metal-free-microalgae recovery. Bioresour Technol 239:190–196. https://doi.org/10.1016/j.biortech.2017.05.021

Kumari S, Turkar P, Subramanian S (2018) An insight on algal cell disruption for biodiesel production. Asian J Pharm Clin Res 11(2):21–26. https://doi.org/10.22159/ajpcr.2018.v11i2.22481

Lee J-Y, Yoo C, Jun S-Y, Ahn C-Y, Oh H-M (2010) Comparison of several methods for effective lipid extraction from microalgae. Bioresour Technol 101(1):S75–S77. https://doi.org/10.1016/j.biortech.2009.03.058

Lee A, Lewis D, Ashman P (2012) Disruption of microalgal cells for the extraction of lipids for biofuels: processes and specific energy requirements. Biomass Bioenergy 46:89–101. https://doi.org/10.1016/j.biombioe.2012.06.034

Li Q, Jiang X, He Y, Li L, Xian M, Yang J (2010) Evaluation of the biocompatibile ionic liquid 1-methyl-3-methylimidazolium dimethylphosphite pretreatment of corn cob for improved saccharification. Appl Microbiol Biotechnol 87(1):117–126. https://doi.org/10.1007/s00253-010-2484-8

Liang K, Zhang Q, Cong W (2012) Enzyme-assisted aqueous extraction of lipid from microalgae. J Agric Food Chem 60(47):11771–11776. https://doi.org/10.1021/jf302836v

Liu D, Wang P, Wei G, Dong W, Hui F (2013) Removal of algal blooms from freshwater by the coagulation-magnetic separation method. Environ Sci Pollut Res Int 20(1):60–65. https://doi.org/10.1007/s11356-012-1052-4

Mengstie K, Dibabe Y, Sema Y (2023) Review on microalgae potential innovative biotechnological applications. Biosci Biotechnol Res 1:35–43

Michalak I, Chojnacka K (2014) Algal extracts: technology and advances. Eng Life Sci 14(6):581–591. https://doi.org/10.1002/elsc.201400139

Milledge JJ (2011) Commercial application of microalgae other than as biofuels: a brief review. Rev Environ Sci Biotechnol 10(1):31–41. https://doi.org/10.1007/s11157-010-9214-7

Mo W, Soh L, Werber JR, Elimelech M, Zimmerman JB (2015) Application of membrane dewatering for algal biofuel. Algal Res 11:1–12. https://doi.org/10.1016/j.algal.2015.05.018

Molina Grima E, Belarbi EH, Acién Fernández FG, Robles Medina A, Chisti Y (2003) Recovery of microalgal biomass and metabolites: process options and economics. Biotechnol Adv 20(7):491–515. https://doi.org/10.1016/S0734-9750(02)00050-2

Morales M, Aflalo C, Bernard O (2021) Microalgal lipids: a review of lipids potential and quantification for 95 phytoplankton species. Biomass Bioenergy 150:106108. https://doi.org/10.1016/j.biombioe.2021.106108

Ndikubwimana T, Zeng X, Murwanashyaka T, Manirafasha E, He N, Shao W, Lu Y (2016) Harvesting of freshwater microalgae with microbial bioflocculant: a pilot-scale study. Biotechnol Biofuels 9(1):47. https://doi.org/10.1186/s13068-016-0458-5

Neeti K, Gaurav K, Singh R (2023) The potential of algae biofuel as a renewable and sustainable bioresource. Eng Proc 37(1):22

Neto AMP et al (2013) Improvement in microalgae lipid extraction using a sonication-assisted method. Renew Energy 55:525–531. https://doi.org/10.1016/j.renene.2013.01.019

Ninomiya K, Kamide K, Takahashi K, Shimizu N (2012) Enhanced enzymatic saccharification of kenaf powder after ultrasonic pretreatment in ionic liquids at room temperature. Bioresour Technol 103(1):259–265. https://doi.org/10.1016/j.biortech.2011.10.019

Oliveira GA, Carissimi E, Monje-Ramírez I, Velasquez-Orta SB, Rodrigues RT, Ledesma MTO (2018) Comparison between coagulation-flocculation and ozone-flotation for Scenedesmus microalgal biomolecule recovery and nutrient removal from wastewater in a high-rate algal pond. Bioresour Technol 259:334–342. https://doi.org/10.1016/j.biortech.2018.03.072

Pahl SL, Lee AK, Kalaitzidis T, Ashman PJ, Sathe S, Lewis DM (2013) Harvesting, thickening and dewatering microalgae biomass. In: Borowitzka MA, Moheimani NR (eds) Algae for biofuels and energy. Springer, Cham, pp 165–185. https://doi.org/10.1007/978-94-007-5479-9_10

Park JBK, Craggs RJ, Shilton AN (2011) Wastewater treatment high rate algal ponds for biofuel production. Bioresour Technol 102(1):35–42. https://doi.org/10.1016/j.biortech.2010.06.158

Pragya N, Pandey KK, Sahoo PK (2013) A review on harvesting, oil extraction and biofuels production technologies from microalgae. Renew Sust Energ Rev 24:159–171. https://doi.org/10.1016/j.rser.2013.03.034

Quitério E, Grosso C, Ferraz R, Delerue-Matos C, Soares C (2022) A critical comparison of the advanced extraction techniques applied to obtain health-promoting compounds from seaweeds. Mar Drugs 20(11):677. https://doi.org/10.3390/md20110677

Ramesh D (2013) Lipid identification and extraction techniques. In: Biotechnological applications of microalgae: biodiesel and value-added products. CRC Press, Boca Raton, pp 89–97. https://doi.org/10.1201/b14920-8

Ranjith Kumar R, Hanumantha Rao P, Arumugam M (2015) Lipid extraction methods from microalgae: a comprehensive review. Front Energy Res 2:61. https://doi.org/10.3389/fenrg.2014.00061

Rawat I, Ranjith Kumar R, Mutanda T, Bux F (2011) Dual role of microalgae: phycoremediation of domestic wastewater and biomass production for sustainable biofuels production. Appl Energy 88(10):3411–3424. https://doi.org/10.1016/j.apenergy.2010.11.025

Rhea N, Groppo J, Crofcheck C (2017) Evaluation of flocculation, sedimentation, and filtration for dewatering of Scenedesmus algae, vol 60. American Society of Agricultural and Biological Engineers, St Joseph

Routray W, Orsat V (2017) Microwave-assisted extraction of flavonoids. Food Bioprocess Technol 5:409–424. https://doi.org/10.1016/B978-0-12-809380-1.00008-5

Sarwer A, Moussa S, Osman A, Jamil F, Al-Muhtaseb A, Alhajeri N, Rooney D (2022) Algal biomass valorization for biofuel production and carbon sequestration: a review. Environ Chem Lett. https://doi.org/10.1007/s10311-022-01458-1

Sharif KM, Rahman MM, Azmir J, Mohamed A, Jahurul MHA, Sahena F, Zaidul ISM (2014) Experimental design of supercritical fluid extraction – a review. J Food Eng 124:105–116. https://doi.org/10.1016/j.jfoodeng.2013.10.003

Shen Y, Yuan W, Pei J, Wu Q, Mao E (2009) Microalgae mass production methods. Trans ASABE 52(4):1275–1287. https://doi.org/10.13031/2013.27771

Show K-Y, Lee D-J (2013) Algal biomass harvesting. In: Biofuels from algae. Elsevier, Amsterdam, pp 85–110. https://doi.org/10.1016/B978-0-444-59558-4.00005-X

Singh G, Patidar SK (2018) Microalgae harvesting techniques: a review. J Environ Manag 217:499–508. https://doi.org/10.1016/j.jenvman.2018.04.010

Soto L, Dixon C, Wilken L (2017) Enzymatic cell disruption of the microalgae Chlamydomonas reinhardtii for lipid and protein extraction. Algal Res 25:149–159. https://doi.org/10.1016/j.algal.2017.04.004

Vandamme D, Pontes SCV, Goiris K, Foubert I, Pinoy LJJ, Muylaert K (2011) Evaluation of electro-coagulation–flocculation for harvesting marine and freshwater microalgae. Biotechnol Bioeng 108(10):2320–2329. https://doi.org/10.1002/bit.23199

Vandamme D, Foubert I, Fraeye I, Meesschaert B, Muylaert K (2012) Flocculation of Chlorella vulgaris induced by high pH: role of magnesium and calcium and practical implications. Bioresour Technol 105:114–119. https://doi.org/10.1016/j.biortech.2011.11.105

Viegas CV, Hachemi I, Mäki-Arvela P, Smeds A, Aho A, Freitas SP, da Silva Gorgônio CM, Carbonetti G, Peurla M, Paranko J, Kumar N, Aranda DAG, Murzin DY (2015) Algal products beyond lipids: comprehensive characterization of different products in direct saponification of green alga Chlorella sp. Algal Res 11:156–164. https://doi.org/10.1016/j.algal.2015.06.014

Wang X, Liu X, Wang G (2011) Two-stage hydrolysis of invasive algal feedstock for ethanol fermentation. J Integr Plant Biol 53(3):246–252. https://doi.org/10.1111/j.1744-7909.2010.01024.x

Weerheim AM, Kolb AM, Sturk A, Nieuwland R (2002) Phospholipid composition of cell-derived microparticles determined by one-dimensional high-performance thin-layer chromatography. Anal Biochem 302(2):191–198. https://doi.org/10.1006/abio.2001.5552

Wiley PE, Brenneman KJ, Jacobson AE (2009) Improved algal harvesting using suspended air flotation. Water Environ Res 81(7):702–708. https://doi.org/10.2175/106143009x407474

Wu Z, Zhu Y, Huang W, Zhang C, Li T, Zhang Y, Li A (2012) Evaluation of flocculation induced by pH increase for harvesting microalgae and reuse of flocculated medium. Bioresour Technol 110:496–502. https://doi.org/10.1016/j.biortech.2012.01.101

Xu L, Wim Brilman DW, Withag JA, Brem G, Kersten S (2011) Assessment of a dry and a wet route for the production of biofuels from microalgae: energy balance analysis. Bioresour Technol 102(8):5113–5122. https://doi.org/10.1016/j.biortech.2011.01.066

Xu K, Zou X, Wen H, Xue Y, Zhao S, Li Y (2018) Buoy-bead flotation harvesting of the microalgae Chlorella vulgaris using surface-layered polymeric microspheres: a novel approach. Bioresour Technol 267:341–346. https://doi.org/10.1016/j.biortech.2018.07.065

Yoo G, Park WK, Kim CW, Choi YE, Yang JW (2012) Direct lipid extraction from wet Chlamydomonas reinhardtii biomass using osmotic shock. Bioresour Technol 123:717–722. https://doi.org/10.1016/j.biortech.2012.07.102

Zenouzi A, Ghobadian B, Hejazi MA, Rahnemoon P (2013) Harvesting of microalgae dunaliella salina using electroflocculation. J Agric Sci Technol 15(5):879–887. http://jast.modares.ac.ir/article-23-4125-en.html

Zhang X, Wang L, Sommerfeld M, Hu Q (2016) Harvesting microalgal biomass using magnesium coagulation-dissolved air flotation. Biomass Bioenergy 93:43–49. https://doi.org/10.1016/j.biombioe.2016.06.024

Zheng G, Li C, Guo L, Ruo W, Wang S (2012) Purification of extracted fatty acids from the microalgae spirulina. J Am Oil Chem Soc 89(4):561–566. https://doi.org/10.1007/s11746-011-1956-z

Zhou J, Wang M, Saraiva JA, Martins AP, Pinto CA, Prieto MA, Simal-Gandara J, Cao H, Xiao J, Barba FJ (2022) Extraction of lipids from microalgae using classical and innovative approaches. Food Chem 384:132236. https://doi.org/10.1016/j.foodchem.2022.132236

Pretreatment and Fractionation of Algae Biomass for Value-Added Extraction

5

Shalini Sharma, Sudarshan Sahu, and Shailendra Kumar Arya

Contents

S. Sharma · S. Sahu
Department of Biotechnology Engineering, University Institute of Engineering and Technology, Panjab University, Chandigarh, India

S. K. Arya (✉)
University Institute of Engineering and Technology (UIET), Panjab University, Chandigarh, India
e-mail: skarya@pu.ac.in

S. K. Arya et al. (eds.), *Value Added Products From Bioalgae Based Biorefineries: Opportunities and Challenges*,
https://doi.org/10.1007/978-981-97-1662-3_5

Abstract

Algae biomass is a valuable and sustainable resource with enormous potential for use in a variety of industries. The methods of pretreatment and fractionation are essential in releasing the wide range of components present in algal biomass, facilitating their extraction and conversion into valuable products. In order to produce value-added products for a variety of industries, this chapter attempts to give a thorough review of the pretreatment and fractionation methods used on algae biomass. The first section of this chapter describes the varied makeup of algal biomass, emphasizing the abundance of lipids, proteins, carbohydrates, pigments and bioactive chemicals found in it. It then explores the pretreatment techniques—physical, chemical and biological—that are used to alter the intricate structure of algal biomass. These pretreatment techniques are essential for improving the components of algal biomass that are valuable to access and remove. This chapter also clarifies the many fractionation strategies that are applied to isolate particular fractions or chemicals from pretreatment algal biomass. These cutting-edge techniques for separation include solvent extraction, chromatography, membrane separation and other recently developed technologies that all help to extract valuable substances including proteins, polysaccharides, bioactive pigments and omega-3 fatty acids. The value-added applications of these extracted components in various sectors are summarized in the next section. It outlines their use in biofuels, bioplastics, medicines, cosmetics, functional foods and environmental remediation, highlighting how they contribute to creative and sustainable solutions. The assessment concludes by highlighting the prospects for the future and possible developments in fractionation and pretreatment methods for algae biomass. It envisions the integration of cutting-edge methods to significantly improve the extraction efficiency and diversity of value-added products from algal biomass, including genetic engineering, sophisticated bioprocessing and optimized biorefinery concepts. The analysis concludes by highlighting the critical role that fractionation and pretreatment play in maximizing the potential of algal biomass and laying the groundwork for the creation of a lucrative and diversified range of products that can be used in a variety of industrial sectors to build a sustainable bioeconomy.

Keywords

Algal biomass · Macroalgae · Microalgae · Pretreatment techniques · Fractionation · Biorefinery · Value-added products

Abbreviations

DESs	Deep eutectic solvents
ILs	Ionic liquids
PEF	Pulsed electric field
SFC	Supercritical fluid chromatography

SFE Supercritical fluid extraction
TEA Techno economic analysis
TLC Thin-layer chromatography

5.1 Introduction

The term 'algal biomass' describes the total mass or number of algae, which are photosynthetic organisms that can grow in varying range of environment from freshwater to extreme conditions such as saltwater or wastewater, among other aquatic conditions. Algal biomass, which is usually expressed in terms of weight or volume, is the total amount of algae present in a specific area or volume. Algae are a broad category of basic, photosynthetic organisms that can be as big as multicellular seaweeds or as tiny as single-celled creatures. Because they are primary producers in aquatic ecosystems, they use photosynthesis to turn carbon dioxide and sunlight into organic matter. They use carbonic anhydrase as a CO_2 compensatory mechanism to catalyse the dehydration of HCO_3^-. The majority of marine algae use the Calvin-Benson cycle (C3) to ingest CO_2; however, some species also use the Hatch-Slack cycle to improve photosynthetic efficiency. They are classified as macroalgae or microalgae based on their size. While macroalgae are multicellular and can only be seen with the use of a microscope, microalgae are minuscule and typically unicellular (Uma et al. 2023). The high lipid content and fast growth rate of algae have made them a promising source of renewable biomass for biofuels and other bioproducts. Scientists and businesses investigate ways to grow and gather algae biomass in order to extract oils or make valuable substances like biofuels, bioplastics, animal feed and others (Wang et al. 2023). Various types of algae are utilized for the extraction of different compounds, using different fractionation techniques and pretreatment methods that enhance the ultimate yield the of the final product. Comprehensive extraction of multi-algal product biorefinery is imagined as an assured method to improve the economic viability of algal goods in the worldwide market. Because of the intricate construction of the algal cell wall, a variety of pretreatment procedures are required, including physicochemical, chemical, biological and physical methods for cell disruption. Building a sustainable biorefinery for the circular economy requires understanding the pretreatment of the microalgal biomass in order to achieve the goal of each molecule's solubility. Pectin and cellulose make up the strong cell walls of microalgae, which prevent hydrolytic enzymes and further processing. The purpose of pretreatments for microalgal biomass is to modify the composition of the materials found in the cell wall, such as cellulose and starch, and to increase the solubility of the materials. These pretreatment and fractionation methods are generally product specific and to obtain the better results, these need to be highly optimized (Ventura et al. 2017). The development of algal biomass valorization involves four key steps: choosing an appropriate microalgal strain and growing it; harvesting the microalgal biomass once it has grown; producing bioenergy from the obtained biomass; and, finally, using suitable pretreatment techniques to increase the biomass's conversion efficiency to the necessary bioenergy product (de Carvalho et al. 2020). Here, an extensive examination outlines various methods along with their associated challenges, followed by strategies to mitigate these issues. Additionally, the study delves into the analysis of bioproduct quality and the examination of various attributes. Furthermore, it discusses future directions and emerging trends comprehensively.

5.2 Algae Biomass: Overview and Characteristics

It is critical to comprehend the makeup of algal biomass so as to (1) select the best genera for the production of a particular fraction; (2) identify the technique to be utilized for specific product formation, that is identifying valuable compounds and optimizing extraction processes; (3) to analyse the nutrient uptake ability of particular algal species; (4) to ensure the safety, quality and nutritional content of algal-based food and feed products; and (5) to analyse the algal ability of carbon sequestration and climate mitigation. Thus, to ensure all these parameters, the physical and chemical characteristics of the algae are analysed prior to their further treatment.

5.2.1 Physical Characteristics of Algae

Algae are a class of plants that have been around since prehistoric times. After being originally described by Linnaeus in 1753, A. L. de Jussieu (1789) distinguished the algae from the rest of the plant kingdom and classed the plants according to their current classification. They are generally aquatic, autotrophic and occasionally terrestrial. It is known as thallophytes because the plant body varies from unicellular to multicellular structures without vasculature and minimal differentiation into different tissue systems. These plants lack genuine roots, stems and leaves. Seaweeds range in size from single cells as small as 1 μ to massive organisms that can reach heights of over 60 m.

Algal classification can be well understood from Fritsch classification (Karaduman et al. 2022). Fritsch classified the whole of the algae into 11 classes on the basis of type of pigments, nature of reserve food material, mode of reproduction etc.

5.2.2 Chemical Composition and Nutritional Value

The protein and fat contents of macroalgae can vary from 7 to 31% and 2 to 13% of dry weight respectively. Additionally, macroalgae contain a significant amount of carbohydrates (up to 32–60% of dry weight). In terms of micronutrient content, macroalgae are rich in vitamins, particularly those belonging to the B-group (i.e. B1, B12) and the lipophilic vitamins A and E (tocopherol). The abundance of vitamin B12 in macroalgae-based products makes them ideal as dietary supplements for vegans, who are thought to be susceptible to vitamin B12 insufficiency. The most important microelements found in seaweeds are typically potassium, sodium, magnesium and calcium, which together make up over 97% of the total mineral content. Small concentrations of additional microelements, such as zinc, manganese, copper and iron, range from 0.001 to 0.094% of seaweeds dry weight (Biris-Dorhoi et al. 2020).

Three additional factors are crucial to consider when assessing the composition of microalgae in addition to the traditional protein-carbohydrate-lipid form: the pace of cell development, the presence of important compounds like carotenoids and the makeup of the cell wall. Two clear benefits of industrial strains are high content of a desired fraction and rapid growth rate (=high productivity). Also, the algal biomass lacks lignin content, hence making it easier to be hydrolysed.

5.2.3 Challenges in Utilizing Algae Biomass

Based on the area of research, there are various challenges that need to be encountered while dealing with algal biomass. Various algae biorefineries do come across various problem while degrading the algal components into value-added products. These issues can be common to all refinery units or can be sometimes specific to each one of them. Thus, before studying the pretreatment methods thoroughly, there needs to be done a proper study on the problems that can be faced later in production process and the cost associated with the production process needs to be analysed via TEA (techno economic analysis). Also, the cost of pre-treating microalgae is far less than that of pre-treating terrestrial plant biomass.

Issues that are generally common to all biorefinery units are as follows:

- The most significant characteristic of microalgae is their capacity for photosynthetic activity, which is effectively achieved in outdoor cultivation methods. However, the irregularity of natural sunshine is a hindrance. These restrictions have led to issues with productivity for the large-scale sector.
- Drying process of algal biomass is tedious and consumes high energy that ultimately adds on to the production cost (Yun et al. 2019).
- In most situations, the biomass productivity of mixotrophic farming—which makes use of organic carbon sources—outperforms that of photoautotrophic cultivation. Algae grown heterotrophically use only organic carbon sources; in contrast, algae grown mixotrophically can use both organic and inorganic carbon sources at the same time. The use of two pathways for organic and inorganic carbon raises questions since it may result in a poorer yield or productivity of algae growth than the total of the conditions.
- Furthermore, in mixotrophic environments, some species experience photosynthetic inhibition.
- There can also be a problem with contamination-related decreased productivity. Single-strain cultivation is generally challenging to maintain outside. The culture system is usually an open pond or semi-enclosed kind where contamination is

inevitable in order to maximize the utilization of photo-energy. Target cells must therefore compete with other microorganisms, which will reduce overall output (Park et al. 2022)

5.3 Pretreatment Techniques for Algae Biomass

An essential first stage in turning algal biomass into a range of useful products, such as food supplements, biofuels and bioplastics, is pretreatment. Pretreatment aims to break down the algae's inflexible cell walls and increase the accessibility of the intracellular components for further processing. Algal biomass must be broken down or disturbed in order to convert, gather and process the lipids and carbohydrates it contains. Numerous studies have demonstrated that glucose, followed by galactose, mannose, xylose and other sugar molecules, is the primary sugar molecule found in the majority of algae species. Various methods of pretreatment can be utilized such as physical/mechanical, chemical, thermal and biological (Kumar et al. 2016) based on the final product that is to be synthesized; as there is no single 'magic potion' that can be utilized for all the biomasses. In order to produce better results than applying a single pretreatment process, studies were also conducted by combining the aforementioned pretreatment techniques, also known as combined approaches. However, most methods can be classified into four categories according to the goal: (1) drying, (2) thermochemical processing, (3) cell disruption and (4) targeted fractionation as shown in Fig. 5.1. All the above-mentioned processes will be discussed further in brief.

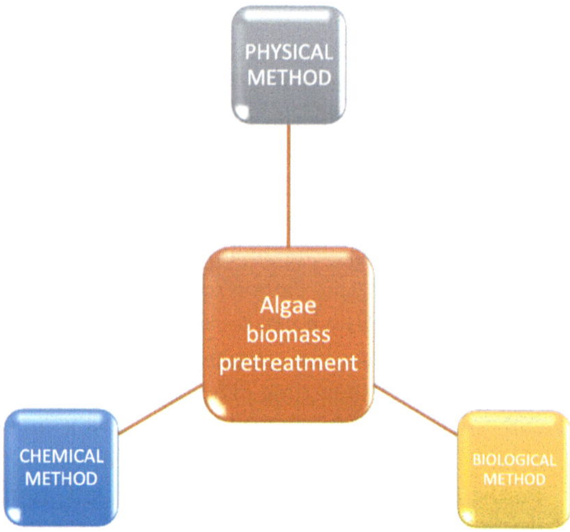

Fig. 5.1 Different methods for pretreatment of algae biomass

5.3.1 Physical/Mechanical Method

Mechanical pretreatment includes physical methods aimed at size reduction (milling) or cell wall disruption by provoking physical damage (ultrasound, microwave), the latter being more commonly used for microalgae biomass.

Ultrasound, grinding, microwaves and the use of high temperatures are examples of mechanical or physical pretreatment techniques that convert algal biomass into nutritional molecules that are easily absorbed. Due to its high energy input requirements, this algal biomass pretreatment process is not economically practical.

5.3.1.1 Milling

One popular mechanical technique that has a high disruption efficacy in a single run is milling (de Carvalho et al. 2020). Shear pressures are created as the algal biomass is stirred quickly by beads. Algal cells are broken apart by the beads' physical impact with the biomass. Shaking vessels have the potential to replace beads, although their effectiveness is not as great. Large amounts of heat are released during the operation, necessitating the use of energy to cool the machinery.

High-pressure homogenization (HPH) is an additional procedure that entails applying extremely high pressure to the suspended algal culture and causing a quick drop in pressure to mechanically disturb the algal cells. It was demonstrated that cell wall stiffness is negatively related to cell wall disruption efficiency. Microalgae species such as *P. cruentum* and *A. platensis*, characterized by less rigid cell walls, exhibited a protein release of 57% in response to high-pressure homogenization (HPH) treatment. Conversely, microalgae species with more rigid cell walls, such as *N. oculata* and *C. vulgaris*, demonstrated a slightly lower protein release of approximately 47% under the same intensity treatment (Soto-Sierra et al. 2018; Farooqui et al. 2021). Large-scale applications can make use of this technique because it releases less heat than ball milling does. The algae cells' viscosity and velocity, along with a high-pressure homogenizer, make up the cell disruption efficacy. The desired result in high-pressure homogenization requires multiple runs (Jagaba et al. 2021). A few algal species that have undergone cell wall disruption analysis using this technique are *Porphyridium centum*, *Arthrospira platensis*, *Nannochloropsis oculata*, *Chlorella vulgaris* and *Haematococcus pluvialis* (Yaashikaa et al. 2023).

5.3.1.2 Sonication

Sonication is the process of application of sound waves to a sample for a wide range of applications. When biomass is exposed to the sound waves of ultrahigh frequency, that is,100 kHz to 1 MHz, it results in disintegration of the polymeric constituents in the biomass. Biomass is usually suspended in an aqueous medium and subjected to sound-wave treatment. Furthermore, because oscillating bubbles are similar in size to most microalgae cell walls, given the right conditions they can greatly increase mechanical resonance and destroy cells. Without the need for expensive dehydration processes, continuous ultrasonication is a viable method for releasing lipids from microalgae species with robust cell walls (Ha et al. 2020).

5.3.1.3 Microwaves

Microwaves high efficiency in the disruption of molecules, rapid pretreatment, low power consumption and suitability for a large group of microalgal species are some of the advantages of microwave pretreatment technique. However, maintenance costs and irradiation of heat-sensitive components are flaws inhibiting the commercialization process. Hence, microwave-assisted pretreatment techniques are often employed for the extraction biomolecules. The pretreatment's main benefit is that the high temperature accelerates the rate of reaction and uses less chemicals overall. However, this pretreatment is typically used extensively in conjunction with other pretreatment techniques (Saini et al. 2015). While the microwave enhances the substrate's susceptibility to enzymatic hydrolysis, sonication produces acoustic waves that cause cavitation in the cells which ultimately lead to cell disruption.

5.3.1.4 Extrusion

Extrusion pretreatment is also a physical pretreatment approach that uses heat, compression, mixing and shear forces to physically disrupt and chemically modify biomass as it passes through the extruder, forming steam-expanded pellets at the system outlet. Extrusion pretreatment has several benefits, including improved monitoring and control, economy, excellent adaptability to different process changes, products with no sugar degradation and high continuous output (Azmi et al. 2020).

5.3.1.5 Pulsed Electric Field (PEF)

The principle behind pulsed electric field is to expose microalgal cell wall to a very high electric field for brief periods of time (pulses). The creation of pores (electroporation) on the cell membranes is possible when the applied electric field is usually greater than a certain threshold. One prerequisite for this process is that the microalgae suspension should be non-conductive. Hence, salt-containing marine algae must be treated to remove salts before PEF treatment. Compared to conventional and other physical methods, PEF has several benefits, including a reduced solvent requirement, lesser energy consumption and reduced production costs. PEF efficiency is influenced by the microalgae culture's growth stage and cell size. Larger cells are extremely sensitive to electroporation, but smaller cells are more resistant. Consequently, larger *Haematococcus pluvialis* cells may help with PEF treatment, since astaxanthin ethanol extraction was found to be 96% effective (Costa et al. 2020; Martínez et al. 2019). Due to its minimal impact on cell integrity, pulsed electric field pretreatment is frequently referred to as a moderate or selective approach. It can target the carbohydrates and lower molecular weight proteins found inside of cells. PEF-assisted techniques are preferable for macromolecule extraction.

5.3.2 Chemical Pretreatment

One of the simplest, most economical and fastest ways to obtain the desired molecules is the chemical pretreatment process. To solubilize the polymers and promote biodegradations, this pretreatment consists of adding alkaline, organic solvents and acid/acidic oxidative agents. Furthermore, hydrolysing starch or cellulose derived from microalgae is somewhat simpler to do in milder circumstances. In the sections that follow, various chemical pretreatment techniques used for microalgae are discussed.

5.3.2.1 Acid Hydrolysis

Acidic pretreatments rely on the use of acids as a catalyst to accelerate the breakdown of cellulose. Concentrated and diluted acids are the two divisions included in this. Because concentrated acids cause equipment corrosion and create inhibitory components that grow at greater temperatures and acid concentrations, they are less preferred than diluted acids. Sulphuric acid, hydrochloric acid, nitric acid and phosphoric acid are typically used in these pretreatments. For the conversion of lignocellulosic structures into soluble sugars, moderate temperatures and diluted acid are needed. These days, preparation of biomass is frequently done with diluted sulphuric acid (Abou El-Souod et al. 2021). To prevent the enzymatic hydrolysis process, acid pretreatment promotes the breakdown of the cellulosic matrix found in the cell wall and the depolymerization of hemicelluloses, which includes the hydrolysis of starch molecules into simpler molecules. Thus, acid pretreatment is the process of dissolving the algae's hard cell walls using acids like hydrochloric or sulphuric acid. The release of sugars and other useful components is facilitated by the acids' hydrolysis of hemicellulose and disruption of lignin. However, because it can be challenging to dissolve some algae species' stiff structure, the efficacy of acid pretreatment may be limited (Bhosle et al. 2010). The main drawback of acid hydrolysis is the creation of undesirable sugar and lignin derivatives, including 5-hydroxymethylfurfural, furfurals and phenolic chemicals, which have antimicrobial properties. These harmful substances slow down the rate at which yeast grows and converts during fermentation.

5.3.2.2 Alkaline Hydrolysis

Alkaline pretreatment weakens the structure of lignin and solubilizes hemicellulose, improving the accessibility of proteins and carbohydrates. Bases such as sodium hydroxide (NaOH) or ammonia (NH_3) are used in this process. Alkaline treatment can help extract bioactive chemicals from algal biomass and improve its digestion. The alkaline solutions act by breaking down hemicellulose and lignin present in the cell walls, thereby reducing the structural complexity and enhancing the accessibility of intracellular constituents. The effectiveness of alkali pretreatment depends on various factors, including the concentration of the alkaline solution, temperature, duration of treatment and the type of algal biomass being processed. components (Zhang et al. 2021). Bhosle et al demonstrated that boiling *Spirogyra* sp. and *Nostoc* commune for 15 min in 500 ml of 0.5N NaOH solution enhanced their biosorption capability as 5.12 and 5.82 mg/g respectively. Before applying these alkalis for hydrolysis, the process optimization needs to be done that includes, its concentrations and temperature. The kind of strain, length of the treatment, timing of other treatments and recuperation period are all factors that affect how effective the treatment is. Due to the reaction's severe conditions—high pH and temperature, denaturation of protein and continuous pigment release from the cell suspension—the container eventually deteriorates. Since disposing of them presents a significant risk, the discharge of toxic compounds is also a major concern. Acetate and propionate are two inhibitors that have also been reported in trace levels (Mandotra et al. 2019). This results in limited carbohydrate extraction; hence, it would be advantageous to include inhibitor-resistant fermentative strains.

5.3.2.3 Ionic Liquids (ILs)

Ionic liquids are low-melting-point organic salts that effectively dissolve and alter the structure of algal biomass by forming hydrogen bonds with them. Ionic liquid pretreatment can make it easier to extract lipids, proteins and carbohydrates from algae. Because ionic liquids may establish hydrogen bonds with biological polymers, such as cellulose, which is typically found in the cell walls of microalgae, they can interact with them. The hydrogen bonds that develop between the cellulose microfibrils are broken by means of this characteristic of ionic liquids. Ionic liquid treatment is advantageous due to its low vapour pressure, fast reaction time, reusability and high-performance yield. By extracting the microalgae biomass without changing its molecular structure, Ionic Liquids (ILs) make it possible to produce products with additional value (Chen et al. 2018; Lu et al. 2019). Gyeong-Uk Kim et al demonstrated the use of ionic liquids in *Chlamydomonas Mexicana* for enhanced biodiesel production. It showed lipid extraction: 99% by using 4 g of IL/g of microalgae; Speed: 150 rpm; Time: 1.5 h. 1-ethyl-3-methylimidazolium acetate, 1-butyl-3-methylimidazolium chloride, 1-ethyl-3-methylimidazolium chloride, choline-based ionic liquids etc. can be employed for this purpose.

5.3.2.4 Deep Eutectic Solvents (DESs)

Deep eutectic solvents (DES) are a class of green solvents formed by the combination of two or more solid or liquid components, typically a hydrogen bond donor (HBD) and a hydrogen bond acceptor (HBA), at certain molar ratios to create a eutectic mixture with a lower melting point than either of the individual components. These solvents possess several properties that make them attractive alternatives to conventional organic solvents such as low toxicity, low volatility, tunable properties, solvent for biomass processing and versatility in reactions. The most common components used in the formation of DES include choline chloride, urea, organic acids, amides, sugars and alcohols. When combined at specific molar ratios, these components form a eutectic mixture with a melting point significantly lower than that of the individual constituents, resulting in a liquid or viscous solvent at room temperature. DES have gained attention as potential alternatives to conventional organic solvents in various industries due to their environmentally benign nature, tunable properties and applicability in diverse chemical processes. Ongoing research focuses on exploring and optimizing the properties and applications of DES in fields ranging from green chemistry to industrial processes and material sciences (Chen and Wan 2018).

5.3.3 Biological Pretreatment

The microbial-driven pretreatment methods are more encouraged by the research communities to increase the algal biomass following biofuel production. The microbial techniques are eco-friendly and environmentally friendly. Microbes require minor energy input and support sustainable developmental goals. In a broader context, there are two approaches: either utilizing microbes directly to process biomass, or producing hydrolytic enzymes and subsequently digesting the biomass with these enzymes. Some methods of biological pretreatment include enzymatic hydrolysis, microbial treatment, algae-assisted pretreatment and combined biological methods. Each of these has their own merits and demerits, but this process as a whole is much more ecofriendly and sustainable.

5.3.3.1 Microbial Pretreatment

The main biological pretreatment technique for algae that enhances downstream processing is whole-cell pretreatment. The microbial cell pretreats the algal cell in whole-cell pretreatment. The microbial cell is the only extract or metabolite utilized in this procedure. Heterotrophic feeders, such as bacteria and fungi, are linked to algae in a variety of biological niches, including freshwater and marine environments. By upsetting the cells and dissolving polymeric materials (lipids, proteins and carbohydrates) into low molecular weight components, these microorganisms can proliferate on algae. The released chemicals are eventually consumed by the microorganisms to meet their metabolic requirements (Bhushan et al. 2023). As a result, during colonization, these bacteria have constantly created more efficient ways to obtain the carbon and nutrients they require. In nature, symbiotic relationships exist between microorganisms and algae, wherein the former provide nutrients and coenzymes in return for the latter's oxygen and dissolved organic matter. Enzymatic catalysis by both anaerobic and aerobic microbial strains can break down the algal cells, releasing components for biofuel. Various bacteria, fungi such as Firmicutes, Proteobacteria, Bacteroidetes, Streptomyces, *Vibrio*, *Pseudoalteromonas*, *Anaeromyces*, *Neocallimastix*, *Orpinomyces* and *Piromyces* have been reported for this purpose (Aydin et al. 2017; Llamas et al. 2021).

5.3.3.2 Enzymatic Pretreatment

The most energy-efficient method for breaking down algal biomass is enzymatic hydrolysis. Because of their great selectivity and specificity, less or no harmful or inhibiting by-products are produced. They have a significant advantage when it comes to large-scale application because of the shorter incubation period (faster reaction rate) and lower dosage because of the biological pretreatment's increased accuracy (Demuez et al. 2015). Hydrolytic enzymes are typically produced by fungus and bacteria. Enzyme operational parameters must match the ideal strain requirement in order to achieve high activity. Enzymes for large-scale industrial uses are created by fermentation. Major enzymes used for algal biomass treatment are α-amylase, amyloglucosidase, lysozyme, pectinase, xylanase etc. These enzymes can be used as single unit or in mixed co-culture form. To enhance their catalytic ability, these enzymes can be chemically modified or even immobilized (Rodrigues et al. 2013).

5.4 Fractionation Processes for Algae Biomass

Once the process of pretreatment is done that aims at preparing the algae biomass for further processing by altering its structure or composition, the next step is the fractionation that involves the separation or isolation of different fractions or components present in the pretreated algae biomass. The main objective of fractionation is to isolate specific compounds or groups of compounds from the complex mixture of the pretreated biomass. Thus, to conclude it can be stated that, fractionation entails extracting or isolating particular components or fractions from the pretreated biomass, whereas pretreatment concentrates on changing the structure or content of algal biomass to make it better suited to further processing. By changing the physical or chemical characteristics of the biomass before fractionation, pretreatment makes it easier to extract or isolate specific components later on in the process. For the purpose of effectively using algal biomass for a range of commercial, industrial or biotechnological uses, both phases are necessary.

The term 'fractionation' describes the process of separating various constituents or fractions within algal biomass into discrete and valuable streams. Using a set of criteria related to their functional, physical or chemical characteristics, each ingredient or fraction of the complex mixture of biomass is separated out in this process. Fractionation makes it possible to separate and extract useful chemicals from algal biomass for a variety of uses. There are various methods of fractionating the pretreated algal biomass into several components as per the final product requirements. These methods include Solvent Extraction, Centrifugation and Filtration, Chromatography, Enzymatic Treatments, Flocculation and Sedimentation, Membrane Separation Technique and Precipitation.

5.4.1 Solvent Extraction

One well-known method for extracting microalgae biomolecules is to employ organic solvents. This type of extraction is often linked to an earlier stage of cell disruption in order to improve the extraction yield by facilitating the solvent's access to inner cell components. Traditional organic solvent-based methods such as the Bligh and Dyer (B&D) method, Folch method and Soxhlet method are used to extract and recover microalgae bio-compounds, primarily lipids. The Folch and B&D procedures are two extensively used lipid extraction techniques that are somewhat comparable. Folch is a two-step process that starts with homogenizing lipids in a 2:1 ratio of methanol to chloroform. In the second phase, which involves washing the crude oil extract with at least five times the volume of water, the homogenate is filtered and the lipids are separated (Corrêa et al. 2021). The homogenate produced in the chloroform/methanol mixture for the B&D method needs to be a monophasic system that forms a bi-phase when combined with water. Lipids are present in the chloroform phase, while non-lipid components, methanol and water make up the other phase. Franz von Soxhlet developed the Soxhlet method in 1879. It uses an instrument called a Soxhlet to extract lipids from biomass by exhausting reflux in an organic solvent, most often n-hexane or petroleum ether (Guldhe et al. 2016).

5.4.2 Centrifugation and Filtration

The fundamental idea behind centrifugation-based particle separation is the differential sedimentation of molecules based on their densities and the force of gravity (g-force), which separates components based on differences in particle size, density, or shape. Centrifugation uses centrifugal force to separate components based on their density, while filtration separates particles based on size through porous membranes or filters. This method can isolate different fractions by collecting components with distinct sizes or densities. Once the pretreated biomass is subjected to any further extraction procedure, it is subjected to different cycles of centrifugation at specific rotating speeds and temperatures for a fixed amount of time that help to separate out the different components based on their density gradients. Also, filtration process uses filters of varied sizes such as dialysis filtration that help to separate one individual components from other based on the differences in their sizes (Farooqui et al. 2021).

5.4.3 Chromatography

A mixture of molecules can be separated using chromatographic procedures based on their molecular weights, adsorption (liquid–solid), partition (liquid–solid), affinity and other distinctions in their features (Corrêa et al. 2021). The efficacy of various chromatography techniques varies according on the type of molecule that needs to be separated. The goal of using chromatography, which is a quantitative analysis method in and of itself, is to achieve an acceptable separation in a reasonable amount of time. To that aim, several chromatography techniques have been developed. Several techniques are available for chromatography, such as gel permeation chromatography, high-pressure liquid chromatography, paper chromatography, gas chromatography, thin-layer chromatography (TLC), column chromatography and ion exchange chromatography (Coskun 2016). Based on the molecules to be separated, the respective technique can be employed.

5.4.4 Precipitation

This method is employed for the separation of different protein molecules and is generally followed by chromatographic methods or membrane filtration so as to achieve high purity levels. It is a low-cost technique and also it is feasible for the large scale. It causes the precipitation or separation of specific compounds from the solution by the addition of specific reagents or changes in environmental conditions (such as temperature or pH), allowing for their separation from the solution. After centrifugation, the protein rich and protein poorer phases are developed, which can then be easily separated by shifting the pH using any acid such as sulphuric acid or any base such as sodium hydroxide (Cutshaw et al. 2023).

5.4.5 Flocculation and Sedimentation

Flocculation utilizes additives or chemicals to induce the aggregation of components, forming larger particles that settle down, allowing for their separation from the solution by sedimentation. It aggregates and separates components based on differences in density or chemical properties. Chemical flocculants that are commonly available and reasonably priced, including aluminium and iron salts, have found extensive application in various sectors. These chemical flocculants' fractionating efficacy is dependent on their solubility and electronegativity as well as operating parameters including algal solution pH and dose and algal cell density. (Farooqui et al. 2021). For the purpose of algal fractionation, aluminium sulphate ($Al_2(SO_4)_3$), commonly referred to as 'alum,' is a kind of inorganic flocculant. It is a very acceptable inorganic chemical flocculant for the bulk recovery of biomass because of its extremely low cost. When it comes to algal fractionation, organic polymers are commonly used. Cationic polymers are preferred over anionic polymers because they neutralize the negative charges on the algal cell's surface, which eventually cause the cells to destabilize and flocculate. Deacetylated chitin, or chitosan, is a less toxic, biodegradable and economical cationic polymer. However, the effectiveness and dosage of chitosan are dependent on a number of variables, including temperature, cell density, ionic strength, pH and the stage of growth of the culture. pH is one of the most important elements out of all of them. In general, the flocculation of chitosan is advised in an acidic environment (Bracharz et al. 2018). This process is then followed by sedimentation, and the key concept behind this method is the separation of solid or liquid particles based on their density difference. It is a low-cost method having potential utilization as a first stage to minimize energy input. Gravity sedimentation is usually done to serve this purpose.

5.4.6 Fractionation Via Advanced Separation Technologies

Advanced separation technologies play a crucial role in the fractionation of algae biomass, allowing for precise separation and isolation of different components or fractions with improved efficiency and selectivity. Supercritical Fluid Extraction (SFE), Supercritical Fluid Chromatography (SFC), Membrane Separation Techniques, Electrically Driven Membrane Processes etc. have also been employed. SFE uses supercritical fluids (commonly carbon dioxide) as the solvent. Under

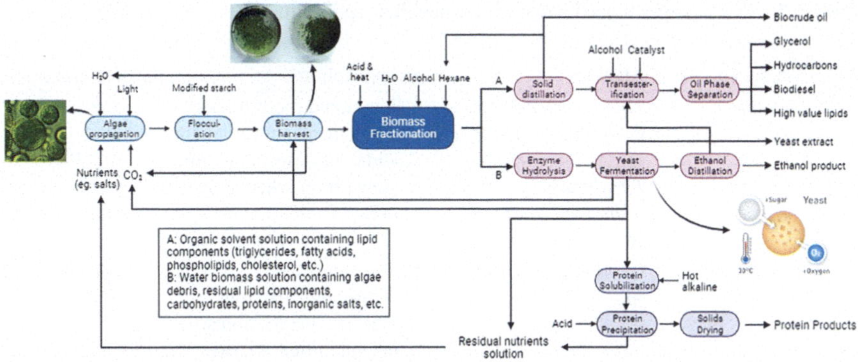

Fig. 5.2 A complete representation of whole process of valorization of algal mass into value-added products

specific temperature and pressure conditions, carbon dioxide reaches a supercritical state, exhibiting both liquid and gas properties. It selectively extracts compounds based on pressure and temperature adjustments, while SFC employs a supercritical fluid as the mobile phase. It separates components based on differences in polarity, size or chemical affinity (Farooqui et al. 2021). Membrane Separation Techniques use membranes with specific pore sizes to separate components based on size or molecular weight. Nanofiltration allows for the separation of compounds based on their molecular size, while reverse osmosis separates solvents from solutes based on pressure differences.

Any of the above method can be employed as per the type of compound one wish to separate as shown in Fig. 5.2.

5.5 Value-Added Extraction from Algae Biomass Components

Isolating and extracting particular components or fractions with high market value, functionality or uses is the process of value-added extraction from algal biomass. Algal biomass is rich in many important components, and the biomass can be significantly valued for a variety of industries by extracting these compounds. The extracted molecules have the potential to be further refined, formed or applied in a variety of industries. This might lead to the creation of novel and sustainable products in the food, healthcare, energy and biotechnology sectors.

Bioactive proteins, lipids & fatty acids, polysaccharides, secondary metabolites, vitamins, pigments can be extracted from the algal biomass using various techniques that are specific for each compound as shown in Table 5.1.

Table 5.1 Various extractions of algal genera and their applications

Sl No.	Component	Algal sp.	Application	Reference
1	Bioactive proteins	*Aphanizomenon* sp., *Chlorella* sp., *Dunaliella* sp., *Arthrospira* sp., *Tetraselmis* sp., *Nannochloropsis* sp., etc.	Functional foods, dietary supplements, essential amino acids, fluorescent labelling, cancer biomarker, aquafeed, animal feed	Jönsson et al. (2020), Koyande et al. (2019)
2	Polysaccharides	*Gelidium* sp., *Gracilaria* sp	Nutraceuticals, cosmeceuticals, drug delivery, dietary fibres, anti-microbial activity, carrageenan—canned and processed food, agar-capsule manufacture and as foods	Diaz et al. (2023)
3	Lipids and fatty acids	*Chlorella* sp., *Nannochloropsis* sp., *Odontella aurita*, *Parietochloris incise*, *Phaedactylum tricornutum*, *Diacronema vlkianum*	Biodiesel feedstock, oleochemicals, dietary supplements, aquafeed, rich source of DHA & EPA, phytosterols-therapeutics	Galasso et al. (2019), Uma et al. (2023)
4	Vitamins	*Chlorella pyrenoidosa*, *Arthrospira platensis*, and *Spirulina* sp.	Nutritional balance, reduces blood cholesterol, immuno-modulatory agent, improves cardio health	Edelmann et al. (2019)
5	Carotenoids	*Dunaliella salina*, *Nannochloropsis* sp., *Spirulina platensis*, *Tetraselmis suecica*	Used as synergistic compounds in preventing cellular damage in cancer patients, repress cardiovascular disorders, reduce cholesterol in blood streams, used as antioxidants and are aidful in retinol functions	Galasso et al. (2019)

5.6 Future Trends and Applications in Potential Fuel/Food/Agriculture/Pharma Fields

Algal biomass has a wide range of uses and products that show off its adaptability and potential as a sustainable and renewable resource for many industries, including food and nutrition, energy, healthcare, cosmetics and environmental solutions. The use of algae in a variety of industries is still being explored and expanded by ongoing research, which helps to create sustainable and creative goods. Algae show great promise as feedstocks for the industrial production of viable metabolites, which may eventually displace artificial food and feed additives. These feasible compounds have numerous uses in the oleochemical, plastics and polymers, pharmaceutical, cosmeceutical, nutraceutical and agricultural sectors. Microalgae yield high-value chemicals that are mostly utilized as medicinal ingredients. Algae has numerous opportunities in different aspects of life as mentioned below.

5.6.1 Biofuels Production and Utilization

Microalgae are a diverse group of single-celled organisms that may be able to provide us with a number of options for meeting our needs for liquid fuel for transportation. Algal species can develop in a variety of aquatic conditions, including saturated saline and freshwater. More than 40% of the world's carbon fixation is attributed to algae, most of which are marine microalgae that utilize CO_2 efficiently. Algae have an extremely high rate of biomass production; some species can double in as little as six hours, and many can double twice a day. Various biofuels such as bioethanol, biodiesel and biobutanol can be derived from algae using approaches such as transesterification, fermentation and saccharification. These products are environmentally friendly and are future fuels (Hannon et al. 2010).

5.6.2 In Food and Agriculture Sector

One of the biggest concerns of the twenty-first century is food insecurity, which is only going to become worse as population increases and climate change continues to pressurize our already fragile food production systems. Here algae come to the rescue as they can be used as a rich source of energy and can be consumed as protein source. Algae have a lot of potential as dietary supplements, even in modest doses, since they can supply important macronutrients like fatty acids and amino acids. The protein percentage of dry biomass in a number of algae that are safe for human consumption might vary from 27 to 70%. Algae are a great source of omega-3 fatty acids because they are the main manufacturers of these goods, including the vital compounds eicosapentaenoic acid (EPA) and docosahexaenoic acid (DHA). Algae are a great source of micronutrients like vitamins and minerals, just like terrestrial plants. Algae are a great option for adding more protein to the human diet since they produce vital lipids and fatty acids and have a high protein content along with crucial vitamins (Diaz et al. 2023). Also, algae are capable of thriving in analogous to crop species that enhance the plants' nutrient uptake efficiency, hence can be employed as biofertilizers.

5.6.3 In Pharmaceuticals

There are still many ongoing studies on the bioactive substances found in algae to create medications that target a range of illnesses, including cancer, inflammation and neurological diseases. Algae are a rich and varied source of pharmacologically active natural products and nutraceuticals. Highly bioactive secondary metabolites that could provide valuable clues for the creation of novel pharmacological requirements can be found in marine algae. The properties of algae such as the antioxidant properties, anticancer activity and antiviral properties make them a promising candidate for pharmaceutical industry. Numerous algae species have been discovered to reduce oxidative damage by scavenging free radicals and active oxygen, which reduces the risk of cancer, according to a study. Sargassum species and cyanobacteria *S. platensis* have the highest levels of antioxidants, which contribute to their anticancer efficacy, in aqueous extracts of algae. Phycobilin proteins are another example. Red algae called Carragren is used to fight HIV, DENV, HSV-1, HSV-2, HPV and HRV. Red algae (*Callophyllis variegate*, *Agardhiella tenera*) and its galactan are employed as anti-HIV-1, anti-HIV-2, DENV and HAV. PUFA has a major role in human body metabolism. These vital and healthful fatty acids are extracted using bryophyta algae or chlorophytes. N-6 derived from the aforementioned, PUFA is a fatty acid high in diet. Algae are a well-known bioactive chemical that are highly beneficial to the pharmaceutical sector because of their high fatty acid content (Eleanor 2020).

5.7 Conclusion

As civilization has advanced, people have come to understand the value of goods made from natural resources. Algal biomass has a significant competitive advantage in the worldwide market when renewability becomes a critical component for the industrial production of food, feed and fuel. The state-of-the-art technologies and new developments in the valorization of algal biomass for innovative biotechnological products are conceptualized in this thorough assessment. We have endeavoured to incorporate the most recent technological developments and innovative ways of valorization that have advanced the algal biorefinery framework and brought it one step closer to the blue-bioeconomy. The integration of economical extraction and fractionation techniques for several coproducts is the present barrier preventing more progress in this area. The optimal solvents and/or technologies were determined by examining various biocompounds and evaluating various extraction techniques in this chapter; nevertheless, additional research appears to be needed to enhance the currently known outcomes. In fact, a thorough analysis should be conducted on the newly developed systems that use solvents other than traditional ones, like ionic liquids, which have a higher solvent power. In order to treat the biomass with aqueous solutions rather than organic volatile and poisonous solvents, researchers must properly handle the solvent formulations being employed. Furthermore, a thorough investigation should be conducted into the relationship between the chemical structures of the chemicals extracted from the microalgae biomass and the solvents and methods used. Researchers should be focused on developing new goods and markets, but industry should play a major role in this as well, working with academics.

References

Abou El-Souod GW, Morsy EM, Hassan LHS, El-Sheekh MM (2021) Efficient saccharification of the microalga Chlorella vulgaris and its conversion into ethanol by fermentation. Iran J Sci Technol Trans A Sci 45(3):767–774. https://doi.org/10.1007/s40995-021-01097-1

Aydin S, Yıldırım E, Ince O, Ince B (2017) Rumen anaerobic fungi create new opportunities for enhanced methane production from microalgae biomass. Algal Res 23:150–160. https://doi.org/10.1016/j.algal.2016.12.016

Azmi AA, Sankaran R, Show PL, Ling TC, Tao Y, Munawaroh HSH, Kong PS, Lee DJ, Chang JS (2020) Current application of electrical pre-treatment for enhanced microalgal biomolecules extraction. In: Bioresource technology, vol 302. Elsevier, Amsterdam. https://doi.org/10.1016/j.biortech.2020.122874

Bhosle C, Vishwakarma A, Tupkar C (2010) Effect of pretreatment of algal biomass on bioadsorption of manganese. Int J Eng Sci Technol 2:12

Bhushan S, Jayakrishnan U, Shree B, Bhatt P, Eshkabilov S, Simsek H (2023) Biological pretreatment for algal biomass feedstock for biofuel production. J Environ Chem Eng 11(3):109870. https://doi.org/10.1016/j.jece.2023.109870

Biris-Dorhoi ES, Michiu D, Pop CR, Rotar AM, Tofana M, Pop OL, Socaci SA, Farcas AC (2020) Macroalgae—a sustainable source of chemical compounds with biological activities. Nutrients 12(10):1–23. https://doi.org/10.3390/nu12103085

Bracharz F, Helmdach D, Aschenbrenner I, Funck N, Wibberg D, Winkler A, Bohnen F, Kalinowski J, Mehlmer N, Brück TB (2018) Harvest of the oleaginous microalgae Scenedesmus obtusiusculus by flocculation from culture based on natural water sources. Front Bioeng Biotechnol 6:200. https://doi.org/10.3389/fbioe.2018.00200

Chen Z, Wan C (2018) Ultrafast fractionation of lignocellulosic biomass by microwave-assisted deep eutectic solvent pretreatment. Bioresour Technol 250:532–537. https://doi.org/10.1016/j.biortech.2017.11.066

Chen Z, Wang L, Qiu S, Ge S (2018) Determination of microalgal lipid content and fatty acid for biofuel production. Biomed Res Int 2018:1503126. https://doi.org/10.1155/2018/1503126

Corrêa PS, Morais Júnior WG, Martins AA, Caetano NS, Mata TM (2021) Microalgae biomolecules: extraction, separation and purification methods. Processes 9(1):1–40. https://doi.org/10.3390/pr9010010

Coskun O (2016) Separation techniques: chromatography. N Clin Istanb 3(2):156–160. https://doi.org/10.14744/nci.2016.32757

Costa JAV, Freitas BCB, Moraes L, Zaparoli M, Morais MG (2020) Progress in the physicochemical treatment of microalgae biomass for value-added product recovery. Bioresour Technol 301:122727. https://doi.org/10.1016/j.biortech.2019.122727

Cutshaw A, Frost H, Uludag-Demirer S, Liu Y, Liao W (2023) Protein extraction, precipitation, and recovery from Chlorella sorokiniana using mechanochemical methods. Energies 16(12):4809. https://doi.org/10.3390/en16124809

de Carvalho JC, Magalhães AI, de Melo Pereira GV, Medeiros ABP, Sydney EB, Rodrigues C, Aulestia DTM, de Souza Vandenberghe LP, Soccol VT, Soccol CR (2020) Microalgal biomass pretreatment for integrated processing into biofuels, food, and feed. Bioresour Technol 300:122719. https://doi.org/10.1016/j.biortech.2019.122719

Demuez M, Mahdy A, Tom As-Pej E, Gonz Alez-Fern Andez C, Ballesteros M (2015) Enzymatic cell disruption of microalgae biomass in biorefinery processes. Biotechnol Bioeng 112:1955–1966. https://doi.org/10.1002/bit.25644/abstract

Diaz CJ, Douglas KJ, Kang K, Kolarik AL, Malinovski R, Torres-Tiji Y, Molino JV, Badary A, Mayfield SP (2023) Developing algae as a sustainable food source. Front Nutr 9:1029841. https://doi.org/10.3389/fnut.2022.1029841

Edelmann M, Aalto S, Chamlagain B, Kariluoto S, Piironen V (2019) Riboflavin, niacin, folate and vitamin B12 in commercial microalgae powders. J Food Compos Anal 82:9. https://doi.org/10.1016/j.jfca.2019.05.009

Eleanor MG (2020) The role of algae in pharmaceutical development. J Pharm Nanotechnol 4:1–8

Farooqui A, Tripathi G, Moheet K, Dubey P, Ahmad S, Husain A, Shamim A, Mahfooz S (2021) Algal biomass: potential renewable feedstock for bioenergy production. Springer, Cham, pp 85–113. https://doi.org/10.1007/978-981-16-1888-8_5

Galasso C, Gentile A, Orefice I, Ianora A, Bruno A, Noonan DM, Sansone C, Albini A, Brunet C (2019) Microalgal derivatives as potential nutraceutical and food supplements for human health: a focus on cancer prevention and interception. Nutrients 11(6):1226. https://doi.org/10.3390/nu11061226

Guldhe A, Singh B, Ansari FA, Sharma Y, Bux F (2016) Extraction and conversion of microalgal lipids. In: Green energy and technology. Springer, Cham, pp 91–110. https://doi.org/10.1007/978-3-319-12334-9_6

Ha GS, El-Dalatony MM, Kurade MB, Salama ES, Basak B, Kang D, Roh HS, Lim H, Jeon BH (2020) Energy-efficient pretreatments for the enhanced conversion of microalgal biomass to biofuels. Bioresour Technol 309:123333. https://doi.org/10.1016/j.biortech.2020.123333

Hannon M, Gimpel J, Tran M, Rasala B, Mayfield S (2010) Biofuels from algae: challenges and potential. Biofuels 1(5):763–784

Jagaba AH, Kutty SRM, Lawal IM, Abubakar S, Hassan I, Zubairu I, Umaru I, Abdurrasheed AS, Adam AA, Ghaleb AAS, Almahbashi NMY, Al-dhawi BNS, Noor A (2021) Sequencing batch reactor technology for landfill leachate treatment: a state-of-the-art review. J Environ Manag 282:111946. https://doi.org/10.1016/j.jenvman.2021.111946

Jönsson M, Allahgholi L, Sardari RRR, Hreggviosson GO, Karlsson EN (2020) Extraction and modification of macroalgal polysaccharides for current and next-generation applications. Molecules 25(4):930. https://doi.org/10.3390/molecules25040930

Karaduman FR, Türk Çulha S, Horzum N (2022) Algal nanofibers: current status and recent developments. Mater Today Commun 33:104248. https://doi.org/10.1016/j.mtcomm.2022.104248

Koyande AK, Show PL, Guo R, Tang B, Ogino C, Chang JS (2019) Bio-processing of algal biorefinery: a review on current advances and future perspectives. Bioengineered 10(1):574–592. https://doi.org/10.1080/21655979.2019.1679697

Kumar G, Sivagurunathan P, Thi NBD, Zhen G, Kobayashi T, Kim SH, Xu K (2016) Evaluation of different pretreatments on organic matter solubilization and hydrogen fermentation of mixed microalgae consortia. Int J Hydrog Energy 41(46):21628–21640. https://doi.org/10.1016/j.ijhydene.2016.05.195

Llamas M, Magdalena JA, Greses S, Tomás-Pejó E, González-Fernández C (2021) Insights on the microbial communities developed during the anaerobic fermentation of raw and pretreated microalgae biomass. Chemosphere 263:127942. https://doi.org/10.1016/j.chemosphere.2020.127942

Lu H, Yu X, Li H, Tu ST, Sebastian S (2019) Lipids extraction from wet Chlorella pyrenoidosa sludge using recycled [BMIM]Cl. Bioresour Technol 291:1221819. https://doi.org/10.1016/j.biortech.2019.121819

Mandotra SK, Sharma C, Srivastava N, Ahluwalia AS, Ramteke PW (2019) Current prospects and future developments in algal bio-hydrogen production: a review. Biomass Convers Biorefinery 13:8575–8592. https://doi.org/10.1007/s13399-021-01414-z

Martínez JM, Gojkovic Z, Ferro L, Maza M, Álvarez I, Raso J, Funk C (2019) Use of pulsed electric field permeabilization to extract astaxanthin from the nordic microalga Haematococcus pluvialis. Bioresour Technol 289:121694. https://doi.org/10.1016/j.biortech.2019.121694

Park WK, Min K, Yun JH, Kim M, Kim MS, Park GW, Lee SY, Lee S, Lee J, Lee JP, Moon M, Lee JS (2022) Paradigm shift in algal biomass refinery and its challenges. Bioresour Technol 346:126358. https://doi.org/10.1016/j.biortech.2021.126358

Rodrigues RC, Ortiz C, Berenguer-Murcia Á, Torres R, Fernández-Lafuente R (2013) Modifying enzyme activity and selectivity by immobilization. Chem Soc Rev 42(15):6290–6307. https://doi.org/10.1039/c2cs35231a

Saini A, Aggarwal NK, Sharma A, Yadav A (2015) Prospects for irradiation in cellulosic ethanol production. Biotechnol Res Int 2015:1–13. https://doi.org/10.1155/2015/157139

Soto-Sierra L, Stoykova P, Nikolov ZL (2018) Extraction and fractionation of microalgae-based protein products. Algal Res 36:175–192. https://doi.org/10.1016/j.algal.2018.10.023

Uma VS, Usmani Z, Sharma M, Diwan D, Sharma M, Guo M, Tuohy MG, Makatsoris C, Zhao X, Thakur VK, Gupta VK (2023) Valorisation of algal biomass to value-added metabolites: emerging trends and opportunities. Phytochem Rev 22(4):1015–1040. https://doi.org/10.1007/s11101-022-09805-4

Ventura SPM, Nobre BP, Ertekin F, Hayes M, Garciá-Vaquero M, Vieira F, Koc M, Gouveia L, Aires-Barros MR, Palavra AMF (2017) Extraction of value-added compounds from microalgae. In: Microalgae-based biofuels and bioproducts: from feedstock cultivation to end-products. Elsevier, Amsterdam, pp 461–483. https://doi.org/10.1016/B978-0-08-101023-5.00019-4

Wang Y, Ma C, Kong D, Lian L, Liu Y (2023) Review on application of algae-based biochars in environmental remediation: progress, challenge and perspectives. J Environ Chem Eng 11(6):111263. https://doi.org/10.1016/j.jece.2023.111263

Yaashikaa PR, Keerthana Devi M, Senthil Kumar P, Rangasamy G, Rajendran S, Xiao L (2023) A review on pretreatment methods, photobioreactor design and metabolic engineering approaches of algal biomass for enhanced biohydrogen production. Int J Hydrog Energy 48(55):21110–21127. https://doi.org/10.1016/j.ijhydene.2022.10.092

Yun JH, Cho DH, Heo J, Lee YJ, Lee B, Chang YK, Kim HS (2019) Evaluation of the potential of Chlorella sp. HS2, an algal isolate from a tidal rock pool, as an industrial algal crop under a wide range of abiotic conditions. J Appl Phycol 31(4):2245–2258. https://doi.org/10.1007/s10811-019-1751-z

Zhang K, Zhang F, Wu YR (2021) Emerging technologies for conversion of sustainable algal biomass into value-added products: a state-of-the-art review. Sci Total Environ 784:147024. https://doi.org/10.1016/j.scitotenv.2021.147024

Biochemical Conversion of Algae Biomass: Enzymatic Hydrolysis and Fermentation

6

Josimayra Almeida Medeiros, João Victor Ferro, Gabrielly Karla Silva Santos, João Victor Oliveira Nascimento da Silva, Brígida Maria Villar da Gama, Carlos Eduardo De Farias Silva, Larissa Rodrigues Macário, and Albanise Enide da Silva

Contents

Abstract

Due to the high availability of carbohydrates in cellular composition of algae and the absence of lignin in most species, the application of their hydrolysates as raw material in the production of biomolecules in biorefineries has stimulated the development of studies that aim to reduce challenges related to green production, in parallel with economic obstacles, when compared to other biomass generation

J. A. Medeiros · J. V. Ferro · G. K. S. Santos · J. V. O. N. da Silva · B. M. V. da Gama · C. E. De Farias Silva (✉) · L. R. Macário · A. E. da Silva
Technology Center, Federal University of Alagoas, Maceió, Alagoas, Brazil
e-mail: carlos.eduardo@ctec.ufal.br

S. K. Arya et al. (eds.), *Value Added Products From Bioalgae Based Biorefineries: Opportunities and Challenges*,
https://doi.org/10.1007/978-981-97-1662-3_6

routes. Classified as third-generation raw material, micro and macroalgae are highly efficient photosynthetic organisms that have biomasses of great economic power. However, when applied to industry, depending on the product, their applications require a pre-treatment step, which changes according to their cell wall characteristics and macromolecular components, followed by an enzymatic hydrolysis process to utilize carbohydrate or lipid fraction, mainly. In this context, this chapter aimed to discuss the main characteristics of this biomass type to fermentation applications (bioethanol, biodiesel, biogas and lactic acid) from hydrolysate obtained by an enzymatic process.

Keywords

Third generation biomass · Depolymerization · Polysaccharides · Enzymes · Ethanol · Biogas

Abbreviations

°C	Degree Celsius
ACod	Anaerobic co-digestion
ATCC	American-type culture collection
$CaCO_3$	Calcium carbonate
CO_2	Carbon dioxide
Cu	Copper
FAME	Fatty acid methyl esters
g	Gram
H_2	Hydrogen
H_2SO_4	Sulphuric acid
HCl	Chloridric acid
L	Litre
mg	Milligram
Min	Minutes
PHB	Polyhydroxybutyrate
PVA	Polyvinyl alcohol
VS	Volatile solids
w/v	Weight per volume

6.1 Introduction

The global energy demand grows over time, and environmental issues from fossil fuels are discussed making essential the investment in renewable sources (Sudhakar et al. 2019). Micro and macroalgae are biomasses that have been widely studied due to their composition, which contain considerable concentrations of carbohydrates, lipids, proteins, pigments, vitamins and minerals, having high biotechnological potential.

Micro and macroalgae are promising biomass for obtaining biomolecules, in particular, biofuels can be mentioned, such as biogas through anaerobic digestion process, bioethanol through alcoholic fermentation, biodiesel with the synthesis of lipid reserves and biohydrogen with photobiological transformations. They have several advantages compared to other methods of obtaining biofuels reducing the greenhouse effect and fossil fuels dependency, since they capture carbon dioxide to perform photosynthesis, and don't need arable land for cultivation (Dębowski et al. 2020; Sudhakar et al. 2019).

Additionally, the nutritional composition of algae, such as polysaccharides and their derivatives (oligosaccharides generally obtained by enzymatic fractionation), are not completely fermented by the intestinal microbiota of humans and animals, making them a potential prebiotic source and also makes them applicable as therapeutics having antioxidant, antihypertensive, anticancer and immunomodulatory responses. Brown macroalgae and some species of microalgae, such as *Chlorella* sp., are highlighted for this type of application (Krishna Perumal et al. 2023; Patel et al. 2021; Camacho et al. 2019).

However, algal large-scale application still faces some barriers since to access intracellular compounds, treatment processes are necessary, such as a pretreatment (cell/tissue disruption) and enzymatic hydrolysis, making the process relatively expensive. Therefore, it is necessary to promote studies that optimize the process in order to improve its economic feasibility, making its applicability more accessible (Sudhakar et al. 2019; Tawfik et al. 2022).

Therefore, this chapter will address the importance of algae depolymerization (micro and macro) towards application in fermentative processes, especially biofuels, detailing the pre-treatment process, enzymatic hydrolysis and fermentations carried out with the algal hydrolysate.

6.2 Algal Biomass and Its Value for the Fermentation Study Field

Algae are primarily photoautotrophic organisms that during their growth are capable of metabolizing nitrogen, phosphorus and other nutrients from the environment, in addition to capturing atmospheric CO_2 and light energy, generating biomass with great economic potential, which can be applied in biotechnological processes (Sudhakar et al. 2019). These biomasses are considered sustainable and renewable raw materials to produce biomolecules, such as biofuels, for example, bioethanol, biogas, biobutanol and biohydrogen, through aerobic and anaerobic processes as shown in Fig. 6.1 (Dębowski et al. 2020; Sudhakar et al. 2019).

Algae are basically classified into micro and macroalgae. Microalgae are formed by microorganisms, and considering industrial applications, mainly chlorophytes (eukaryotic) and cyanobacteria (prokaryotic), while macroalgae are formed by seaweed subdivided into green, red and brown algae (Tazikeh et al. 2022).

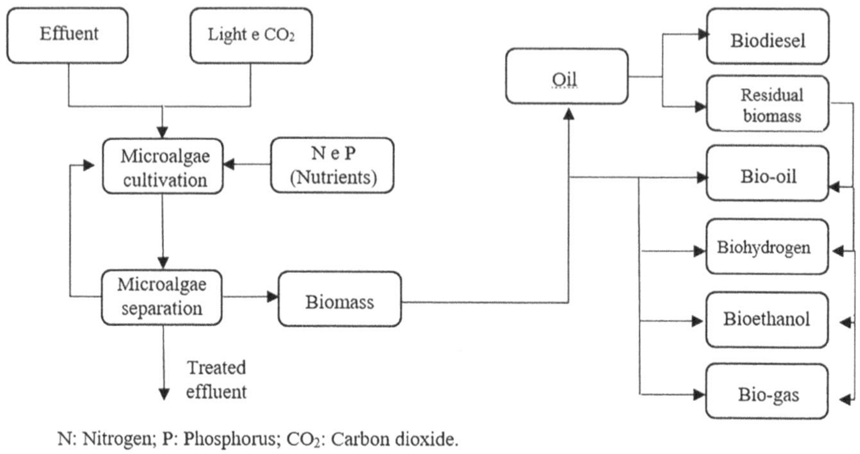

Fig. 6.1 Flowchart of the process for obtaining biofuels from microalgae biomass. (Adapted from Arun et al. 2021)

In relation to microalgae, they have a carbohydrate composition favourable for fermentative processes (up to 60% in dry cell weight can be reached; mainly starch—energy reserve and cellulose—component of the cell wall). Despite an average composition of lipids (5–60%), carbohydrates (8–30%) and proteins (40–60%), as dry cell weight (Kumar et al. 2022), they have biochemical plasticity and can produce a biomass rich in carbohydrates, proteins or lipids, depending on environmental (temperature, pH, salinity etc.), nutritional (availability of carbon, nitrogen, phosphorus etc.) and operational conditions (type of reactor, operating mode, cultivation time etc.) of cultivation (Bentahar et al. 2019).

Furthermore, microalgae have a high potential to be used in industry since their duplication time is less than 24 h and they do not compete with arable land and can be cultivated in wastewater (Sudhakar et al. 2019; Mahmood et al. 2023).

On the other hand, macroalgae have complex sugars in their composition, phycocolloids, which in their simple form can be used in fermentations. Therefore, to be used in the production of third-generation fuels, these algae require pretreatment/hydrolysis to break down the phycocolloids into simpler compounds such as glucose, and galactose among others. Some advantages of this biomass application are fast growth and great availability, do not require fresh water nor do not depend on agricultural land, and do not compete with food crops (Ramachandra and Hebbale 2020).

The efficiency and viability of the fermentative process will depend on the effectiveness of algal biomass formation, and the cost of the depolymerization process combined with the fermentative process yield. Therefore, it is essential to design the most profitable process.

The industrial application of this process still faces some challenges in relation to its economic feasibility, as the cost of this production route is higher when compared to other biomass types, requiring optimization of the cultivation system and reducing the cost involved in the pre-treatment process and enzymes applied in the depolymerization step (Tawfik et al. 2022; Tazikeh et al. 2022).

Microalgae cultivation can be carried out in open, closed or hybrid manner in different types of bioreactors, with open ponds and closed photobioreactors being the most applied (Mahmood et al. 2023; De Andrade et al. 2022). Furthermore, the presence of sugars is extremely important for fermentative processes, so microalgae must be induced to accumulate energy reserves (carbohydrate—starch, mainly or lipids) through nutrient limitation such as nitrogen (De Sá Filho et al. 2023; Jacob-Lopes et al. 2022).

The use of industrial effluents has high potential for cultivating them since they contain high amounts of chemical components such as nitrogen and phosphorus, essential nutrients for the metabolism and growth of microalgae (Sathya et al. 2022; Ali et al. 2021). On the other hand, macroalgae are generally cultivated in open sea, thus obtaining sustainable and abundant cultivation, given that the seas and oceans cover more than 70% of the surface of the planet (Tan et al. 2020).

Greater fermentation efficiency can be obtained if the pre-treatment stage is well conducted, which can be chemical, physical, biological processes or a combination of these, followed by the hydrolysis process, generally, an enzymatic process (Dave et al. 2019; Ramachandra and Hebbale 2020).

6.3 Pre-Treatment Step and Enzymes Applied for Biomass Depolymerization

6.3.1 Pre-Treatment of Algal Biomass

Algae contain a diverse range of organic and chemical substances. These substances are produced by algae as reserve or structural components (Chan et al. 2022). Pretreatment techniques are pointed out as a necessary step for algae cell/tissue disruption to access these substances of interest.

The effectiveness of pretreatment methods depends on the characteristics of biomass, i.e. the toughness and structure of the cell wall, and the macromolecular composition of cells (González-Fernández et al. 2012). Algal cell wall is rigid and requires pre-treatment, which can be chemical applying acids, basics, solvents and catalysts; physical employing high temperature and/or pressure, ultrasonication and mixing; or biological which requires the action of microorganisms such as fungi and bacteria or enzymes. These methods can be used separately or combined to obtain simple sugars (Fig. 6.2). No single method is universally applicable as algal characteristics are diverse, and pretreatment conditions depend on the biomass applied (Bhatia et al. 2023).

The main algal biofuels produced by biochemical conversion include biodiesel, bioethanol, biohydrogen and biogas. Each of them presents its benefits and limitations, which should be carefully considered (González-Fernández et al. 2012; Bhushan et al. 2023). Overall, the economy of the pre-treatment process is based on the energy requirement and costs with chemicals, microorganisms' multiplication

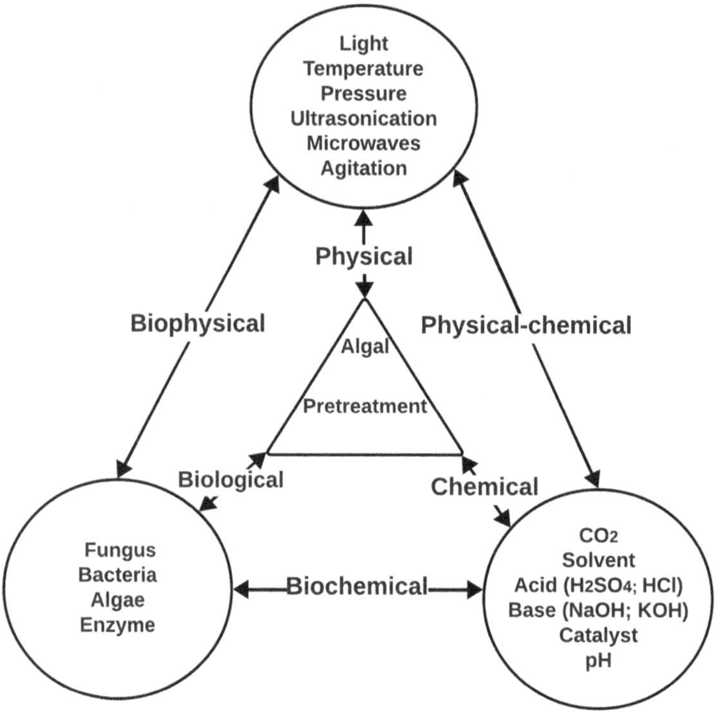

Fig. 6.2 Pretreatment methods applied to algal biomass

and enzymes applied. When the process is analysed from a comparative perspective of efficiency and costs, it is worth highlighting the purpose and characteristics relating to the cell wall of the algal species used in view of these different complexities in breaking down and extracting the targeted raw material.

For example, different species of microalgae showed different responses to sonication or mechanical cell rupture, because cell wall can be different in composition and morphology, for example, the cyanobacterium *Arthrospira platensis* is surrounded by a peptidoglycan layer while the eukaryotic microalga *Chlorella vulgaris* has a more robust cellulose cell wall (Williamson et al. 2023).

Physical methods, generally, are not able to completely hydrolyse algal biomass, and are used for cell/tissue disruption, only (as a pretreatment). Among the possible options for this type of treatment, high-pressure homogenization, ball milling, microwaves, ultrasonication, electrocoagulation and hydrodynamic cavitation stand out. (Shetty and Krishnakumar 2023; Sirohi et al. 2021).

Considering macroalgae, the pretreatment is an essential part. Mechanical pretreatment is a popular technique that involves the use of blades, knives and hammers to chip, grind, mill and shred biomass into small particles. This process increases the biomass reaction surface and releases simpler but still complex sugars, which are completely saccharified by enzymatic hydrolysis, thereby optimizing bio-energy production. The use of microwaves and ultrasound waves can increase enzyme accessibility. But the most common pretreatment is heating the biomass (100–250 °C).

When chemical methods are applied, generally, they perform both pretreatment and acidic hydrolysis and the main parameters to be considered are biomass load, treatment time, catalyst concentration and temperature, although a chemical pretreatment only to de-structure the cells/tissues with very dilute acid can be applied (De Farias Silva and Bertucco 2018). Pre-treatment with dilute acid cannot only disrupt but also hydrolyse the intracellular starch content present in microalgae if the concentration is slightly higher (De Farias Silva and Bertucco 2018). However, some results for specific species were not very effective, for example, Sirohi et al. (2021) explain the inefficiency of chemical methods when imposed on microalgae of the genus *Nannochloropsis*, justified by the presence of the biopolymer algaenan. This behaviour can also be related given the recommendations for the application of chemical methods to act on components such as cellulose, hemicellulose or other recalcitrant polymers (Bhatia et al. 2022). For macroalgae, the chemical methods used include acid or alkaline pretreatment (Thompson et al. 2019).

In relation to biological processes, the key factors such as pH, temperature, biomass concentration, biomass particle size, treatment time, enzyme action (when enzymes are used directly) and microbial growth (when applying microorganisms) influence the pretreatment process (more applied to macroalgae) (Bhushan et al. 2023). Recognized for their green technology characteristics (promoting reductions in environmental impacts) and economic cost, given the possible and wide use of fungi of the genus *Trichoderma* and *Penicillium*, to produce enzymes, treatments assisted by these microorganisms for cell wall degradation of macroalgae have come to being studied, as well as the species *Trametes hirsuta* grown in macroalgal biomass reached effective degradation rate (Tapia-Tussell et al. 2018).

It is worth highlighting the benefit of these optimizations aimed at the application of enzymes during pre-treatment processes is related to the economic viability factors of processes for obtaining biofuels that require more complex depolymerization/saccharification steps to promote the breakdown of constituents present in the biomass when it has many polymers in its composition, increasing operational costs. In these cases, the action of enzymes from the cellulase group is of paramount importance, especially endo-glucanases, exo-glucanases and β-glucosidases responsible for the cleavage stages of cellulose fibres and hydrolysis of oligosaccharides during depolymerization, which is considered the limiting factor in process of cellulose breakdown (Singh et al. 2023; Gong et al. 2022).

As explained by Ferreira et al. (2021) and Reis et al. (2023), the commercial cost of cellulase is stipulated at around 3 and 8 EUR kg^{-1}, thus considering uses of 10 and 30 mg g^{-1} to obtain 70–90% of hydrolysis rate, pricing in the order of 0.095 EUR kg^{-1} and 0.18 EUR kg^{-1}, respectively, for the hydrolysis process of biomass.

Therefore, among cost optimization strategies, the production of enzymes through fungal strains can be seen, promoting the addition of green technologies to the process, reductions in transport and preparation expenses (Ellilä et al. 2017) and reduction of unproductive cellulase-lignin-like bonds through the addition of extra protein such as bovine serum albumin (BSA) or soy protein (Florencio et al. 2016; Ázar et al. 2019).

According to the arguments of Millati et al. (2023) and Zabed et al. (2020), biohydrogen production through anaerobic digestion processes with the integration of methane and productive division into two stages, production of biohydrogen and biomethane, offers the possibility of using the first stage of this process as a pretreatment phase of the subsequent anaerobic digestion with increased yield by 123% when compared to monodigestive stages.

There is also evidence of the application of emerging technologies being found in research by Rashid et al. (2019), Sadatshojaei et al. (2020), Fajardo et al. (2020), Wicker et al. (2021) and Lim et al. (2022), starting from the insertion of optimization, monitoring and collaboration processes with genetic engineering and conditioning of symbiotic consortia. Included here as examples:

- Use of life cycle assessments during the pre-implementation process;
- Monitoring of optimal factors and conditions using flow and light sensors, online monitoring of taxonomic structure etc.;
- Generation of modified strains by methods such as electroporation, artificial transposons, silicon carbide whiskers, among others;
- Co-cultivation with species of bacteria and fungi characterized by high adaptability and promotion of autotrophic-heterotrophic symbiosis, mainly for microalgae.

6.3.2 Enzyme Selection in the Pre-Treatment Step

Enzyme-assisted disruption is used to hydrolyse specific bonds in the cell wall constituents (Fig. 6.3). Hence, a detailed understanding of the cell wall composition and structure, as well as of the substrate and hydrolysis reaction specificity of the added enzyme(s) is required (Mienis et al. 2023).

Generally, an enzymatic treatment could be used with another type of pretreatment. Enzymatic pretreatment has a high selectivity and scale-up is relatively easy. For example, cellulase pretreatment at lower temperatures showed very low solubilization. So, it can be concluded that temperature is one of the main driving parameters for cellulase pretreatment. When compared to thermal pretreatment, there was a small increase in biomass solubilization (Kendir and Ugurlu 2020). Enzymatic treatment can even outperform microwave and ultrasonication in terms of extraction yield. However, some disadvantages of a disruption process include long process periods, resulting in a lower production capacity than mechanical or chemical disruption, and the possibility of product inhibition. Depending on the intensity of the ultrasound, the combined use of ultrasound may increase or reduce viscosity, with the impact being either permanent or transient. In addition, ultrasound helps with processes including homogenization, cell dispersion, emulsification, intracellular

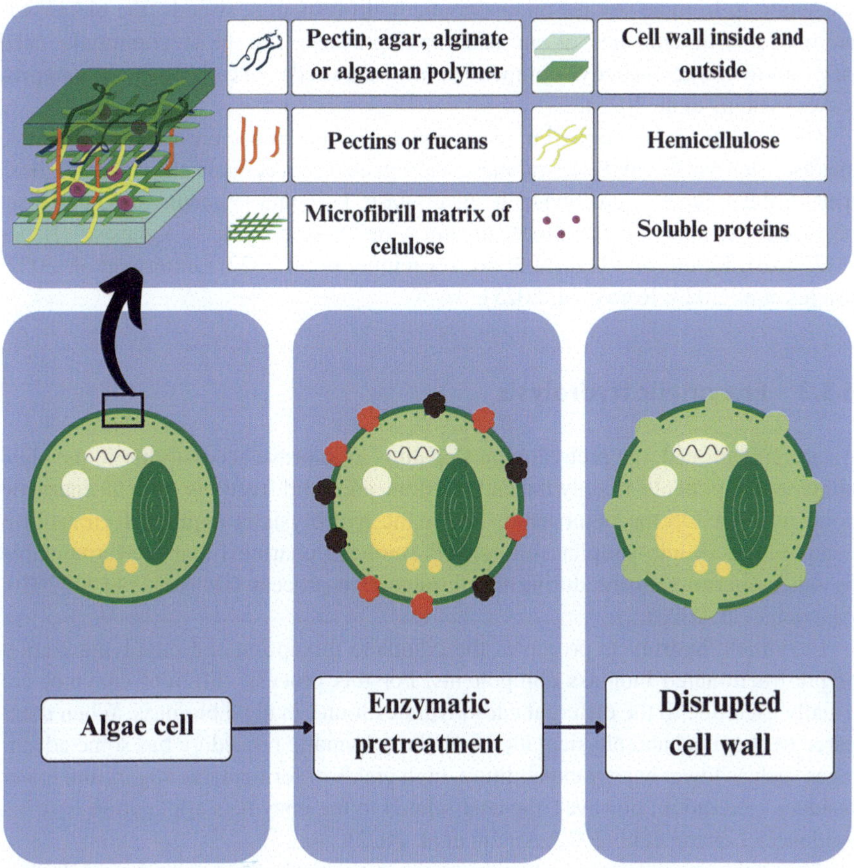

Fig. 6.3 Structure and action of enzymes in algal cell wall. (Based on Mienis et al. 2023 and Velazquez-Lucio et al. 2018)

content extraction, maturation, drying, enzyme activation and/or deactivation and degassing (Singla and Sit 2021).

From the pretreatment of green microalgae, the following suggested order of cell wall robustness can be proposed for some species like *Chlorella pyrenoidosa* that has harder cell wall than *Chlamydomonas reinhardtii*. *C. pyrenoidosa* is composed mainly of cellulose and *C. reinhardtii* of hydroxyproline-rich glycoproteins (Duan et al. 2017). The cell wall of *Chlorella* strains also contains a chitin-like layer. Besides the organic layers, some microalgae cell walls have inorganic components, like the cell wall of the diatom *Phaeodactylum tricornutum,* which contains a silica shell (Nitsos et al. 2020).

Considering macroalgae, specifically seaweed, cell wall mostly contains cellulose and hemicellulose, which make the cell walls of seaweeds chemically complex. In general, the enzymatic disruption of seaweeds depends on the ability of enzymes to hydrolyse seaweed cell wall components, disrupting its structural complexity by forming an enzyme–substrate binding complex and releasing the cell content into the solution. In most studies on enzymatic hydrolysis of seaweeds, the biomass is pretreated with enzymes as an additional process. The most commonly used enzymes to extract bioactive compounds from seaweeds are carbohydrases and proteases (Mienis et al. 2023).

The enzymes more applied in macroalgae and microalgae pre-treatment include agarase, alcalase®, amylase, arabinase, carrageenanase, celluclast®, cellulase, dextrinase, flavorzyme®, glucanase, hemicellulase, kojizyme™, neutrose®, protease, pectinase, termamyl®· ultraflo®, umamizyme™, viscozyme®, xylanase (Bader et al. 2020; Kendir and Ugurlu 2020; Jeyakumar et al. 2022; Quitério et al. 2022; Mienis et al. 2023; Rojo et al. 2023).

6.3.3 Enzymatic Hydrolysis

As aforementioned, the pretreatment step aims to improve accessibility to the algal biomass components, mainly the carbohydrate and lipid fractions from an energetic point of view. On the other hand, enzymatic hydrolysis is required to transform complex sugars into simpler components (monosaccharides) that are fermentable by other microorganisms during the fermentative process (De Souza et al. 2020; Alavijeh et al. 2020).

Enzymatic hydrolysis promotes the complete disruption and depolymerization of pre-fractionated biomass components. For this process, different enzymes are usually used due to the different carbohydrates found in algal biomass. When compared to chemical and physical methods, the enzymatic procedure has some advantages such as lower energy expenditure, high yields of fermentable sugars and lower residues generation; however, the cost related to the enzyme's application is still a bottleneck (Zhang et al. 2022; Araújo et al. 2022).

Although it is a relatively expensive process, the application of enzymatic hydrolysis is still required for a series of processes, such as in the case of obtaining bioactive peptides through protein hydrolysis and there is a need for neutralization, as occurs when the acidic method is applied (Akaberi et al. 2019; Trigueros et al. 2021).

Several studies were carried out with the aim of optimizing this method to improve its efficiency and feasibility. For example, Shokrkar and Keighobadi (2020) studied the effects of hydrodynamic conditions (agitation speed, types of impellers and presence of baffles) on the hydrolysis of microalgae through simulations and experimental tests, aiming at the extraction of glucose using cellulases and Sedighi et al. (2019) who extracted and purified protein hydrolysates from *Dunaliella salina* and *Spirulina platenses* using the enzymes trypsin and chymotrypsin.

On the other hand, González-Gloria et al. (2023) studied the application of hydrothermal pretreatment methods (physicochemical pretreatment method) based on the application of hot water and a tubular steam explosion reactor to improve the efficiency of enzymatic hydrolysis (using cellulases and hemicellulases) with the biomass of the macroalgae *Sargassum* spp. to obtain bioethanol. They achieved a saccharification efficiency of 90% and a biochemical ethanol yield of 70%.

Acid pretreatment process (diluted H_2SO_4) of the macroalgae *Ulva lactuca* and *Enteromorpha intestinalis* associated with enzymatic hydrolysis was evaluated, using a purified enzyme obtained from the bacterium *Vibrio parahaemolyticus.* The results showed that the combination of these methods promoted a greater release of monosaccharides for macroalgae (*Ulva lactuca*: 261.76 mg g^{-1}; *Enteromorpha intestinalis:* 289.89 mg g^{-1}). Following the same line of sequential processes, Romero-Vargas et al. (2023a) used the industrial residue of *Eucheuma spinosum,* after the extraction of carrageenan, to produce fermentable sugars with applicability in obtaining bioproducts. The authors combined a thermochemical pre-treatment, with HCl, and enzymatic hydrolysis, with a commercial Cellic CTec® enzyme cocktail. Under optimized conditions, a reducing sugar concentration of 21.4 g L^{-1} was obtained, and the hydrolysate was used to produce polyhydroxybutyrate (PHB), achieving a yield of up to 0.26 g of PHB g^{-1} of reducing sugar consumed. Other studies are shown in Table 6.1.

6.4 Algal Hydrolysate Fermentation

Regarding extraction or production of biomolecules resulting from algae, considered in an algal biorefinery, biochemical, thermochemical and mechanical conversion mechanisms are used in the industrial sector (Aparicio et al. 2020; Raikova et al. 2022; Liu et al. 2022).

Pointing out the presence of carbohydrates as a cellular component or reserve of algae with emphasis on the forms of starch and cellulose or phycocolloids, mainly in the form of sugars, biofuels can be mentioned as a promising fermentative application such as bioethanol and biogas (Seon et al. 2020). Furthermore, oleaginous microalgae can be used in biodiesel production. This biofuel production flowchart is shown in Fig. 6.4.

Given the absence of lignin in the composition of the cell wall in most of micro and macroalgae species, the conversion of carbohydrates into sugars can be configured, with microalgae being mainly a source of starch and cellulose, and for macroalgae, a diversity of polymers can be found. In Table 6.2, structural components of macroalgae are mentioned (Humbird et al. 2011; Jayakumar et al. 2022).

The main fermentative processes that have been discussed in the literature for algal hydrolysates application are the production of ethanol, biodiesel and biogas, in addition to lactic acid fermentation and the combination with nanoparticle technology.

6.4.1 Bioethanol

Regarding bioethanol production, the process involves the use of carbohydrases to de-structure the biomass into simple sugars (monosaccharides) and to ferment the hydrolysate with yeasts or bacteria (*Saccharomyces cerevisiae* and *Zymomonas mobilis* strains, mainly).

Table 6.1 Enzymatic hydrolysis of micro and macroalgae

Biomass	Experimental procedure	Main results	Enzymes	References
Scenedesmus almeriensis	The biomass was pretreated using a pulsed electric field. Hydrolysis occurred for 180 min in flasks with a temperature of 50 °C and pH 8. The enzymes were added in a proportion of 3% in relation to the biomass (dry weight).	The pretreatment reduced the hydrolysis time to 60 min, reaching 80% of maximum hydrolysis yield.	Alcalase and flavourzyme	Akaberi et al. (2019)
Chlorella sorokiniana	Hydrolysis occurred in Eppendorf tubes with 1.25 g of the reaction mixture composed of 1% biomass at 50 °C, 200 rpm for 24 h. To evaluate grinding as a pretreatment, a vibrating mill was used for 90 min at an amplitude of 1.5 mm.	Microalgae biomass had a starch accumulation of 26% (w/w) after nitrogen starvation. Cellulases were not effective in breaking the microalgae cell wall; however, the milling process allowed high hydrolysis yields with fungal amylases.	The enzymes manufactured in the laboratory were produced by the fungi *Trichoderma reesei* Rut C30 (producer of cellulase) and *Aspergillus awamori* (producer of β-glucosidases and amylases)	De Souza et al. (2020)
Chlorella vulgaris	The biomass cell wall was disrupted using a 75 mL horizontal bead mill operated in batch recirculation mode at 25 °C and 2039 rpm. Hydrolysis was carried out with an enzyme dosage of 2% (v/w) with cellulase and protease at 45 °C, while for the other enzymes at 37 °C. Hydrolysis time was 24 h.	Increased recovery yield of lipids, carbohydrates and proteins from 75%, 31% and 40%, to 88%, 74% and 68%, respectively, when ball-milled biomass was applied.	Cellulase, protease, lipase and phospholipase A1.	Alavijeh et al. (2020)

Chlorella sp.	The biomass was pre-treated by a acid-hydrothermal method in a high-pressure reactor, which was placed in an oil bath to allow the reaction. Enzymatic hydrolysis was carried out on the pretreated biomass paste, which had its pH adjusted to 7 and then α-amylase was added in different dosages and the reaction took place at 85 °C. In the case of adding glucoamylase, the pH of the biomass suspension was adjusted to 4.5 and the reaction temperature was 65 °C. Then, fermentation was carried out to produce ethanol using *Saccharomyces cerevisiae.*	Production of up to 11 g. L^{-1} of ethanol. Showing a 7.1-fold increase in ethanol production, when compared to the hydrolysis without any pretreatment.	Thermostable α-amylase and glucoamylase	Ngamsirisomsakul et al. (2019)
Chlorella	The dry and properly diluted biomass (60 g.L^{-1}) was used in enzymatic hydrolysis. The pH of the suspension was adjusted to 6 and then, when the reaction medium was at 95 °C, the alpha-amylase enzyme was added. Glucoamylase was introduced into the medium when the broth was cooled to 50 °C. The hydrolysis process took place for 5 h at 400 rpm. Then the glucose concentration was analysed and fermentation was carried out using *S. cerevisiae* in a medium with pH 5 for 30 h, at the end of the process the alcohol content was determined.	The maximum glucose conversion was of 80% and occurred through the addition of 3% (v/v) of glucoamylase, in a reaction medium with pH 5 and a temperature of 80 °C for 5 h. Hydrolysis process reached its highest productivity rate (6.83 g/(L.h)) in the first hour of reaction. Ethanol was produced with a maximum biochemical yield of 94.21% in 30 h of fermentation.	Alpha-amylase and glucoamylase	Megawati et al. (2022)

(continued)

Table 6.1 (continued)

Biomass	Experimental procedure	Main results	Enzymes	References
Sargassum muticum	Biomass pretreatment was carried out using the non-isothermal autohydrolysis process in a pressurized Parr reactor. A biomass load of 14.3% (w/v) with water was applied under a stirring of 150 rpm and heating followed by cooling to room temperature. Enzymatic hydrolysis was carried out in the solid and liquid phases, in an orbital shaker at 150 rpm, 48.5 °C and pH 4.85. Simultaneous saccharification and fermentation occurred in an orbital incubator at 35 °C and 150 rpm of agitation. At the end of the process, the samples were analysed for the concentration of sugars and ethanol.	With the pretreatment it was possible to recover in the solid phase, more than 90% of the glucan and in enzymatic hydrolysis up to 94% of this material was converted into glucose. A maximum ethanol yield of 80% was achieved.	Cellulase, carbohydrase, pectinase and β-glucosidase.	Del Río et al. (2019)
Gelidium sesquipedale	Enzymatic hydrolysis occurred in an agitated jacketed reactor with 7.5 g of macroalgae residue (MR) and 150 mL of reaction medium, with an enzyme ratio (w/w) varying between 0.25% and 8%. The process was carried out at 50 °C, at different pHs (5.6 and 8) during more of 24 h.	Release of phenolic compounds bound to the cell wall and the polysaccharide fraction in the form of monomers and oligomers. Higher hydrophobic content and amino acids in the protein fraction were available for hydrolysis. Protein extraction yield was close to 30%. Cellulase had greater hydrolytic capacity in algae.	Cellulase, endo-1,3-β xylanase and protease.	Trigueros et al. (2021)

Ulva lactuca	The residual biomass from oil extraction by solvent (n-hexane) was pretreated with acid optimized the type of acid, temperature and reaction time. Enzymatic hydrolysis was initially carried out using 1 g of biomass in 20 mL of citrate buffer (pH 4.8) and occurred for 48 h at 50 °C and 150 rpm. After that, the parameters related to biomass concentration, enzyme load, temperature and incubation time were optimized. Ethanol production was carried out and the parameters pH, temperature and time were optimized.	A maximum yield of fermentable sugars of 13.48 $g.L^{-1}$ was achieved and the bioethanol concentration obtained was 2.8 ± 0.12 $g.L^{-1}$.	Cellulase	Gengiah et al. (2023)
Dictyota dichotoma	The dry biomass was pretreated by an ultrasonic processor with a pulse maintained at 0.5 and variations were made in the amplitude percentage, operating time and biomass concentration. Enzymatic hydrolysis was carried out in an Erlenmeyer flask coupled to a rotary shaker at 150 rpm and 50 °C for several days. Solid-liquid ratios of 2.1 and 4.3% (w/v) in citrate buffer were used. Reducing sugar was analysed.	The best results for sugar concentration were achieved with a pretreatment time of 40 min and an amplitude of 40% for algae suspension at 4.3% w/v.	A cellic CTec 2 enzyme cocktail was used which is a mix formed by cellulase, β-glucosidase and hemicellulase.	Romero-Vargas et al. (2023b)

(continued)

Table 6.1 (continued)

Biomass	Experimental procedure	Main results	Enzymes	References
Warty gracilaria	The biomass was pretreated with acidic solution, in a conical tube, for 1 h and then by ultrasonicaton with an on-off cycle of 10 s at this stage, parameters were optimized (the type and concentration of acid, ultrasonic amplitude, temperature, time, among others). For enzymatic hydrolysis, the pretreated biomass had the pH adjusted to 5% and 20% (v/w) of enzyme mix was used (in relation to biomass). The process took place in a shaking incubator at 50 °C and 180 rpm for 48 h finally, reducing sugar were analysed.	Higher yields of reducing sugar in the pretreatment (60.4%) and subsequent hydrolysis (76.3%), under optimal conditions were reached. Sugar yields close to those obtained with hydrothermal pretreatment followed by enzymatic hydrolysis.	Cellic® CTec2, Cellic® HTec2 and Viscozyme® L	Mi-Ra and Jeong (2021)

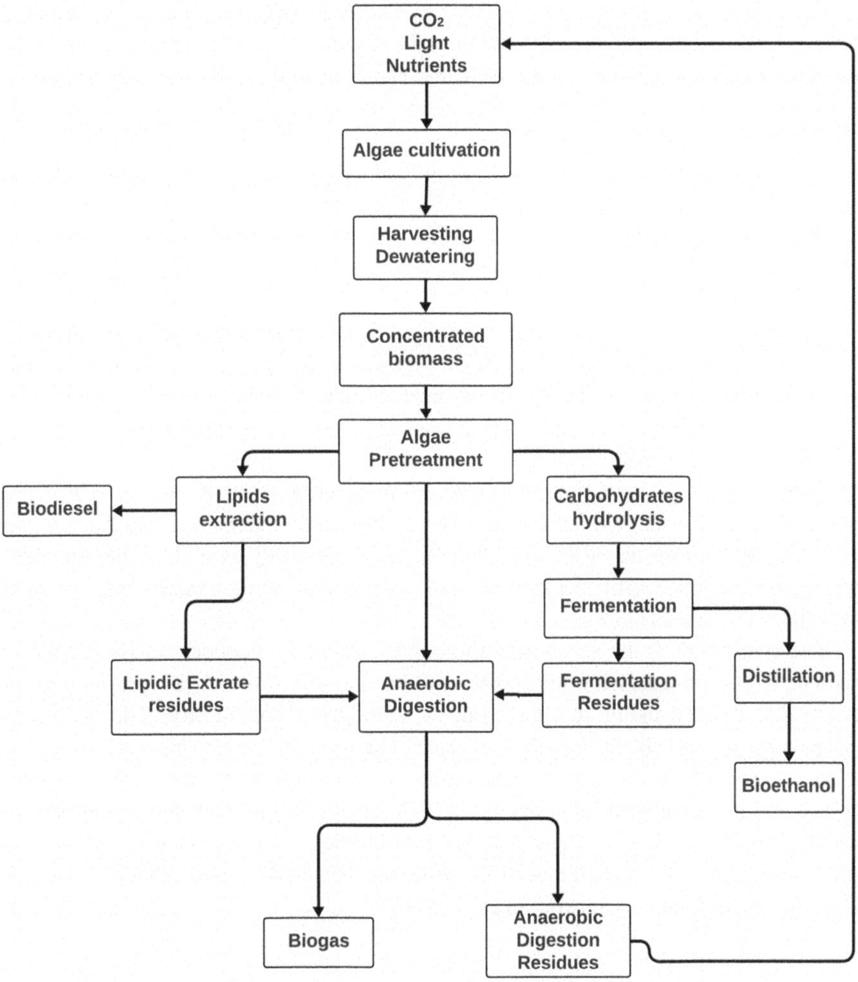

Fig. 6.4 Biorefinery concept based on biofuels production from algal biomass

Table 6.2 Sugars found in macroalgal composition

Macroalgae	Available sugars
Green algae	D-glucose, L-rhamnose, D-psicose, D-tagatose, L-iduronic acid
Red algae	D-galactose, 3,6-anhydro-D-galactose, agarobiose
Brown algae	L-fucose, D-mannitol, mannuronic acid, glucuroni acid
Pentoses can be also present	L-arabinose, D-xylose, D-ribose

Adapted from Nagarajan et al. (2022)

For example, the microalga *Chlorella pyrenoidosa* was submitted to a combined process of enzymatic hydrolysis using alpha-amylase and glucoamylase since the energy reserve is composed of starch, mainly. The highest glucose concentration obtained was subsequently fermented using the yeast *Saccharomyces cerevisiae*. (Megawati et al. 2022).

Crude enzymes are an alternative, produced by most of the fungus on a selected substrate which generally have a cocktail of cellulase, hemicellulose and xylanase (Cekmecelioglu and Demirci 2020). Crude enzymes from *Aspergillus fischer* were effective in the hydrolysis of *C. pyrenoidosa* for bioethanol production (Bhushan et al. 2021).

An efficient saccharification of microalgal *Chlamydomonas reinhardtii* biomass by *Trichoderma harzianum* crude enzymes composed of pectinase, amylase, dextrinase, cellulase and protease confirms the technical feasibility of enzymatic saccharification of algal biomass and provided a promising alternative source of hydrolytic enzymes for ethanol production (Bader et al. 2020).

Experimental investigation on bioethanol production from macro-algae utilizing neutralized acid-pretreated biomass from *Padina tetrastromatica* and *Sargassum swartzii* was demonstrated by applying enzymatic hydrolysis with alpha-amylases and fermentation with *Saccharomyces cerevisiae* and *Zymomonas mobilis* (Jeyakumar et al. 2022).

The evaluation of the saccharification and fermentation process of two different seaweeds for an eco-friendly bioethanol production showed the production of renewable energy using marine biomass with great potentiality. The seaweeds *Sargassum* sp. and *Gracilaria sp* were used. The reducing sugar (monosaccharides) produced was markedly higher in the biomass after obtained from two-stage hydrolysis (acid pre-treatment and enzymatic hydrolysis applying cellulases and pectinases). This study clearly suggested that pretreatment of biomass with dilute acid greatly improves the efficiency of enzyme saccharification and ethanol was produced by fermentation (Saravanan et al. 2018).

6.4.2 Biodiesel

Sustainable industrial processes make enzymes more attractive as catalysts (Chapman et al. 2018). As a result of their adaptability, lipases are the most common enzymes involved in conducting biotransformation methods such as esterification, transesterification and hydrolysis (Ferreira et al. 2021). Micro and macroalgae have the potential as lipids for biodiesel production (Chhandama et al. 2023).

Many microalgae species possess a cellulosic cell wall, which is resistant to mechanical disruption and prevents diffusion during solvent extraction processes for lipid recovery. It is for this reason that extraction contributes 30–50% of the overall process cost of biodiesel from microalgae (Orr et al. 2016). Enzyme-assisted extraction (pretreatment) has been reported to allow nearly 99% FAME (fatty acid methyl esters) recovery and reduced energy input. It is speculated that the cost of enzymes represents about 22% of the total process cost (Wang et al. 2017). Xu et al.

(2021) have shown that an enzyme mixture of cellulase, protease, lysozyme and pectinase extracted 73% lipids from *Nannochloropsis* biomass (Vishwakarma et al. 2023).

Macroalgal biomass has also been considered to produce biodiesel. Some species have been shown with high lipid content and quality and potential for this application: *Ulva* spp., *Sargassum* spp., *Cladophora* spp. *and Spirogyra* spp. (Ruangrit et al. 2023), *Ulva lactuca*, *Padina boryana* and *Ulva intestinalis* (Abomohra et al. 2021).

6.4.3 Biogas

The constraints of employing microalgae as an energy-efficient source for biodiesel synthesis are discussed because processes such as harvesting and drying require a lot of energy. Because of this, there are more affordable and innovative ways to use microalgae for the wet production of biofuel by applying anaerobic digestion with the biomass residue after lipid extraction (Vargas-Estrada et al. 2022).

After lipid extraction or fermentation, anaerobic co-digestion (ACod) of microalgae wastes has received little attention. Researchers agree that there is needed to incorporate and develop a microalgal biorefinery concept (Fig. 6.4). As a result, it is commonly assumed that ACod will rapidly acquire popularity. The fundamental advantage of microalgae waste ACod is that it maximizes the economic and energy advantages of algal biomass (Vargas-Estrada et al. 2022).

Kendir and Ugurlu (2020) studied the effect of treating *Porphyridium cruentum* with mixed enzymes (cellulases, proteases and viscozyme). The results showed that the effect of mixed enzyme treatment was better than the single enzyme treatment, demonstrating a synergistic effect. Proteases were responsible for degrading cell proteins. Viscozyme and cellulase were a carbohydrase mix that degraded cell wall and intracellular components. The enzymatic pretreatment provided a reduction in lag phase and maximized methane production.

There are a lot of studies based on the use of enzymatic pretreatment for biogas production. A one-step multi-enzyme pre-treatment composed of Cellic® CTec 2, Termamyl® SC and Dextrozyme® GA in pre-treated *Chlorella* sp. biomass was used as a substrate for biohydrogen production. A hydrogen yield of 43.16 mL-H_2/gVS (volatile solids) was obtained under optimal conditions, which was 39.63% greater than that observed with non-treated biomass (Sriyod et al. 2021).

Also, macroalgal biomass can be used for biogas production. The use of cellulase, amylase and an enzyme mixture (cellulase, amylase and xylanase) was applied to *Rhizoclonium* sp. biomass. Among them, the mixed enzyme cocktail had a synergistic effect. Cellulase hydrolyses cellulose, while amylase and xylanase hydrolysed other components. Additionally, in the enzyme treatment process, the type and dosage of the enzyme were adjusted optimizing the process (Wu et al. 2022).

Biogas production from *Sargassum fulvellum* biomass was improved after the chemical and biological treatments. The alkali pretreatment of macroalgae increased the degradation of macroalgal biomass and improved the methane yield significantly. On the other hand, the biological pretreatment (with enzymes) through the supplementation of cellulolytic enzymes enhanced biomass biodegradation, volatile fatty acids utilization, and methane formation in a thermophilic anaerobic digestion system. Thus, enzymatic pretreatment of macroalgae can replace chemical and mechanical pre-treatments that require high energy and maintenance costs and are often accompanied by equipment and facility corrosion (Farghali et al. 2021).

Seaweeds have the potential for dual use, in heavy metal biosorption and biofuel production. Three dominant seaweeds were collected representing the three macroalgal phyla, namely *Ulva* spp. *(Chlorophyte), Gracilaria* spp. *(Rhodophyte) and Sargassum* spp. *(Phaeophyte). Sargassum* spp. showed the highest cumulative copper (Cu^{2+}) biosorption with 80% removal efficiency, which increased to 94.6% after process optimization. Biogas production by anaerobic digestion of Cu-sorbed biomass from *Sargassum* spp. was obtained (Abomohra et al. 2021).

6.4.4 Lactic Acid and Applicability of Nanoparticles

The application of fermentation of macroalgae hydrolysates to produce lactic acid is also evaluated by several scientists based on the insertion of lactic acid bacteria capable of converting sugars into lactic acid configured as the main product in homofermentative or heterofermentative processes (in addition to the production of other products such as ethanol) (Peinemann et al. 2019; Thygesen et al. 2021).

As examples, using microalgae, Pagnini et al. (2023) studied the performance of *Tisochrysis lutea* hydrolysate imposed on the fermentative process carried out by multiple *Lactobacillus* (*L. plantarum* ATCC 8014, *Lactobacillus delbrueckii subsp. bulgaricus* LB28A, and *L. casei* LB28B). After digestibility analyses, better resistance and high growth rate were noticed by *L. plantarum strain* during the pretreatment process with pepsin and pepsin-pancreatic in addition to better yield in the production of lactic acid (about 5 g L^{-1}) and Chen et al. (2020) who used hydrolysate of *Chlorella vulgaris* ESP-31 pretreated in acid (diluted sulphuric acid—4% w/v) at 120 °C for 20 min and neutralized in $CaCO_3$. After fermentation testing different configurations, excellent conditions were observed under batch operation mode with immobilized *L. plantarum* in polyvinyl alcohol (PVA), obtaining results equivalent to 37.93 g L^{-1} for lactic acid production with a yield of 0.91 g g^{-1}.

Considering macroalgae, Chai et al. (2021) promoted studies on the application of the hydrolysate of the red macroalgae *Eucheuma denticulatum* to obtain lactic acid through thermal processes at 120–150 °C and a microwave-assisted autohydrolysis (MAA) system (pretreatments), enzymatic hydrolysis and lactic fermentation by *Bacillus coagulans* ATCC 7050. In the pre-treatment conditions applying MAA at 120 °C for 50 min, promoting a glucose concentration of 19.3 g L^{-1} after 48 h of hydrolysis and a lactic acid production yield of 98.6%.

Nagarajan et al. (2022) detected lactic acid at a concentration of 36.8 g L^{-1} and a yield of 0.91 g g^{-1} when analysing hydrolysates of *Ulva* sp. pre-treated in acid (diluted sulphuric acid—4% w/v) at 120 °C for 20 min and fermented with *Lactobacillus plantarum* immobilized in polyvinyl alcohol (PVA). Other species of micro and macroalgae such as *Gracilaria* sp. (Lin et al. 2020), *Kappaphycus alvarezii* (Sudhakar and Dharani 2022) and *Chlorella sorokiniana* (Morón-Ortiz et al. 2024) were also applied to lactic acid production.

In addition, technological innovations were studied and gradually used in the industrial sector regarding nanoparticle application. For example, in the production of biohydrogen conditioned in photofermentation processes (based on the use of substrates not suitable for dark fermentation). The basis of this process stands out photosynthetic bacteria capable of capturing light energy and, in addition to the presence of nitrogenase enzymes, converting it into hydrogen using silica nanoparticles and green algae from species such as *Chlorella vulgaris*, *Chlorella protothecoides* and *Chlamydomonas reinhardtii* (Giannelli and Torzillo 2012; Sengmee et al. 2017; Martinez-Burgos et al. 2021).

Promoting greater coverage on the subject, some nanoparticles and their respective advantages in fermentative processes in the production of biofuels conditioned in algae hydrolysates are mentioned below:

- Carbon nanoparticles: they have functionalities in redox functions and electron transfer kinetics through the application of nanotubes (tubular fullerenes) resulting in an increase in the rates of biohydrogen or graphene oxide as a nanocatalyst in biodiesel production (Liu et al. 2012; Arya et al. 2021; Nematian et al. 2020);
- Metallic nanoparticles: characterized by offering binding conditions between oxidoreductase enzymes and magnetic nanoparticles which increase biomass growth and substrate optimization (gold and silver nanoparticles) and improve the generation of biogas from waste effluents (iron nanoparticles) (Arya et al. 2021; Xiang et al. 2019);
- Ceramic nanoparticles: their particularity is a chemical inertness and high heat resistance for immobilized enzymes or bacteria/other microrganisms, where, in the latter case, increasing in biohydrogen generation applying ceramic nanoparticles with silica or silicon carbide (Boshagh et al. 2019);
- Semiconductor nanoparticles: characterized by having the effect of quantum confinement and attributes between metals and non-metals. They were identified as promoters of both increasing in biomass and lipid yield and in the stability of immobilized cellulase enzyme (Sheikh et al. 2023).

6.5 Conclusions and Future Perspectives

Micro and macroalgae biomass is a sustainable and renewable raw material for the production of biofuels and has many applications in several fields of biotechnology. However, there are still some operational and production bottlenecks. The fractionation of this type of biomass is necessary in a biorefinery context to convert the maximum biomass fractions, mainly carbohydrates and lipids. Enzymatic hydrolysis still presents a high cost when considering biofuels, in addition to the fact that biomass also requires a pretreatment step to show efficiency during the fermentation process. Pretreatment methods can be chemical, physical or biological and are designed to provide better accessibility to the enzymes during the saccharification process. Carbohydrates found in micro and macroalgae are generally different and even when similar, are present in different fractions, which requires different types of enzymes for effective saccharification. They can be used to produce ethanol or lactic acid. The lipid fraction can also be applied for the biodiesel production. Biomass resulting from the lipid fraction extraction process can be used to produce biogas. Much remains to be studied to consolidate third-generation biomass in a viable way on a large scale, effectively contributing to a more sustainable production.

References

Abomohra AE-F, El-Hefnawy ME, Wang Q, Huang J, Li L, Tang J et al (2021) Sequential bioethanol and biogas production coupled with heavy metal removal using dry seaweeds: towards enhanced economic feasibility. J Clean Prod 316:128341. https://doi.org/10.1016/j.jclepro.2021.128341

Akaberi S, Gusbeth C, Silve A, Senthilnathan DS, López EN, Molina-Grima E, Müller G, Frey W (2019) Effect of pulsed electric field treatment on enzymatic hydrolysis of proteins of *Scenedesmus almeriensis*. Algal Res 43:43. https://doi.org/10.1016/j.algal.2019.101656

Alavijeh RS, Karimi K, Wijffels RH, Van Den Berg C, Eppink M (2020) Combined bead milling and enzymatic hydrolysis for efficient fractionation of lipids, proteins and carbohydrates of *Chlorella vulgaris* microalgae. Bioresour Technol 309:309. https://doi.org/10.1016/j.biortech.2020.123321

Ali S, Paul Peter A, Chew KW, Munawaroh HSH, Show PL (2021) Resource recovery from industrial effluents through the cultivation of microalgae: a review. Bioresour Technol 337:125461. https://doi.org/10.1016/j.biortech.2021.125461

Aparicio E, Rodríguez-Jasso RM, Lara A, Loredo-Treviño A, Aguilar CN, Kostas ET, Ruiz HA (2020) Biofuels production of third generation biorefinery from macroalgal biomass in the Mexican context: an overview. In: Sustainable seaweed technologies—cultivation, biorefinery, and applications. Elsevier, pp 393–446. https://doi.org/10.1016/B978-0-12-817943-7.00015-9

Araújo RRL, Silva CEF, De Sá Filho MLF, Almeida RMRG (2022) Bioethanol from microalgae. In: 3rd generation biofuels. Elsevier, pp 439–462. https://doi.org/10.1016/B978-0-323-90971-6.00020-6

Arun J, Gopinath KP, Sivaramakrishnan R, SundarRajan P, Malolan R, Pugazhendhi A (2021) Technical insights into the production of green fuel from CO2 sequestered algal biomass: a conceptual review on green energy. Sci Total Environ 755:142636. https://doi.org/10.1016/j.scitotenv.2020.142636

Arya I, Poona A, Dikshit PK, Pandit S, Kumar J, Singh HN, Jha NK, Rudayni HA, Chaudhary AA, Kumar S (2021) Current trends and future prospects of nanotechnology in biofuel production. Catalysts 11:1308. https://doi.org/10.3390/catal11111308

Ázar RISL, Morgan T, Barbosa MHP, Guimarães VM, Ximenes E, Ladisch M (2019) Impact of protein blocking on enzymatic saccharification of bagasse from sugarcane clones. Biotechnol Bioengineering 116:1584–1593. https://doi.org/10.1002/bit.26962

Bader AN, Sanchez Rizza L, Consolo VF, Curatti L (2020) Efficient saccharification of microalgal biomass by Trichoderma harzianum enzymes for the production of ethanol. Algal Res 48:101926. https://doi.org/10.1016/j.algal.2020.101926

Bentahar J, Doyen A, Beaulieu L, Deschênes JS (2019) Acid whey permeate: an alternative growth medium for microalgae Tetradesmus obliquus and production of β-galactosidase. Algal Res 41:101559. https://doi.org/10.1016/j.algal.2019.101559

Bhatia SK, Ahuja V, Chandel N, Gurav R, Bhatia RK, Govarthanan M, Tyagi VK, Kumar V, Pugazendhi A, Banu JR (2022) Advances in algal biomass pretreatment and its valorisation into biochemical and bioenergy by the microbial processes. Bioresour Technol 358:127437. https://doi.org/10.1016/j.biortech.2022.127437

Bhatia SK, Rajesh Banu J, Singh V, Kumar G, Yang YH (2023) Algal biomass to biohydrogen: pretreatment, influencing factors, and conversion strategies. Bioresour Technol 368:128332. https://doi.org/10.1016/j.biortech.2022.128332

Bhushan S, Rana MS, Bhandari M, Sharma AK, Simsek H, Prajapati SK (2021) Enzymatic pretreatment of algal biomass has different optimal conditions for biogas and bioethanol routes. Chemosphere 284:131264. https://doi.org/10.1016/j.chemosphere.2021.131264

Bhushan S, Jayakrishnan U, Shree B, Bhatt P, Eshkabilov S, Simsek H (2023) Biological pretreatment for algal biomass feedstock for biofuel production. J Environ Chem Eng 11:109870. https://doi.org/10.1016/j.jece.2023.109870

Boshagh F, Rostami K, Moazami N (2019) Biohydrogen production by immobilized Enterobacter aerogenes on functionalized multi-walled carbon nanotube. Int J Hydrog Energy 44:14395–14405. https://doi.org/10.1016/j.ijhydene.2018.11.199

Camacho F, Macedo A, Malcata F (2019) Potential industrial applications and commercialization of microalgae in the functional food and feed industries: a short review. Mar Drugs 17:312. https://doi.org/10.3390/md17060312

Cekmecelioglu D, Demirci A (2020) Production of cellulase and xylanase enzymes using distillers dried grains with solubles (DDGS) by Trichoderma reesei at shake-flask scale and the validation in the benchtop scale bioreactor. Waste Biomass Valor 11:6575–6584. https://doi.org/10.1007/s12649-020-00934-5

Chai CY, Tan IS, Foo HCY, Lam MK, Tong KTX, Lee KT (2021) Sustainable and green pretreatment strategy of Eucheuma denticulatum residues for third-generation l-lactic acid production. Bioresour Technol 330:124930. https://doi.org/10.1016/j.biortech.2021.124930

Chan SS, Khoo KS, Chew KW, Ling TC, Show PL (2022) Recent advances biodegradation and biosorption of organic compounds from wastewater: microalgae-bacteria consortium—a review. Bioresour Technol 344:126159. https://doi.org/10.1016/j.biortech.2021.126159

Chapman J, Ismail A, Dinu C (2018) Industrial applications of enzymes: recent advances, techniques, and outlooks. Catalysts 8:238. https://doi.org/10.3390/catal8060238

Chen PT, Hong ZS, Cheng CL, Ng IS, Lo YC, Nagarajan D, Chang JS (2020) Exploring fermentation strategies for enhanced lactic acid production with polyvinyl alcohol-immobilized *Lactobacillus plantarum* 23 using microalgae as feedstock. Bioresour Technol 308:123266. https://doi.org/10.1016/j.biortech.2020.123266

Chhandama MVL, Ruatpuia JVL, Ao S, Chetia AC, Satyan KB, Rokhum SL (2023) Microalgae as a sustainable feedstock for biodiesel and other production industries: prospects and challenges. Energy Nexus 12:100255. https://doi.org/10.1016/j.nexus.2023.100255

Dave N, Selvaraj R, Varadavenkatesan T, Vinayagam R (2019) A critical review on production of bioethanol from macroalgal biomass. Algal Res 42:101606. https://doi.org/10.1016/j.algal.2019.101606

De Andrade FP, De Farias Silva CE, Medeiros JA, De Sá Filho MLF, Santos GKS (2022) Consortium between microalgae and other microbiological groups: a promising approach to emphasizing the sustainability of open cultivation systems for wastewater treatment. J Water Process Eng 50:50. https://doi.org/10.1016/j.jwpe.2022.103211

De Farias Silva CE, Bertucco A (2018) Severity factor as an efficient control parameter to predict biomass solubilization and saccharification during acidic hydrolysis of microalgal biomass. Bioenergy Res 11:491–504. https://doi.org/10.1007/s12155-018-9913-4

De Sá Filho MLF, De Farias Silva CE, De Oliveira AMM, De Andrade FP, Medeiros JA, Tonholo J (2023) Tetradesmus obliquus to treat groundwater contaminated with nitrate towards a semicontinuous process. Energ Ecol Environ 8:262–272. https://doi.org/10.1007/s40974-023-00274-9

De Souza MF, Rodrigues MA, Freitas SP, Bom EPS (2020) Effect of milling and enzymatic hydrolysis in the production of glucose from starch-rich *Chlorella sorokiniana* biomass. Algal Res 50:50. https://doi.org/10.1016/j.algal.2020.101961

Dębowski M, Zieliński M, Kazimierowicz J, Kujawska N, Talbierz S (2020) Microalgae cultivation technologies as an opportunity for bioenergetic system development—advantages and limitations. Sustain For 12:9980. https://doi.org/10.3390/su12239980

Del Río PG, Domínguez E, Domínguez VD, Romaní A, Domingues L, Garrote G (2019) Third generation bioethanol from invasive macroalgae *Sargassum muticum* using autohydrolysis pretreatment as first step of a biorefinery. Renew Energy 141:728–735. https://doi.org/10.1016/j.renene.2019.03.083

Duan Z, Tan X, Guo J, Kahehu CW, Yang H, Zheng X et al (2017) Effects of biological and physical properties of microalgae on disruption induced by a low-frequency ultrasound. J Appl Phycol 29:2937–2946. https://doi.org/10.1007/s10811-017-1178-3

Ellilä S, Fonseca L, Uchima C, Cota J, Goldman GH, Saloheimo M, Sacon V, Siika-aho M (2017) Development of a low-cost cellulase production process using Trichoderma reesei for Brazilian biorefineries. Biotechnol Biofuels 10:30. https://doi.org/10.1186/s13068-017-0717-0

Fajardo C, De Donato M, Carrasco R, Martínez-Rodríguez G, Mancera JM, Fernández-Acero FJ (2020) Advances and challenges in genetic engineering of microalgae. Rev Aquac 12:365–381. https://doi.org/10.1111/raq.12322

Farghali M, Ap Y, Mohamed IMA, Iwasaki M, Tangtaweewipat S, Ihara I et al (2021) Thermophilic anaerobic digestion of *Sargassum fulvellum* macroalgae: biomass valorization and biogas optimization under different pre-treatment conditions. J Environ Chem Eng 9:106405. https://doi.org/10.1016/j.jece.2021.106405

Ferreira RG, Azzoni AR, Freitas S (2021) On the production cost of lignocellulose-degrading enzymes. Biofuels Bioprod Biorefin 15:85–99. https://doi.org/10.1002/bbb.2142

Florencio C, Badino AC, Farinas CS (2016) Soybean protein as a cost-effective lignin-blocking additive for the saccharification of sugarcane bagasse. Bioresour Technol 221:172–180. https://doi.org/10.1016/j.biortech.2016.09.039

Gengiah K, Rajendran N, Al-Ghanim KA, Govindarajan M, Gurunathan B (2023) Process and technoeconomic analysis of bioethanol production from residual biomass of marine macroalgae *Ulva lactuca*. Sci Total Environ 868:868. https://doi.org/10.1016/j.scitotenv.2023.161661

Giannelli L, Torzillo G (2012) Hydrogen production with the microalga *Chlamydomonas reinhardtii* grown in a compact tubular photobioreactor immersed in a scattering light nanoparticle suspension. Int J Hydrog Energy 37:16951–16961. https://doi.org/10.1016/j.ijhydene.2012.08.103

Gong C, Cao L, Fang D, Zhang J, Kumar Awasthi M, Xue D (2022) Genetic manipulation strategies for ethanol production from bioconversion of lignocellulose waste. Bioresour Technol 352:127105. https://doi.org/10.1016/j.biortech.2022.127105

González-Fernández C, Sialve B, Bernet N, Steyer J (2012) Impact of microalgae characteristics on their conversion to biofuel. Part I: focus on cultivation and biofuel production. Biofuels Bioprod Biorefin 6:105–113. https://doi.org/10.1002/bbb.338

González-Gloria KD, Rodríguez-Jasso RM, Rosero-Chasoy G, Shiva KET, Aparício E, Sánchez A, López-Sandin I, Ruiz HÁ (2023) Scale-up of hydrothermal processing: liquid hot water and pilot-scale tubular steam explosion batch reactor for bioethanol production using macroalgae *Sargassum spp* biomass. Bioresour Technol 369:369. https://doi.org/10.1016/j.biortech.2022.128448

Humbird D, Davis R, Tao L, Kinchin C, Hsu D, Aden A, Schoen P, Lukas J, Olthof B, Worley M, Sexton D, Dudgeon D (2011) Process design and economics for biochemical conversion of lignocellulosic biomass to ethanol: dilute-acid pretreatment and enzymatic hydrolysis of corn stover. National Renewable Energy Lab, Golden, p 147. https://doi.org/10.2172/1013269

Jacob-Lopes E, Zepka LQ, Severo IA, Maroneze MM, editors. (2022) 3rd Generation biofuels: disruptive technologies to enable commercial production. Woodhead Publishing, an imprint of Elsevier, Cambridge, MA/Kidlington

Jayakumar M, Vaithilingam SK, Karmegam N, Gebeyehu KB, Boobalan MS, Gurunathan B (2022) Fermentation technology for ethanol production: current trends and challenges. In: Biofuels and bioenergy, pp 105–131. https://doi.org/10.1016/B978-0-323-90040-9.00015-1

Jeyakumar N, Hoang AT, Nižetić S, Balasubramanian D, Kamaraj S, Lakshmana Pandian P et al (2022) Experimental investigation on simultaneous production of bioethanol and biodiesel from macro-algae. Fuel 329:125362. https://doi.org/10.1016/j.fuel.2022.125362

Kendir ÇE, Ugurlu A (2020) Enhanced biogas production of red microalgae via enzymatic pretreatment and preliminary economic assessment. Algal Res 50:101979. https://doi.org/10.1016/j.algal.2020.101979

Krishna Perumal P, Huang CY, Chen CW, Anisha GS, Singhania RR, Dong CD et al (2023) Advances in oligosaccharides production from brown seaweeds: extraction, characterization, antimetabolic syndrome, and other potential applications. Bioengineered 14:2252659. https://doi.org/10.1080/21655979.2023.2252659

Kumar A, Yan B, Tao J, Li J, Kumari L, Oba BT et al (2022) Influence of waste plastic on pyrolysis of low-lipid microalgae: a study on thermokinetics, behaviors, evolved gas characteristics, and products distribution. Renew Energy 185:416–430. https://doi.org/10.1016/j.renene.2021.12.087

Lim HR, Khoo KS, Chia WY, Chew KW, Ho SH, Show PL (2022) Smart microalgae farming with internet-of-things for sustainable agriculture. Biotechnol Adv 57:107931. https://doi.org/10.1016/j.biotechadv.2022.107931

Lin H-TV, Huang MY, Kao TY, Lu WJ, Lin HJ, Pan CL (2020) Production of lactic acid from seaweed hydrolysates via lactic acid bacterial fermentation. Fermentation 6(1):37. https://doi.org/10.3390/fermentation6010037

Liu Z, Lv F, Zheng H, Zhang C, Wei F, Xing XH (2012) Enhanced hydrogen production in a UASB reactor by retaining microbial consortium onto carbon nanotubes (CNTs). Int J Hydrog Energy 37:10619–10626. https://doi.org/10.1016/j.ijhydene.2012.04.057

Liu JJ, Dickson R, Niaz H, Van Hal JW, Dijkstra JW, Fasahati P (2022) Production of fuels and chemicals from macroalgal biomass: current status, potentials, challenges, and prospects. Renew Sust Energ Rev 169:112954. https://doi.org/10.1016/j.rser.2022.112954

Mahmood T, Hussain N, Shahbaz A, Mulla SI, Iqbal HMN, Bilal M (2023) Sustainable production of biofuels from the algae-derived biomass. Bioprocess Biosyst Eng 46:1077–1097. https://doi.org/10.1007/s00449-022-02796-8

Martinez-Burgos WJ, De Souza CE, Pedroni Medeiros AB et al (2021) Hydrogen: current advances and patented technologies of its renewable production. J Clean Prod 286:124970. https://doi.org/10.1016/j.jclepro.2020.124970

Megawati BZAS, Damayanti A, Putri RDA, Triwibowo B, Prasetiawan H, Aji SPK, Prawisnu A (2022) Bioethanol production from glucose obtained from enzymatic hydrolysis of chlorella microalgae. Mater Today Proc 63:5373–5378. https://doi.org/10.1016/j.matpr.2022.03.551

Mienis E, Vandamme D, Foubert I (2023) Enzyme-assisted disruption of oleaginous microalgae to increase the extraction of lipids: nannochloropsis as a case study. Curr Opin Food Sci 51:101034. https://doi.org/10.1016/j.cofs.2023.101034

Millati R, Wikandari R, Ariyanto T, Hasniah N, Taherzadeh MJ (2023) Anaerobic digestion biorefinery for circular bioeconomy development. Bioresour Technol Rep 23:101315. https://doi.org/10.1016/j.biteb.2022.101315

Mi-Ra P, Jeong GT (2021) Production of reducing sugar in *Gracilaria verrucosa* using physiochemical pretreatment and subsequent enzymatic hydrolysis. Algal Res 60:60. https://doi.org/10.1016/j.algal.2021.102531

Morón-Ortiz Á, Mapelli-Brahm P, León-Vaz A, Benitez-González AM, León R, Meléndez-Martínez AJ (2024) Ultrasound-assisted extraction of carotenoids from phytoene-accumulating Chlorella sorokiniana microalgae: effect of milling and performance of the green biosolvents 2-methyltetrahydrofuran and ethyl lactate. Food Chem 434:137437. https://doi.org/10.1016/j.foodchem.2023.137437

Nagarajan D, Chen CY, Ariyadasa TU, Lee DJ, Chang JS (2022) Macroalgal biomass as a potential resource for lactic acid fermentation. Chemosphere 309:136694. https://doi.org/10.1016/j.chemosphere.2022.136694

Nematian T, Shakeri A, Salehi Z, Saboury AA (2020) Lipase immobilized on functionalized superparamagnetic few-layer graphene oxide as an efficient nanobiocatalyst for biodiesel production from *Chlorella vulgaris* bio-oil. Biotechnol Biofuels 13:15. https://doi.org/10.1186/s13068-020-01688-x

Ngamsirisomsakul M, Reungsang A, Liao Q, Kongkeitkajorn MB (2019) Enhanced bio-ethanol production from *Chlorella sp.* biomass by hydrothermal pretreatment and enzymatic hydrolysis. Renew Energy 141:484–492. https://doi.org/10.1016/j.renene.2019.04.008

Nitsos C, Filali R, Taidi B, Lemaire J (2020) Current and novel approaches to downstream processing of microalgae: a review. Biotechnol Adv 45:107650. https://doi.org/10.1016/j.biotechadv.2020.107650

Orr VCA, Plechkova NV, Seddon KR, Rehmann L (2016) Disruption and wet extraction of the microalgae *Chlorella vulgaris* using room-temperature ionic liquids. ACS Sustain Chem Eng 4:591–600. https://doi.org/10.1021/acssuschemeng.5b00967

Pagnini C, Sampietro G, Santini G, Biondi N, Rodolfi L (2023) *Tisochrysis lutea* as a substrate for lactic acid fermentation: biochemical composition, digestibility, and functional properties. Food Secur 12(6):1128. https://doi.org/10.3390/foods12061128

Patel AK, Singhania RR, Awasthi MK, Varjani S, Bhatia SK, Tsai ML et al (2021) Emerging prospects of macro- and microalgae as prebiotic. Microb Cell Factories 20:112. https://doi.org/10.1186/s12934-021-01601-7

Peinemann JC, Demichelis F, Fiore S, Pleissner D (2019) Techno-economic assessment of non-sterile batch and continuous production of lactic acid from food waste. Bioresour Technol 289:121631. https://doi.org/10.1016/j.biortech.2019.121631

Quitério E, Grosso C, Ferraz R, Delerue-Matos C, Soares C (2022) A critical comparison of the advanced extraction techniques applied to obtain health-promoting compounds from seaweeds. Mar Drugs 20:677. https://doi.org/10.3390/md20110677

Raikova S, Allen MJ, Chuck CJ (2022) Hydrothermal liquefaction of macroalgae for the production of renewable biofuels. Biofuels Bioprod Biorefin 13:1483–1504. https://doi.org/10.1002/bbb.2047

Ramachandra TV, Hebbale D (2020) Bioethanol from macroalgae: prospects and challenges. Renew Sust Energ Rev 117:109479. https://doi.org/10.1016/j.rser.2019.109479

Rashid TU, Salem KS, Islam MM, Zaman A, Khan MN, Luna IZ, Haque P, Rahman MM (2019) Recent developments of microbial fuel cell as sustainable bio-energy sources. In: Microbial catalysts, vol 2. Nova Science Publishers, Inc., Hauppauge, pp 57–115

Reis CER, Libardi Junior N, Bento HBS, De Carvalho AKF, De Souza Vandenberghe LP, Soccol CR, Aminabhavi TM, Chandel AK (2023) Process strategies to reduce cellulase enzyme loading for renewable sugar production in biorefineries. Chem Eng J 451:138690. https://doi.org/10.1016/j.cej.2022.138690

Rojo EM, Filipigh AA, Bolado S (2023) Assisted-enzymatic hydrolysis vs chemical hydrolysis for fractional valorization of microalgae biomass. Process Saf Environ Prot 174:276–285. https://doi.org/10.1016/j.psep.2023.03.067

Romero-Vargas A, Gallé A, Blandino A, Romero-García LI (2023a) Use of macroalgal waste from the carrageenan industry as feedstock for the production of polyhydroxybutyrate. Biofuels Bioprod Biorefin 17:1290–1302. https://doi.org/10.1002/bbb.2508

Romero-Vargas A, Muñoz I, Marzo C, Diaz AB, Romero-García LI, Blandino A (2023b) Ultrasound pretreatment to enhance the enzymatic hydrolysis of *Dictyota dichotoma* for sugar production. Algal Res 71:71. https://doi.org/10.1016/j.algal.2023.103083

Ruangrit K, Chaipoot S, Phongphisutthinant R, Duangjan K, Phinyo K, Jeerapan I et al (2023) A successful biorefinery approach of macroalgal biomass as a promising sustainable source to produce bioactive nutraceutical and biodiesel. Biomass Conv Bioref 13:1089–1099. https://doi.org/10.1007/s13399-021-01310-6

Sadatshojaei E, Wood DA, Mowla D (2020) Third generation of biofuels exploiting microalgae. In: Sustainable green chemical processes and their allied applications. Springer, pp 575–588. https://doi.org/10.1007/978-3-030-42284-4_21

Saravanan K, Duraisamy S, Ramasamy G, Kumarasamy A, Balakrishnan S (2018) Evaluation of the saccharification and fermentation process of two different seaweeds for an ecofriendly bioethanol production. Biocatal Agric Biotechnol 14:444–449. https://doi.org/10.1016/j.bcab.2018.03.017

Sathya K, Nagarajan K, Carlin Geor Malar G, Rajalakshmi S, Raja LP (2022) A comprehensive review on comparison among effluent treatment methods and modern methods of treatment of industrial wastewater effluent from different sources. Appl Water Sci 12:70. https://doi.org/10.1007/s13201-022-01594-7

Sedighi M, Jalili H, Darvish M, Sadeghi S, Ranaei-Siadat SO (2019) Enzymatic hydrolysis of microalgae proteins using serine proteases: a study to characterize kinetic parameters. Food Chem 284:334–339. https://doi.org/10.1016/j.foodchem.2019.01.111

Sengmee D, Cheirsilp B, Suksaroge TT, Prasertsan P (2017) Biophotolysis-based hydrogen and lipid production by oleaginous microalgae using crude glycerol as exogenous carbon source. Int J Hydrog Energy 42:1970–1976. https://doi.org/10.1016/j.ijhydene.2016.10.089

Seon G, Kim HS, Cho JM et al (2020) Effect of post-treatment process of microalgal hydrolysate on bioethanol production. Sci Rep 10:16698. https://doi.org/10.1038/s41598-020-73816-4

Sheikh ZUD, Bajar S, Devi A et al (2023) Nanotechnology based technological development in biofuel production: current status and future prospect. Enzym Microb Technol 171:110304. https://doi.org/10.1016/j.enzmictec.2023.110304

Shetty K, Krishnakumar G (2023) Strategies to enhance lipid productivity in microalgae: role of cell disruption and media composition. Fuel 345:128240. https://doi.org/10.1016/j.fuel.2023.128240

Shokrkar H, Keighobadi A (2020) Effect of fluid hydrodynamic situations on enzymatic hydrolysis of mixed microalgae: experimental study and simulation. Energy 241:241. https://doi.org/10.1016/j.energy.2021.122804

Singh R, Saati AA, Faidah H, Bantun F, Jalal NA, Haque S, Rai AK, Srivastava M (2023) Prospects of microbial cellulase production using banana peels wastes for antimicrobial applications. Int J Food Microbiol 388:110069. https://doi.org/10.1016/j.ijfoodmicro.2022.110069

Singla M, Sit N (2021) Application of ultrasound in combination with other technologies in food processing: a review. Ultrason Sonochem 73:105506. https://doi.org/10.1016/j.ultsonch.2021.105506

Sirohi R, Ummalyma SB, Sagar NA, Sharma P, Awasthi MK, Badgujar PC, Madhavan A, Rajasekharan R, Sindhu R, Sim SJ (2021) Strategies and advances in the pretreatment of microalgal biomass. J Biotechnol 341:63–75. https://doi.org/10.1016/j.jbiotec.2021.09.010

Sriyod K, Reungsang A, Plangklang P (2021) One-step multi enzyme pretreatment and biohydrogen production from Chlorella sp. biomass. Int J Hydrog Energy 46:39675–39687. https://doi.org/10.1016/j.ijhydene.2021.09.232

Sudhakar MP, Dharani G (2022) Evaluation of seaweed for the production of lactic acid by fermentation using *Lactobacillus plantarum*. Bioresour Technol Rep 17:100890. https://doi.org/10.1016/j.biteb.2021.100890

Sudhakar MP, Kumar BR, Mathimani T, Arunkumar K (2019) A review on bioenergy and bioactive compounds from microalgae and macroalgae-sustainable energy perspective. J Clean Prod 228:1320–1333. https://doi.org/10.1016/j.jclepro.2019.04.287

Tan IS, Lam MK, Foo HCY, Lim S, Lee KT (2020) Advances of macroalgae biomass for the third generation of bioethanol production. Chin J Chem Eng 28:502–517. https://doi.org/10.1016/j.cjche.2019.05.012

Tapia-Tussell R, Avila-Arias J, Domínguez Maldonado J, Valero D, Olguin-Maciel E, Pérez-Brito D, Alzate-Gaviria L (2018) Biological pretreatment of Mexican Caribbean macroalgae consortiums using Bm-2 strain (*Trametes hirsuta*) and its enzymatic broth to improve biomethane potential. Energies 11(3):494. https://doi.org/10.3390/en11030494

Tawfik A, Niaz H, Qadeer K, Qyyum MA, Liu JJ, Lee M (2022) Valorization of algal cells for biomass and bioenergy production from wastewater: sustainable strategies, challenges, and techno-economic limitations. Renew Sust Energ Rev 157:112024. https://doi.org/10.1016/j.rser.2021.112024

Tazikeh S, Zendehboudi S, Ghafoori S, Lohi A, Mahinpey N (2022) Algal bioenergy production and utilization: technologies, challenges, and prospects. J Environ Chem Eng 10:107863. https://doi.org/10.1016/j.jece.2022.107863

Thompson TM, Young BR, Baroutian S (2019) Advances in the pretreatment of brown macroalgae for biogas production. Fuel Process Technol 195:106151. https://doi.org/10.1016/j.fuproc.2019.106151

Thygesen A, Tsapekos P, Alvarado-Morales M, Angelidaki I (2021) Valorization of municipal organic waste into purified lactic acid. Bioresour Technol 342:125933. https://doi.org/10.1016/j.biortech.2021.125933

Trigueros E, Sanz MT, Filipigh A, Beltrán S, Riano P (2021) Enzymatic hydrolysis of the industrial solid residue of red seaweed after agar extraction: extracts characterization and modeling. Food Bioprod Process 126:356–366. https://doi.org/10.1016/j.fbp.2021.01.014

Vargas-Estrada L, Longoria A, Arenas E, Moreira J, Okoye PU, Bustos-Terrones Y et al (2022) A review on current trends in biogas production from microalgae biomass and microalgae waste by anaerobic digestion and co-digestion. Bioenergy Res 15:77–92. https://doi.org/10.1007/s12155-021-10276-2

Velazquez-Lucio J, Colla LM, Rodríguez-Jasso RM, Sáenz-Galindo A, Cervantes-Cisneros DE, Aguilar CN, Fernandes BD, Ruiz HA (2018) Microalgal biomass pretreatment for bioethanol production: a review. Biofuel Research Journal 17:780–791. https://doi.org/10.18331/BRJ2018.5.1.5

Vishwakarma R, Dey S, Samuchiwal S, Malik A (2023) A biphasic photobioreactor system for consecutive extraction of lipids and carotenoids from pre-hydrolysed microalgae and evaluation of its biodiesel potential. Environ Res 226:115681. https://doi.org/10.1016/j.envres.2023.115681

Wang Y, Lee Y-Y, Santaus TM, Newcomb CE, Liu J, Geddes CD, Zhang C, Hu Q, Li Y (2017) In situ enzymatic conversion of nannochloropsis oceanica IMET1 biomass into fatty acid methyl esters. Bioenerg Res 10(2):438–448. https://doi.org/10.1007/s12155-016-9807-2

Wicker RJ, Kumar G, Khan E, Bhatnagar A (2021) Emergent green technologies for cost-effective valorization of microalgal biomass to renewable fuel products under a biorefinery scheme. Chem Eng J 415:128932. https://doi.org/10.1016/j.cej.2021.128932

Williamson E, Ross IL, Wall BT, Hankamer B (2023) Microalgae: potential novel proteins for sustainable human nutrition. Trends Plant Sci 23:S1360–S1385. https://doi.org/10.1016/j.tplants.2023.08.006

Wu Z, Peng K, Zhang Y, Wang M, Yong C, Chen L et al (2022) Lignocellulose dissociation with biological pretreatment towards the biochemical platform: a review. Mater Today Bio 16:100445. https://doi.org/10.1016/j.mtbio.2022.100445

Xiang Y, Yang Z, Zhang Y, Xu R, Zheng Y, Hu J, Li X, Jia M, Xiong W, Cao J (2019) Influence of nanoscale zero-valent iron and magnetite nanoparticles on anaerobic digestion performance and macrolide, aminoglycoside, β-lactam resistance genes reduction. Bioresour Technol 294:122139. https://doi.org/10.1016/j.biortech.2019.122139

Xu J, Zhao F, Su X (2021) Direct extraction of lipids from wet microalgae slurries by super-high hydrostatic pressure. Algal Res 58:102412. https://doi.org/10.1016/j.algal.2021.102412

Zabed HZ, Akter S, Yun J, Zhang G, Zhang Y, Qi X (2020) Biogas from microalgae: technologies, challenges and opportunities. Renew Sust Energ Rev 117:109503. https://doi.org/10.1016/j.rser.2019.109503

Zhang Y, Kang X, Zhen F, Wanga Z, Kong X, Sun Y (2022) Assessment of enzyme addition strategies on the enhancement of lipid yield from microalgae. Biochem Eng J 177:108198. https://doi.org/10.1016/j.bej.2021.108198

Thermochemical Conversion of Algae Biomass: Pyrolysis and Gasification

7

Jyoti Kaushal, Madhu Khatri, and Shailendra Kumar Arya

Contents

J. Kaushal · M. Khatri · S. K. Arya (✉)
Department of Biotechnology Engineering, University Institute of Engineering and Technology, Panjab University, Chandigarh, India
e-mail: skarya@pu.ac.in

S. K. Arya et al. (eds.), *Value Added Products From Bioalgae Based Biorefineries: Opportunities and Challenges*,
https://doi.org/10.1007/978-981-97-1662-3_7

Abstract

Algae biomass typically includes cyanobacteria, macro and microalgae and is considered as one of the most easily available and important biomasses for the production of bio-fuels. Conversion of algae biomass into valuable by-products through thermochemical conversion is considered as the most straightforward method of all other techniques available. Thermochemical conversion involves the application of high temperature to the biomass for transformation into energy-providing by-products by accelerating the chemical conversion process. This chapter discusses the two main techniques that are used in thermochemical process for conversion of algal biomass namely pyrolysis and gasification. Gasification is used for the production of energy-rich combustible gas from algal biomass whereas pyrolysis leads to the production of bio-oil at very high temperatures.

Keywords

Algae biomass · Thermochemical conversion · Pyrolysis · Gasification · Bio-oil · Char · Syngas

Abbreviations

°C	Degree Celsius
ER	Equivalence ratio
g/kcal	Gram per kilo calorie
L/ha	Litre per hectare
MJ/Kg	Mega joules per kilogram
MPa	Mega Pascal
SWG	Supercritical water gasification
ZSM-5	Zeolite Socony Mobil-5

7.1 Introduction

Algae comprise a large and diversified group of eukaryotic living organisms that are able to perform photosynthesis on their own. Algae can be unicellular (*Prototheca, Chlorella* etc.) or multicellular (brown algae, green algae etc.) and are classified as micro or macroalgae broadly (Mohsenpour et al. 2021). Algae possess various important characteristics including high growth rate, lower water consumption, high CO_2 mitigation efficiency, economic farming and diverse species that make them a suitable candidate for energy harvesting and industrial applications. The three core components that make up an algal specie comprise of mainly proteins (as high as 71%), lipids (as high as 23%) and carbohydrates (as high as 57%) provide an opportunity to convert the algae biomass into an energy source including bio-ethanol, biobutanol, biodiesel, biohydrogen etc. (Kumar et al. 2020). Fuels derived from algae biomass are categorized as solid fuels (Biochar), liquids (biodiesel, bio-oil, bio-crude or bioethanol) and gases (syngas). This conversion is dependent on the methodology carried out for the conversion of biomass and various process parameters (Saber et al. 2016).

The three main strategies used for the conversion of algae biomass to energy-rich fuels include the biochemical method, which generally uses biocatalyst-like enzymes for the conversion process, a straightforward chemical conversion or a thermochemical conversion process, which utilizes heat from high temperature for speeding up the chemical conversion process (Raheem et al. 2015). Out of these three methods for algae biomass conversion, the thermochemical conversion process is the most straightforward one owing to its ease and simplicity and lesser time consumption as compared to the other two methods whereas in the chemical conversion the by-products need to be further purified and refined whereas in the biochemical method, time consumption increases owing to the need of the biomass to undergo fermentation requiring several days. This chapter extensively elaborates the thermochemical process of biomass conversion where two main aspects of the technology namely pyrolysis and gasification have been explained in detail. These two strategies are extensively used for the effective biomass conversion of algae to energy-efficient fuels.

7.2 Algae Biomass

The identification of algal species to produce biomass suitable for energy production and biofuels is the prerequisite step. Substantial efforts have therefore been made in this context so as to find algae species that are easy to grow, and have faster growth rates. Apart from a faster growth rate, the algal species must possess the potential to tolerate a wide range of temperature, salinity, sunlight and various biotic factors like attack from herbivores or other weeds in order to minimize the time required for their cultivation.

Algae species are classified into macroalgae which are filamentous in nature and micro-algae, which includes phytoplankton. Macroalgae, generally referred as seaweeds are further subdivided into three key categories grounded on their type of pigmentations. These include Rhodophyceae (Red seaweed), Phaeophyceae (Brown seaweeds) and Chlorophyceae (Green seaweed). Microalgae in contrast are unicellular organisms sub-divided into four classes. These include Chlorophyceae (green algae), Bacillariophyceae (diatom), Cyano-phyceae (blue algae) and Chrysophyceae (golden algae) (Demirbas and Demirbas 2011). Algal biomass typically consists of proteins (as high as 71%), lipids (as high as 23%) and carbohydrates (as high as 57%) on dry weight bases.

7.3 Cultivation of Algae

There are various factors that affect the growth and cultivation of algal biomass for their efficient utilization, which have been discussed further in the chapter. The main factors (external and internal) include water, light, space, carbon requirement, nitrogen, phosphorous and other additional nutrients.

7.3.1 Water

One simple component required by algae for their growth is water rich with essential minerals and salts. The requirement for water puts algal biomass into two main categories of semiaquatic and aquatic species. An approximate of 1.5 L/ha water is required for the effective cultivation and farming of algae that survive in aquatic environments, which therefore requires an open pond system. Wastewater from industrial and domestic areas can be utilized in the cultivation of such algal species thereby simultaneously providing the solution for bioremediation of waste water as well (Majid et al. 2014).

7.3.2 Light

All algae are photosynthetic organisms and therefore require light as one of the key factors for their growth. In comparison to the other higher plant species, algae require lower intensity of light for performing photosynthesis and for their appropriate development. Sunlight is the main source of light for algae. Approximately 27% of the photosynthetically active radiations referred to as PAR are converted into carbohydrates by algae and the rate of growth of algae can be established with the following equation:

$$P = aE.I$$

where P represents the production rate of dry mass of algae measured in $g/m^2/day$, a represents the conversion coefficient in g/kcal, E gives the photosynthetic efficiency and I signifies energy in light ($kcal/m^2/day$). The light source can be an artificial light source or natural sunlight, the latter being the cheapest source and most suitable for an open pond system.

7.3.3 Space

Requirement of arable land for the growth of algae is redundant unlike other organisms since algae can grow in varied other areas like ponds, reactors or water bodies thereby rendering the budget for the provision of space unconcerned.

7.3.4 Carbon

Efficient production and growth of algae require a large amount of carbon. Approximate 1.6 kg of carbon is required for the production of 1 kg algal biomass, which accounts for as high as 60% of the total cost required for the cultivation of algae. Carbon for algal cultivation is obtained from either the atmosphere, carbonates or carbon from industrial smoke.

7.3.5 Nitrogen and Phosphorous

Nitrogen is required for constructing proteins and nucleic acids within the algae thereby playing an important role in algal metabolism. Phosphorous is made available in the form of phosphates so as to ensure efficient availability to the growing algal biomass (Majid et al. 2014).

7.4 Thermochemical Method for Algae Biomass Conversion

Thermochemical methodology is considered as the most straightforward and easy technique for the efficient conversion of algal biomass in comparison to other techniques. Thermochemical conversion process is more flexible in the manner that it can handle different types of algal biomass and is therefore versatile in comparison to biochemical process. Moreover, it is direct, simple and fast. The process involves the decomposition of the organic matter at a very high temperature with the conversion of the algal biomass into oils or gas. Thermochemical methodology has further four more subtypes which are stated in Fig. 7.1.

7.5 Pyrolysis

Pyrolysis is the thermochemical process that is carried out without the presence of oxygen and with or without the use of a catalyst. Different temperatures can be used for the biomass decomposition with different time intervals for the recovery of low

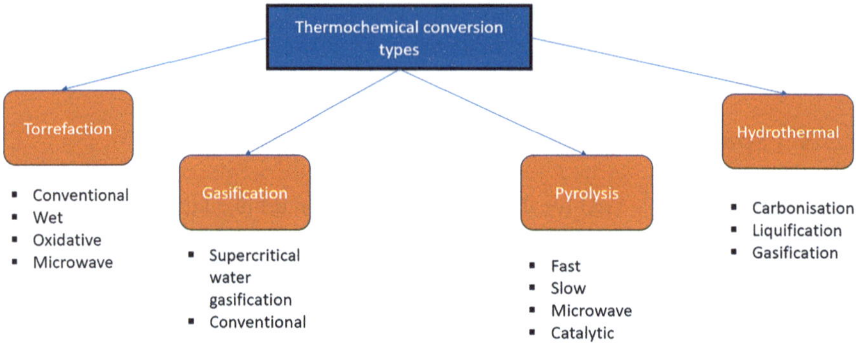

Fig. 7.1 Thermochemical conversion process types

to medium caloric value fuels. The end products after the process of pyrolysis can be a solid like bio-char, a liquid like bio-oil, or various gases (Sirajunnisa and Surendhiran 2016).

Algal species like *Chlorella vulgaris* and *C. protothecoides*, *Scenedesmus* sp., *Nannochloropsis* sp. and *Microcystis aeruginosa* have been mainly recognized as the main species generally considered suitable for conversion to energy-efficient fuels. Pyrolysis carried out for the conversion of algal biomass is gaining much attention due to the end products, which are produced in the form of solid, liquids, or gases. The temperature requirement for solid biochar formation is lower in comparison to that required for liquid end products like bio-oils, which typically occurs at high temperatures ranging between 450 and 550 °C (Adeniyi et al. 2018). Further, the gaseous by-products require even higher temperatures. The higher the temperature, the lower is the time required for conversion of the biomass. Based on these temperature and time differences, the pyrolysis process is divided into four subcategories namely; Fast, slow, microwave or catalytic pyrolysis with temperatures ranging from 300 to 800 °C and time ranging from 10 to 120 min. Pyrolysis assisted by a microwave requires a temperature of 800 °C, which is the highest temperature for this process whereas the lowest temperature is required for catalytic pyrolysis (Auersvald et al. 2019).

Three main steps involved in the conversion of algal biomass are dehydration, devolatilization and decomposition. The first step of dehydration occurs at a temperature $\leq$200 °C. The de-volatilization step is the focal step in the entire process, which occurs at a temperature between 200 and 500 °C whereas the decomposition, where the end product is achieved is carried out at temperatures above 550 °C. Among the main components of the algal biomass, the decomposition of the proteins and carbohydrates takes place at temperature $\leq$400 °C; however, the lipids tend to degrade at higher temperatures generally above 500 °C (Pourkarimi et al. 2019).

Figure 7.2 further gives the description of all the different pyrolysis processes discussed above.

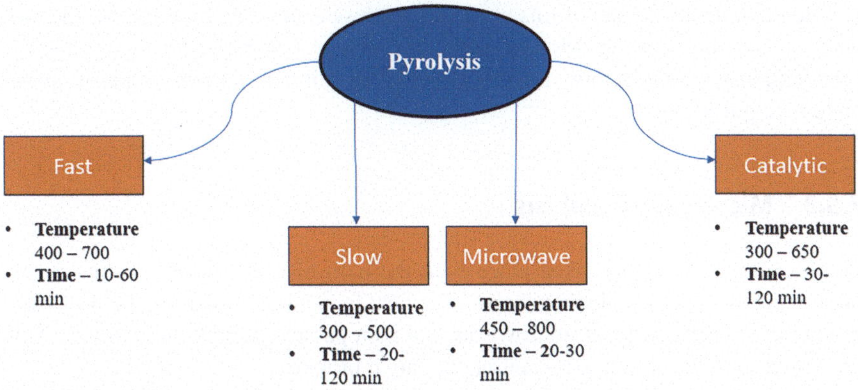

Fig. 7.2 Types of pyrolysis processes

7.5.1 Fast Pyrolysis

The temperature range in the fast pyrolysis process is quite high as given in Fig. 7.2, with a lower time-span required for biomass conversion. The process is carried out in the absence of air/oxygen. Firstly, the biomass collected is dried for the process of grinding into small particles of size ≤1 mm and these particles are then put into the reactor for further processing. Due to the small size, the biomass particles heat up quite rapidly and are processed into small chain molecules and gaseous substances, which are then cooled down to form bio-oils and other liquids. The reason why this method of pyrolysis is getting more attention is because in order for the yield of bio-oils to increase, the algal biomass needs to be treated at high temperature followed by instant cooling in the minimum time possible and these conditions are fulfilled by this procedure of pyrolysis. The oil yields are in the range of 60–75% following fast pyrolysis, which is quite high (Jahirul et al. 2012). The yield of liquid fuels accounted for the conversion of *Chlorella protothecoides* or *Microcystis aeruginosa* algal biomass was 18% and 24% stated by Miao et al. (2004).

7.5.2 Slow Pyrolysis

This process of pyrolysis requires a low temperature of heating and a longer time interval in comparison to the fast pyrolysis process. This increased time required for this process is the reason why this type of pyrolysis causes an increased yield of biochar and certain gases that are non-condensable including H_2, CH_4, CO_2 and C_2H_4. The rate of reaction is typically slow in this case. It requires a pre-heating of the algal biomass in order to dry it out and remove any moisture from it which helps in breaking of the bond. Further, the biomass decomposes into carbohydrates, proteins and lipids leading to the formation of biofuels. The yield of bio-oils from algal biomass in case of slow pyrolysis varies from 20% to 43% in weight and in the case of MJ/Kg, it varies from 1.2 to 33.62 (Grierson et al. 2009).

The main disadvantage of this pyrolysis process is the time required for the biomass conversion to occur. Moreover, the energy input required for this process is very high because of the low heat. It is therefore suitable in cases where more oil yield is required only.

7.5.3 Microwave Pyrolysis

The heating process that takes place with the help of microwave is called as the microwave-assisted pyrolysis process. The heat is generated with the help of electromagnetic radiation from microwaves that then passes through the biomass. This process creates an even heating among the algal biomass.

The microwave pyrolysis process depends on ionic conduction and dipolar polarization. The first one affects the capability of heat production. The main components are the same as those required in a conventional pyrolysis process with just the addition of a microwave oven. Transparent materials like borosilicate glass or quartz are used to make this oven for the pyrolysis process (Borges et al. 2014). Microwave-assisted pyrolysis again requires a high temperature and a low residence time for the reaction to take place. The microwaves are difficult to penetrate within the biomass, therefore, the algal biomass is mixed with enhancers that help in the absorption of microwaves like activated carbon, water and certain catalysts like $MgCl_2$ and charcoal (Motasemi and Afzal 2013). There is significant enhancement following this procedure in the quickness of pyrolysis because of the increased absorption of microwaves.

The yield of oil in this case ranges from 18% to 60% in weight and 21.9–42 MJ/Kg. The use of a catalyst in this process further enhances the process and improves the oil yield and physiochemical characteristics.

7.5.4 Catalytic Pyrolysis

The main drawbacks of the energy fuels generated in the fast pyrolysis process are the high viscosity of the fuels, high release of oxygenated compounds and low value of heating. In order to improve the oil quality, the addition of a catalyst was done in the fast pyrolysis process which was called catalytic pyrolysis. Catalysts help in breaking the high molecular weight compounds into smaller ones and enhance the rate of the process (Aysu and Sanna 2015). The temperature in this process ranges from 300 to 500 °C and the ratio of the catalyst to algal biomass is between 0.2 and 5. The yield of biofuels is in the range of 20–55% in this case. The main catalysts that are used in this process include ZSM-5 zeolites and Na_2CO_3 (Campanella and Harold 2012).

The bio-oils produced from this process have a low acidity and produce lower oxygenated compounds. The main components required for this process include a biomass feeding system, a reactor, a regenerator for catalyst, a cooling system, a cleaning process and a system to separate the gases generated from char.

7.6 Reactor Types Used in Pyrolysis Process

The two types of reactors generally used in a pyrolysis process include the fluidized bed reactor for the process of fast pyrolysis and the fixed bed reactor is used for the other three types. The reactor types have the capability to improve the outcome of biomass conversion yield.

7.6.1 Fixed Bed Reactors

The three main components of a fixed bed reactor comprise a cylindrical tube, a cooling system and a cleaning process (Campanella and Harold 2012). Firstly, the algal biomass is dehydrated and then ground to small particles which are then loaded into the tube of the reactor. An externally heated furnace is used to provide heat to the reactor system. An inert gas like N_2, Ar, or He is passed through the biomass inside the reactor in order to create a neutral and inert environment. The components that are volatile are made to flow out of the reactor with a carrier flow. The other gases that are non-volatile and condensable are passed onto the cooling system where the condensation into the liquids and oils takes place. This entire process is quite straightforward and is suitable for small-scale projects of algal biomass conversion (Aravind et al. 2020).

7.6.2 Fluidized Bed Reactors

The main principle behind a fluidized bed reactor is the ability of a fluid in the reactor with high velocity to be able to make a liquid continuum of a fixed bed in the reactor. The reactor is assembled with a fixed bed of algal biomass maintained in an inert environment. The mixing of the algal biomass with the inert gas in the chamber is responsible for the heating inside the reactor. The reactor contains a cyclone that does the task of separating of solids from gaseous particles. The other processes are typically the same as that of the fixed bed reactor. The main advantage of this reactor system is the temperature control mechanism, which provides the reactor the ability to handle greater rates of heat transfer. This reactor has the capability to be scaled up for larger operations as well (Aravind et al. 2020).

7.7 Factors Affecting the Process of Pyrolysis

As stated above, the pyrolysis process has three major steps involved, which include dehydration, decomposition and devolatilization. The main products obtained from pyrolysis are either solids like bio-char, or liquids like bio-oils or syngas (Xiong et al. 2017). Various factors have been identified, which play an important role in the performance of the process of pyrolysis including the operation conditions, the composition of the feedstock, reactor type etc. which have been discussed further in the chapter.

7.7.1 Temperature

The final outcome of the pyrolysis process is dependent on the temperature of the process. The process of dehydration takes place below 200 °C. The decomposition occurs above 550 °C and the devolatilization between 200 and 500 °C (Adeniyi et al. 2018). The secondary cracking occurs at a temperature above 500 °C resulting in a decreased yield of bio-oil. Any further rise in temperature results in a rise in the content of carbon present in the yield thereby leading to the formation of aromatic hydrocarbon molecules.

The composition of pyrolysis products can therefore vary with changes in temperature conditions. An increase in temperature typically decreases the solid char production and an increase in the yield of gaseous products (Ly et al. 2016). In the same manner, the increase in the heating rate has a direct influence on the final composition of the product from pyrolysis. Low rate of heating enhances the rate of biochar production. The rise in microwave radiation increases the yield of bio-oil. On the contrary, the production of biochar decreases with an increase in microwave radiation (Du et al. 2011).

7.7.2 Residence Time

The holding/residence time is another significant factor that has a direct influence on the yield of bio-char since the longer is the time period, the more is the production of bio-char. It signifies the type of pyrolysis process depending on the time period and distinguishes it between fast and slow. In the case of slow pyrolysis, the time taken is longer in comparison to the fast process. The production of char is higher in case the time is increased since there is complete re-polymerization of the algal biomass with increased residence time. Moreover, the rate of heating also affects the yield of biochar and bio-oil. Longer holding time with a low rate of heating produces a high yield of char whereas short holding time and high rate of heating means high oil yield.

7.7.3 Composition of Feedstock

The quality and the structure of the product in the pyrolysis process also depend on the composition of the algal feedstock. The amount of carbon highly depends on the percentage of carbohydrates found in the algal biomass (Ji et al. 2015). It has been observed that the composition and structures of the pyrolysis yield are different in the case of algal biomass when compared to the lignocellulosic biomass. The nutrients in the pyrolysis products of algal biomass are extremely enriched and therefore can also be infields as a fertilizer.

7.7.4 Size of Particles

Biomass particle size also affects the bio-oil's properties and yield. The greater is the particle size the more is the yield of bio-char whereas, lower is the yield of bio-oil. Large size of particle requires more amount of heat as well as activation energy in comparison to small biomass particles. However, the small particles demand a higher handling cost and increased time period of the process. It is therefore more suitable to choose a particular size of particle to ensure that the process of pyrolysis is systemized and proves to be profitable. Oil is produced when the algal biomass is lower in size (≤1 mm) whereas when particle size increases, bio-char is produced (Kocer et al. 2018).

7.8 Gasification

This process of thermochemical conversion is used for the conversion of algae biomass into gaseous energy fuels. The biomass is converted into syngas like CO, CO_2, H_2, or CH_4. Apart from these chief gases, other gases like C_2H_4, C_3H_8 and C_2H_6 are also generated from this process (Arun et al. 2021). The syngas produced can further be used to produce fuels required for the transportation process. The gasification process allows the biomass to undergo treatments like dehydration followed by devolatilization and then finally oxidation and gas formation (Cheah et al. 2016).

Drying of the algal biomass allows for it to be fully utilized as a source of energy production. Following this process, the biomass is made to go through the process of pyrolysis, which helps in the production of bio-char, tar and gases. After this process, the stage of oxidation is carried out where all the volatile material and char are oxidized to Co, CO_2 and H_2O along with the production of heat. Lastly, the gasification step is carried out also called the reduction step which is carried out in an environment free from oxygen. The biomass oxidized above is made to react with gasifying agents like air or steam in order to permanently change it into syngas.

There are typically two types of gasification processes namely the conventional process and the supercritical water gasification. Both of these types have been discussed below in detail.

7.8.1 Conventional Gasification

In the conventional gasification process, the algal biomass is firstly dried and then later subjected to the process of pyrolysis. Approximately 90% of the moisture content is removed from the drying process before the gasification can happen. This process requires a high temperature since the water requires high latent heat of vaporization (Chen et al. 2015). Conventionally, the gasification process occurs within the range of temperature 800–1000 °C with pressure ranging from 0.1 to 1 MPa.

The biomass is firstly dried for the removal of moisture as much as possible. Secondly, the biomass is then treated by the pyrolysis process for breakage of the complex structure of biomass. Thirdly, oxidation process takes place. In the final process, the biomass undergoes gasification for conversion of the products obtained from pyrolysis to small molecules yielding high amount of energy. Final products of syngas, bio-oil, biochar and tar are delivered from the completion of this process.

7.8.2 Supercritical Water Gasification

SWG or supercritical water gasification is the process in which the biomass with high water content (up to around 80%) can undergo the hydrothermal process. Therefore, the need for drying of the biomass is rendered redundant. The algal biomass in this way can be converted into energy-providing gas directly. The thermal productivity of SWG process is higher than that received from the conventional process. The main concern is the setting up of the system for this process since the techniques required for its commercial running are still in the stage of development (Haiduc et al. 2009). The temperature of its running process can vary from 375 to as high as 700 °C in some instances. The range of pressure can also vary from 22 to 36 MPa (Khoo et al. 2013). This process decomposes the algal biomass into small molecules. Once again, the steps required for the conversion of biomass from this process include the pyrolysis step followed by hydrolysis, condensation and then finally dehydrogenation. The main products of the SWG process are gases like CO, H_2, CO_2 etc. and certain secondary by-products like ashes, char and tar. If there is any catalyst present in the system, the outcome of SWG process can be affected; however, the main dependency is on the algal biomass composition.

7.9 Reactor Types Used in Gasification Process

Three types of reactors have been mainly utilized in the gasification thermochemical process, which include the fixed bed, fluidized bed and the catalytic/non-catalytic type reactors. All the reactor types have been discussed further in the chapter.

7.9.1 Fixed Bed Reactors

A fixed bed reactor employs a system in which the raw algal biomass is fixated in a reactor unit. The main components include the system where the feed of algal biomass is entered, a chamber for gasification, a system for reforming syngas and finally a cooling system. The algal biomass that has been partly decomposed following the process of pyrolysis is made to enter the gasifier chamber for the gasification process to take place. Oxygen is also made available in this chamber. The syngas produced is then separated and is made to undergo a series of treatment (drying and desulphurization).

Depending on the way the gas flows in the gasifier, the fixed bed reactor is subdivided further into three types, which include the updraft, the downdraft and the cross-flow gasifiers. The updraft gasifier gives a better thermal efficiency with low formation of slag and lower pressure drop. On the other hand, the downdraft type gives products with low tar content (Caputo et al. 2016). This process can be used for the generation of power in small-scale operations. The cross-flow type gasifier favours the production of syngas and is a more flexible system in the processing of different biomass.

7.9.2 Fluidized Bed Reactor

The fluidized bed reactor puts across an agent of gasifying along with the algal biomass in order to convert it into a fluid consistency. This process renders the biomass susceptible to the procurement of uniform temperature and maintenance of good contact between the solid biomass and the gas. This type of reactor is preferred for large-scale industrial applications because of its flexibility in handling diverse algal biomass types and higher thermal inertia (Ebadi and Hisoriev 2019). In this type of reactor, the main components again consist of a feeding system, a chamber for gasification, a syngas cleaning system and lastly a system for the supply of air. The gasification chamber helps in the maintenance of pressure and temperature. Preheated air serves as a fluidising gas, which is heated externally with the help of an external heating agent before it enters the chamber of gasification. The biomass is continuously supplied in this chamber through a system of feeding process comprising of mainly two systems. These systems include the gate valves with a hopper for the purpose of blocking the flow of air and a feeder that is connected to gasifier. The primary cleaning for the separation of char is done with the use of a cyclone. The syngas is collected with the help of a scrubber attached to the gasifiers end (Qian et al. 2013).

7.9.3 Catalytic/Non-Catalytic Reactors

The algal biomass in this type of reactor is converted into the energy-efficient fuels in a gasification process in the presence/absence of a catalyst. The process of gasification is dependent on various parameters like the type of gasifier, rate of heating, type of biomass and other operating conditions like pressure, temperature, oxidizing agents etc. (Cortright et al. 2002).

7.10 Factors Affecting the Process of Gasification

Similar to the pyrolysis process, the gasification process also is affected by various different factors that influence the end products and yields following the procedure. These factors are discussed further in the chapter.

7.10.1 Temperature

With the increase in temperature of gasification process, there is an increase in the production of H_2 but there is a decrease in the yield of CO_2 and CH_4 in biomass. It therefore leads to reduced production of char and tar in the end. High temperature is the most influential factor to cause improvement in the efficiency of gasification, production of methanol and yield of hydrogen. Higher the temperature, the higher is the yield of hydrocarbon in algal biomass however, the oxygenated contents like nitrogen and certain phenolic components suffer a loss (Hirano et al. 1998). However, with an excessive increase in heat and temperature, the consumption of energy and cost of operation also increases which is not considered feasible. It is therefore better if the optimization of the temperature conditions is done beforehand so as to improve efficiency of the process.

7.10.2 Equivalence Ratio

Increase in the ER (equivalence ratio) causes improvement in temperature, quality of fuel and conversion efficiency. An increase in ER slightly decreases the composition of CH_4, CO and H_2 whereas a minute enhancement in CO_2 of the biomass. Improvement in the CO_2 is due to an increase in the supply of oxygen during the process of oxidation (Yang et al. 2013).

7.10.3 Pressure

The conventional process of gasification occurs at a pressure within 0.1–1 MPa while in SWG, pressure ranges from 22.1 to 36 MPa (Chen et al. 2015). The carbon gasification rate of carbon reduces significantly with a rise in pressure. The moderate pressure range lying equivalent to the supercritical pressure condition (22.1 MPa) gives a better condition of gasification.

7.10.4 Time Period

The holding or retention time is an imperative parameter that ensures complete combustion of the algal biomass. This parameter also guarantees an increase in the efficiency of the process, yield of product and amount of energy. An extended time of the reaction is vital in case the reaction rate has been fixed and kept constant. To reduce the consumption of energy and improve productivity, it is important to optimize reaction time.

7.10.5 Loading Rate of Catalyst

A catalyst improves the product quality and yield as well as process efficiency. The gasification rate is sensitive towards the structure of catalyst and it affects the composition of the products (Tiong and Komiyama 2019).

7.11 Conclusion and Future Prospectives

Algal biomass is a promising, abundantly available material for the purpose of renewable energy production. The diverse algal species available can be efficiently used as bio-stock for the purpose of energy production in the form of bio-oil, char and syngas. The thermochemical method for conversion of algal biomass is a straightforward and easy method for biomass conversion to valuable products. Among the many thermochemical methods, this chapter discusses the applicability and process of pyrolysis and gasification. Pyrolysis is the thermochemical process that is carried out without the presence of oxygen and with or without the use of a catalyst. Gasification is used for the conversion of algae biomass into gaseous energy fuels. The biomass is converted into syngas like CO, CO_2, H_2, or CH_4. Apart from these chief gases, other gases like C_2H_4, C_3H_8, C_2H_6 are also generated from this process.

Fixed bed reactor is appropriate for small-scale applications whereas the fluidized bed reactors have the possibility of scaling up for industrial applications in both the thermochemical methods of conversion.

Among the macro and microalgae biomass available for energy production, the application of microalgae has recently gained much attention. However, the upstream and downstream processes are still unable to tap the full potential of the microalgae biomass. Moreover, the need to provide a cost-effective and efficient methodology for microalgae cultivation needs to be standardized to provide quality biofuels. There still exists a gap in knowledge in this area for taping bioenergy from microalgae biomass. Characterized research is required to fill this knowledge gap.

References

Adeniyi OM, Azimov U, Burluka A (2018) Algae biofuel: current status and future applications. Renew Sust Energ Rev 90:316–335

Aravind S, Kumar PS, Kumar NS, Siddarth N (2020) Conversion of green algal biomass into bioenergy by pyrolysis. A review. Environ Chem Lett 18:829–849

Arun J, Gopinath KP, Sivaramakrishnan R, SundarRajan P, Malolan R, Pugazhendhi A (2021) Technical insights into the production of green fuel from CO2 sequestered algal biomass: a conceptual review on green energy. Sci Total Environ 755:142636

Auersvald M, Shumeiko B, Vrtiška D, Straka P, Staš M, Šimáček P et al (2019) Hydrotreatment of straw bio-oil from ablative fast pyrolysis to produce suitable refinery intermediates. Fuel 238:98–110

Aysu T, Sanna A (2015) Nannochloropsis algae pyrolysis with ceria-based catalysts for production of high-quality bio-oils. Bioresour Technol 194:108–116

Borges FC, Xie Q, Min M, Muniz LAR, Farenzena M, Trierweiler JO et al (2014) Fast microwave-assisted pyrolysis of microalgae using microwave absorbent and HZSM-5 catalyst. Bioresour Technol 166:518–526

Campanella A, Harold MP (2012) Fast pyrolysis of microalgae in a falling solids reactor: effects of process variables and zeolite catalysts. Biomass Bioenergy 46:218–232

Caputo G, Dispenza M, Rubio P, Scargiali F, Marotta G, Brucato A (2016) Supercritical water gasification of microalgae and their constituents in a continuous reactor. J Supercrit Fluids 118:163–170

Cheah S, Jablonski WS, Olstad JL, Carpenter DL, Barthelemy KD, Robichaud DJ et al (2016) Effects of thermal pretreatment and catalyst on biomass gasification efficiency and syngas composition. Green Chem 18(23):6291–6304

Chen WH, Lin BJ, Huang MY, Chang JS (2015) Thermochemical conversion of microalgal biomass into biofuels: a review. Bioresour Technol 184:314–327

Cortright RD, Davda RR, Dumesic JA (2002) Hydrogen from catalytic reforming of biomass-derived hydrocarbons in liquid water. Nature 418(6901):964–967

Demirbas A, Demirbas MF (2011) Importance of algae oil as a source of biodiesel. Energy Convers Manag 52(1):163–170

Du Z, Li Y, Wang X, Wan Y, Chen Q, Wang C et al (2011) Microwave-assisted pyrolysis of microalgae for biofuel production. Bioresour Technol 102(7):4890–4896

Ebadi AG, Hisoriev H (2019) Gasification of algal biomass (*Cladophora glomerata* L.) with CO2/H2O/O2 in a circulating fluidized bed. Environ Technol 40(6):749–755

Grierson S, Strezov V, Ellem G, Mcgregor R, Herbertson J (2009) Thermal characterisation of microalgae under slow pyrolysis conditions. J Anal Appl Pyrolysis 85(1–2):118–123

Haiduc AG, Brandenberger M, Suquet S, Vogel F, Bernier-Latmani R, Ludwig C (2009) SunCHem: an integrated process for the hydrothermal production of methane from microalgae and CO2 mitigation. J Appl Phycol 21:529–541

Hirano A, Hon-Nami K, Kunito S, Hada M, Ogushi Y (1998) Temperature effect on continuous gasification of microalgal biomass: theoretical yield of methanol production and its energy balance. Catal Today 45(1–4):399–404

Jahirul MI, Rasul MG, Chowdhury AA, Ashwath N (2012) Biofuels production through biomass pyrolysis—a technological review. Energies 5(12):4952–5001

Ji X, Liu B, Chen G, Ma W (2015) The pyrolysis of lipid-extracted residue of Tribonema minus in a fixed-bed reactor. J Anal Appl Pyrolysis 116:231–236

Khoo, HH, Koh, CY, Shaik, MS, & Sharratt, PN (2013). Bioenergy co-products derived from microalgae biomass via thermochemical conversion–life cycle energy balances and CO2 emissions. Bioresource technology, 143, 298–307

Kocer AT, Inan B, Ozcimen D (2018) A comparison of bioethanol and biochar production from various algal biomass samples and sweet sorghum energy crop. Environ Res Technol 1(1):43–50

Kumar M, Sun Y, Rathour R, Pandey A, Thakur IS, Tsang DC (2020) Algae as potential feedstock for the production of biofuels and value-added products: opportunities and challenges. Sci Total Environ 716:137116

Ly HV, Kim SS, Choi JH, Woo HC, Kim J (2016) Fast pyrolysis of Saccharina japonica alga in a fixed-bed reactor for bio-oil production. Energy Convers Manag 122:526–534

Majid M, Shafqat S, Inam H, Hashmi U, Kazi AG (2014) Production of algal biomass. In: Biomass and bioenergy: processing and properties. Springer, pp 207–224

Miao X, Wu Q, Yang C (2004) Fast pyrolysis of microalgae to produce renewable fuels. J Anal Appl Pyrolysis 71(2):855–863
Mohsenpour SF, Hennige S, Willoughby N, Adeloye A, Gutierrez T (2021) Integrating micro-algae into wastewater treatment: a review. Sci Total Environ 752:142168
Motasemi F, Afzal MT (2013) A review on the microwave-assisted pyrolysis technique. Renew Sust Energ Rev 28:317–330
Pourkarimi S, Hallajisani A, Alizadehdakhel A, Nouralishahi A (2019) Biofuel production through micro-and macroalgae pyrolysis–a review of pyrolysis methods and process parameters. J Anal Appl Pyrolysis 142:104599
Qian K, Kumar A, Patil K, Bellmer D, Wang D, Yuan W, Huhnke RL (2013) Effects of biomass feedstocks and gasification conditions on the physiochemical properties of char. Energies 6(8):3972–3986
Raheem A, Azlina WW, Yap YT, Danquah MK, Harun R (2015) Thermochemical conversion of microalgal biomass for biofuel production. Renew Sust Energ Rev 49:990–999
Saber M, Nakhshiniev B, Yoshikawa K (2016) A review of production and upgrading of algal bio-oil. Renew Sust Energ Rev 58:918–930
Sirajunnisa AR, Surendhiran D (2016) Algae–a quintessential and positive resource of bioethanol production: a comprehensive review. Renew Sust Energ Rev 66:248–267
Tiong L, Komiyama M (2019) Supercritical water gasification of microalga Chlorella vulgaris over supported Ru. J Supercrit Fluids 144:1–7
Xiong X, Iris KM, Cao L, Tsang DC, Zhang S, Ok YS (2017) A review of biochar-based catalysts for chemical synthesis, biofuel production, and pollution control. Bioresour Technol 246:254–270
Yang KC, Wu KT, Hsieh MH, Hsu HT, Chen CS, Chen HW (2013) Co-gasification of woody biomass and microalgae in a fluidized bed. J Taiwan Inst Chem Eng 44(6):1027–1033

Sustainable Approaches to Algal Biofuels: Opportunities, Key Challenges and Current Status

8

Anuradha Devi, Christina Saran, Luiz Fernando R. Ferreira, Sikandera I. Mulla, and Ram Naresh Bharagava

Contents

A. Devi · C. Saran · R. N. Bharagava (✉)
Department of Environmental Microbiology (DEM), School of Earth and Environmental Sciences (SEES), Babasaheb Bhimrao Ambedkar University, Lucknow, Uttar Pradesh, India

L. F. R. Ferreira
Postgraduate Program in Genomic Sciences and Biotechnology, Catholic University of Brasília, Brasília, Brazil

S. I. Mulla
Department of Biochemistry, School of Applied Sciences, REVA University, Bangalore, India

S. K. Arya et al. (eds.), *Value Added Products From Bioalgae Based Biorefineries: Opportunities and Challenges*,
https://doi.org/10.1007/978-981-97-1662-3_8

Abstract

As fossil fuel reserves dwindle and the concerns over pollution, global warming and soaring fuel prices exaggerate, the perseverance for renewable energy sources becomes paramount. Algae-based biofuels emerge as an excellent alternative to conventional fossil fuels, leveraging their inherent photosynthetic efficiency, rapid growth rate, minimal land and water requirements in comparison to conventional crops, and superior quality and quantity of biomass yield rich in carbohydrates and lipids. The produced biomass can be used for producing diverse array of biofuels like bioethanol, biomethane, biohydrogen and biodiesel. However, despite their potential opportunistic limitations and challenges persist, notably in production scalability and research focus limited to selection and lab scale cultivation of algal species. This chapter comprehensively explore the opportunities, challenges and economic viability of algal biofuels. From delving into the generations of biofuels to highlighting algae as a third-generation feedstock, it scrutinizes cultivation and harvesting techniques, elucidates various biofuel production opportunities, and addresses engineering for production scale-up. Also, the chapter seeks to address critical limitations inhibiting widespread adoption by evaluating the economic landscape and challenges. By spotlighting these facets, it aspires to catalyse further research and technological advancements in algal biofuel generation, propelling the field towards a sustainable, impactful future.

Keywords

Microalgae · Biomass · Biofuels · Biodiesel · Bioethanol · Third generation feedstock

Abbreviations

ABE	Acetone-Butanol-Ethanol
AMOPS	Aquatic Microbial Oxygenic Photoautotroph
ASP	Aquatic Species Program
COD	Chemical oxygen demand
EESI	Environmental and Energy Study Institute
EPS	Extracellular polymer substances
FAME	Fatty Acid Methyl Ester
GHGs	Greenhouse gases
IRR	Internal Rate of Return
LCA	Life cycle assessment
LED	Light emitting diodes
MFSP	Minimum Fuel Selling Price
NER	Net energy ratio
NPV	Net Present Value
NREL	National Renewable Energy Laboratory

PAR	Photosynthetic active radiation
PBR	Photobioreactor
RITE	Research Institute of Innovative Technology for the Earth
SDG	Sustainable Development Goals
TCI	Total capital investment
USD	United States dollar

8.1 Introduction

Biofuels have attracted considerable interest as an eco-friendly substitute of traditional fossil fuels because of their capacity to decrease greenhouse gas emissions and reduce reliance on finite resources. Within the array of options for biofuel development, microalgae emerge as a particularly encouraging contender (Zabermawi et al. 2022). These self-sustaining, solitary microorganisms possess exceptional attributes that render them exceptionally well-suited for biofuel production.

Microalgae are found everywhere, flourishing in a wide range of locations, including marine and fresh waters, deep ocean regions and rocky coastlines (Punia et al. 2021; Kashyap et al. 2023). Their ability to thrive in diverse environments can be attributed to their uncomplicated single-celled structure. Whether prokaryotic (Cyanophyceae and Chlorophyta) or eukaryotic (like Bacillariophyta), various species of microalgae show substantial promise in the realm of biofuel production. Their capacity to adapt and their versatility make them a superb option for biofuel source material (Taher et al. 2011). Microalgae offer a wide array of applications beyond biofuel production. They have a diverse range of uses, including bioplastics, bioenergy, animal feed, human food, pharmaceuticals, nutraceuticals, vitamins, cosmetics, fertilizers and agar production (Orejuela-Escobar et al. 2021; Saran et al. 2023).

This versatility makes them a valuable resource with relevance to numerous industries. A notable characteristic of microalgae is their capacity to amass substantial levels of lipids. This lipid-rich biomass can be effectively transformed into biodiesel via the process of transesterification, rendering microalgae an economically appealing choice for biofuel generation. Numerous investigations have verified the viability of producing biodiesel from microalgae with minimal expenses. Biofuels are currently in production worldwide, and specific nations are taking the lead in this endeavour. Notably, the United States stands at the forefront of biofuel production, closely followed by countries like Brazil, Indonesia, China and others (Pahl 2008). This emphasizes the increasing importance of biofuels within the broader global energy context. Microalgae also contribute to ecological advantages. They exhibit effectiveness in eliminating phosphorus, nitrogen and harmful substances from wastewater, serving as valuable agents for environmental remediation. Furthermore, microalgae are environmentally conscious, non-harmful, devoid of sulphur and undergo natural decomposition. Over the course of time, substantial endeavours in research and development have been dedicated to uncovering the potential of microalgae in the context of energy generation.

The US Aquatic Species Program (ASP) and the Japanese Research Institute of Innovative Technology for the Earth (RITE) programme serve as noteworthy instances of such initiatives (Borowitzka 2012). While some initial projects were abandoned due to unfavourable economic predictions, research persists in other locations, with a focus on diverse aspects of microalgae application. In conclusion, microalgae present a compelling solution for the sustainable production of biofuels owing to their exceptional adaptability, versatility, substantial lipid accumulation and their beneficial impact on the environment. Biofuels such as bioethanol, biomethane, bioethanol and biohydrogen are also produced using microalgal biomass which is discussed in this chapter (Khan et al. 2018). In exploring the realm of microalgal biofuels, this chapter embarks on illuminating their extraordinary adaptability, multifaceted versatility and remarkable impact on environmental sustainability. This chapter unravels the diverse array of biofuels like bioethanol, biomethane, biohydrogen and biodiesel and also sheds light on the inherent limitations and challenges that have constrained microalgal biofuel generation to lab-scale. Moreover, it uncovers the innovative strategies, breakthroughs and potential solutions poised to elevate microalgae's role in sustainable energy, marking a pivotal step towards widespread commercial viability and environmental impact. From their extensive lipid accumulation capabilities to their role in environmental remediation, the narrative of this chapter extends far beyond biofuel production, unveiling a narrative of innovation and sustainability.

As the world seeks sustainable energy alternatives, the intricate potentials of microalgae in virtue of biofuels emerge as not just a solution but a transformative force. This chapter stands as a testament to the flourishing possibilities, charting a course towards a more sustainable and eco-friendly energy.

8.2 Generations of Biofuels

Biofuels are hydrocarbon fuels that are produced by using organic matter (can be living or once living material). These can be in the state of solid (manure, wood and dried plant materials), liquid (bioethanol and biodiesel) or gaseous (biogas). The generations of biofuels are divided into four generations based on the raw material used (Fig. 8.1).

The different quantity of oil contents present in different feedstock of all biofuel generations is given in Table 8.1.

8.2.1 First-Generation Biofuels

First-generation biofuels are primarily produced by sugar, starch, animal fats and vegetable oils (Dimitriou et al. 2018). These biofuels include fuels like biodiesel from esterified of immaculate waste vegetables plant oil or waste vegetable oils,

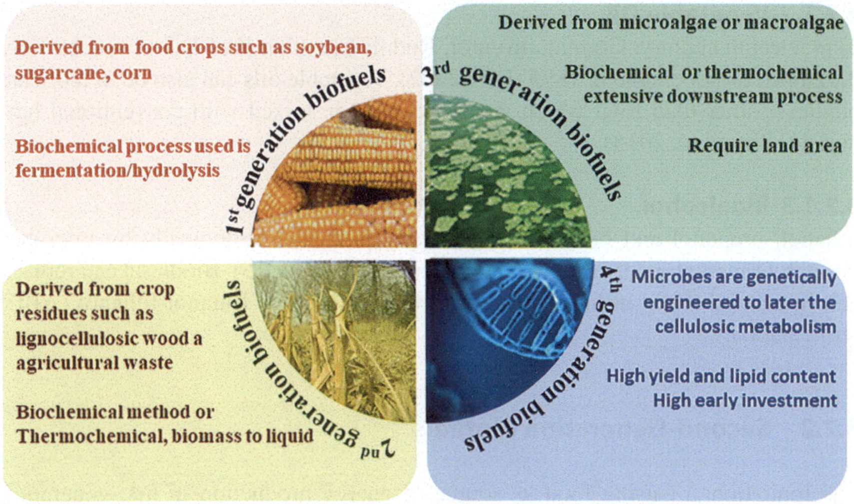

Fig. 8.1 Generations of biofuel with their sources and yielding technologies

Table 8.1 Comparative tabulation of sources and their oil contents different crops and microalgae stating that microalgae being excellent choice for biodiesel production

Crop/Microalgae	Oil content yield (L/acre)	References
Sunflower	386.07	Zheljazkov et al. (2008)
Soybean	181.68	Gouda et al. (2022)
Jatropha	788.33	Muhit et al. (2014)
Corn	68.13	Muhit et al. (2014)
Rapeseed	480.69	Muhit et al. (2014)
Microalgae	19,000–57,000	Bajhaiya et al. (2010), Muhit et al. (2014)

bioethanol from sugar, starch crops and derivative of ethanol (t-butyl ether of ethanol). Three main biofuels regarding this generation are written below (Ullah et al. 2015) First-generation bioalcohols (bioethanol) are produced via yeast (*Saccharomyces cerevisiae*) fermentation of plant sugars and starches obtained from crops such as sugarcane (*Saccharum* sp.), sugar beet and corn (*Zea mays*) (Ebabhi 2012).

8.2.1.1 Biodiesel

Biodiesels are fatty esters commonly classified as monoalkyl esters composed of different variety of vegetable oils (hemp, soybean and canola) and animal fats undergo simple transesterification process. In transesterification process oils are chemically reacted with methanol and sodium hydroxide to form biodiesel and glycerol in the ratio of 10:1 respectively. This biodiesel can be easily used for diesel engine when mixed with conventional diesel in definite composition (Knothe and Razon 2017).

8.2.1.2 Vegetable Oil

In nineteenth century, German inventor, Rudolf Christian Karl Diesel successfully developed diesel engine in 1895 (EPA 2002). Vegetable oils can also be to fabricate biodiesel other than food which works good when mixed with conventional biodiesel (Ullah et al. 2015).

8.2.1.3 Bioalcohol

Ethanol, propanol and butanol are generally produced biologically by microbes through fermentation of cellulose or sugar (Ullah et al. 2015). Biodiesel can replace gasoline and directly utilized in gasoline engine Acetone-Butanol-Ethanol (ABE) fermentation.

8.2.2 Second-Generation Biofuels

The limitations of using food as source of energy production in first-generation biofuels are countered by second-generation biofuel technologies. Other than raw material used as food crops in first-generation the in second-generation biomass produced from higher plants is more efficient and non-competitive for biofuel production (Alalwan et al. 2019). This sustainable approach is fulfilled by production of biofuel breaking down plant matter into crude oil material as more bimolecular components like carbohydrate are present in plant biomass compared to food crops (Chundawat et al. 2011). The genetic modifications are constantly attempted to reduce the impurities (lignin and hemicelluloses) and increasing the yield and reduce the cost in order to compete the cost of fossil fuels without subsidy (Wang et al. 2019).

8.2.2.1 'Plant Waste' Biofuels

Plant biomass used for the production of second-generation biofuel. In order to remove natural deposits from fields, more nitrate manures will need, which will increase nitrous oxide emissions, nitrate overburden and the destruction of biodiversity on land, in freshwater and in the oceans. Additionally, top soil losses are probably to increase (Bhatia et al. 2017). The removal of dead and dying trees from managed forests has caused widespread biodiversity disasters, which have decreased carbon sequestration in forests (Ullah et al. 2015).

However, the preparation of land for food and for common biological communities would be under tremendous strain if millions of hectares of land are planted in permanent crops for bioenergy. The feed for second-generation biofuel can be offered by bio-diverse meadow. It is stated that biomass contains more intensely formed molecule by healthy bio-diverse system. The monocultures are considerably complex to handle in terms of technical hurdles other than multiple feedstocks. To breakdown compound plant material biomass, a mixture of distinctive enzymes is applied to break down which is more complicated than breaking down one particular feedstock (Ullah et al. 2015). Research and development are invested in the favour of genetically modifying monoculture as they result in more yields.

Additionally, the majority of land requirement predictions along with decreasing per-hectare grain yields will put additional strain on the land in order to generate the same quantity of agro fuels. With the aforementioned information in mind, it can be said that there is no proof that large-scale second-generation biofuel would be environmentally beneficial or sustainable (Bhatia et al. 2017).

8.2.2.2 Cellulosic Ethanol

Cellulosic ethanol, derived from nonedible crop wastes like rice hulls, sawdust and citrus peels, offers a sustainable alternative to traditional biofuels. This ethanol is produced by breaking down complex carbohydrates into simple glucose using enzymes, then fermenting it to create ethanol. Lignin, as a by-product, holds potential as a carbon-neutral fuel source (Tarasov et al. 2018). Compared to first-generation biofuels, this second-generation variant emits up to 90% fewer greenhouse gases. Successful demonstrations in Canada generated 700,000 litres of bioethanol annually, inspiring similar projects globally. The Fischer-Tropsch or Biomass-To-Liquids process holds promise for producing bio-synthetic liquid fuels from biomass (Balan et al. 2013).

8.2.3 Third-Generation Biofuels

The worldwide demand of fossil fuels due to ever growing population results in diminishing non-renewable resource (fossil fuels). The demand can be countered by replacing non-renewable fossil fuels to renewable biofuels (Arutyunov and Lisichkin 2017). The first- and second-generation biofuels have some challenges, firstly the first-generation biofuels lead the debate of food versus fuel. When it comes to second-generation biofuels the efficiency of cellulose conversion becomes the drawback. The enzyme can convert ethanol from sugar with an efficiency of 90%, while ethanol conversion from cellulose has only 40% efficiency (Ullah et al. 2015). Apart from this finding huge area of land for the production of feedstock has become more difficult due to the increasing demand for biofuels throughout the world. Developing countries are looking to take advantage of trade and investment possibilities, which may put far greater strain on limited land resources than they would if energy security were the only goal of their policies. There have been demands for the use of third-generation biofuel sources, which need a great deal of fewer acres and can be used for decreasing carbon dioxide emissions into the environment, in order to alleviate the issues frequently connected with land-based biofuel feedstock (Ullah et al. 2015). Specifically, the biofuel made from diatoms, algae and cyanobacteria, together known as Aquatic Microbial Oxygenic Photoautotroph (AMOPS) has been more sustainable resource that could meet the world's fossil fuel needs without affecting the food supply in developing countries (Ye et al. 2020).

Biofuel made from algae seems to have the finest chances as it is the only renewable energy source that might simultaneously supply the world's need for transportation fuels, combat carbon build-up and global warming. This draws interest among many researchers to perform algaculture (farming algae) for producing biofuels (Adeniyi et al. 2018).

8.2.3.1 Algae as Third-Generation Feedstock

Algal biomass is an asset for the bioenergy sector as they are fastest growing and are a crucial source of biomass. There are various kinds of fuel that may be produced with particular species of algae (Adeniyi et al. 2018). Algae do not need fertile soil or food crops to thrive, and they may even be produced in saline water or wastewater. Algae can serve as a more efficient and disadvantage-free alternative to oil-based fuels. Algae are among the organisms with the highest rate of growth, and nearly half of their weight is made up of lipid content. This lipid content can be used for making various kinds of fuels like in transport section (Demirbas and Demirbas 2011).

Algae can be cultivated nearly anywhere and grows 20–30 times faster than food crops while storing up to 30 times more fuel per unit than other biofuel sources including canola, soybean, palm oil, jatropha. According to studies, they could produce approx. 60% of their biomass composed of carbohydrate and lipid content (Ullah et al. 2015; Adeniyi et al. 2018). In fact, when compared to food crops algae could yield up to about 94,000 litres per ha (10,000 gallons per acre) while corn can produce about 560 litres per ha only 960 gallons per acre) annually (Hussian and Ellatif 2018).

Algae-oriented energy is more manageable and stable in comparison to other forms of renewable energy sources like geothermal, wind, solar and tidal energy. The advantage over other forms of energy is the ability to generate more biofuel with less arable land and better water utilization than land-based biomass (Adeniyi et al. 2018). The main focus is shifted to growing microalgae for the large-scale production which are smaller in size (>0.4 mm in diameter), able to do photosynthesis including the diatoms and cyanobacteria. In contrast to macroalgae, such as seaweed fewer researches are going on most likely because of the specific asset's accessibility (Ullah et al. 2015). However, the preference to microalgae is mostly due to their faster growth, simpler structure and high lipid content. There are more than 100,000 different types of algae, each having differing amounts of the three main types of molecules: proteins, carbohydrates and lipids. Algal strains with high carbohydrate and lipid content yield starches that utilized make ethanol, while the remaining proteins could be converted to bioactive compounds and animal fodder (Khan et al. 2018).

8.2.4 Fourth-Generation Biofuels

Third-generation biofuels are promising due to low cultivation costs and high oil production. However, their energy-intensive lipid extraction and drying process often fuelled by fossil fuels, increase greenhouse gas emissions. Moreover, they are not economically competitive with fossil fuels, with prices ranging from $0.44 to $8.76/L (Aron et al. 2020).

To address these limitations, fourth-generation biofuels use genetically modified microalgae, which utilize more carbon dioxide for photosynthesis, act as carbon sinks and enhance biofuel production. Algae strains, like *Phaeodactylum tricornutum, Chlamydomonas reinhardtii* and *Thalassiosira pseudonana*, have been genetically modified to grow faster and adapt to poor nutrient conditions, improving lipid productivity and carbon dioxide sequestration (Gautam et al. 2015). Although fourth-generation biofuels aim to have a low environmental impact, research is still in the early stages, mainly focusing on the environmental effects of genetic modification.

Genetically modified microalgae can be grown in closed or open systems. Closed systems are more secure but costlier, while open systems are at risk of leakage and contaminations. Concerns about genetically modified microalgae include potential environmental issues like changes in the natural environment, toxicity, horizontal gene transfer and competition with native species (Aron et al. 2020).

For example, naturally toxic dinoflagellates can lead to eutrophication and sea surface discoloration. Some studies suggest that the environmental impact of cultivating genetically modified microalgae in open systems may not harm the environment or native species. However, the introduction of genetically modified algae offers customization while introducing uncertainty and potential ecological risks to the ecosystem (Aron et al. 2020).

8.3 Potential of Algae as Third-Generation Biofuel

Microalgae are microscopic (from 1 to 100 μm) eukaryotic photosynthetic organisms can be unicellular or multicellular. Microalgae can survive a variety of environments, including freshwater and saline water (Udayan et al. 2021; Ajala et al. 2021). The advantages of employing microalgae for biofuel production over other biofuel feedstocks include greater growth rates, photosynthetic efficiency, huge surface to volume ratios and reduced land area utilization.

The constraints of the downstream process can be managed due to the simpler structure of microalgae. Microalgal biomass may be used to make a variety of products, including biofuels (biodiesel, biohydrogen, bioethanol and bioelectricity), nutraceuticals, bioactive substances, biopolymers, fodder and more (Khan et al. 2018; Devi et al. 2023). The need for sustainable and alternative fossil fuels is driven by the ever-increasing demand for fossil fuels (diesel, gasoline, natural gas and coal) as well as their depletion (Enamala et al. 2018). Talking about biofuels, the lipid content of microalgal biomass is mostly used to produce biodiesel. In comparison to oil crops, the rate of lipid generation in microalgae is approximately 30% higher. Therefore, the biofuel produced from microalgae is likewise sulphate-free and safe. Utilizing carbon dioxide during growth lowers atmospheric greenhouse gas concentrations (Gautam et al. 2015). According to Falkowski et al. (1988), they effectively use carbon dioxide for photosynthesis and are responsible for around 40% of the world's carbon fixation. Biodiesel, bioethanol, biogas and biohydrogen are the primary biofuels produced by algae. Using different processing techniques, such as transesterification, lipids can subsequently be transformed into biofuels, such as biodiesel.

This supports both environmental and energy sustainability goals by providing a sustainable solution by biofuel generation. Some of the promising algae species for the generation of biofuels are *Chlorella vulgaris, Scenedesmus obliquus, Chlorella pyrenoidosa, Trachydiscus minutus, Scenedesmus rubescens, Tetraselmis suecica and Parachlorella kessleri* (Sarkar et al. 2021). High lipid content may be found in microalgae such as *Chlamydomonas polypyrenoideum, Neochloris oleoabundans, Chlorella protothecoides, Scenedesmus obliqus*, etc. (Arutselvan et al. 2022). Among all the microalgae species *Chlorella* is labelled as the dominant in terms biodiesel production due to its high biomass productivity and lipid production. Algal species with high carbohydrate content include *Dunaliella salina, Anabaena cylindrica, Spirogyra* sp., *Porphyridium cruentum, Scenedesmus dimorphus* and *Prymnesium parvum* (Sharma et al. 2021).

8.4 Cultivation and Harvesting of Algal Biomass

Cultivation of algal biomass is subjected to particular use (for remediation purpose, agricultural, industrial, aquaculture, etc.). The key concept of cultivating and harvesting effective algal biomass in regulated manner depends on temperature, carbon dioxide levels, light source (artificial/natural) and nutrients.

8.4.1 Cultivation Methods of Algae for Production of Biomass

The cultivation of algal biomass can be done in open ponds or closed photobioreactors having different design variations (e.g. tubular, flat plate, airlift, vertical column, flat panel, etc.) (Leite et al. 2013). Nitrogen, phosphorus and potassium are essential nutrients for algal growth. Silica, iron and trace elements are considered crucial marine nutrients. The wastewater with organic and inorganic waste serves as economically suitable for supporting algal growth. The major advantages and limitations of some cultivation systems with tier net energy gain are given in Table 8.2.

8.4.1.1 Photobioreactor (PBRs) Cultivation System

In PBR systems a transparent receiver encloses the cultures. PBRs can have many forms and sizes: fermenter-like containers, flat panels, tubes and plastic bags as well as others. Vertical tubes are a widely used technology because of their high surface to volume ratio, low cost, and relatively low cost (Suali and Sarbatly 2012).

There are several advantages of using PBRs such as high production per unit area, resistance to contamination by herbivores or wild algal strains and parameter controls, such as light intensity, pH and temperature. The PBR may be used in an indoor or outdoor environment with natural light, artificial light or a combination of the two. The use of optical fibres to bring external light energy within is an intriguing variant on lighting systems (Pulz and Scheibenbogen 2006).

Table 8.2 Microalgal cultivation systems with their advantages, limitations and met energy gain outputs

Cultivation systems	Advantages	Limitations	NEG	Reference
Raceway pond (open cultivation system)	Relatively cost effective, easy maintenance, low energy inputs, easy to clean, good for mass cultivation	Large land area is needed, biomass productivity is low, poor circulation of air and CO_2, light penetration is poor, easily get contaminated	26,666,657 (MWh/month)	Jorquera et al. (2010)
Column photobioreactor (closed cultivation system)	High mass transfer, minimum energy use, low chances of contamination, low shear stress Reduced photo-inhibition, high potentials for scalability, controlled culture conditions, good for immobilization of algae, reduced photo-oxidation	Limited illumination area, cost intensive, sophisticated construction Decrease of illumination surface area upon scale-up	–	
Flat plate photobioreactor (closed cultivation system)	High biomass yields, easy sterilization, minimal oxygen accumulation, easily controllable temperature, effective light penetration, cost-effective, low maintenance, well-suited for immobilizing algae, expansive illuminated surface area, appropriate for outdoor cultivation	Small degree of hydrodynamic stress, some degree of wall growth	26,666,650 (MWh/month)	Jorquera et al. (2010)
Tubular photobioreactor (closed cultivation system)	Large illumination surface area, suitable for outdoor cultures, relatively cheap, good biomass productivities	Some degree of wall growth, fouling requires large land space, gradients of pH, dissolved oxygen and CO_2 along the tubes	26,666,299 (MWh/month)	Jorquera et al. (2010)

Any common light source, such as tungsten or fluorescent bulbs, can provide artificial light. Since light emitting diodes (LEDs) limit light to photosynthetic active radiation (PAR) and even allows the study of the effects of various wavelengths and intensities on these microorganisms, their use is increasing due to their low heat generation, lower power consumption and specificity of emitted light wavelength. According to a recent study, the lipid profile, biomass and lipid production may all be significantly impacted by different wavelengths (Benner et al. 2022). According to Das et al. (2011), a strain of *Nannochloropsis* that was isolated locally exhibited distinct lipid profiles, greater growth rate and lipid production when exposed to blue light (470 nm) in comparison to white, red (680 nm) or green light (550 nm).

The primary challenge lies in sourcing affordable sterile carbon dioxide (carbon dioxide). Certain studies have demonstrated the effectiveness of utilizing carbon dioxide emissions from industrial smokestacks for algae cultivation. To enhance economic viability, experts propose locating algae farms for biofuel production near power plants, where they can help absorb pollutants (Qin et al. 2019).

8.4.1.2 Open Ponds Cultivation System

Open pond cultivation encompasses the use of shallow, cost-effective basins that come in various designs, including raceway, circular, inclined and mixed. These ponds are relatively easy to set up, either directly on the ground or with covers to minimize water loss and optimize solar energy utilization while mitigating contamination risks (Suali and Sarbatly 2012; Qin et al. 2019). The use of low-cost, low-energy paddle wheels ensures efficient mixing, supporting aeration and nutrient dispersion.

However, the shallow depth and large surface area make evaporation a significant concern, limiting its suitability to areas with accessible low-cost water sources like marine water or wastewater to address environmental concerns related to potable water usage. The operational and maintenance costs for this system remain relatively low, making it an economical choice for biomass production (Leite et al. 2013). Despite its cost-efficiency, open ponds are not without drawbacks. They are susceptible to rapid algal invasions, which can occur more quickly than in closed systems. Additionally, invasions by fungi, bacteria and algae grazers can lead to contamination, affecting the quality of the biomass produced. In regions with colder climates, limited sunlight can pose temperature maintenance issues, while in hotter areas, elevated temperatures and humid weather may impact the cultivation process (Leite et al. 2013). Nonetheless, open pond cultivation plays a pivotal role in the commercial production of algal biomass, with approximately 98% of production being directed towards high-value nutritional products that fetch significantly higher prices compared to biofuels. This approach allows algae to thrive in marginal lands with saline groundwater, doubling their biomass volume overnight and yielding 15–25 dry tons per hectare annually. Remarkably, around half of an alga's weight is comprised of oil, in stark contrast to oil palm trees, which yield only about 20% of their weight in oil for biofuel production (Qin et al. 2019).

8.4.2 Harvesting Methods of Algae for Production of Biomass

The process of producing microalgal biodiesel is often quite similar to that of producing first-generation biodiesel. The process begins with the production and harvesting of biomass, followed by the extraction of lipids and transesterification to create Fatty Acid Methyl Ester (FAMES) also known as biodiesel. Harvesting microalgal cells can be difficult unlike oil seed plants because it is more difficult to reach the microscopic cells floating in water than it is for macroscopic plants, oil extraction becomes more intricate than the centuries-old pressing method that is often employed for oil seeds (Singh and Patidar 2018). Also, dewatering is required before using biomass from algal cultures since they are extremely diluted typically autotrophic growth and heterotrophic growth is 1% and 10% of culture media (Leite et al. 2013). Harvesting can be done in single or multistep phase, with each phase changing based on the ultimate total solids concentration that is required. Typically, the initial stage yields just green slurry (biomass with water), and more drying could be required. Of course, the final use of the biomass will determine the selection of harvesting technique.

Like in case of nutraceuticals it might need to be harvested using physical methods in order to prevent chemical contamination and preserve the natural composition (Muhammad et al. 2021). For fish feed, continuous centrifugation is the recommended technique for harvesting. On contrary the quality value of harvesting technique will compensate to cost and energy intensity giving an advantage of longer shelf life. But when it comes to low-value, high-demand products like biodiesel, harvesting and dewatering techniques are crucial. Centrifugation and tangential filtration are two examples of energy-intensive harvesting processes that can account for 20–57% of the total cost of biomass and jeopardize the overall net energy ratio (Singh and Patidar 2018). Therefore, one of the biggest challenges in developing a practical method for producing biofuels from microalgae is figuring out how to collect the biomass efficiently and affordably (Singh and Patidar 2018). Centrifugation, filtering, flocculation, biofilm formation and sedimentation are only a few of the possible techniques along with the effectiveness of many of these varies depending on the algal species. Appropriate harvesting practices may vary depending on the cultivation technique (Muhammad et al. 2021). Open pond systems are ideal for producing biofuels for a variety of reasons, as previously mentioned, but they also have a higher probability of contaminations. Therefore, it is likely that species-specific approaches will only work well with cultures cultured in photobioreactors where some degree of species control is feasible.

8.4.2.1 Centrifugation Harvesting Method

Centrifugation is effective for small-scale or lab scale studies but considered energy-intensive for low-value products, impacting the Net Energy Ratio (NER) (Gupta et al. 2019).

However, it is suggested that increasing flow (volumetric throughput) and accepting lower capture efficiency can reduce energy consumption. Dassey and Theegala (2013) found an 82% energy decrease by collecting only 28% of algal biomass, leading to a harvesting cost of $0.864/L oil.

8.4.2.2 Filtration Harvesting Method

Filtration is a potentially efficient harvesting method, primarily suitable for larger microalgae or those that grow in filamentous structures. However, its effectiveness hinges on maintaining a near-homogeneous monoculture of the desired species. For many microalgae, this method faces limitations due to their small size and the presence of extracellular material. These characteristics make microalgae prone to quickly clogging filtration systems in experimental trials, rendering filtration less practical for harvesting algae (Muhammad et al. 2021).

8.4.2.3 Flocculation Harvesting Method

Flocculation is a long-standing method for removing algae and suspended particles from water. By adding certain compounds (flocculants), it causes algae to clump into flocks that can settle naturally. Flock formation depends on mixing speed, particle size and is a physico-chemical process (Hallenbeck 1943). To facilitate microalgae harvest, flocculation agents neutralize the negative charges on cell walls, causing aggregation and settling. Various chemical flocculants have been investigated, with the ideal one being non-toxic, recyclable, inexpensive and efficient at low concentrations. Traditional options like alum and alkali aren't suitable due to cost and toxicity (Singh and Patidar 2018). However, recent research suggests potential cost savings as flocculant requirements vary with cell density. A study proposes using a mixture of calcium and magnesium hydroxides for cost-effective flocculation concentration at less than $10.00 per ton of algal biomass having cell density (6–10^7 cells/mL). Flocculants demand may vary by algal species based on characteristics like cell size and extracellular polysaccharide production. In some cases, ammonia can recover up to 99% of the biomass and be recycled for nitrogen, but its applicability to various algae species remains uncertain (Leite et al. 2013).

8.4.2.4 Bioflocculation Harvesting Method

Bioflocculation is a potential but understudied substitute of conventional flocculation harvesting technique. Kim et al. (2011) reported that certain algal strains may be flocculated by adding a bacterial culture, whereas Olguín (2012) reported that certain strains have the inherent potential to auto-flocculate under certain conditions. This implies that new compounds may be discovered that have the potential to function as flocculants and circumvent some of the drawbacks associated with the chemical flocculants (Singh and Patidar 2018). One such is the recently identified flocculant secreted by *Solibacillus silvestris* cultures, which has been demonstrated to effectively flocculate marine microalgae cultures of *Nannochloropsis oceanica* and is reusable. Similarly, an autofloculating *Scenedesmus* has yielded a bioflocculant (Wan et al. 2013).

8.4.2.5 Biofilm Formation Harvesting Method

Since it is clearly difficult to maintain species that quickly form biofilms as a homogeneous suspension in the culture medium, but it received not much attention in the biofuels research. Nonetheless, a number of recent experiments using two distinct systems have demonstrated that this type of growth mode can provide the simplicity of straightforward mechanical harvesting, resulting in slurries with a dry weight content of 9–16%. In one instance, an open pond system was utilized to grow algae on a rotating drum. The harvesting process was as easy as unspooling and scraping the cotton 'rope' fibre that was employed (Leite et al. 2013). In an alternative method, microalgae were cultivated on a flat surface and water was dripped onto them. When the growth period concluded, the microalgae were collected through straightforward mechanical scraping. Both approaches not only simplified the harvesting process significantly but also achieved substantial biomass production rates with respectable light conversion efficiencies (Singh and Patidar 2018).

8.4.2.6 Sedimentation/Flotation Harvesting Method

In the lack of adequate mixing, several microalgal species exhibit the unusual characteristics of either sedimenting or floating. Again, a high degree of species control throughout culture would be necessary for this strategy to be applicable, even if it might be beneficial in the earliest stages of dewatering and harvesting. Furthermore, these characteristics could make harvesting less expensive, but they might also have a negative effect on the amount of mixing needed since it might be harder to keep algal cells dispersed growth and distribution and throughout culture (Muhammad et al. 2021).

8.5 Biofuel Production Opportunities Using Algal Biomass

Microalgae are recognized as the fastest-growing organisms globally, hold significant promise as a biomass source, particularly for biofuel production. The versatility of algae makes them suitable for various types of biofuels (biodiesel, biohydrogen and bioalcohol) with different species exhibiting specific advantages. Algae cultivation offers several advantages like they can thrive in diverse environments, including sewage or saltwater, without the need for fertile land or food crops. The cultivated biomass is rich in various biomolecules (carbohydrate, lipid and protein) which can be utilized for generation of various biofuels of interest. The energy required for processing algae into biofuel is outweighed by the energy they produce.

Talking about biodiesel most know application and opportunity of microalgal biofuel generation. Biodiesel involves value chain selecting appropriate species based on local environmental conditions, cultivating biomass, extracting oil and processing. The growth of algae is 30 times faster than food crops and the high oil content (about 50% of their weight) make them a highly efficient biodiesel source. Additionally the ability to sequester carbon dioxide during microalgae growth contributes to greenhouse gas reduction. Researches on various biofuels such as biohydrogen, biomethane, biomethane and bioethanol are still going on and are discussed below. Despite of many advantages, the widespread use of algae for biofuel faces challenges, primarily the high cost of cultivation. Large-scale application uncertainties persist, with economic feasibility studies showing varying net revenues. Advantages attributed to algae include their adaptability to various environments, high energy and oil density, quick reproduction and eco-friendliness. Notably, algae are non-toxic, lack sulphur, and are highly biodegradable.

8.5.1 Lipid Extraction and Transesterification for Biodiesel Production

To achieve large-scale biodiesel production in an energy-efficient manner, researchers are actively exploring various conversion techniques. Traditional extraction processes have long relied on hazardous solvents, and the energy-intensive procedures required for their recovery have raised environmental concerns (Kumar et al. 2017). In the interest of both energy conservation and environmental sustainability, avoiding the use of organic solvents in the conventional lipid extraction process from algal biomass has become a priority (Zapata-Boada et al. 2021). In response to this challenge, a sophisticated technique known as solvent switching, as introduced by Boyd et al. (2019), offers a compelling solution. This method allows solvents to be seamlessly interconverted from a polar to a non-polar character by utilizing carbon dioxide, providing a versatile and eco-friendly approach to lipid extraction.

Moreover, the incorporation of environmentally friendly solvents further enhances the sustainability of this process. A notable breakthrough in this field is the demonstration of a wet method for lipid extraction, as showcased by Alassali et al. (2016). This innovative approach eliminates the need for extensive drying of harvested algae before lipid extraction, offering a promising new direction for biodiesel production. Impressively, approximately 80% of the lipids suitable for transesterification can be efficiently obtained from wet algal biomass, even when it retains moisture content as high as 84% (Boyd et al. 2019).

Researchers have explored various techniques aimed at improving lipid extraction yields by facilitating better solvent penetration through cell disruption. Among these techniques, pulsed field electroporation stands out as particularly advantageous due to its relatively low energy requirements, as reported by Alassali et al. (2016).

In the realm of microalgae, these organisms typically produce lipids with carbon chain lengths ranging from 12 to 22 carbons, which can be either saturated or unsaturated. These lipids play diverse roles within microalgae, including contributing to membrane synthesis (polar lipids) and serving as a carbon reserve (neutral lipids). In the context of biodiesel production, there is a preference for saturated fatty acids within the 12–16 carbon range, although the ideal composition may vary depending on specific environmental conditions (Kumar et al. 2017).

In summary, the pursuit of energy-efficient and sustainable biodiesel production has prompted the development of innovative techniques that prioritize eco-friendly practices while maximizing lipid extraction yields from algal biomass (Kumar et al. 2017). Some microalgae reported for biodiesel production are *Chaetomorpha antennina, Gracilaria corticata, Ulva intestinalis* with quantity of 2.1 mL/10 g of biomass, 2 mL/10 g of biomass and 32.3 mg/g of dry weight respectively (Osman et al. 2020).

8.5.2 Biohydrogen Production

The potential of algae as 'microscopic power plants' was initially recognized by Hans Gaffron in 1939. Gaffron observed that, for an unknown reason, algae would occasionally transition from producing oxygen to generating hydrogen, albeit for a brief period. However, it wasn't until 1999 that Professor Tasios Melis and researchers from the National Renewable Energy Laboratory (NREL) made a ground breaking discovery (Melis and Happe 2001). They found that by depriving algae of sulphur and oxygen, it could consistently produce hydrogen over an extended duration. There are three primary methods for extracting hydrogen from algae: biochemical processes, gasification and steam reforming. In biochemical processes, microscopic green algae, such as *Chlamydomonas reinhardtii* or pond scum, play a crucial role. These algae split water molecules with the assistance of enzymes acting as catalysts. Recent advancements in controlling the algae's hydrogen production have sparked interest in scaling up hydrogen production from algae for commercial purposes (Ullah et al. 2015). Gasification of algal biomass is another method, where biomass is converted into a gaseous mixture primarily consisting of hydrogen and carbon monoxide. This process involves applying heat under pressure in the presence of steam and a controlled amount of oxygen. Various techniques are available to separate hydrogen from the resulting syngas (Rostrup-Nielsen and Christiansen 2011).

The third method, steam reformation of methane derived from the fermentation of algal biomass, allows for hydrogen production. Traditional steam reformation techniques can be used to extract hydrogen from methane. Steam reformation is a well-established method employed not only for hydrogen production but also in the industrial synthesis of ammonia. It is recognized as a cost-effective approach (Ullah et al. 2015). This process involves subjecting methane to high temperatures (700–1100 °C) and utilizing a metal-based catalyst, typically nickel. Under these conditions, steam reacts with methane, yielding carbon monoxide and hydrogen. While research on algal hydrogen production has been ongoing for many years, there hasn't been a demonstrated mechanism suitable for scaling up, even at the laboratory level. As a result, this aspect of algal research is not currently being further addressed (Ullah et al. 2015). Some microalgae reported for biohydrogen production are *Laminaria japonica, L. japonica, Laminaria digitate* with amount of *15.8* mL/g of total solids, *23.56* ± 4.5 mL/g, 44.0 ± 1.2 mL/g of volatile solute respectively (Yin et al. 2019; Liu and Wang 2014).

8.5.3 Biogas and Bioethanol

The majority of the early research on producing biofuel from microalgae focused on methane production after the primary use of microalgae, i.e. for wastewater treatment. Although anaerobic digestion of algal biomass is still a possibility, since the first oil shocks in the 1980s, the major focus of algae oil research, and especially biodiesel production, has been on the greater value of liquid transportation fuels derived from microalgae (Ebabhi 2012). Recently, corn and sugarcane have emerged as the most popular commercially feasible sources for producing ethanol fuel. However, as ethanol production grows, so does the demand for corn and sugarcane, raising worries that environmentally sensitive regions may be used again. Additionally, the amount of maize that can be used to make ethanol is significantly reduced, which raises the cost of food for consumers.

The fact that producing ethanol from microalgae has several intrinsic restrictions is emphasized by certain researchers who are investigating the idea of doing so. Either the algae themselves can directly make ethanol through photosynthesis, or else they can store a lot of starch and then metabolize it to produce ethanol. In either scenario, in order to permit its recovery, the ethanol would have to be expelled at very high quantities into the growth medium. Even highly developed genetic engineering methods would face enormous obstacles in overcoming them. A simpler method would be the microalgae's synthesis of starch, followed by yeasts' fermentation, as is done with cane sugar and maize starch to produce fuel ethanol. *Chlorella*, *Scenedesmus*, *Dunaliella*, *Chlamydomonas* and *Tetraselmis* have found to be rich in carbohydrates (above 40% of the dry weight). Various species of genus Chlorella possess high carbohydrate content, for example, *C. vulgaris* has 37–55% of carbohydrates (dry weight) (Dolganyuk et al. 2020). Also, some microalgae reported for biomethane production are *Palmaria palmate, Chaetomorpha antennina, Chaetomorpha antennina* with amount of 320 mL CH_4/g of volatile solute, 0.20 g COD/g COD, 215 mL/g volatile solute respectively (Chai et al. 2021; Tamilarasan et al. 2017) and some microalgae reported for bioethanol production are *Chaetomorpha linum, Saccharina japonica* with amount of 44 g ethanol/100 g glucan 6.65 g/L respectively (Offei et al. 2018; Khambhaty et al. 2012).

8.5.4 Biobutanol

The residual green waste resulting from the extraction of algae oil can be effectively utilized for the synthesis of butanol, a fuel with an energy density comparable to gasoline and superior to both ethanol and methanol. Notably, butanol can be seamlessly employed in most gasoline engines without requiring any engine modifications. Numerous tests have indicated that butanol consumption is on par with gasoline, and when incorporated as a blend with gasoline, it exhibits improved performance and corrosion resistance compared to E85 or ethanol, as reported in research (Keasling et al. 2021).

8.5.5 Jet Fuel

The prospects for the commercial use of jet fuel derived from algae were significantly reinforced when Continental Airlines conducted an inaugural test flight using algae-based biofuel on January 8, 2009. During this test, a twin-engine commercial jet ran on a blend comprising 50% biofuel and 50% conventional aviation fuel. A series of evaluations, including tests performed at an altitude of 38,000 feet (11.6 km) and even a mid-flight engine shutdown, demonstrated that no engine modifications were necessary to accommodate the biofuel. Notably, the biofuel received commendation for its low flash point and the mitigation of freezing point issues that have posed challenges for other biofuels in aviation applications (Keasling et al. 2021).

8.6 Engineering for Scale-Up of Production

The growing research for algal derived products necessitates further in-depth study at large-scale cultivation s methods. Although open pond farming is a common cultivation method, research has indicated that there is a high risk of contamination and little control over growth conditions, particularly when it comes to the influence of outside variables like temperature and sun radiation (Enamala et al. 2018). The open pond culture techniques are less expensive in terms of energy and materials the usage of closed system photobioreactors (PBRs) is preferred since it has control over contamination and growing conditions. Its efficacy is being thoroughly investigated in order to develop batch-PBR for large-scale manufacturing (Benner et al. 2022). The primary constraint encountered in closed batch culture (PBRs) techniques is the extended maturation period of microalgal cells, leading to lengthier processing times and elevated operational expenses. To address these batch operating issues, continuous fed culture techniques can be utilized to substitute above-mentioned issues as it shortens the microalgal cell maturation time and produces high-quality microalgal biomass (Benner et al. 2022).

At large-scale commercial production of bioenergy (biofuels) requires maintaining the biomass productivity yield, which requires careful temperature control between 20 and 30 °C and light intensity levels between 60 and 100 mol/m^2/s. Also, in continuous mode the steady-state operations offer a controlled and precise way of controlling the carbon dioxide supply, culture medium flow rate supply, and light intensities. The parameter adjustments facility when applied to large-scale microalgae biomass production it can decrease the harvesting process downtime (Gautam et al. 2015). Various challenges and limitations for scaling up microalgal cultivation are shown in Fig. 8.2.

The cultivated microalgal biomass needs to harvest using methods such as sedimentation/filtration, centrifugation and flocculation/coagulation. However, these processes need a large amount of energy input (Enamala et al. 2018). On contrary, bio-flocculation method can reduce the energy required to harvest microalgae. Extracellular polymer substances (EPS) can function as bio-flocculation which include proteins and polysaccharides made from microalgae (Alassali et al. 2016). Therefore, in order to provide a high-quality bioenergy production, it is imperative to begin an efficient cultivation procedure. Cost-effective bioenergy processing would come from continuous mode cultivation, which has advantages over fouling and labour saving (Enamala et al. 2018). Moreover, partial harvesting of microalgal biomass during the exponential phase of cultivation may be possible with

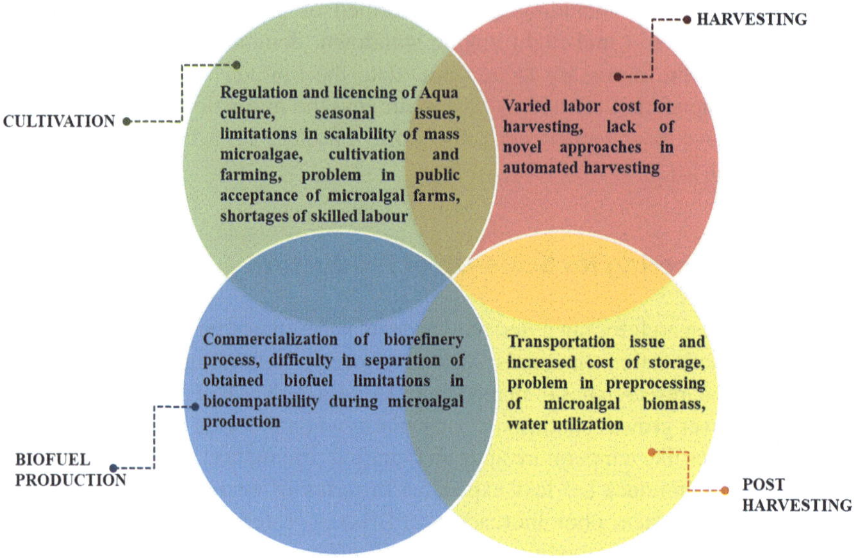

Fig. 8.2 Challenges and limitations in scaling up microalgal cultivation for biofuel generation

continuous culture, reducing the total cultivation time and removing the lag phase while increasing biomass output (Enamala et al. 2018).

The PBR construction cultivation and harvesting plant however needs to take an account on the monetary aspects like installation and maintenance cost, running expenses, microalgal output like lipid content (in case of biodiesel), conversion tax, labour cost, etc. (Benner et al. 2022). The research on effective, economic continuous cultivation and harvesting system should be focussed. Also, the microalgal biomass composition and quality needs to be ensured side-by-side to modify the desired market product and fulfil the requirement of bioenergy.

8.7 Economic Analysis and Challenges

The economic viability of biofuel production hinges on several critical factors, including its design, cost estimation and simulation models. These elements serve as the foundation for research, allowing for adjustments to assess marketability and product potential (Dimitriou et al. 2018). In evaluating the true economic landscape of biofuels, it is essential to move beyond the simplistic comparison of biofuel and fossil fuel prices at the gas pump. A broader perspective is necessary, one that considers the hidden expenses associated with fossil fuel production and consumption (Gupta et al. 2019).

A preliminary examination reveals the existence of numerous concealed costs connected to fossil fuel use, significantly surpassing the price consumers pay at the pump for gasoline or diesel. These hidden costs ultimately need to be covered, either immediately or in the future, often through increased taxation. Direct subsidies to the fossil fuel industry, estimated at around $20 billion (USD) yearly in the United States alone, as reported by Environmental and Energy Study Institute (EESI) in 2011, underscore the complexity of the economic landscape surrounding this issue (Coleman and Dietz 2019). For in-depth economic analysis, the total capital investment (TCI) relies on the complete equipment cost initially. Subsequently, fixed equipment capital costs and variable operating expenses, encompassing feedstock, co-product credits and raw materials, are factored in. Even in operational conditions, variable operating costs, covering expenses for feedstock, co-product credits and raw materials are considered. The quantities of raw materials utilized and co-products generated are determined using the Aspen material balance model (Davis et al. 2014).

When assessing biofuels at market levels, the Minimum Fuel Selling Price (MFSP) becomes crucial to achieve a zero Net Present Value (NPV) with a finite Internal Rate of Return (IRR). This calculation considers a range of expenses and utilizes discounted cash flow analysis (Davis et al. 2014). However, before embarking on large-scale microalgal biofuel production, a comprehensive life cycle assessment (LCA) of the production system is indispensable. This assessment helps identify potential environmental effects, calculates the Net Energy Ratio (NER) and conducts a detailed economic analysis. In reality, the hidden costs associated with fossil fuel utilization far exceed the mentioned figures. In 2005, external effects, excluding those related to climate change, were estimated to cost the United States an astounding $120 billion (Russo et al. 2012; Davis et al. 2014). This estimate doesn't encompass the costs linked to climate change. To gauge the damages, one method is to assess the expenses of adapting to climate change, although this approach falls short of capturing the full costs of mitigation. An initial international study projected these costs to range from $49 to $171 billion annually (USD), with arguments suggesting this is likely an underestimate due to the significant uncertainty surrounding the actual impacts, as noted by Parry et al. (2009).

Thus, conducting a comprehensive economic analysis to ascertain the true competitive cost of biofuels is a formidable task. Precise cost analysis encounters challenges owing to uncertainties in the practical design of algal biodiesel plants. Additionally, determining both the capital and operating costs of Algae Cultivation systems remains a critical challenge. Previous economic studies indicate that even basic open pond systems, including harvesting and algal biomass processing equipment, would cost a minimum of $100,000 per hectare, if not more (Richardson et al. 2014). These costs are compounded by operating expenses. Current commercial algae production is generally regarded as small-scale and inefficient. Still, there is potential for scaling up and leveraging technological advancements to bridge the existing gap. Some recent breakthroughs suggest that profitable oil production costs as low as $0.08–$0.12/pound (about $0.18–$0.26 per kg) may be achievable. The affordability of biofuel production is a critical factor for its sustainability. This cost encompasses feedstock, processing, and capital expenditures. First-generation biofuels, like ethanol and biodiesel, incur production costs of around $0.49 and $0.86/L, respectively, with feedstock, transportation, technology, and labour as significant cost components.

Second-generation biofuels have production costs ranging from $1.15 to $7.85 per gallon equivalent to gasoline, with feedstock and operational expenses playing a substantial role. The capital cost of producing microalgae-based biofuels varies depending on the cultivation method, where open systems are cheaper but demand a significant water supply. Third-generation biofuels are more expensive, with costs ranging from $0.44 to $8.76/L, making them less competitive economically compared to fossil fuels (Davis et al. 2014). Research and development efforts are pivotal in demonstrating the feasibility of mass-culturing algae for optimal oil productivity and cost-effective harvesting.

Despite these formidable challenges, the principal obstacle to the widespread utilization of algae for biofuel production remains the high cost of cultivation (Dimitriou et al. 2018). While there is ample experience in the commercial-scale growth of food-grade algae in various countries, this experience primarily pertains to relatively small open pond systems, frequently not exceeding 5 hectares. Techno-economic feasibility studies have yielded a wide range of net revenue projections, underscoring the uncertainties regarding large-scale application of algae for energy production. The process of cultivating, harvesting and converting algae into fuel and other valuable products in an economically competitive manner is an evolving endeavour. Nevertheless, the potential advantages of algae in the biofuel sector continue to be a subject of great interest and research (Gupta et al. 2019).

8.8 Conclusion and Future Prospectus

In conclusion, microalgae biofuel is facing economic competitiveness challenges compared to earlier generations and offers potential through genetic modification, enabling increased lipid production even in adverse conditions. However, all biofuel generations, including microalgae, still emit greenhouse gases, albeit at reduced levels compared to fossil fuels. The need for substantial energy inputs, often from fossil fuels (responsible for 60% of global GHG emissions and a major driver of global warming), during microalgae processing contradicts Sustainable Development Goals (SDGs) related to greenhouse gas reduction. Given the energy sector's significant role in global GHG emissions, fourth-generation technologies are imperative to enhance GHG uptake by microalgae. First-generation feedstocks are well established but have raised concerns regarding food security due to the diversion of substantial food resources for biofuel production. To align with SDGs like zero hunger and land preservation, second-, third- and fourth-generation feedstocks emerge as promising alternatives, with a particular focus on the latter two to achieve sustainable development goals. Further research and exploration are essential to address the challenges faced by third- and fourth-generation feedstocks.

Microalgae cultivation holds potential in industrial wastewaters and can significantly aid in water purification. These microorganisms play a dominant role in biological treatment, known for their capacity to adsorb toxic heavy metals and utilize nutrients from wastewater, addressing major environmental pollution concerns. Microalgae-based wastewater treatment systems outperform traditional methods in pollutant removal while remaining environmentally friendly. The accumulation of specific bioenergy carrier strains within microalgae promises a future of oil, biodiesel, ethanol, methanol, biogas production, wastewater treatment and GHG reduction.

Efforts should now focus on assessing nutrient consumption rates, exploring diverse wastewater environments to identify indigenous oil-producing microalgal strains and developing sustainable practices for the cultivation of microalgae in wastewater industries. These steps will be essential in realizing the full potential of microalgae as a renewable and sustainable source of biofuel, addressing the challenges and uncertainties in the journey towards fourth-generation biofuels.

Acknowledgments Ms. Anuradha Devi is highly thankful to the 'University Grants Commission' (UGC), New Delhi, India for providing the financial support for her research work.

References

Adeniyi OM, Azimov U, Burluka A (2018) Algae biofuel: current status and future applications. Renew Sust Energ Rev 90:316–335

Ajala EO, Ajala MA, Akinpelu GS, Akubude VC (2021) Cultivation and processing of microalgae for its sustainability as a feedstock for biodiesel production. NJTD 18(4):322–343

Alalwan HA, Alminshid AH, Aljaafari HA (2019) Promising evolution of biofuel generations. Subject review. Renew Energy Focus 28:127–139

Alassali A, Cybulska I, Brudecki GP, Farzanah R, Thomsen MH (2016) Methods for upstream extraction and chemical characterization of secondary metabolites from algae biomass. Adv Tech Biol Med 4(1):2379–1764

Aron NSM, Khoo KS, Chew KW, Show PL, Chen WH, Nguyen THP (2020) Sustainability of the four generations of biofuels–a review. Int J Energy Res 44(12):9266–9282

Arutselvan C, Seenivasan HK, Oscar FL, Ramya G, Chi NTL, Pugazhendhi A, Thajuddin N (2022) Review on wastewater treatment by microalgae in different cultivation systems and its importance in biodiesel production. Fuel 324:124623

Arutyunov VS, Lisichkin GV (2017) Energy resources of the 21st century: problems and forecasts. Can renewable energy sources replace fossil fuels. Russ Chem Rev 86(8):777

Bajhaiya AK, Mandotra SK, Suseela MR, Toppo K, Ranade S (2010) Algal biodiesel: the next generation biofuel for India. Asian J Exp Biol Sci 4:728–739

Balan V, Chiaramonti D, Kumar S (2013) Review of US and EU initiatives toward development, demonstration, and commercialization of lignocellulosic biofuels. Biofuels Bioprod Biorefin 7(6):732–759

Benner P, Meier L, Pfeffer A, Krüger K, Oropeza Vargas JE, Weuster-Botz D (2022) Lab-scale photobioreactor systems: principles, applications, and scalability. Bioprocess Biosyst Eng 45(5):791–813

Bhatia SK, Kim SH, Yoon JJ, Yang YH (2017) Current status and strategies for second generation biofuel production using microbial systems. Energy Convers Manag 148:1142–1156

Borowitzka MA (2012) Energy from microalgae: a short history. In: Algae for biofuels and energy. Springer, Dordrecht, pp 1–15

Boyd BJ, Bergström CA, Vinarov Z, Kuentz M, Brouwers J, Augustijns P et al (2019) Successful oral delivery of poorly water-soluble drugs both depends on the intraluminal behavior of drugs and of appropriate advanced drug delivery systems. Eur J Pharm Sci 137:104967

Chai WS, Bao Y, Jin P, Tang G, Zhou L (2021) A review on ammonia, ammonia-hydrogen and ammonia-methane fuels. Renew Sust Energ Rev 147:111254

Chundawat SP, Beckham GT, Himmel ME, Dale BE (2011) Deconstruction of lignocellulosic biomass to fuels and chemicals. Annu Rev Chem Biomol Eng 2:121–145

Coleman C, Dietz E (2019) FactSheet|fossil fuel subsidies: a closer look at tax breaks and societal costs. LaShier B, Stolark J, Laporte A (eds). The Environmental and Energy Study Institute (EESI)

Das P, Lei W, Aziz SS, Obbard JP (2011) Enhanced algae growth in both phototrophic and mixotrophic culture under blue light. Bioresour Technol 102(4):3883–3887

Dassey AJ, Theegala CS (2013) Harvesting economics and strategies using centrifugation for cost effective separation of microalgae cells for biodiesel applications. Bioresour Technol 128:241–245

Davis R, Kinchin C, Markham J, Tan E, Laurens L, Sexton D et al (2014) Process design and economics for the conversion of algal biomass to biofuels: algal biomass fractionation to lipid-and carbohydrate-derived fuel products (No. NREL/TP-5100-62368). National Renewable Energy Lab (NREL), Golden

Demirbas A, Demirbas MF (2011) Importance of algae oil as a source of biodiesel. Energy Convers Manag 52(1):163–170

Devi A, Verma M, Saratale GD, Saratale RG, Ferreira LFR, Mulla SI, Bharagava RN (2023) Microalgae: a green eco-friendly agents for bioremediation of tannery wastewater with simultaneous production of value-added products. Chemosphere 336:139192

Dimitriou I, Goldingay H, Bridgwater AV (2018) Techno-economic and uncertainty analysis of biomass to liquid (BTL) systems for transport fuel production. Renew Sust Energ Rev 88:160–175

Dolganyuk V, Belova D, Babich O, Prosekov A, Ivanova S, Katserov D et al (2020) Microalgae: a promising source of valuable bioproducts. Biomol Ther 10(8):1153

Ebabhi AM (2012) Bioethanol (bio-fuel) production from agro-waste using yeast isolates from Nigeria. Dissertation, University of Lagos (Nigeria)

Enamala MK, Enamala S, Chavali M, Donepudi J, Yadavalli R, Kolapalli B et al (2018) Production of biofuels from microalgae-a review on cultivation, harvesting, lipid extraction, and numerous applications of microalgae. Renew Sust Energ Rev 94:49–68

EPA (2002) Clean alternative fuels: biodiesel. United States Environmental Protection Agency; EPA 420-F-00-032. http://www.epa.gov

Falkowski PG, Kolber Z, Fujita Y (1988) Effect of redox state on the dynamics photosystem II during steady-state photosynthesis in eucaryotic algae. Biochim Biophys Acta (BBA)-Bioenerg 933(3):432–443

Gautam K, Pareek A, Sharma DK (2015) Exploiting microalgae and macroalgae for production of biofuels and biosequestration of carbon dioxide—a review. Int J Green Energ 12(11):1122–1143

Gouda SP, Jasha H, Kumar P, Dhakshinamoorthy A, Rashid U, Rokhum SL (2022) Microwave-assisted biodiesel production using UiO-66 MOF derived nanocatalyst: process optimization using response surface methodology. Catalysts 12(11):1312

Gupta SK, Dhandayuthapani K, Ansari FA (2019) Techno-economic perspectives of bioremediation of wastewater, dewatering, and biofuel production from microalgae: an overview. In: Phytomanagement of polluted sites, Elsevier, pp 471–499

Hallenbeck LW (1943) A study of the effects of mixing speed on settling rates. Doctoral dissertation, M. Sc. thesis, Civil Engineering, New York University

Hussian M, Ellatif A (2018) The role of microalgae in renewable energy production: challenges and opportunities. In: Marine ecology—biotic and abiotic interactions

Jorquera O, Kiperstok A, Sales EA, Embiruçu M, Ghirardi ML (2010) Comparative energy life-cycle analyses of microalgal biomass production in open ponds and photobioreactors. Bioresour Technol 101(4):1406–1413

Kashyap BK, Saran C, Solanki MK, Divvela PK (2023) Emerging frontiers of microbes as liquid waste recycler. In: Current research trends and applications in waste management. Singapore, Springer Nature Singapore, pp 3–35

Keasling J, Garcia Martin H, Lee TS, Mukhopadhyay A, Singer SW, Sundstrom E (2021) Microbial production of advanced biofuels. Nat Rev Microbiol 19(11):701–715
Khambhaty Y, Mody K, Gandhi MR, Thampy S, Maiti P, Brahmbhatt H et al (2012) Kappaphycus alvarezii as a source of bioethanol. Bioresour Technol 103(1):180–185
Khan MI, Shin JH, Kim JD (2018) The promising future of microalgae: current status, challenges, and optimization of a sustainable and renewable industry for biofuels, feed, and other products. Microb Cell Factories 17(1):1–21
Kim DG, La HJ, Ahn CY, Park YH, Oh HM (2011) Harvest of Scenedesmus sp. with bioflocculant and reuse of culture medium for subsequent high-density cultures. Bioresour Technol 102(3):3163–3168
Knothe G, Razon LF (2017) Biodiesel fuels. Prog Energy Combust Sci 58:36–59
Kumar SJ, Kumar GV, Dash A, Scholz P, Banerjee R (2017) Sustainable green solvents and techniques for lipid extraction from microalgae: a review. Algal Res 21:138–147
Leite GB, Abdelaziz AE, Hallenbeck PC (2013) Algal biofuels: challenges and opportunities. Bioresour Technol 145:134–141
Liu H, Wang G (2014) Fermentative hydrogen production from macro-algae Laminaria japonica using anaerobic mixed bacteria. Int J Hydrog Energy 39(17):9012–9017
Melis A, Happe T (2001) Hydrogen production. Green algae as a source of energy. Plant Physiol 127(3):740–748
Muhammad G, Alam MA, Mofijur M, Jahirul MI, Lv Y, Xiong W et al (2021) Modern developmental aspects in the field of economical harvesting and biodiesel production from microalgae biomass. Renew Sust Energ Rev 135:110209
Muhit IB, Baidya D, Nahid N (2014) Prospect of algal biodiesel production in Bangladesh: overview from developed countries. IOSR-JMCE 11(1):49–54
Offei F, Mensah M, Thygesen A, Kemausuor F (2018) Seaweed bioethanol production: a process selection review on hydrolysis and fermentation. Fermentation 4(4):99
Olguín EJ (2012) Dual purpose microalgae–bacteria-based systems that treat wastewater and produce biodiesel and chemical products within a biorefinery. Biotechnol Adv 30(5):1031–1046
Orejuela-Escobar L, Gualle A, Ochoa-Herrera V, Philippidis GP (2021) Prospects of microalgae for biomaterial production and environmental applications at biorefineries. Sustain For 13(6):3063
Osman ME, Abo-Shady AM, Elshobary ME, Abd El-Ghafar MO, Abomohra AEF (2020) Screening of seaweeds for sustainable biofuel recovery through sequential biodiesel and bioethanol production. Environ Sci Pollut Res 27:32481–32493
Pahl G (2008) Biodiesel: growing a new energy economy. Chelsea Green Publishing
Parry M, Arnell N, Berry P, Dodman D, Fankhauser S, Hope C et al (2009) Adaptation to climate change: assessing the costs. Environment 51(6):29–36
Pulz O, Scheibenbogen K (2006) Photobioreactors: design and performance with respect to light energy input. In: Bioprocess and algae reactor technology, apoptosis. Springer, pp 123–152
Punia H, Tokas J, Malik A, Kumar N (2021) Enzymes as nanoadditives: a promising alternative for biofuel production. In: Nanomaterials. Academic Press, pp 631–662
Qin L, Alam MA, Wang Z (2019) Open pond culture systems and photobioreactors for microalgal biofuel production. In: Microalgae biotechnology for development of biofuel and wastewater treatment. Springer, pp 45–74
Richardson JW, Johnson MD, Zhang X, Zemke P, Chen W, Hu Q (2014) A financial assessment of two alternative cultivation systems and their contributions to algae biofuel economic viability. Algal Res 4:96–104
Rostrup-Nielsen J, Christiansen LJ (2011) Concepts in syngas manufacture, vol 10. World Scientific
Russo D, Dassisti M, Lawlor V, Olabi AG (2012) State of the art of biofuels from pure plant oil. Renew Sust Energ Rev 16(6):4056–4070
Saran C, Purchase D, Saratale GD, Saratale RG, Ferreira LFR, Bilal M et al (2023) Microbial fuel cell: A green eco-friendly agent for tannery wastewater treatment and simultaneous bioelectricity/power generation. Chemosphere 312:137072

Sarkar RD, Singh HB, Kalita MC (2021) Enhanced lipid accumulation in microalgae through nanoparticle-mediated approach, for biodiesel production: a mini-review. Heliyon 7(9):e08057
Sharma N, Jaiswal KK, Kumar V, Vlaskin MS, Nanda M, Rautela I et al (2021) Effect of catalyst and temperature on the quality and productivity of HTL bio-oil from microalgae: a review. Renew Energy 174:810–822
Singh G, Patidar SK (2018) Microalgae harvesting techniques: a review. J Environ Manag 217:499–508
Suali E, Sarbatly R (2012) Conversion of microalgae to biofuel. Renew Sust Energ Rev 16(6):4316–4342
Taher H, Al-Zuhair S, Al-Marzouqi AH, Haik Y, Farid MM (2011) A review of enzymatic transesterification of microalgal oil-based biodiesel using supercritical technology. Enzyme Res 2011:468292
Tamilarasan K, Kavitha S, Banu JR, Arulazhagan P, Yeom IT (2017) Energy-efficient methane production from macroalgal biomass through chemo disperser liquefaction. Bioresour Technol 228:156–163
Tarasov D, Leitch M, Fatehi P (2018) Lignin–carbohydrate complexes: properties, applications, analyses, and methods of extraction: a review. Biotechnol Biofuels 11(1):1–28
Udayan A, Pandey AK, Sharma P, Sreekumar N, Kumar S (2021) Emerging industrial applications of microalgae: challenges and future perspectives. SMAB 1(4):411–431
Ullah K, Ahmad M, Sharma VK, Lu P, Harvey A, Zafar M, Sultana S (2015) Assessing the potential of algal biomass opportunities for bioenergy industry: a review. Fuel 143:414–423
Wan C, Zhao XQ, Guo SL, Alam MA, Bai FW (2013) Bioflocculant production from Solibacillus silvestris W01 and its application in cost-effective harvest of marine microalga Nannochloropsis oceanica by flocculation. Bioresour Technol 135:207–212
Wang H, Pu Y, Ragauskas A, Yang B (2019) From lignin to valuable products–strategies, challenges, and prospects. Bioresour Technol 271:449–461
Ye W, Xia A, Chen C, Liao Q, Huang Y, Zhu X, Zhu X (2020) Sustainable carbon capture via halophilic and alkaliphilic cyanobacteria: the role of light and bicarbonate. BRJ 7(3):1195–1204
Yin Y, Hu J, Wang J (2019) Fermentative hydrogen production from macroalgae Laminaria japonica pretreated by microwave irradiation. Int J Hydrog Energy 44(21):10398–10406
Zabermawi NM, Alsulaimany FA, El-Saadony MT, El-Tarabily KA (2022) New eco-friendly trends to produce biofuel and bioenergy from microorganisms: an updated review. Saudi J Biol Sci
Zapata-Boada S, Gonzalez-Miquel M, Jobson M, Cuellar-Franca RM (2021) Integrating techno-economic, environmental, and safety criteria in solvent screening for extraction processes: the case of algae lipid extraction. ACS Sustain Chem Eng 10(1):472–485
Zheljazkov VD, Vick BA, Ebelhar MW, Buehring N, Baldwin BS, Astatkie T, Miller JF (2008) Yield, oil content, and composition of sunflower grown at multiple locations in Mississippi. Agron J 100(3):635–642

Bioenergy Generation and Cogeneration from Algae-Based Biorefineries

9

Christina Saran, Anuradha Devi, Ganesh Dattatraya Saratale, Luiz Fernando R. Ferreira, and Ram Naresh Bharagava

Contents

C. Saran · A. Devi · R. N. Bharagava (✉)
Department of Environmental Microbiology (DEM), School of Earth and Environmental Sciences (SEES), Babasaheb Bhimrao Ambedkar University, Lucknow, Uttar Pradesh, India

G. D. Saratale
Department of Food Science and Biotechnology, Dongguk University, Seoul, Republic of Korea

L. F. R. Ferreira
Postgraduate Program in Genomic Sciences and Biotechnology, Catholic University of Brasília, Brasília, Brazil

S. K. Arya et al. (eds.), *Value Added Products From Bioalgae Based Biorefineries: Opportunities and Challenges*,
https://doi.org/10.1007/978-981-97-1662-3_9

Abstract

Algae are a newly recognised renewable energy source derived from living creatures. They are very remarkable energy and are attracting interest due to their potential to increase their market share in the biofuel industry. Algal-based biorefineries have drawn a lot of attention as a feedstock for biofuels; but, to meet the financial requirements of an algal biorefinery, high-value co-products from the algae fraction must be produced. Additionally, algal-based wastewater treatment is thought to be a lower energy alternative to traditional wastewater treatment, making it a good fit for procedures used in algae biorefineries. The food, medicine and beauty sectors may make use of the many biomolecules found in algae, including proteins, carbohydrates, lipids, pigments, vitamins and antioxidants. Researchers from all across the world are focusing their attention on these aspects. As of their high biomass output, low cultivation area requirements and ability to produce important metabolites, algae can be a good alternative to other renewable resources. In this chapter we focus primarily on the potentials, qualities and implication of algae as a sustainable renewable resource.

Keywords

Algae · Biorefinery · Bioenergy · Cogeneration · Renewable resources

Abbreviations

TAG	Triglyceride
GHG	Greenhouse gas
CN	Cetane number
SHF	Separated hydrolysis and fermentation
SSF	Solid state fermentation
CO_2	Carbon dioxide
CH_4	Methane
C	Carbon
N	Nitrogen
P	Phosphorus
VS	Volatile solids
COD	Chemical oxygen demand
PBRs	Photo bioreactors
L	Litre
mL	Millilitre
Kg	Kilogram
g	Gram
m^3	Cubic meters

9.1 Introduction

The world's environmental problems have gotten worse due to the population's fast rise. Global warming and ecological imbalance are caused by major forms of environmental pollution, including contamination of the air, land and water. Significant environmental harm is caused by a variety of harmful compounds that contribute to pollution, such as heavy metals, chemical fertilisers, nuclear waste, hydrocarbons, pharmaceutical and pesticides by-products (Ahmad et al. 2016). Science faces a great task in solving this issue, which calls for in-depth understanding and innovative technical solutions. Numerous physicochemical and biological treatment techniques had remained created to eliminate or mitigate pollutants from impacted areas. Osmosis, evaporation, membrane filtration, precipitation, electrochemical treatment and ion exchange are distinctive physicochemical remediation techniques for wastewater treatment (Alazaiza et al. 2022). Nevertheless, the majority of these techniques are not ecologically sustainable nor financially viable.

In recent times, biological wastewater treatment had gained popularity over chemical treatments. The most effectual kind of wastewater treatment is chemical; however, the effluent contains synthetic chemicals that harm the environment and living things in many ways. Biological wastewater treatment is therefore the utmost eco-friendly choice (Kashyap et al. 2023). One of the most promising methods is biological remediation, or bioremediation, because of its long-term efficacy, cheap cost and ecological friendliness. Microorganisms are used in bioremediation to detoxify or lower concentrations of inorganic or organic contaminants in the surrounding water. To eliminate environmental pollutants, bioremediation has employed a variety of microbes, such as bacteria, fungus and algae (Shahid et al. 2020).

Specifically, bioremediation happens via two primary processes: biosorption and bioaccumulation. Through their cell walls, living things absorb the toxins through a process known as bioaccumulation, which is an active kind of bioremediation. Meanwhile, utilising inert biomass or live algae, biosorption is a novel method for removing heavy metal ions from acidic liquids. On the other hand, physicochemical processes carried out by living or dead microbial cells are involved in passive bioremediation. Unconventional materials that can be employed as affordable, dependable and secure wastewater treatment instruments have been found as a result of the investigation of resources that can be utilised to lower the levels of heavy metals in environment. In light of this, biological materials have become more affordable and ecologically benign than their predecessors. The usage of algae as biosorbents had been emphasised due to their versatility and capacity to withstand contaminated water (Alazaiza et al. 2022).

Despite a prodigious deal of study on the usage of different microorganisms to bioremediate contaminated settings, algae-based bioremediation technology has drawn a lot of interest as a practical method because of its benefits. Microalgae-based bioremediation may eliminate contaminants, and the biomass generated during the procedure can be processed into additional bioproducts with added value (Kalra et al. 2021). The most promising approach for treating industrial wastewater over the long run is the biological one. Additionally, it could help to advance the creation of supplementary circular economic methods.

The idea of algae as a feedstock for biofuel generation has gained a lot of interest because of the devastating impacts of using fossil fuels due to global warming, as well as the growing costs of oil and the fast reduction of natural oil sources. Future generations will not have access to fossil fuel resources. Renewable energy resources, which emit little to no carbon dioxide (CO_2) and can supply a sizable amount of the world's energy demands at a price comparable to present petroleum prices, can meet a sizable fraction of the world's energy needs (Pahunang et al. 2021). Compared to other feedstocks, microalgae provide more economic and environmental benefits. There is also a sufficient supply of microalgae to fulfil growing demand with no negative environmental effects. Biodiesel, bioethanol and biohydrogen have previously been produced using microalgae (Yap et al. 2021). In many industries, but particularly in transportation, alternative fuels have gained significant importance. The public is interested in biofuels made from algae as a possible remedy for all of these issues, both in terms of production and consumption. Numerous studies had been conducted on the advantages of using algal biomass as a source of biofuel and to produce bioproducts with added value (Saravanan et al. 2021; Jiang et al. 2021).

The emergence of algae-based biorefineries holds hope for the escalating environmental issues. This budding field not only addresses resource depletion and pollution issues but also offers a strong avenue for sustainable energy generation. Within this background, the pivotal role of algae in biofuel generation and bioremediation cannot be overstated. While previous research has illuminated the potential of algae in mitigating contaminants and yielding valuable products, our exploration probes deeper. Our work seeks to loosen the intricacies of harnessing the unique properties of algae, anticipating their multifaceted role in bioenergy generation and cogeneration from biorefineries. Through a fusion of innovative techniques and comprehensive understanding, we aim to elucidate the untapped potential of microalgae, posing specific inquiries about their efficiency, scalability and ecological impact.

In this pursuit, this chapter endeavours to carve a path towards sustainable energy solutions. The aim of this chapter is to explore the advancements in the development of environmentally friendly technologies by analysing their abilities in bioremediation and biofuel generation. Also, the concept of biorefineries based on algae is transforming the energy sector and restoring environmental equilibrium.

9.2 Overview of Algal-Based Biorefineries

The justifiable conversion of biomass in a variety of commercial products that is, materials, chemicals, feed, food and energy is known as biorefinery. Biorefining is a technology that is expanding globally. Implementing biotechnological techniques and procedures, biorefineries had the possibility to manufacture a variety of goods. As an example, new applications of biotechnology include the modification of algae to generate fatty acids for petrochemicals, alternative fuels and detergents (Poggi-Varaldo et al. 2014).

A biorefinery it is a process in which it turns a single biomass into many products. Most often, microalgae are used as feedstocks for producing biofuels (Saran et al. 2023). Algae, however, also contain a variety of valuable elements that could be processed or turned into high-value products in addition to biofuels (Sivaramakrishnan et al. 2022).

According to Ferreira (2017) a biorefinery is a unified bio-based industry that uses a range of technologies to process biomass sustainably and produce a range of marketable products which includes: biofuels, chemicals food and feed ingredients, fibres, biomaterials, power and heat. The goal is to maximise value-added along with the three pillars of sustainability: the environment, the economy and society.

9.2.1 Algae Used as Biofuel Feedstock

Triglyceride (TAG) concentration, which makes up to 70% of lipid content, and biomass productivity are key factors in manufacture of biodiesel, one kind of biofuel, from algae (Bigogno et al. 2002). The majority of research published is related to lipid rather than TAG as the feedstock for the manufacture of biodiesel, most likely due to more TAG in lipid content. According to the transesterification hypothesis, mass of the feedstock and the amount of biodiesel generated in 1:1 ratio; this indicates that the smaller the feedstock mass, the less biodiesel is produced. According to a review study by Sun et al. (2011), the cost of producing algal oil ranges from $10.87 to $13.32 per gallon (in US dollars), whereas the cost of producing triglycerides is $3 per gallon. Genetics is one aspect that influences the production of algae. As a result, selecting the right algae for production is crucial.

TAG that develops an alga for storing energy is represented as lipids, which make up a portion of biomass. TAG, which is essentially glycerol esterified with three fatty acids, combines with alcohol to produce biodiesel, with glycerol remaining as a by-product. Nonetheless, biomass output offers an indicator of how well photosynthesis and agricultural processes are working. Thus, the cost of producing biodiesel is influenced by the biomass yield. Additionally, the eco-design of an effective and sustainable production chain may be supported by choosing a microalgal species that is suited for culture while maintaining the growth rates and high lipid contents (Suali and Sarbatly 2012).

The method of cultivating algae is the initial step in turning them into biofuel. Algae could be full-grown in an open system like a raceway pond and closed system like in a photobioreactor. After that, microalgal cells are extracted by filtering or centrifugation with the help of flocculants. The gathered biomass is either biochemically transformed into biodiesel or processed to extract its oil. Algal biomass has a heating value of 24 MJ/kg (Clarens et al. 2010), which makes it appropriate for generating syngas, bio-oil fuel and other products.

Due to the consequences of large-scale production, selecting the right biomass and end products is essential to the design of a biorefinery. It is necessary to consider the features, possible applications and initial availability of biomass. Many solutions are accessible, and each has benefits and drawbacks. The biorefinery can process a variety of biomass materials, such as short-rotation woody crops, wood wastes, agricultural wastes, bagasse, sawdust, waste paper, organic residues aquatic plants and algae and animal wastes (Sun et al. 2011).

Any biomass conversion process must have precise and thorough classification of its feedstocks, intermediates, and end-products in order to comprehend the interactions between the different reaction products and components of biomass at every stage of the process. The biorefineries can be categorised as marine or lignocellulosic based on the biomass feedstock.

9.2.2 Concept of Biorefineries

The word 'biorefinery' comes from renewable biomass which serves as the raw material feedstock as well as the conversion procedures that are frequently used to treat and refine raw materials. The biorefinery strategy comprises multiple-step procedures, the first of which often involves pretreatment (biomass treatment for subsequent processing) after feedstock selection. Biomass is treated chemically and biologically (Conteratto et al. 2021). The phrase 'biorefinery' has several meanings, all leading to the same conclusion. A straightforward biorefinery model for the National Renewable Energy Laboratory (NREL) has been developed. It is based on three distinct platforms: biochemical, thermochemical and microbiological platforms to support various product paths (Bose et al. 2022).

According to Juan et al. (2011), the biorefinery idea seen as viable way to reduce greenhouse gas emissions (GHGs). Additionally, the biorefinery idea minimises the accumulation of biomass waste and facilitates the maximum utilisation of algae. The process of combining several fractions into the required single fraction is the main bottleneck in the microalgae biorefinery. This may be fixed by developing processes in an economical way with algae as the promising organisms to generate high-value chemicals together with biofuels (Sivaramakrishnan et al. 2019).

A biorefinery is an establishment that combines machinery and processes for converting biomass into fuels, electricity and chemicals with value added. The modern petroleum refinery, which turns petroleum into a variety of fuels and goods, is comparable to the biorefinery. A biorefinery maximises the value received after the biomass feedstock by maximising the numerous mechanisms in biomass and their mediation by creating several products. One or more low-volume, high-value chemical compounds and a low-value, high-volume liquid transportation fuel, i.e. bioethanol or biodiesel, might be produced by a biorefinery. It has the capacity to produce heat and power simultaneously, potentially enough for it to sell to the nearby utility in addition to using it for its own usage. According to (Biernat and Grzelak 2015) high-value goods boost profitability, fuel helps satisfy energy demands and power production lowers energy prices and lowers GHGs from conventional power plant facilities.

The biorefinery's goal is to maximise profits and benefits by minimising waste and optimising resource usage. Entire-scale, highly effective, integrated biorefineries enable the production of high-value products at a reasonable price. Adaptability is essential, it is necessary to define flexibility as the ability to pick from a variety of processes and, within a particular process, the ability to choose the best technology and operating conditions based on the exclusive properties of biomass stream is available. This makes it possible to produce a wide range of worthwhile, commercially viable goods (De Jong and Jungmeier 2015).

9.3 Biofuel Generation from Algae

Renewable energy produced from natural or biological sources is called bioenergy. One common term for bioenergy is 'sustainable and renewable energy'. It has lately grown in importance as a field of study for scientists worldwide. According to Hariz and Takriff (2017), bioenergy is a fantastic short- to medium-term way for reducing global warming and generate sustainable energy. Algae have drawn a lot of interest in the last decade for their potential as a feedstock for biofuel. When compared to fossil fuels, algae biofuels effectively lower CO_2 emissions, safeguarding the environment through preserving equilibrium between CO_2 consumption and production. Algae take up CO_2 released during the burning of biodiesel. It has been demonstrated that burning biofuels releases less CO_2 than burning fossil fuels (Ahmad et al. 2022).

Biofuels generated from biomass offer a number of benefits, such as their ability to regenerate and their little effect on pollution and global warming. Compared to fossil fuels, algal biofuels have a number of benefits, including the following: (a) they can be easily obtained using common algal biomass; (b) when burned, they subsidise to the CO_2 cycle; (c) environmentally friendly in nature; (d) they are advantageous to consumers, the economy and the environment; and (e) they support long-term sustainability and they are biodegradable (Ahmad et al. 2022). Depending as the feedstocks utilised, viability of production and the state of technical advancement, they are divided into four generations (Alalwan et al. 2019).

Since conventional crops are key source of human food, the quantity of feedstock available for first-generation biofuels made from them is restricted (Rodionova et al. 2017).

Land-based food production is still in competition with second-generation biofuels are product of agricultural waste or by-products (Chowdhury and Loganathan 2019). Because they provide a substantial contribution to first- and second-generation biofuels, which are generated using edible and non-edible resources, respectively, biofuels from algae and microalgae are classified as third-generation biofuels. According to Hussain et al. (2021), the principal biofuels produced from algae are bioethanol, biodiesel, biohydrogen and biogas. Utilising genetically modified microorganisms, fourth-generation biofuels are made from bacteria, fungi, yeast and microalgae. There have been several attempts to boost potential production of biofuels derived using algae (Chowdhury and Loganathan 2019) which is shown Table 9.1.

Table 9.1 Biofuel generation from different algae species

Type of biofuel	Algae species	Conversion techniques	Conversion condition	Biofuel yield	References
Biodiesel	*Ulva fasciata*	Catalytic transesterification	Molar ratio of methanol: oil, 9:1, time: 6 h, temperature: 80–100 °C	88%	Abomohra et al. (2018)
	Chaetomorpha antennina	Transesterification	Chloroform: ethanol solvent 1:20 (w/v)	2.1 mL/10 $g_{biomass}$	Atadashi et al. (2010)
	Gracilaria corticata	Transesterification	Hexane: ether solvent 1:20 (w/v)	2 mL/10 $g_{biomass}$	
	Enteromorpha compressa	Base transesterification	Base: 1%NaOH, methanol: oil ratio 9:1, temperature: 60 °C, time: 70 min	90.6%	Kraan (2013)
Bioethanol	*Chaetomorpha linum*	Wet oxidation method	Temperature: 200 °C	44 g ethanol/100 g glucan	Kumar et al. (2021); Schultz-Jensen et al. (2013)
	Saccharina japonica	Low acid pretreatment	Acid: 0.06% (w/w) sulphuric acid temperature: 170 °C, time: 15 min	6.65 g/L	Offei et al. (2018)
	Saccharina japonica	Thermal acid hydrolysis	Acid: 40 mM H_2SO_4, temperature: 121 °C, time: 60 min	7.7 g/L	Kumar et al. (2021)
	Laminaria digitata	Oven drying	Temperature: 70°C, time: 72 h	13.6 ± 0.2 μL/gDS	Jang et al. (2012)
	Ulva linza	Mild acid hydrolysis	Acid condition: 3% H_2SO_4	12.01%	Palupi et al. (2022)

(continued)

Table 9.1 (continued)

Type of biofuel	Algae species	Conversion techniques	Conversion condition	Biofuel yield	References
Biohydrogen	*Laminaria japonica*	Microwave	Temperature: 160 °C, time: 30 min	15.8 mL/g TS	Robak and Balcerek (2018)
	Laminaria japonica	Ultrasonic	Sonication frequency: 20 kHz	23.56 ± 4.5 mL/g	Wu et al. (2022)
	Laminaria digitate	Hydrothermal	Temperature: 140 °C, time: 20 min	44.0 ± 1.2 mL/g VS	Yin et al. (2019)
	Chaetomorpha antennina	Surfactant-aided microwave pretreatment	Microwave power: 0.36 kW, ammonium dodecyl sulphate: 0.0035	74.5 mL/g COD	Kumar et al. (2019)
	Ulva reticulata	Microwave-H_2O_2 alkali pretreatment	Microwave power: 0.36 kW, H_2O_2 dosage:24 mg H_2O_2/g biomass, pH: 10	87.5 mL H_2/g COD	Kumar et al. (2020)
Biomethane	*Palmaria palmata*	Anaerobic digestion	Semi continuous anaerobic digestion	320 mL CH4/ gVS	Tamilarasan et al. (2019)
	Chaetomorpha antennina	Ozone disperser pretreatment	Disperser gravitational force:1613 g, treatment time: 30 min, ozone dosage: 0.00049 g O_3/g TS	0.20 g COD/g COD	Tamilarasan et al. (2017)
		Thermo-chemo disperser	Disperser gravitational force: 1613 g, temperature: 80°C, NaOH:1 N, pH: 11	215 mL/g VS	
	Laminaria digitata	Heat	Temperature: 104 °C	Dried biomass: 97.66 m^3 CH_4 Fresh biomass: 67.24 m^3 CH_4	Ding et al. (2020)

9.3.1 Biomass Conversion Technologies

Biobased goods are very important to many sectors, but before the industry can be successfully commercialised on a big scale, there are still a number of technological, strategic and commercial obstacles that need to be resolved. Therefore, every planned biorefinery should be concerned with the development and use of the best technologies available for all processes, as well as for pretreatment and storage (Alalwan et al. 2019). Algal biofuels can be produced through a number of processes, including (a) physico-chemical conversion, which involves transesterification or esterification for the producing biodiesel; (b) biochemical conversion, which involves fermentation process to produce bioethanol and anaerobic digestion for production of biomethane/biogas; and (c) thermochemical conversion, which involves gasification, pyrolysis and combustion for the production of biohydrogen and bioenergy as shown Fig. 9.1.

9.3.1.1 Biodiesel Production

Biodiesel can substitute petroleum diesel as a sustainable fuel made from waste lipids and renewable biomass. Algal biomass's lipid content has an impact on the fuel's generation, quality and acceptability in place of diesel derived from petroleum (Devi et al. 2023). Additionally, biodiesel made from algae biomass has reduced emissions of GHGs and particulate matter and is sulphur-free, just as diesel derived from petroleum (Ahmad et al. 2022). However, algal biodiesel's poor performance at low temperatures limits its oxidation stability. Milano et al. 2016, looked at the process of turning microalgal strains with the high oleic acid concentrations in their fatty acids into biodiesel. Microalgae including *Nannochloropsis oculata, Chlorella* sp., *Scenedesmus* sp., *Botryococcus* sp. and *Picochlorum* sp. are used to make biodiesel. More oleic acid is present in *Saccharomyces cerevisiae*, which enhances the oxidative stability of biodiesel (Milano et al. 2016; Devi et al. 2023).

Through a process called transesterification, microalgae are the primary source of biodiesel. Glycerol and biodiesel are produced when lipids, particularly triacyl glycerides, are transesterified in the existence of an alkali or an acid. A critical step in the manufacturing biodiesel is choosing the right microalgae strain (Chye et al. 2018). The calorific value, melting temperature, viscosity and cetane number (CN) all affect how well biodiesel performs in engines. One suggested technique to raise the quality of biodiesel is the bioengineering of algal strains (Nigam and Singh 2011).

Because of its higher viscosity than diesel oil, crude oil derived from algae is not appropriate for direct engine usage. Transesterification is a chemical procedure that is needed to improve the fluidity and reduce the viscosity of algal oil. Any reaction between an alcohol and a fatty acid (or organic acid) that produces an ester is known as esterification. On the other hand, transesterification is replacing the alkoxy group in an ester with alcohol to generate fatty acid-alcohol esters. Diglycerides, monoglycerides and glycerol are the by-products of the breakdown of triglycerides. Short, straight-chain esters are produced during the transesterification of big, branching triglyceride molecules (Devi et al. 2022, 2023).

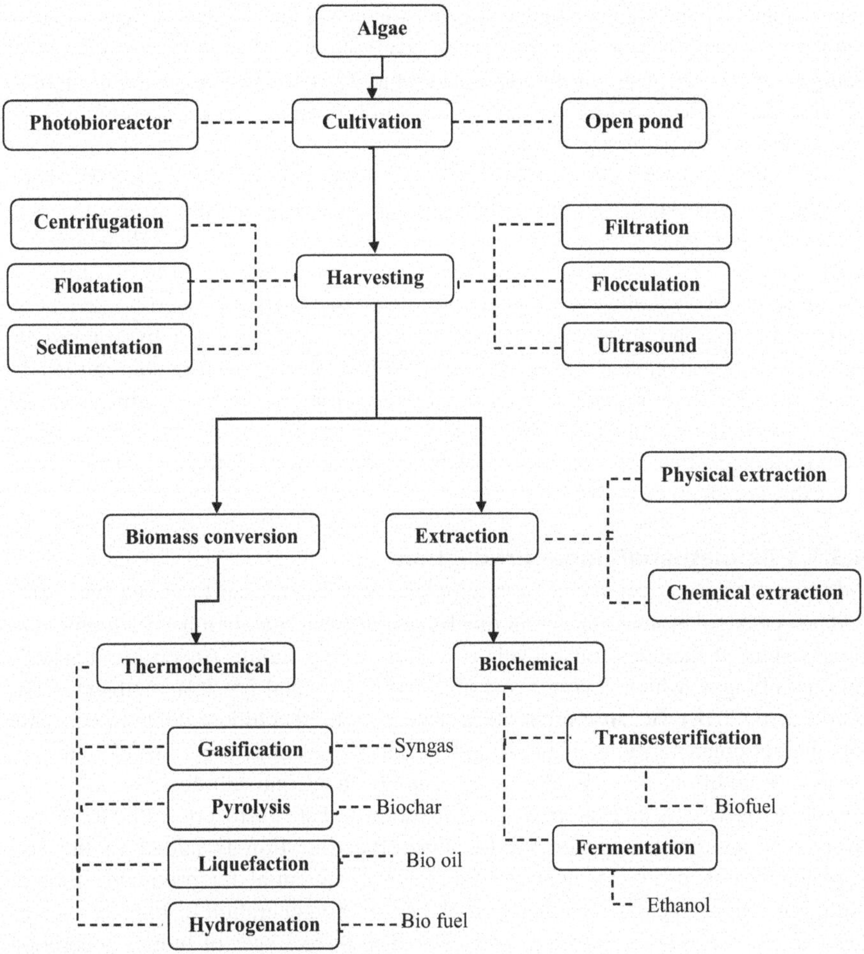

Fig. 9.1 Summarisation of algae conversion into biofuel

9.3.1.2 Bioethanol Productions

The most efficient biofuel that can presently be made from third-generation microalgal biomass is bioethanol. The manufacture of bioethanol using microalgae biomass has many benefits, two of which are that it does not need large tracts of arable land and that it helps environment by removing CO_2 from the atmosphere. Microalgal cell walls are rich in lipids and carbohydrates, including starch, xylans, mannans, cellulose and sulphated glycans. These complex constituents are reduced enzymatically or chemically to simple sugars, which are then anaerobically transformed into bioethanol (Constantino et al. 2021). The method of producing bioethanol from microalgae involves many steps, such as selecting and growing the algal biomass, pretreatment of algal biomass produces, liquefaction, saccharification, anaerobic fermentation and distillation to purify the bioethanol (Kim et al. 2017).

Pure ethanol is an effective fuel substitute for petrol since it has a higher-octane number and vaporisation heat than petrol. By volume, it has the same amount of energy as 66% petrol. According to Zhang et al. (2014), microalgae are thought to be a more efficient feedstock for the synthesis of bioethanol than conventional crops like maize, soybeans or sugarcane.

According to Constantino et al. (2021), a number of algal species have been known as important for the generation of bioethanol, like *Spirogyra* sp., *Chlorococcum* sp., *Chlorella sorokiniana, Sargassum* sp., *Gelidiumamansii, Laminaria* sp., *Gracilaria* sp. and *Prymnesium parvum.* According to different research, *Porphyridium cruentum* marine algae showed high glucose conversion yields (89.8%) and freshwater *P. cruentum* yields (85%) when hydrolysed by enzymes. In freshwater *P. cruentum*, separate hydrolysis and fermentation (SHF) produced a bioethanol yield of 65.4%, whereas simultaneous saccharification and solid-state fermentation (SSF) produced a higher bioethanol yield (70.3%). These findings suggest that *P. cruentum* is a viable candidate for bioethanol and can thrive in freshwater (Kim et al. 2017).

9.3.1.3 Biomethane/Biogas Production

Additionally, biogas produced by microalgae may be utilised to create fuel cells, and power. Algal biomass is a great fuel for anaerobic fermentation, which produces biogas since it includes trace levels of cellulose and lignin. Anaerobic digestion produces biogas, which is composed of 25–45% CO_2 and 55–75% methane (CH_4) (Chye et al. 2018). Because microalgae have decreased cellulose content, no lignin and greater concentrations of lipids and carbohydrates, they are a useful source of biogas. In addition, they develop more quickly than lignocellulosic biomass, are simpler to harvest, need less space for cultivation and are easily converted to biogas. Moreover, solid waste generated by anaerobic digestion is added to the soil (Constantino et al. 2021). Seaweed has a lot of potential for producing biogas. Numerous algae species have been employed in the generation of biogas, such as *Scenedesmus, Euglena, Spirulina* and *Ulva.* The three stages of biogas generation include methanogenesis, acetogenesis and hydrolysis (Ahmad et al. 2022).

However, producing biogas is undesirable due to a number of limitations. The process is complicated, for instance, by the possibility of eutrophication and the production of dangerous compounds. Ammonia, which is produced when the carbon: nitrogen (C:N) ratio is low, is toxic to anaerobic organisms like bacteria. Moreover, sodium ions inhibit microbial activity. Therefore, it is advised to use salt-tolerant bacteria for anaerobic digestion of microalgae biomass in order to produce biogas (Klassen et al. 2015).

The generation of nitrogen in microalgal cultures cultivated in various environments led to an increase in the C/N ratio to 24–26. In contrast to other species, *Chlamydomonas reinhardtii* produced 65% more biogas (69.823 mL/g VS (volatile solids)). *Parachlorella kessleri* and *Scenedesmus obliquus* produced 70.639 mL/g of VS and 58.636 mL/g of VS biogas, respectively, in these circumstances (Klassen et al. 2015). In a further investigation, *Arthrospira platensis* was utilised to generate 4 gVS/L/day of biogas when carbon-rich substrates were present in varying C:N ratios. The treatment of palm oil mill effluent (POME) and the synthesis of biomethane were achieved by the anaerobic co-digestion of freshwater microalgae *Chlorella* sp. with oil palm. According to Ahmad et al. (2014), the greatest specific biogas generation rate was recorded at 0.128–0.129 m^3/kg/COD/day. The co-cultivation of POME and *N. oculata* marine algae produced a maximum specific biogas production rate of 1.13–1.14 m^3/kg/COD/day, according to study. The specific biogas production rate was similar at 1.13–1.16 m^3 kg/COD/day, while the biomethane yield was 1.3-fold lower in the absence of *N. oculata* co-cultivation (Ahmad et al. 2015).

9.3.1.4 Biohydrogen

Biohydrogen has high calorific value and clean oxidation characteristics have garnered significant interest in it as a possible energy source. Biological processes are more energy-efficient, less harmful to the environment and capable of producing biohydrogen under ambient temperatures than thermochemical methods. The synthesis of biohydrogen using microalgae has surfaced as a viable method for producing green energy (Singh and Das 2020). Microalgae may produce biohydrogen in two different ways: light-dependent and light-independent. In the first group, cyanobacteria and microalgae bio-photolyse water, whereas photosynthetic bacteria primarily do photo-fermentation.

Anaerobic bacteria ferment organic compounds in the second category by a process known as dark fermentation (Budiman and Wu 2018). Microalgae may make biohydrogen under anaerobic circumstances if they are exposed to water and light. Due to its ability to create water as a by-product and lack of GHGs emission, biohydrogen is regarded as an efficient energy source (Chye et al. 2018). Under dark, anaerobic fermentation conditions, the proton-lowering enzyme hydrogenase divides water to create hydrogen. Hydrogenase was initially identified in microalgae (Cuellar-Bermudez et al. 2015).

The restricted reaction time resulting from oxygen production, which quickly deactivates the hydrogenase enzyme, is a drawback of this method. Further technological advancements are necessary to enhance the hydrogenase enzyme's oxygen tolerance through improved methodologies (Cuellar-Bermudez et al. 2015; Chye et al. 2018). In addition to producing readily commercialised by-products like butyric, lactic and acetic acid, dark fermentation is low-cost, ecologically benign process. The lack of aeration and light source in this procedure could lower additionally costs. Consequently, combined algae–bacteria system offers a more cost-effective and sustainable method of raising the production efficiency of biohydrogen via dark fermentation.

9.4 Cogeneration in Algal-Based Biorefineries

Cogeneration of algal-based biorefineries refers to the concurrent generation of multiple valuable products or forms of energy from the cultivation and processing of algae. This typically includes the production of biofuels such as biodiesel or bioethanol, along with the generation of bio-electricity and utilisation of excess heat generated during various biorefinery processes. The key idea is to maximise resources efficiency and sustainability by extracting multiple benefits from the same biomass and process heat, thereby enhancing the overall economic and environmental viability of the biorefinery.

9.4.1 Cogeneration Potential for Bioenergy

Algal fuels are a sustainable way of decreasing CO_2 emissions worldwide. One benefit that makes algae a good substrate for biofuels is that, in comparison to plants, their biomass has a lower lignin content (Kavitha et al. 2022). It is feasible to transform algae's carbohydrates into bioethanol and lipids into biodiesel by the conversion of algae into various biofuels (Tazikeh et al. 2022; Sun et al. 2022). Algae such as *Eucheuma cottonii* act as a sugar feedstock for bioethanol production and *Palmaria palmata* acid hydrolysis yields glucose and galactose, which may be fermented into bioethanol (Torres et al. 2019; Offei et al. 2018).

Algal lipid content has the potential to produce biodiesel; however, extracting of lipid fraction which makes up a smaller portion of the biomass requires eliminating all of the watery portions of the biomass, or dehydration. A transesterification or hydrogenolysis procedure must be carried out in order to transform the algal lipids into biodiesel. There are serious concerns about the feasibility of commercialising algal biodiesel, and it is possible that manufacturing the fuel on its own will not be feasible in the absence of biorefineries that produce co-products (Bhatia et al. 2020). A high concentration of lipids may be recovered from some microalgae and cyanobacteria using various methods. According to Moretto et al. (2022), strains of *Sphaerospermopsis torques-reginae*, *Arthrospira* sp., and *Chlorella* sp. have favourable fatty acid profiles and the ability to produce biodiesel. Mamo and Mekonnen (2020) used a chemical process aided by a microwave digester to extract oil from algae (*Scenedesmus*), producing 20.8% biodiesel of its dry weight.

Goat bone debris was employed to create a nano-catalyst for the transesterification process. The findings of this study point to the possible use of animal bone debris as a nano-catalyst and *Scenedesmus* algal strain as a feedstock for biodiesel. Brown algae are biggest and most filamentous macroalgae that have the ability to produce biofuel. Based on published research, the orders Fucales and Laminariales account for nearly all of the world's biomass generated by brown algae, and brown algae are more productive than remaining algae. Because brown algae are significant sources of chemicals (oils and carbohydrates) that may be utilised for thermochemical and biochemical conversions, there is significance in the sustainable generation of biofuels (Kouhgardi et al. 2023).

Kouhgardi et al. (2023) state that drop-in biofuel which can be manufactured on large or industrial scale at a minimum cultivation cost has previously been made from brown algae. This fuel may be utilised in engines that are now in use. In relation to other biofuels like biogas and biohydrogen, Chen et al. (2022) pretreated *Laminaria japonica* with microwave gamma irradiation and achieved a biohydrogen yield of 15.8 mL/g and Kumar et al. (2018) pretreated *Ulva reticulata* using a surfactant-assisted disperser and obtained 45.5 mL of biohydrogen. Anaerobic digestion used for production of biogas from algal biomass, whereas microbial fermentation and sugar hydrolysis produce biomethane. Regarding hydrogen, during photosynthesis the algal cells produce and store it already (Bhatia et al. 2020).

Breaking through rigid cell wall that keeps the soluble organics within primary obstacle to using algal biomass for the manufacture of biofuel. Obtaining a standard pretreatment approach is very challenging due to the very specific variations in the cell wall components of algal biomass. According to findings in the literature, trilaminar layers give microalgal biomasses like *Chlorella* spp. and *Nanochlorum* spp. strong resistance to the cell wall. In their description of the current pretreatments used to break down algal biomass, Kavitha et al. (2022) focused on increasing the yield of biofuel intended for pilot-scale use. The process known as liquefaction is aided by these pretreatments in breaking down the cell walls and releasing the intracellular organics into the aqueous phase. The liquefaction of algal biomass can be accomplished by a variation of pretreatment techniques, which includes individual (e.g. thermal, dispersion, biological, ultrasonic and chemical), combined (chemo-mechanic and thermo-chemical) and phase-separated techniques. To boost liquefaction, biodegradation and ultimately the generation of biofuel, phase-separated or combination pretreatments are advised (Kavitha et al. 2022).

An important disadvantage of biological pretreatments for algae biomass is the expense of the enzymes, which is enzyme-mediated pretreatment. Furthermore, pretreatment procedure with live microorganisms may produce by-products such as propionates, butyrates and formates that interfere with the downstream process. Sometimes the inoculated bacteria may use the by-products that are produced; other times, the bacteria capture the released soluble organics, which reduces the amount of biofuel that is produced. The answer lies in biotechnology; certain fermentation and by-product problems may be resolved by using genetically engineered microorganisms for algae pretreatment (Da Rosa et al. 2023).

Although producing biofuel from algal biomass has gained attention recently, the process is challenging due to the inflexible and complicated algae cell walls, even though producing ethanol from algal biomass offers the advantage of higher output efficiency. Algal biofuel production has advanced, but due to the energy- and money-intensive pretreatments required, it is still in its infancy. Currently, the most effective liquefaction techniques for producing algal biofuels with reduced energy and cost inputs are phase-separated and synergistic pretreatments.

9.4.2 Cogeneration Potential as Biomass Utilisation

The natural process of algal blooms affects water treatment, industry and the environment overall negatively. They arise in aquatic habitats. The overabundance of nutrients in the water, mostly from human sources, is what causes this phenomenon. Excessive biomass can serve numerous uses, increasing value of an initially difficult resource (Biswas et al. 2018). Algal biomass leftovers from industrial operations are a further issue that has to be addressed and applied. Under these conditions, algal biomass wastes could be used in biotechnological applications, i.e. bioenergy, biofuels and biosorbents (Zhong et al. 2020). The remainder of algal biomass from blooms or industrial processing may be used as a culture medium to cultivate various microorganisms, revealing a biotechnological use in the field of microbiology. Because of their rich composition of monosaccharides and polysaccharides also in other nutrients, the species *Enteromorpha prolifera* and *Laminaria japonica* were employed in the study done by (Yuan et al. 2021) to produce a culture medium.

Considering cyanobacteria that create cyanotoxins can be found in bloom processes, these processes can be harmful. The cyanobacterial (*Microcystis aeruginosa*) strain that was isolated from a bloom found on the banks of the Salto Grande reservoir in São Paulo was studied by Passos et al. (2021). It was able to conclude from this analysis that the biomass had a poor calorific value. After acid hydrolysis and fermentation, the biomass of the species *Eucheuma denticulatum* (Spinosum) demonstrated a 75% conversion efficiency in bioethanol for this investigation. Using a thermogravimetric capacity analysis, Zhong et al. (2020) assessed the viability of generating bioenergy from the leftover algae biomass of the *Enteromorpha clathrata* species, which is a by-product of a green tide phenomenon.

According to Ayala-Parra et al. (2017) study on biogas generation using a strain of the microalga *Chlorella sorokiniana* for fuel production and nutrient recovery. This alga showed a significant capacity for producing biodiesel, creating a residue that can be converted back into methane (CH_4) and some other soluble nutrients. For use in the bioenergy industry, Chen et al. (2020) illustrated the necessity of learning more about the leftovers left behind from the burning of algal biomass (slag). The species *Chlorella vulgaris*, which the authors of this study employed, had a significant dross propensity as a result of the high phosphorus content.

Equipment may corrode and melting crystals may develop as a result of the slag produced during combustion lowering the melting temperature of the ashes. The aforementioned studies highlight the significance of doing research on the generation of bioenergy from leftover biomass and the optimisation of industrial processes. The biomass leftover created by the algae-based biodiesel business can be transformed into biochar, a vegetable-based charcoal or organic residues resulting from thermal treatments (such as pyrolysis) in low-oxygen environments. Biochar can be used as a biosorbent or to remediate soil.

Although these molecules have several active sites that enable the adsorption of various substances, biosorbents have many benefits. Chemical or physical techniques can be used to assess the adsorption effectiveness of biochar (Gümüş 2022). Green aniline was eliminated by Gümüş (2022) using the leftovers of the algae species *Gongolaria barbata* (Stackhouse). Aniline removal in this investigation yielded 99% biochar efficiency and a reuse capacity greater than 71.6%. Zhu and Zou (2021) provide another usage for biochar: they used it to eliminate methyl orange with good adsorption results. The leftover biomass of *Spirulina* microalga from a phycocyanin manufacturing facility was used to create this biochar.

9.5 Environmental and Economics Implication

Algae-based biorefineries have both environmental and economic implication.

9.5.1 Environmental Benefits

Algal biorefineries offer several environmental benefits for sustainable approach which includes (a) algal refineries are major sink of CO_2 and helps in its mitigation from environment. Algae could capture and use CO_2 during photosynthesis helping to reduce GHGs emission. These algal refineries are particularly very useful when established near industrial area or power plants to capture carbon emission. (b) Algae can help in wastewater treatment by absorbing access nutrients such as Nitrogen (N) and Phosphorus (P) which can otherwise lead to eutrophication in natural water resources. (c) Also, algae cultivation does not need any pesticides or fertilisers to enhance their growth leading to less or no chemical usage. (d) The algal biomass production in biorefineries produce less waste and harmful environmental impacts. As these are closed loop system where waste (algal biomass) can be utilised to bioenergy generation and wastewater generated can be utilised human use. (e) Many renewable products such as biofuels, animal feed, food additives and bioactive compounds can be generated using the algal biomass (Chen et al. 2022).

Here, we conclude the discussion of algae's environmental advantages in this chapter. Currently, the combustion of fossil fuels like coal, natural gas and petrol provides 88% of the world's energy needs. These fossil fuels will eventually run out since they are not renewable. As a result, more innovative and promising approaches are needed to reduce CO_2 emissions and provide energy security (Milano et al. 2016). Biofuels are a possible supply for future energy needs. Liquid biofuels have proliferated on a worldwide scale in recent years. First-generation biofuels are now widely available and well-established in the United States, Brazil, and the European Union. These biofuels are mostly made from food and oil crops, such as vegetable oils, animal fats, sugarcane and beetroot. Due to direct competition with food for crop usage and agricultural land, the impact of these biofuels on the transportation sector is currently restricted (Adenle et al. 2013). According to Alam et al. (2012), the majority of the second-generation biofuels produced from biomass sources are made from agricultural residue, forest harvesting residue, wood processing residue and non-edible parts of food crops. Non-food crops like jatropha, mahua, tobacco seeds and miscanthus are also produced. As a result, they have less of an impact on the environment than the first generation and are not directly competing with arable land. Second-generation biofuels, however, have an inadequate conversion rate, and the conversion technologies are currently not commercially viable. Third-generation biofuels, or biofuel derived from microalgae, can be produced on a big scale. Microalgae are resistant organisms that can withstand severe environments while evolving quickly. Microalgae growth can be accelerated by carefully choosing the proper species under suitable conditions. Microalgae are often grown in sizable open ponds or photo-bioreactors (PBRs) to generate biomass, which is then collected and processed further to create biofuels.

9.5.2 Economics Viability

The economic viability of algal based biorefineries could be influenced by various factors, including the cost of algal cultivation, processing technologies, etc. Algal biorefineries aim to convert algae into various valuable products. (a) The cost of growing and harvesting algae is a significant factor. Factors like land, water, nutrients and labour expenses could impact overall viability. (b) Efficient and cost-effective extraction and conversion technologies are essential. Innovations in extraction and processing methods can reduce costs and improve profitability. (c) Government policies, incentives and regulations related to biofuels, food and environmental sustainability can influence the viability of algal biorefineries. (d) Algal biorefineries often focus on sustainable and environmental value but may also have associated costs. (d) Algal biorefineries could influences economics of scale and production costs. Larger operation may achieve lower costs per unit of product. Economics viability in alga biorefineries varies depending on these factors and the specific products and markets they target. It is important for operators to conduct thorough feasibility studies and adapt their strategies based on changing conditions to improve their chances of success (Conteratto et al. 2021).

9.6 Conclusion

Algal-based biorefineries represent a promising avenue for sustainable bioenergy generation and cogeneration. The cultivation and conversion of algae offer a wide range of environmental as well as economic benefits, and opportunities to address the growing demand for biofuel and other bioproducts. Algal-based biorefineries have emerged as a vital component of the transition towards a more sustainable and environmentally friendly energy and chemical production landscape. These biorefineries utilise algae as a versatile feedstock, allowing them to produce biofuel, chemicals and valuable bioproducts.

The best possibility for producing energy sustainably is through algae biorefineries. In contrast to conventional biofuel sources like maize or sugarcane, algae may grow in a variety of habitats, reducing their conflict with food crops. Because of their high growth and yield rates, effective CO_2 absorption, capacity to create biofuel, biobased compounds and even nutritional supplements, algal biorefineries are attractive. Algae-derived biorefineries, when compared to solar or wind energy, offer a consistent biomass supply that can be processed into a variety of products (biofuels and electricity generation), providing adaptability beyond the production of electricity. Furthermore, wastewater may be used to grow algae, which lowers the need for freshwater resources by cultivating them on non-arable land. To compete on a wide scale with alternative sources of energy that are environmentally friendly, algae biorefineries still need to solve issues including cost-effectiveness, scalability and technical innovation.

In this chapter, we conclude that algal-based biorefineries hold great promise in the field of sustainable bioenergy generation and cogeneration. These innovative systems harness the rapid growth and high lipid content of algae to produce biofuels such as biodiesel and bioethanol, as well as other valuable co-products like biofertiliser, animal feed and specialty chemicals. Algal biorefineries offer several benefits, including their ability to utilise non-arable land and wastewater for cultivation, thereby reducing competition with food crops and addressing the water pollution issue.

Also, the cogeneration aspect of algal biorefineries enhances their overall efficiency. By simultaneously producing both bioenergy and valuable co-products, these systems not only reduce waste but also increase economic viability. The flexibility of algal biorefineries to adapt to different feedstocks and environmental conditions makes them a sustainable and adaptable solution for clean energy and resource production. However, it is important to acknowledge that there are many challenges to overcome in terms of scalability, cost-effectiveness and technological optimisation. Research and development efforts in the field are ongoing to address these issues and unlock the full potential of algae-based biorefinery.

Acknowledgments The first author (Christina Saran) is highly thankful to the 'University Grants Commission' (UGC), New Delhi, India for providing the financial support for her research work.

References

Abomohra AEF, El-Naggar AH, Baeshen AA (2018) Potential of macroalgae for biodiesel production: screening and evaluation studies. J Biosci Bioeng 125(2):231–237

Adenle AA, Haslam GE, Lee L (2013) Global assessment of research and development for algae biofuel production and its potential role for sustainable development in developing countries. Energy Policy 61:182–195

Ahmad A, Shah SMU, Othman MF, Abdullah MA (2014) Enhanced palm oil mill effluent treatment and biomethane production by aerobic and anaerobic co-cultivation of Chlorella sp. Can J Chem Eng 92:1636–1642

Ahmad A, Shah SMU, Othman MF, Abdullah MA (2015) Aerobic and anaerobic co-cultivation of *Nannochloropsis oculata* with oil palm empty fruit bunch for enhanced biomethane production and palm oil mill effluent treatment. Desalin Water Treat 56(8):2055–2065

Ahmad A, Buang A, Bhat AH (2016) Renewable and sustainable bioenergy production from microalgal co-cultivation with palm oil mill effluent (POME): a review. Renew Sust Energ Rev 65:214–234

Ahmad A, Banat F, Alsafar H, Hasan SW (2022) Algae biotechnology for industrial wastewater treatment, bioenergy production, and high-value bioproducts. Sci Total Environ 806:150585

Alalwan HA, Alminshid AH, Aljaafari HA (2019) Promising evolution of biofuel generations. Subject review. Renew Energy Focus 28:127–139

Alam F, Date A, Rasjidin R, Mobin S, Moria H, Baqui A (2012) Biofuel from algae—is it a viable alternative? Procedia Engineering 49:221–227

Alazaiza MY, Albahnasawi A, Ahmad Z, Bashir MJ, Al-Wahaibi T, Abujazar MSS et al (2022) Potential use of algae for the bioremediation of different types of wastewater and contaminants: production of bioproducts and biofuel for green circular economy. J Environ Manag 324:116415

Atadashi IM, Aroua MK, Aziz AA (2010) High quality biodiesel and its diesel engine application: a review. Renew Sust Energ Rev 14(7):1999–2008
Ayala-Parra P, Liu Y, Field JA, Sierra-Alvarez R (2017) Nutrient recovery and biogas generation from the anaerobic digestion of waste biomass from algal biofuel production. Renew Energy 108:410–416
Bhatia L, Bachheti RK, Garlapati VK, Chandel AK (2020) Third-generation biorefineries: a sustainable platform for food, clean energy, and nutraceuticals production. Biomass Convers Biorefinery:1–16
Biernat, K., & Grzelak, P. L. (2015). Biorefinery systems as an element of sustainable development. Biofuels–status and perspective, Intech 30(1), 427–444
Bigogno C, Khozin-Goldberg I, Boussiba S, Vonshak A, Cohen Z (2002) Lipid and fatty acid composition of the green oleaginous alga *Parietochloris incisa*, the richest plant source of arachidonic acid. Phytochemistry 60(5):497–503
Biswas G, Pokkatt PP, Ghosh A, Kamila B, Adhikari K, Dutta S (2018) Valorization of waste micro-algal biomass–collected from coke oven effluent treatment plant and evaluation of sorption potential for fluoride removal. Water Sci Technol 78(1):132–146
Bose D, Bhattacharya R, Rizvi A, Poonia A, Saraf D, Ghodke PK (2022) Biorefineries: an integrated approach for sustainable energy production. In: Thermochemical and Catalytic Conversion Technologies for Future Biorefineries, vol 2. Springer, Cham, pp 185–212
Budiman PM, Wu TY (2018) Role of chemicals addition in affecting biohydrogen production through photofermentation. Energy Convers Manag 165:509–527
Chen C, Huang Y, Qin S, Huang D, Bu X, Huang H (2020) Slagging tendency estimation of aquatic microalgae and comparison with terrestrial biomass and waste. Energy 194:116889
Chen Y, Yin Y, Wang J (2022) Biohydrogen production using macroalgal biomass of *Laminaria japonica* pretreated by gamma irradiation as substrate. Fuel 309:122179
Chowdhury H, Loganathan B (2019) Third-generation biofuels from microalgae: a review. Curr Opin Green Sustain Chem 20:39–44
Chye JTT, Jun LY, Yon LS, Pan S, Danquah MK (2018) Biofuel production from algal biomass. In: Bioenergy and biofuels, vol 621. CRC Press, Boca Raton (Taylor & Francis Group)
Clarens AF, Resurreccion EP, White MA, Colosi LM (2010) Environmental life cycle comparison of algae to other bioenergy feedstocks. Environ Sci Technol 44(5):1813–1819
Constantino A, Rodrigues B, Leon R, Barros R, Raposo S (2021) Alternative chemo-enzymatic hydrolysis strategy applied to different microalgae species for bioethanol production. Algal Res 56:102329
Conteratto C, Artuzo FD, Santos OIB, Talamini E (2021) Biorefinery: a comprehensive concept for the sociotechnical transition toward bioeconomy. Renew Sust Energ Rev 151:111527
Cuellar-Bermudez SP, Garcia-Perez JS, Rittmann BE, Parra-Saldivar R (2015) Photosynthetic bioenergy utilizing CO2: an approach on flue gases utilization for third generation biofuels. J Clean Prod 98:53–65
Da Rosa MDH, Alves CJ, dos Santos FN, de Souza AO, Zavareze EDR, Pinto E et al (2023) Macroalgae and microalgae biomass as feedstock for products applied to bioenergy and food industry: a brief review. Energies 16(4):1820
De Jong E, Jungmeier G (2015) Biorefinery concepts in comparison to petrochemical refineries. In: Industrial biorefineries & white biotechnology. Elsevier, pp 3–33
Devi A, Ferreira LFR, Saratale GD, Mulla SI, More N, Bharagava RN (2022) Microbe-assisted phytoremediation of environmental contaminants. In: Advances in microbe-assisted phytoremediation of polluted sites. Elsevier, pp 3–26
Devi A, Verma M, Saratale GD, Saratale RG, Ferreira LFR, Mulla SI, Bharagava RN (2023) Microalgae: a green eco-friendly agents for bioremediation of tannery wastewater with simultaneous production of value-added products. Chemosphere 139192:139192
Ding L, Cheng J, Lin R, Deng C, Zhou J, Murphy JD (2020) Improving biohydrogen and biomethane co-production via two-stage dark fermentation and anaerobic digestion of the pretreated seaweed Laminaria digitata. J Clean Prod 251:119666

Ferreira AF (2017) Biorefinery concept. In: Biorefineries: targeting energy, high value products and waste valorisation. Springer, Cham, pp 1–20

Gümüş F (2022) Utilization of algal waste biomass-derived biochar prepared by a microwave-assisted method for aniline green adsorption. Water Air Soil Pollut 233(9):364

Hariz HB, Takriff MS (2017) Palm oil mill effluent treatment and CO 2 sequestration by using microalgae—sustainable strategies for environmental protection. Environ Sci Pollut Res 24:20209–20240

Hussain F, Shah SZ, Ahmad H, Abubshait SA, Abubshait HA, Laref A et al (2021) Microalgae an ecofriendly and sustainable wastewater treatment option: biomass application in biofuel and bio-fertilizer production. A review. Renew Sust Energ Rev 137:110603

Jang JS, Cho Y, Jeong GT, Kim SK (2012) Optimization of saccharification and ethanol production by simultaneous saccharification and fermentation (SSF) from seaweed, *Saccharina japonica*. Bioprocess Biosyst Eng 35:11–18

Jiang L, Li Y, Pei H (2021) Algal–bacterial consortia for bioproduct generation and wastewater treatment. Renew Sust Energ Rev 149:111395

Juan JC, Kartika DA, Wu TY, Hin TYY (2011) Biodiesel production from jatropha oil by catalytic and non-catalytic approaches: an overview. Bioresour Technol 102(2):452–460

Kalra R, Gaur S, Goel M (2021) Microalgae bioremediation: a perspective towards wastewater treatment along with industrial carotenoids production. J Water Process Eng 40:101794

Kashyap BK, Saran C, Solanki MK, Divvela PK (2023) Emerging frontiers of microbes as liquid waste recycler. In: Current research trends and applications in waste management. Springer Nature, Singapore, pp 3–35

Kavitha S, Gondi R, Kannah RY, Kumar G, Banu JR (2022) A review on current advances in the energy and cost-effective pretreatments of algal biomass: enhancement in liquefaction and biofuel recovery. Bioresour Technol 369:128383

Kim HM, Oh CH, Bae HJ (2017) Comparison of red microalgae (*Porphyridium cruentum*) culture conditions for bioethanol production. Bioresour Technol 233:44–50

Klassen V, Blifernez-Klassen O, Hoekzema Y, Mussgnug JH, Kruse O (2015) A novel one-stage cultivation/fermentation strategy for improved biogas production with microalgal biomass. J Biotechnol 215:44–51

Kouhgardi E, Zendehboudi S, Mohammadzadeh O, Lohi A, Chatzis I (2023) Current status and future prospects of biofuel production from brown algae in North America: progress and challenges. Renew Sust Energ Rev 172:113012

Kraan S (2013) Mass-cultivation of carbohydrate rich macroalgae, a possible solution for sustainable biofuel production. Mitig Adapt Strateg Glob Chang 18:27–46

Kumar MD, Tamilarasan K, Kaliappan S, Banu JR, Rajkumar M, Kim SH (2018) Surfactant assisted disperser pretreatment on the liquefaction of *Ulva reticulata* and evaluation of biodegradability for energy efficient biofuel production through nonlinear regression modelling. Bioresour Technol 255:116–122

Kumar D, Eswari AP, Park JH, Adishkumar S, Banu JR (2019) Biohydrogen generation from macroalgal biomass, *Chaetomorpha antennina* through surfactant aided microwave disintegration. Front Energy Res 7:78

Kumar MD, Kannah RY, Kumar G, Sivashanmugam P, Banu JR (2020) A novel energetically efficient combinative microwave pretreatment for achieving profitable hydrogen production from marine macro algae (*Ulva reticulate*). Bioresour Technol 301:122759

Kumar D, Pugazhendi A, Bajhaiya AK, Gugulothu P (2021) Biofuel production from macroalgae: present scenario and future scope. Bioengineered 12(2):9216

Mamo TT, Mekonnen YS (2020) Microwave-assisted biodiesel production from microalgae, Scenedesmus species, using goat bone–made nano-catalyst. Appl Biochem Biotechnol 190(4):1147–1162

Milano J, Ong HC, Masjuki HH, Chong WT, Lam MK, Loh PK, Vellayan V (2016) Microalgae biofuels as an alternative to fossil fuel for power generation. Renew Sust Energ Rev 58:180–197

Moretto JA, de Souza AO, Berneira LM, Brigagão LGG, de Pereira CMP, Converti A, Pinto E (2022) Microwave-assisted extraction of fatty acids from cultured and commercial phytoplankton species. Appl Sci 12(5):2407

Nigam PS, Singh A (2011) Production of liquid biofuels from renewable resources. Prog Energy Combust Sci 37(1):52–68

Offei F, Mensah M, Thygesen A, Kemausuor F (2018) Seaweed bioethanol production: a process selection review on hydrolysis and fermentation. Fermentation 4(4):99

Pahunang RR, Buonerba A, Senatore V, Oliva G, Ouda M, Zarra T et al (2021) Advances in technological control of greenhouse gas emissions from wastewater in the context of circular economy. Sci Total Environ 792:148479

Palupi B, Fa'iqoh NA, Neysella ARP, Fachri BA, Putri DKY, Nulhakim L, Septiyana M (2022) Microwave assisted hydrolysis Ulva sp. using HCl to produce bioethanol raw materials. JOBC 2(2):97–112

Passos LS, Almeida ÉC, de Pereira CMP, Casazza AA, Converti A, Pinto E (2021) Chemical characterization of *Microcystis aeruginosa* for feed and energy uses. Energies 14(11):3013

Poggi-Varaldo HM, Munoz-Paez KM, Escamilla-Alvarado C, Robledo-Narváez PN, Ponce-Noyola MT, Calva-Calva G et al (2014) Biohydrogen, biomethane and bioelectricity as crucial components of biorefinery of organic wastes: a review. Waste Manag Res 32(5):353–365

Robak K, Balcerek M (2018) Review of second-generation bioethanol production from residual biomass. Food Technol Biotechnol 56(2):174

Rodionova MV, Poudyal RS, Tiwari I, Voloshin RA, Zharmukhamedov SK, Nam HG et al (2017) Biofuel production: challenges and opportunities. Int J Hydrog Energy 42(12):8450–8461

Saran C, Purchase D, Saratale GD, Saratale RG, Ferreira LFR, Bilal M et al (2023) Microbial fuel cell: a green eco-friendly agent for tannery wastewater treatment and simultaneous bioelectricity/power generation. Chemosphere 312:137072

Saravanan A, Kumar PS, Govarthanan M, George CS, Vaishnavi S, Moulishwaran B et al (2021) Adsorption characteristics of magnetic nanoparticles coated mixed fungal biomass for toxic Cr (VI) ions in aquatic environment. Chemosphere 267:129226

Schultz-Jensen N, Thygesen A, Leipold F, Thomsen ST, Roslander C, Lilholt H, Bjerre AB (2013) Pretreatment of the macroalgae *Chaetomorpha linum* for the production of bioethanol–comparison of five pretreatment technologies. Bioresour Technol 140:36–42

Shahid MJ, AL-surhanee AA, Kouadri F, Ali S, Nawaz N, Afzal M et al (2020) Role of microorganisms in the remediation of wastewater in floating treatment wetlands: a review. Sustain For 12(14):5559

Singh H, Das D (2020) Biohydrogen from microalgae. In: Handbook of microalgae-based processes and products. Academic Press, pp 391–418

Sivaramakrishnan R, Suresh S, Incharoensakdi A (2019) Chlamydomonas sp. as dynamic biorefinery feedstock for the production of methyl ester and ε-polylysine. Bioresour Technol 272:281–287

Sivaramakrishnan R, Suresh S, Kanwal S, Ramadoss G, Ramprakash B, Incharoensakdi A (2022) Microalgal biorefinery concepts' developments for biofuel and bioproducts: current perspective and bottlenecks. Int J Mol Sci 23(5):2623

Suali E, Sarbatly R (2012) Conversion of microalgae to biofuel. Renew Sust Energ Rev 16(6):4316–4342

Sun A, Davis R, Starbuck M, Ben-Amotz A, Pate R, Pienkos PT (2011) Comparative cost analysis of algal oil production for biofuels. Energy 36(8):5169–5179

Sun J, Norouzi O, Mašek O (2022) A state-of-the-art review on algae pyrolysis for bioenergy and biochar production. Bioresour Technol 346:126258

Tamilarasan K, Kavitha S, Banu JR, Arulazhagan P, Yeom IT (2017) Energy-efficient methane production from macroalgal biomass through chemo disperser liquefaction. Bioresour Technol 228:156–163

Tamilarasan K, Banu JR, Kumar MD, Sakthinathan G, Park JH (2019) Influence of mild-ozone assisted disperser pretreatment on the enhanced biogas generation and biodegradability of green marine macroalgae. Front Energy Res 7:89

Tazikeh S, Zendehboudi S, Ghafoori S, Lohi A, Mahinpey N (2022) Algal bioenergy production and utilization: technologies, challenges, and prospects. J Environ Chem Eng 10(3):107863

Torres MD, Flórez-Fernández N, Domínguez H (2019) Integral utilization of red seaweed for bioactive production. Mar Drugs 17(6):314

Wu J, Wang H, Liu Y, Xu B, Du B, Yang Y (2022) Effect of ultrasonic irradiation on the physicochemical and structural properties of *Laminaria japonica* polysaccharides and their performance in biological activities. Molecules 28(1):8

Yap JK, Sankaran R, Chew KW, Munawaroh HSH, Ho SH, Banu JR, Show PL (2021) Advancement of green technologies: a comprehensive review on the potential application of microalgae biomass. Chemosphere 281:130886

Yin Y, Hu J, Wang J (2019) Fermentative hydrogen production from macroalgae *Laminaria japonica* pretreated by microwave irradiation. Int J Hydrog Energy 44(21):10398–10406

Yuan Y, Chu D, Fan J, Zou P, Qin Y, Geng Y et al (2021) Ecofriendly conversion of algal waste into valuable plant growth-promoting rhizobacteria (PGPR) biomass. Waste Manag 120:576–584

Zhang W, Zhang J, Cui H (2014) The isolation and performance studies of an alginate degrading and ethanol producing strain. Chem Biochem Eng Q 28(3):391–398

Zhong L, Zhang J, Ding Y (2020) Energy utilization of algae biomass waste enteromorpha resulting in green tide in China: pyrolysis kinetic parameters estimation based on shuffled complex evolution. Sustain For 12(5):2086

Zhu H, Zou H (2021) Characterization of algae residue biochar and its application in methyl orange wastewater treatment. Water Sci Technol 84(12):3716–3725

Value-Added Chemicals from Algae: Extraction and Refining

10

Sandeep Kaur Saggu and Shiv Kumar

Contents

Abstract

As the world struggles with the pressing challenges of climate change, resource depletion, and environmental degradation, the search for sustainable alternatives has intensified. In this context, algae have emerged as a fascinating and versatile

S. K. Saggu
Department of Biotechnology, Kanya Maha Vidyalaya, Jalandhar, Punjab, India

S. Kumar (✉)
Department of Microbiology, Guru Gobind Singh Medical College, Baba Farid University of Health Sciences, Faridkot, Punjab, India

Department of Microbiology, RT-PCR Laboratory, Civil Hospital, Kapurthala, Punjab, India

S. K. Arya et al. (eds.), *Value Added Products From Bioalgae Based Biorefineries: Opportunities and Challenges*,
https://doi.org/10.1007/978-981-97-1662-3_10

resource with the potential to address these challenges. Algae, a diverse group of photosynthetic microorganisms, stand out as a sustainable resource for applications such as biofuels, bioplastics, pigments, proteins, and polysaccharides. This comprehensive overview explores the vast potential of algae in producing value-added chemicals. The versatility of algae as a source for these compounds positions them as crucial players in diverse industries, including cosmetics, pharmaceuticals, and skin care. Additionally, the environmental sustainability of algae cultivation is emphasized, underlining their minimal land requirement and ability to thrive in diverse water sources. This chapter delves deeper into the extraction and refining processes involved in harnessing value-added chemicals from algae, highlighting the innovative techniques that emphasize the future of sustainable chemistry. In conclusion, this chapter presents algae as a versatile and sustainable solution for the development of value-added products.

Keywords

Algae · Seaweed · Value-added chemicals · Bioactive compounds · Extraction · Refining

10.1 Introduction

Algae have emerged as a promising source of value-added chemicals with the increasing demand for sustainable and eco-friendly solutions. Algae, simple aquatic photosynthetic organisms, have the potential to transform various industries by providing a renewable alternative to traditional chemical sources (Kumar et al. 2021).

Algae are a large and diverse group of eukaryotic photosynthetic organisms that include species ranging from unicellular microalgae such as *Chlorella*, *Chlamydomonas*, and *Euglena* to multicellular macroalgae (Heimann and Huerlimann 2015). Some brown algae grow very large, for example, *Macrocystis pyrifera* (giant kelp), whereas some multicellular algae are found in complex forms. Seaweeds are complex marine algal forms composed of a large number of marine algae whereas Charophytes form complex freshwater algal structures which include *Spirogyra* and members of the order Charales (Delwiche and Cooper 2015; Baweja et al. 2016). Brown algae in particular have been utilized as a source of secondary metabolites with a variety of biological activities. Algae play a significant role in freshwater and marine food chains. Their production of novel compounds with varied biological efficacies, in terms of both pharmacological and chemical properties, has garnered significant interest (Kumar et al. 2021).

This chapter provides a comprehensive analysis to obtain value-added bioactive compounds from algae by exploring the extraction and refining procedures, and their potential applications across various industries.

10.2 Value-Added Chemicals from Algae

The size of the worldwide carotenoids market reached US$1.8 billion in 2021 and it is anticipated to attain US$2.7 billion by 2031, with a projected compound annual growth rate (CAGR) of 3.9% from 2022 to 2031 (https://www.alliedmarketresearch.com/carotenoids-market, accessed on December 24, 2023). The growing demand is attributed to the increasing consumer awareness of the health benefits offered by various carotenoids, coupled with a trend toward healthier and more natural food products (Gebregziabher et al. 2023). The surge in nutraceuticals has incorporated carotenoids primarily due to their established antioxidant properties. Carotenoids are commonly recognized for industrial applications in food products, cosmetics as vitamin supplements, health food products, and as additives for poultry, livestock, fish, and crustaceans (Yaakob et al. 2014).

Carotenoids, bioactive compounds present in plants, animals, and microorganisms, are fat-soluble organic compounds and serve as food colorants, for example, β-carotene, lycopene, lutein, astaxanthin, and bixin. Some carotenoids such as alpha-carotene, beta-carotene, and beta-cryptoxanthin exhibit provitamin-A activity (Blaner 2020). Their consumption has been linked to reinforcing the immune system and mitigating the risk of chronic degenerative diseases, including cancer, cardiovascular diseases, macular degeneration, cataracts, and various others (Eggersdorfer and Wyss 2018). While carotenoids are present in a broad range of plants, their cellular concentrations are typically low. Achieving high production levels necessitates extensive cultivation areas, specific agricultural methodologies, and variations based on geography and season, as well as the expenses associated with raw materials (Zhang 2018). In some cases, manual harvesting is also required, justifying the elevated cost of these molecules.

On the contrary, the carotenoids found in fish oil originate from microalgae, making it reasonable to cultivate microalgae for carotenoid production (Spolaore et al. 2006). Furthermore, algae play a pivotal role in providing high-quality nutrition for animals, ranging from aquaculture to farm animals. Comprehensive assessments of both nutritional content and potential toxicity have affirmed the suitability of algal biomass as a valuable supplement or alternative to traditional animal feed sources (Harun et al. 2010). Reports indicate that the structures of over 600 distinct carotenoids have been derived from a 40-carbon polyene chain (Del Campo et al. 2007). It has been observed that a higher intake of carotenoids reduces the chance of occurrence of coronary heart disease, macular degeneration, and some types of cancer, as well as enhanced resistance against viral, bacterial, fungal, and parasitic infections (Mitra et al. 2021). Various high-value carotenoids as well as bioactive molecules having anti-infective and/or anticancer properties have been reported to be produced by marine as well as freshwater microalgae and macroalgae (Table 10.1).

Astaxanthin, a carotenoid, is often regarded as a potent vitamin E analog owing to its inherent antioxidant properties (Nakagawa et al. 2011; Jafari et al. 2021). Astaxanthin serves as a valuable commercial product in aquaculture, functioning as a source of pigmentation (Lim et al. 2018; Elbahnaswy and Elshopakey 2023). The

Table 10.1 Value-added compounds from algal sources and their biological activities

Compound	Major algal source	Bioactivity	References
Astaxanthin	*Haematococcus pluvialis*	Antioxidant, protective against UVA-induced DNA alterations in human skin fibroblasts (1BR-3), and effective in treating patients with functional dyspepsia, caused by *helicobacter pylori*	Kupcinskas et al. (2008), Ambati et al. (2014) and Aneesh et al. (2022)
Lutein	*Chlorella* spp., *Scenedesmus* spp., *Muriellopsis* spp., *Rhodophyta* spp., *Spirulina* spp.	Antioxidant	Pereira et al. (2021) and Patel et al. (2022)
Beta-carotene	*Dunaliella* spp., *Dunaliella bardawil*	Precursor of vitamin A (retinol), antioxidant, anticancer, protective against age-related macular degeneration, and used to treat skin melisma	Becker (2004), Jayappriyan et al. (2013) and Wang et al. (2022)
Chlorophyll a, b, and their derivatives	Green algae (Chlorophyta)	Repairs cells, increases hemoglobin in the blood, and promotes cell growth. It inhibits myeloma cell multiplication	Becker (2004) and Harun et al. (2010)
Phycobiliproteins	*Spirulina* spp., *Aphanizomenon flos-aquae*, *Arthrospira* spp.	Antioxidant, anti-inflammatory, nephron-protective, anticancer, and effective against hemolysis induced by peroxyl radicals in human erythrocytes	Becker (2004), Chen and Wong (2008), Pagels et al. (2019), Dagnino-Leone et al. (2022) and Tounsi et al. (2023)
Polyunsaturated fatty acids (PUFAs)	*Phaeodactylum* spp., *Thraustochytrium* spp., *Nannochloropsis* spp., and *Schizochytrium* spp.	Regulates carbohydrate and lipid metabolism, effective against various human diseases including heart diseases, essential for brain development	Li et al. (2019) and Khalid et al. (2022)
Beta-1,3-glucan	*Chlorella* sp., *Poterioochromonas malhamensis*	Antibacterial, anti-viral, anticancer, and immune-stimulant	Iwamoto (2004), Ma et al. (2021), Vetvicka et al. (2021) and Bell et al. (2022)

(continued)

Table 10.1 (continued)

Compound	Major algal source	Bioactivity	References
Lophocladine	*Laphocladia* spp.	Exhibit cytotoxicity to NCI-H460 human lung tumor and MDA-MB-435 breast cancer cell lines	Gross et al. (2006), Cheriyamundath and Sirisha (2020) and Rengasamy et al. (2020)
Tolytoxin (6-hydroxy-7-O-methylscytophycin B)	*Scytonema* spp.	Antifungal antibiotic and highly toxic to mice	Yaakob et al. (2014)
Aurilides B and C	*Lyngbya majuscula*	Cytotoxic against human lung tumor and mouse neuroblastoma cell line	Han et al. (2006)
Sulfolipids	Blue-green algae, *Anabaena* 7120	Inhibits HIV-1, herpes and pneumonia viruses, and human lymphoblastoid cell lines	Gustafson et al. (1989) and Archer et al. (1997)

algal source of natural astaxanthin can be *Haematococcus* sp., *Chlorococcum* sp., and *Chlorella zofingiensis* (Nishshanka et al. 2022).

Lutein, a carotenoid, can be obtained from algae and used as a natural colorant of food products, cosmetics, and drugs. It has been reported that lutein plays an important role in delaying chronic diseases, stimulating immune response, and reducing the progression of atherosclerosis, macular degeneration, and cataracts (Mitra et al. 2021). At present, *Tagetes patula* (marigold) is the predominant source for the worldwide production of lutein. However, chlorophycean microalgae biomass, including species like *Chlorella* sp., *Scenedesmus* sp., and *Muriellopsis* sp., is also recognized as a distinguished source of lutein (Xie et al. 2021).

Beta-carotene, a carotenoid, is a pigment of high demand and it has various industrial and pharmaceutical applications such as food coloring agents, dietary supplements, and additives in multivitamin and cosmetics preparations (Farkas et al. 2020; Meléndez-Martínez et al. 2022). Beta-carotene has pro-vitamin A and potent antioxidant properties making them useful for making health food products and animal feed. The global beta-carotene market size reached US$314 billion in 2022, and it is anticipated to attain US$380 billion by the end of 2028, with a projected CAGR of 3.9% from 2023 to 2028 (https://www.marketdataforecast.com/market-reports/beta-carotene-market, accessed on December 29, 2023). *Dunaliella* sp., a halotolerant, photosynthetic, unicellular eukaryote found in marine and marshy environments, is recognized as a natural source for the commercial production of beta-carotene. *Dunaliella* sp. produces various other carotenoids such as α-carotene, lutein, zeaxanthin, violaxanthin, cryptoxanthin, and neoxanthin, which have numerous clinical, pharmaceutical, and cosmetic industries along with dietary supplements for cattle, poultry, and aquaculture (Mehariya et al. 2021; Barbosa et al. 2023).

Chlorophyll, a green pigment essential for photosynthesis, is composed of a porphyrin ring and a phytol tail of a long carbon chain. Dietary chlorophyll is predominantly structurally modified lipophilic derivatives of chlorophyll a and chlorophyll b (Hayes and Ferruzzi 2020). Sodium and copper derivatives of chlorophyll are often used as coloring agents in food or beverages (Fernandes et al. 2020). Chlorophyll has been explored as a potential pigment source in cosmetics, with a focus on utilizing brown and red algae, which are predominantly employed in the cosmetic industry. Moreover, various chlorophyll derivatives were found to promote cell growth and repair, inhibit tumor growth, and increase hemoglobin in the blood (Hayes and Ferruzzi 2020).

Phycobiliproteins (PBPs), water-soluble fluorescent proteins, are auxiliary photosynthetic pigment that helps in light harvesting during photosynthesis. Based on UV-visible absorption maxima, PBPs are classified as phycocyanin, phycoerythrins, phycoerythrocyanin, and allophycocyanin (Dagnino-Leone et al. 2022). Commercially, it is used as a natural dye, and due to diverse biological activities such as antioxidant, antibacterial, and anticancer, it has various applications in the cosmetics, food, and pharmaceutical industries (Merlo et al. 2021). At present, cyanobacteria, as well as micro- and macroalgae belonging to the Rhodophyta phylum, serve as the primary sources of PBPs (Kovaleski et al. 2022). The major algal sources of PBPs are *Spirulina* spp., *Arthrospira* spp., *Aphanizomenon flos-aquae*, *Porphyridium purpureum*, *Phormidium* spp., and *Pyropia* spp. (Dagnino-Leone et al. 2022).

Polyunsaturated fatty acids (PUFAs) have various health benefits and daily dietary intake of PUFAs has been reported to provide immunity against various pathogens and reduce cardiovascular diseases in humans (Saini et al. 2021). The most prevalent PUFAs include eicosapentaenoic acid (EPA), docosahexaenoic acid (DHA), α- and γ-linolenic acid, arachidonic acid, and docosapentaenoic acid. Natural PUFAs are usually extracted from fish as fish oil but it has poor oxidative stability, taste, and smell. Docosahexaenoic acid (DHA) is the most common PUFA obtained from algae and commercialized as an omega-3 fatty acid (Barta et al. 2021). The grey matter of the human brain contains DHA as a major structural fatty acid which shows the importance of algal DHA for humans from the developmental to the adult stage (Khalid et al. 2022). The major sources of DHA are marine microalgae and some algae-related heterotrophs such as *Schizochytrium mangrove*, *Amphidium caryerea*, and *Thrautocytrium aureum*. *Phaeodactylum* spp., *Thraustochytrium* spp., *Nannochloropsis* spp., and *Schizochytrium* spp. have been reported as the major algal source of DHA and EPA (Li et al. 2019).

Beta-1,3-glucans, commonly referred to as glucans, act as defense molecules that play a role in initiating host defense reactions against pathogenic substances (Legentil et al. 2015). *Chlorella* sp. serves as a major algal source of beta-1,3-glucan. It has various health benefits for humans due to its antibacterial, immunogenic, anti-viral, and anti-tumor activities (Bell et al. 2022). It accelerates health recovery for cancer patients undergoing radiotherapy and/or chemotherapy. Beta-1,3-glucans have also been found effective for patients having HIV infection as it show anti-viral activity (Vetvicka et al. 2021).

Algae have also been reported to produce various pharmaceutical and nutraceutical compounds. Various algae-derived compounds show antibacterial, antifungal, anti-viral, and anticancer activities which make them important and valuable for different industries such as pharmaceutical, skin care, and food industries (Table 10.1). The macroalgae *Laphocladia* sp. has been reported to produce lophocladine alkaloids, which show cytotoxic effects on various human cancer cell lines (Gross et al. 2006). Similarly, *Lyngbya majuscula*, a marine cyanobacterium, has the ability to produce aurilides which show cytotoxic effects against mouse neuroblastoma and human lung cancer cell lines (Han et al. 2006). Some microalgae, such as *Arthrospira* spp., *Chlorella* spp., and *Nannochloropsis oculata*, have got attention in the skin care market due to their skin-tightening properties.

Microalgae have garnered significant worldwide interest in recent years due to the valuable natural products they produce, their capacity to remediate effluents, and their potential as renewable energy crops. Algae cultivation requires minimal land, and some species can thrive in brackish or saline water, reducing the competition for freshwater resources. This makes algae an environmentally friendly and sustainable alternative to traditional chemical sources, such as fossil fuels (Ishika et al. 2017). Microalgal biofuels show promise as alternatives to fossil fuels, given the inherent efficiency of microalgae in converting solar energy into chemical energy. Additionally, microalgae offer a substantially higher potential yield of oils suitable for biofuel production compared to terrestrial crops (Fig. 10.1) (Shuba and Kifle 2018).

10.3 Extraction Techniques

Efficient and environmental-friendly extraction techniques are important to explore the potential of algae for the production of valuable chemicals. Common methods include mechanical extraction, solvent extraction, and enzyme-assisted extraction.

10.3.1 Mechanical Extraction Method

Mechanical extraction involves physically breaking down the cell walls of algae to release intracellular components. Numerous mechanical methods are employed for the extraction of value-added chemicals from algae. Mechanical approaches offer a potent alternative due to their reduced reliance on the specific microalgae species being processed and a lower likelihood of contaminating the extracted product. However, these methods typically demand higher energy inputs compared to chemical or enzymatic alternatives (Ranjith Kumar et al. 2015). Furthermore, the generation of heat during mechanical disruptions poses a risk of damaging the end products, necessitating the incorporation of a cooling system while extracting thermolabile compounds (Ahirwar et al. 2021). The inclusion of a cooling system, however, contributes to increased process costs, encompassing energy and equipment expenses for installation and operation. Several mechanical solvent-free

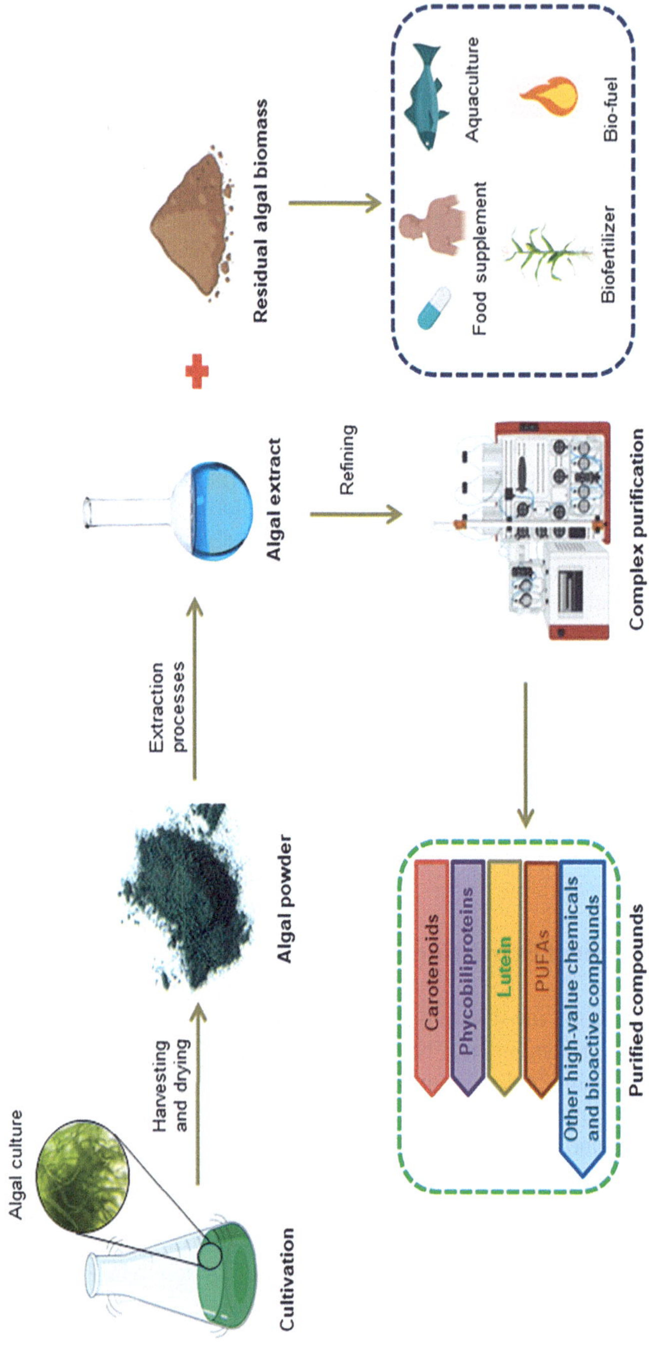

Fig. 10.1 Algae as a biorefinery: source of various value-added chemicals including pharmaceuticals, nutraceuticals, and cosmetics

pretreatment methods for the extraction of valuable chemicals from algal sources are gaining prominence, for example, freeze-thaw, bead milling, expeller press, ultrasonication, microwave-assisted extraction, and application of pulsed electric field (Chittapun et al. 2020; Sommer et al. 2021; Mear et al. 2023). Each of these methods offers distinct advantages and considerations, contributing to the diverse array of mechanical approaches available for value-added compound extraction from algae.

10.3.1.1 Freeze-Thaw

This method stands out as an effective approach for extracting lipids and other chemicals from microalgae biomass, minimizing the loss of volatile and heat-sensitive compounds. In this method, the wet biomass is frozen at ultra-low temperature (−80 °C) or liquid nitrogen temperature (−196 °C), leading to the crystallization of intracellular water. The ice crystals expand during the thawing process leading to cell wall disruption and lysis of frozen cells (Aarthy et al. 2018; Gomes et al. 2020). Typically, this technique is used in combination with other mechanical extraction methods to enhance overall yield efficiency (Aarthy et al. 2018). The number of freeze-thaw cycles affects the yield and quality of products; therefore, it is important to standardize the freeze-thaw cycles for the maximum yield of a particular type of compound extraction (Eilertsen et al. 2014).

10.3.1.2 Expeller Press

The expeller press, commonly referred to as an oil press, is recognized as one of the most simple and well-established techniques for extracting oil from different types of materials including oil seeds and algal biomass (Baskar et al. 2019). When dealing with dried algal biomass, which retains lipid or fatty acids content, an oil press can be employed to extract the oil from algae. The fundamental principle underlying this method involves the application of high mechanical pressure to disrupt the algal cells, subsequently squeezing out the lipid from the algal biomass. While an optimal range of pressure enhances extraction efficiency, excessive pressure can reduce lipid recovery, more heat generation, and operational issues (Ranjith Kumar et al. 2015; Patel et al. 2018). This method is time-consuming and entails a substantial operational cost since it involves the energy-intensive step of drying algae biomass before undergoing the expeller press. Another notable challenge arises from the co-existence of pigments with lipids, and the removal of pigments adds an extra production cost.

10.3.1.3 Bead Beating

Bead beating or bead milling emerges as a mechanical process for cell disruption, employing the principle of grinding algae biomass with fine beads, usually made of steel, zirconia-silica, zirconium oxide, or titanium carbide to induce direct damage to cells (Ranjith Kumar et al. 2015). This method applies to all cell types, including microalgae. Shaking vessels and agitated beads are the two most commonly used bead mill types (Show et al. 2021). Shaking vessels type of bead mill shakes the entire vessel and disrupts the cells. It is typically composed of several containers or wall plates on a vibrating platform. On the other hand, agitated beads consist of a rotating agitator in a fixed vessel containing beads and cell culture. To counteract the heat generated by the rotating agitator, the culture vessels are equipped with cooling jackets to prevent thermal damage of thermolabile compounds (Rajput et al. 2022). Different beads, tailored to various cell types, are utilized. For microalgae cells, an optimal bead diameter of 0.5 mm and a volume fraction of bead loading around 0.5 mm are recommended (Lee et al. 2012). Unlike the expeller press method, bead beating eliminates the need for dewatering of algal slurry, leading to reduced processing costs.

10.3.1.4 Ultrasound-Assisted Extraction

It serves as an innovative alternative to conventional mechanical disruption methods, offering several advantages in terms of simplicity, environmental friendliness, and efficiency. This technique eliminates the challenges associated with wastewater treatment and results in a final product with higher purity (Chemat et al. 2017). Compared to traditional methods, ultrasound-assisted extraction requires minimal energy input, operates at lower temperatures, and is characterized by an easy setup, cost-effective, eco-friendly approach, and a short processing time with high reproducibility (Chemat et al. 2017; Shen et al. 2023).

In liquid cultures, ultrasound-induced cell damage primarily occurs through cavitation and acoustic streaming mechanisms (Tan et al. 2018). Cavitation generates microbubbles under ultrasound, exerting pressure on cells and leading to their disruption. Simultaneously, acoustic streaming promotes mixing of algal cells (Liu et al. 2022). The quick compression-decompression cycles during sonication generate transient and stable cavitation, with the resulting disintegrations producing confined heat shock waves, leading to disruption of microalgae cells (Liu et al. 2022). This cavitation effect leads to the cracking of cell walls and membranes during sonication (Aarthy et al. 2018; Liu et al. 2022). This method is preferred for extracting heat-sensitive compounds because ultrasonic waves generate a relatively lower temperature compared to microwaves. An important advantage is the absence of beads or chemicals, eliminating the need for later removal and reducing costs. However, prolonged ultrasonication may produce free radicals that can potentially impact the quality of the extracted oil and other compounds (Gerde et al. 2012).

10.3.1.5 Microwave-Assisted Extraction

Previously, microwave radiation found limited applications in sample digestion for trace metal measurements and the extraction of organic contaminants. The potential for lipid extraction using microwave irradiation was initially explored in the mid-1980s, demonstrating its efficacy in isolating lipids and pesticides from various sources, surpassing conventional methods (Ganzler et al. 1986). Microwave technology has since paved the way for rapid, safe, and cost-effective lipid extraction methods, eliminating the need for dewatering algal biomass. Among various recognized methods for algal lipid extraction, microwaving remains the simplest and most effective (Ranjith Kumar et al. 2015). When a dielectric substance is exposed to a rapidly oscillating electric field, such as microwaves, heat is generated due to frictional forces arising from molecular movements. This intracellular heating induces water vapor formation, disrupting cells and leading to the electroporation effect, which enhances the extraction of intracellular biomolecules (Mishra and Sharma 2016; Gomes et al. 2020). The rapid generation of heat and pressure expels compounds from the cell matrix, resulting in high-quality extracts and enhanced recovery of target compounds (Meullemiestre et al. 2015). Microwave-pretreated microalgae, with micro-cracks in the cell wall, have been found to yield a higher amount of oil, and microwaves can also be employed for the extraction and transesterification of oils into biodiesel (Bhatia et al. 2021). The economic benefits, including short reaction times, low operational costs, and efficient extraction of algal lipids, position the microwave-assisted method as a preferred option. Reports indicate that the recovery of biodiesel in a microwave-assisted process takes approximately 15–20 min, significantly shorter than the conventional 6-h heating method (Refaat et al. 2008). However, a notable drawback is the maintenance cost associated with microwave-assisted processes, particularly at a commercial scale.

10.3.1.6 Electroporation Method

Electroporation, also termed electropermeabilization, signifies a membrane phenomenon marked by a notable increase in the electrical conductivity and permeability of the cell wall and cell membrane of the algae when subjected to an external electrical field (Geada et al. 2018). Electroporation induces alterations in the cellular membranes and cell walls of the algae, enhancing the efficiency of lipid extraction. This not only reduces the extraction time and solvent usage but also preserves the composition and quality of the extracted fatty acids (Eleršek et al. 2020). Notably, Ranjith Kumar et al. (2015) reported that a solitary electroporation treatment yields the extraction of 92% of the total lipid from the algal biomass, compared to only 62% when employing an equivalent amount of algal biomass without electroporation treatment. This highlights the considerable potential of electroporation as a promising method for elevating lipid extraction efficiency from algal biomass.

10.3.1.7 Osmotic Pressure Method

An innovative and environmentally conscious approach employs osmotic pressure as a cost-effective alternative to traditional extraction methods, presenting a competitive edge (Rakesh et al. 2015). Osmotic pressure disrupts the cell wall of the algae by quickly altering the salt concentration inside and outside the cell, disturbing the osmotic pressure equilibrium across the cell wall of algal cells (Rakesh et al. 2015; Lari et al. 2019). This method induces hyper-osmotic and hypo-osmotic stresses. Hyper-osmotic stress occurs with a higher external salt concentration, causing cells to shrink as internal fluids diffuse outward, resulting in damage to cell envelopes. Conversely, hypo-osmotic stress arises with a lower external salt concentration, prompting fluid influx into the cells to balance osmotic pressure. Excessive stress may cause cells to swell or burst. Hypo-osmotic shock, commonly used for extracting intracellular substances from microorganisms, has shown promising results for oil extraction for biofuel production from various microalgal biomasses, including *Chlamydomonas reinhardtii*, *Botryococcus* spp., *Chlorella vulgaris*, and *Scenedesmus* spp. (Yoo et al. 2012; Ranjith Kumar et al. 2015; Kumar et al. 2019). The osmotic pressure method is considered the most convenient, direct, and efficient technique for lipid extraction from algae (Rakesh et al. 2015; Xu et al. 2021). However, further research is needed to explore lipids and other value-added compounds extraction using this method across various algae, including marine species. Additionally, the feasibility of implementing this technique at laboratory and commercial scales needs thorough examination.

10.3.2 Solvent Extraction Method

Solvent extraction utilizes organic solvents such as chloroform, methanol, and acetic acid to dissolve compounds. Solvent extraction involves the transfer of a compound from one solvent to another due to the varying solubility or distribution coefficient between these two immiscible or partially soluble solvents.

10.3.2.1 Folch Method

Various combinations of organic solvents have been used for the selective extraction of lipids and other value-added compounds from complex organic compound mixtures. The Folch method utilizes a chloroform–methanol mixture (2:1 v/v) to extract lipids from endogenous cells. In this method, homogenized cells are equilibrated with saline solution in a 3:1 ratio (v/v) and thoroughly mixed. The resultant mixture undergoes phase separation, with lipids settling in the upper phase (Folch et al. 1957). This method is among the earliest endeavors in lipid extraction, providing the foundation for the development of subsequent extraction techniques with refinements. Even with some modifications, the Folch method remains in use for the spectrophotometric estimation of algal lipids. Its primary advantage lies in the quick and direct processing of a substantial number of samples but its sensitivity is comparatively lower (Lam et al. 2017). Researchers often choose for the Folch method based on the specific requirements of their studies, appreciating its efficiency in managing large sample volumes.

10.3.2.2 Soxhlet Extraction

Soxhlet extraction (SE) is a method employed to transfer partly soluble components of a solid sample to a liquid phase using a Soxhlet extractor. This technique utilizes nonpolar solvents, such as hexane, for the extraction of neutral lipids. The process involves placing the solid sample in a filter paper thimble within the chamber of the Soxhlet apparatus. The solvent is heated to reflux, flowing into the primary chamber to recover the compounds with lower solubility (Salinas-Salazar et al. 2019; Barbero et al. 2022). The extraction yield of microalgae can be enhanced with an increase in the polarity of the extraction solvent, leading to the recovery of complex lipids and pigments (Baumgardt et al. 2016). This consideration is significant as total lipid extracts with polar solvents often contain complex mixtures having other metabolites along with lipids. SE parameters include solvent selection, particle size of the sample, and extraction time (Garcia-Vaquero et al. 2020; Aravind et al. 2021). SE is typically performed on a laboratory scale, requiring substantial solvent consumption and an extended extraction time.

10.3.2.3 Bligh and Dyer Method

This method facilitates the extraction and partitioning of lipids simultaneously, with the precipitation of proteins occurring at the interface of two liquid phases. Widely employed for lipid extraction, this method shares similarities with the Folch method but distinguishes itself primarily through differences in solvent/solvent and solvent/tissue ratios. In this procedure, lipids are extracted from a homogenized cell suspension using a mixture of one volume of chloroform and two volume of methanol. Following this extraction, lipids from the chloroform phase undergo subsequent processing using various procedures. The gravimetric approach of the Bligh and Dyer method remains a prevalent choice for estimating lipids among algal technologists (Hounslow et al. 2017; Gorgich et al. 2020). This method is consistently applied in pilot-scale and large-scale extraction processes, indicating its enduring relevance in lipid analysis within the scientific community. To enhance the efficacy of this method, researchers have incorporated numerous modifications. Among the most prevalent alterations is the substitution of water with 1 M NaCl, a modification aimed at preventing the binding of acidic lipids to denatured lipids. Jensen et al. (2008) reported that the addition of 0.2 M hydrochloric acid to the salt solution enhances lipid recovery while reducing separation time compared to earlier extraction techniques. Furthermore, the inclusion of 0.5% (v/v) acetic acid in the water phase has been shown to increase the recovery of acidic phospholipids (Weerheim et al. 2002). These modifications collectively signify a continuous effort within the scientific community to refine and optimize lipid extraction processes for improved efficiency and specificity (Ranjith Kumar et al. 2015). Matyash and co-workers (2008) proposed a modified method of the Folch/Bligh and Dyer method for lipid extraction which demonstrates superior recovery of various types of lipids. Methyl *tert*-butyl ether (MTBE) serves as the solvent for lipid extraction, contributing to the attainment of the most accurate lipidome profile. The efficacy of this method is attributed to the formation of a low-density, lipid-containing organic upper phase, facilitating complete extraction (Matyash et al. 2008). Notably, all these methods can be used for lipid extraction from various lipid-containing cells, including algae.

10.3.2.4 Superior Solvent Extraction Methods

The solvent methods described above typically utilize chloroform as the extracting solvent, and while highly effective, the prospect of large-scale lipid extraction using these methods is hindered by associated environmental and health risks. Consequently, researchers have explored alternative, less-toxic substitutes that, although less effective, mitigate these concerns. Ethanol, isopropanol, butanol, MTBE, acetic acid esters, hexane, and various solvent combinations have been investigated for microalgal lipid extraction (Sheng et al. 2011). The selection of solvents depends on the specific class of lipids targeted. Jones et al. (2012) reported the use of 2-ethoxyethanol as an extraction solvent for better yield of lipids compared to commonly used extraction solvents such as chloroform: methanol and hexane. Another approach involves the application of accelerated solvent extraction processes, utilizing heat or pressure, to enhance lipid recovery, shorten processing times, and enable solvent reuse (Cooney et al. 2009; Tang et al. 2016; Lee et al. 2021).

While numerous improved solvent extraction methods and combined solvent/physical extraction systems are under global research scrutiny, it's crucial to acknowledge that organic solvent-based extraction systems pose inherent disadvantages when scaled up for larger industrial applications. Addressing these challenges remains a focal point in ongoing efforts to refine lipid extraction processes.

10.3.2.5 *In Situ* Lipid Hydrolysis and Supercritical *In Situ* Transesterification

Levine et al. (2010) introduced an innovative approach known as in situ lipid hydrolysis and supercritical in situ transesterification for extracting lipids from wet algal biomass. This technique, notable for its potential application in algal biofuel research and development, involves processing of wet algal biomass to extract total lipids and subsequent transesterification for the production of biodiesel. This method employs simultaneous drying and dehydration to transform the wet algal biomass into a solid, facilitating accurate loading of solidified algal biomass in hydrolysis reactions. After cooling, the aqueous phase and solids are separated under light vacuum conditions using an appropriate membrane filter. A comprehensive experiment is conducted to assess the impact of various reaction parameters such as temperature, time, and ethanol loading on composition as well as yield of biodiesel (Levine et al. 2010). However, it is imperative to note that the feasibility of this method on a commercial scale in a large-scale cultivation facility is yet to be explored. Further testing and evaluation are necessary to ascertain its viability in such contexts.

10.3.2.6 Isotonic Extraction Method

This method is an emerging pretreatment alternative for algal lipid extraction in which ionic liquids are used in place of toxic organic solvents. Ionic liquids, defined as non-aqueous solutions of salts maintaining a liquid state at moderate temperatures (0–140 °C), consist of inorganic or organic ions. This unique composition provides synthetic flexibility, enabling the tailoring of the solvent's specific polarity, hydrophobicity, conductivity, and solubility according to specific needs (Cooney et al. 2009). Extensive investigations into ionic liquid extraction from microbial perspectives have been conducted by many research groups, emphasizing its bio-friendly characteristics across various detailed procedures (Orr et al. 2016; Gomes et al. 2019; Vasistha et al. 2021). The economic and technical viability of this approach remains largely unexplored, making it premature to designate it as a superior method for algal oil extraction. Nevertheless, the method exhibits promise as an environmentally friendly alternative to conventional organic solvents. Ongoing research and comprehensive assessments are necessary to ascertain its potential as a sustainable and efficient option for algal lipid extraction.

10.3.3 Enzyme-Assisted Extraction

Enzyme-assisted extraction (EAE) approach involves the utilization of hydrolytic enzymes for efficient cell disruption. Taher and co-workers (2014) introduced cellulase, a cellulolytic enzyme, and proteolytic trypsin to the microalgal biomass, facilitating the extraction of intracellular lipids by breaking down robust polymers on the cell surface structures (Taher et al. 2014). This method stands out for its high specificity and rapidity, although its efficacy is influenced by the lipid class composition and the microalgae type. Furthermore, achieving optimal results necessitates operating at low temperatures with a high level of specificity and selectivity to enhance overall efficiency (Taher et al. 2014). The efficiency of the EAE process relies on several operational parameters, including the particle size of the sample, pH, and temperature of the medium, enzyme concentration, solvent type, agitation rate, and extraction time. These factors significantly impact extraction efficiency (Sanjeewa et al. 2023). Various authors have illustrated that even enzymes without specific targets, such as polysaccharides, possess the ability to break down the cell wall of seaweed. This enhances the extraction of valuable components from seaweed extracts, including water-soluble pigments, proteins, poly- and oligosaccharides, and bioactive compounds (Sanjeewa et al. 2023). Despite its higher cost, this method offers advantages by minimizing the excessive energy typically required for breaking rigid cell walls through mechanical methods.

10.4 Purification/Refining Processes of Algal Value-Added Chemicals

The extracted materials obtained from algae undergo refining processes to isolate specific chemicals with high value. Various methods such as electrophoresis, filtration, centrifugation, precipitation, biphasic separation, and different types of chromatographic techniques are employed to purify the algae-derived compounds. The below described refining processes separate and purify compounds based on physical and chemical properties as well as their chemical interactions with the stationary phase. These refining processes play a crucial role in obtaining pure and high-quality value-added chemicals from algae.

10.4.1 Electrophoresis

In electrophoresis, charged biomolecules such as proteins and macromolecules migrate through an electric field. Unlike gel permeation chromatography (GPC), electrophoresis allows small particles to move through the gel while large particles remain stationary. Two-dimensional electrophoresis achieves high-resolution separations by first sorting molecules based on the charge and then by molecular mass. This technique has been successful in resolving proteins from microalgae such as

Chlorella protothecoides (Dai et al. 2019) and *Porphyridium cruentum* (Tran et al. 2019).

10.4.2 Membrane Separation Processes

Membrane separation methods enable the separation of molecules based on size or charge, such as electromembrane techniques. In traditional membrane filtration, larger molecules are retained, while smaller ones pass through pores indiscriminately, lacking selectivity.

10.4.2.1 Ultrafiltration

Ultrafiltration (UF) is employed to concentrate macromolecules and buffer exchange. It utilizes membranes with reduced pore sizes ranging from 1 to 100 nm and mild temperature conditions for high retention of biomolecules. Despite lacking selectivity compared to chromatographic methods, UF is a low-cost and easily scalable solution for commercial applications. The purification of crude extract from *Spirulina* sp. involved microfiltration and membrane ultrafiltration, resulting in a food-grade product with an increased purity ratio (Chaiklahan et al. 2011). UF was explored as a complementary technique for fucoxanthin purification, attaining a 16% decrease in impurities and highlighting the controllable fouling aspect through ethanol concentration adjustments in the biphasic system (Gómez-Loredo et al. 2015).

10.4.2.2 Electromembrane Filtration

Electromembrane filtration is a better method in comparison to traditional membrane filtration, as it offers selective separation based on charge and molecular weight without the need for pressure. It is a useful tool to harvest *Chlorella* sp. KR-1, showing a quadruple improvement in the concentration factor with the application of electricity, compared to conditions without electricity or using a commercial polyvinylidene fluoride (PVDF) membrane (Kim et al. 2014).

10.4.3 Ultracentrifugation

Centrifugation separates particles based on differential sedimentation under gravity. Ultracentrifugation, reaching forces up to 200,000× g in lab settings, effectively removes impurities from extracts (e.g., chlorophyll from C-phycocyanin extract) (Furuki et al. 2003) and estimates molecular weights of proteins and enzymes from microalgae. However, it is considered outdated for molecular weight estimation due to its energy-intensive nature.

10.4.4 Aqueous Two-Phase Systems

Aqueous two-phase systems (ATPS) involve phases formed by combinations of polymer/polymer or polymer/salt, with alcohol/salt being another common choice for its easy removal. This method is widely used for purifying compounds produced by algae. This method relies on mass transfer mechanisms for equilibrium and phase separation. The phase diagram guides optimal conditions, considering factors like temperature, pH, and phase component concentration. Additional influences on molecule partitioning include crude sample properties, tie line length, phase volume ratio, and additives (Leong et al. 2019).

Liquid biphasic flotation (LBF), a variant of ATPS, combines with solvent sublation (an adsorptive bubble separation technique) for purifying molecules from microalgal biomass. In LBF, target components adhere to bubbles in the aqueous phase, carrying them to the top of an organic solvent-based immiscible phase. Upon bubble rupture, the released molecules are recovered in the upper column zone (Mathiazakan et al. 2016).

10.4.5 Three-Phase Partitioning

Three-phase partitioning (TPP) is widely used to purify proteins using tertiary butanol (*t*-butanol) and ammonium sulfate. Ammonium sulfate induces protein precipitation, and after centrifugation, proteins collect in a middle zone between two non-miscible phases: *t*-Butanol forms the top layer and water forms the bottom layer (Rachana and Lyju 2014). Despite *t*-butanol's usual miscibility in water, the addition of a specific salt, like ammonium sulfate, makes it immiscible. In TPP, *t*-butanol acts as a co-precipitating agent as it enhances protein buoyancy and aids in triphasic formation (Dennison and Lovrien 1997).

10.4.6 Ammonium Sulfate Precipitation

Ammonium sulfate precipitation is a method used to purify proteins, which involves the addition of salt to an aqueous protein solution, resulting in protein aggregation and precipitation (salting out). Ammonium sulfate is often preferred due to its solubility and cost-effectiveness. Studies have optimized parameters for this method, such as concentration, volume, and pH, achieving high purity and recovery rates for proteins like C-phycocyanin (C-PC) from *Spirulina platensis* (Silva et al. 2009). Typically employed as a pre-purification step, ammonium sulfate precipitation is commonly combined with chromatographic techniques (Herrera et al. 1989; Básaca-Loya et al. 2009) or membrane dialysis (Thengodkar and Sivakami 2010; Julianti et al. 2019). Multi-step purification processes, like one for β-glucosidase from *Trichoderma citrinoviride*, show a decrease in yield but significant increases in specific enzyme activity. A combination of techniques such as ammonium sulfate precipitation and chromatography has been used in the purification of C-PC from *S. platensis* (Chen et al. 2016).

10.5 Conclusion and Future Prospects

The extraction and refining of value-added chemicals from algae represent a sustainable and eco-friendly approach to meeting the growing demands of various industries. Algae, with their abundance, rapid growth, and diverse biochemical composition, offer a renewable alternative to conventional sources of chemicals. As technological advancements continue and awareness of environmental sustainability grows, the integration of algae-derived chemicals into mainstream industries holds the potential to reshape the landscape of modern chemistry.

While the potential of algae-derived chemicals is immense, challenges must be addressed to facilitate widespread adoption. Scaling up production, optimizing extraction efficiency, and addressing economic feasibility are key hurdles. Ongoing research and technological advancements are essential to streamline processes, reduce costs, and enhance the overall viability of algae-derived chemicals in the market. Collaboration between researchers, industry stakeholders, and policymakers is crucial for overcoming these challenges and realizing the full potential of algae as a sustainable resource. This comprehensive analysis of a greener and more sustainable future highlights the potential of algae to meet the evolving needs of mankind.

Acknowledgments Not applicable.

Conflict of Interest The authors declare that they have no conflict of interest in the publication.

Authors' Contributions SKS was involved in writing—original draft preparation, methodology, and formal analysis; SK conceived and designed the analysis and was involved in writing—review and editing, supervision, conceptualization, and validation. All authors have read and approved the final version of the manuscript and its contents, and they are aware of the responsibilities connected to the authorship.

Funding Information This research did not receive any specific grant from funding agencies in the public, commercial, or not-for-profit sectors.

Data Availability Statement Not applicable.

References

Aarthy A, Kumari S, Turkar P, Subramanian S (2018) An insight on algal cell disruption for biodiesel production. Asian J Pharm Clin Res 11(2):21–26

Ahirwar A, Meignen G, Khan MJ, Khan N, Rai A, Schoefs B, Vinayak V (2021) Nanotechnological approaches to disrupt the rigid cell walled microalgae grown in wastewater for value-added biocompounds: commercial applications, challenges, and breakthrough. Biomass Convers Biorefinery 13(10):1–26

Ambati RR, Phang SM, Ravi S, Aswathanarayana RG (2014) Astaxanthin: sources, extraction, stability, biological activities and its commercial applications—a review. Mar Drugs 12(1):128–152

Aneesh PA, Ajeeshkumar KK, Lekshmi RK, Anandan R, Ravishankar CN, Mathew S (2022) Bioactivities of astaxanthin from natural sources, augmenting its biomedical potential: a review. Trends Food Sci Technol 125:81–90

Aravind S, Barik D, Ragupathi P, Vignesh G (2021) Investigation on algae oil extraction from algae *Spirogyra* by Soxhlet extraction method. Mater Today Proc 43:308–313

Archer SD, McDonald KA, Jackman AP (1997) Effect of light irradiance on the production of sulfolipids from *Anabaena* 7120 in a fed-batch photobioreactor. Appl Biochem Biotechnol 67:139–152

Barbero GF, Ferreiro-González M, Freitas VCM, Carrera C, Espada-Bellido E, Ruiz-Rodríguez A, Palma M (2022) Extraction of natural products: principles and fundamental aspects. In: Natural product extraction: principles and applications. The Royal Society of Chemistry, pp 66–116

Barbosa M, Inácio LG, Afonso C, Maranhão P (2023) The microalga *Dunaliella* and its applications: a review. Appl Phycol 4(1):99–120

Barta DG, Coman V, Vodnar DC (2021) Microalgae as sources of omega-3 polyunsaturated fatty acids: biotechnological aspects. Algal Res 58:102410

Básaca-Loya GA, Valdez MA, Enríquez-Guevara EA, Gutiérrez-Millán LE, Burboa MG (2009) Extraction and purification of B-Phycoerythrin from the red microalga *Rhodosorus marinus*. Cienc Mar 35:359–368

Baskar G, Kalavathy G, Aiswarya R, Selvakumari IA (2019) Advances in bio-oil extraction from nonedible oil seeds and algal biomass. In: Advances in eco-fuels for a sustainable environment. Woodhead Publishing, pp 187–210

Baumgardt FJ, Zandoná Filho A, Brandalize MV, da Costa DC, Antoniosi Filho NR, Abreu PC, Ramos LP (2016) Lipid content and fatty acid profile of Nannochloropsis oculata before and after extraction with conventional solvents and/or compressed fluids. J Supercrit Fluids 108:89–95

Baweja P, Kumar S, Sahoo D, Levine I (2016) Biology of seaweeds. In: Seaweed in health and disease prevention. Academic Press, pp 41–106

Becker W (2004) Microalgae in human and animal nutrition. In: Handbook of microalgal culture biotechnology and applied phycology. Blackwell Publishing Ltd, Oxford, pp 312–351

Bell V, Silva CRPG, Guina J, Fernandes TH (2022) Mushrooms as future generation healthy foods. Front Nutr 9:1050099

Bhatia SK, Bhatia RK, Jeon JM, Pugazhendhi A, Awasthi MK, Kumar D, Yang YH (2021) An overview of advancements in biobased transesterification methods for biodiesel production: oil resources, extraction, biocatalysts, and process intensification technologies. Fuel 285:119117

Blaner WS (2020) Vitamin A and provitamin A carotenoids. In: Present knowledge in nutrition, 11th edn. Wiley-Blackwell, Cambridge, MA, pp 73–91

Chaiklahan R, Chirasuwan N, Loha V, Tia S, Bunnag B (2011) Separation and purification of phycocyanin from *Spirulina* sp. using a membrane process. Bioresour Technol 102:7159–7164

Chemat F, Rombaut N, Sicaire AG, Meullemiestre A, Fabiano-Tixier AS, Abert-Vian M (2017) Ultrasound-assisted extraction of food and natural products. Mechanisms, techniques, combinations, protocols and applications. A review. Ultrason Sonochem 34:540–560

Chen T, Wong YS (2008) In vitro antioxidant and antiproliferative activities of selenium-containing phycocyanin from selenium-enriched *Spirulina platensis*. J Agric Food Chem 56(11):4352–4358

Chen C, Kao P, Tan CH, Show PL, Cheah WY, Lee W, Chang J (2016) Using an innovative PH-stat CO2 feeding strategy to enhance cell growth and C-phycocyanin production from *Spirulina platensis*. Biochem Eng J 112:78–85

Cheriyamundath S, Sirisha VL (2020) Marine algal-derived pharmaceuticals: potential anticancer agents. In: Encyclopedia of marine biotechnology. Wiley, pp 2691–2724

Chittapun S, Jonjaroen V, Khumrangsee K, Charoenrat T (2020) C-phycocyanin extraction from two freshwater cyanobacteria by freeze-thaw and pulsed electric field techniques to improve extraction efficiency and purity. Algal Res 46:101789

Cooney M, Young G, Nagle N (2009) Extraction of bio-oils from microalgae. Sep Purif Rev 38(4):291–325

Dagnino-Leone J, Figueroa CP, Castañeda ML, Youlton AD, Vallejos-Almirall A, Agurto-Muñoz A, Agurto-Muñoz C (2022) Phycobiliproteins: structural aspects, functional characteristics, and biotechnological perspectives. Comput Struct Biotechnol J 20:1506–1527

Dai L, Reichert CL, Hinrichs J, Weiss J (2019) Acid hydrolysis behavior of insoluble protein-rich fraction extracted from *Chlorella protothecoides*. Colloids Surf A Physicochem Eng Asp 569:129–136

Del Campo JA, García-González M, Guerrero MG (2007) Outdoor cultivation of microalgae for carotenoid production: current state and perspectives. Appl Microbiol Biotechnol 74(6):1163–1174

Delwiche CF, Cooper ED (2015) The evolutionary origin of a terrestrial flora. Curr Biol 25(19):R899–R910

Dennison C, Lovrien R (1997) Three phase partitioning: concentration and purification of proteins. Protein Expr Purif 11(2):149–161

Eggersdorfer M, Wyss A (2018) Carotenoids in human nutrition and health. Arch Biochem Biophys 652:18–26

Eilertsen HC, Huseby S, Degerlund M, Eriksen GK, Ingebrigtsen RA, Hansen E (2014) The effect of freeze/thaw cycles on reproducibility of metabolic profiling of marine microalgal extracts using direct infusion high-resolution mass spectrometry (HR-MS). Molecules 19(10):16373–16380

Elbahnaswy S, Elshopakey GE (2023) Recent progress in practical applications of a potential carotenoid astaxanthin in the aquaculture industry: a review. Fish Physiol Biochem. https://doi.org/10.1007/s10695-023-01143-3

Eleršek T, Flisar K, Likozar B, Klemenčič M, Golob J, Kotnik T, Miklavčič D (2020) Electroporation as a solvent-free green technique for non-destructive extraction of proteins and lipids from *Chlorella vulgaris*. Front Bioeng Biotechnol 8:443

Farkas Á, Bencsik T, Deli J (2020) Carotenoids as food additives. In: Pigments from microalgae handbook, pp 421–447

Fernandes AS, Nass PP, Oliveira Á, Zepka LQ (2020) Chlorophylls as food additives. In: Pigments from microalgae handbook, pp 391–420

Folch J, Lees M, Sloane Stanley GH (1957) A simple method for the isolation and purification of total lipids from animal tissues. J Biol Chem 226(1):497–509

Furuki T, Maeda S, Imajo S, Hiroi T, Amaya T, Hirokawa T, Ito K, Nozawa H (2003) Rapid and selective extraction of phycocyanin from *Spirulina platensis* with ultrasonic cell disruption. J Appl Phycol 15:319–324

Ganzler K, Salgó A, Valkó K (1986) Microwave extraction: a novel sample preparation method for chromatography. J Chromatogr A 371:299–306

Garcia-Vaquero M, Rajauria G, Tiwari B (2020) Conventional extraction techniques: Solvent extraction. In: Sustainable seaweed technologies. Elsevier, pp 171–189

Geada P, Rodrigues R, Loureiro L, Pereira R, Fernandes B, Teixeira JA, Vicente AA (2018) Electrotechnologies applied to microalgal biotechnology–applications, techniques and future trends. Renew Sust Energ Rev 94:656–668

Gebregziabher BS, Gebremeskel H, Debesa B, Ayalneh D, Mitiku T, Wendwessen T, Getachew T (2023) Carotenoids: dietary sources, health functions, biofortification, marketing trend and affecting factors–a review. J Agric Food Res 14:100834

Gerde JA, Montalbo-Lomboy M, Yao L, Grewell D, Wang T (2012) Evaluation of microalgae cell disruption by ultrasonic treatment. Bioresour Technol 125:175–181

Gomes JM, Silva SS, Reis RL (2019) Biocompatible ionic liquids: fundamental behaviours and applications. Chem Soc Rev 48(15):4317–4335

Gomes TA, Zanette CM, Spier MR (2020) An overview of cell disruption methods for intracellular biomolecules recovery. Prep Biochem Biotechnol 50(7):635–654

Gómez-Loredo A, Gonzále-Valdez J, Rito-Palomares M (2015) Insights on the downstream of Fucoxanthin, a microalgal carotenoid, from an aqueous two-phase system stream exploiting ultrafiltration. J Appl Phycol 27:1517–1523

Gorgich M, Mata TM, Martins AA, Branco-Vieira M, Caetano NS (2020) Comparison of different lipid extraction procedures applied to three microalgal species. Energy Rep 6:477–482

Gross H, Goeger DE, Hills P, Mooberry SL, Ballantine DL, Murray TF, Gerwick WH (2006) Lophocladines, bioactive alkaloids from the red alga *Lophocladia* sp. J Nat Prod 69(4):640–644

Gustafson KR, Cardellina JH 2nd, Fuller RW, Weislow OS, Kiser RF, Snader KM, Boyd MR (1989) AIDS-antiviral sulfolipids from cyanobacteria (blue-green algae). J Natl Cancer Inst 81(16):1254–1258

Han B, Gross H, Goeger DE, Mooberry SL, Gerwick WH (2006) Aurilides B and C, cancer cell toxins from a Papua New Guinea collection of the marine cyanobacterium *Lyngbya majuscula*. J Nat Prod 69(4):572–575

Harun R, Singh M, Forde GM, Danquah MK (2010) Bioprocess engineering of microalgae to produce a variety of consumer products. Renew Sust Energ Rev 14:1037–1047

Hayes M, Ferruzzi MG (2020) Update on the bioavailability and chemopreventative mechanisms of dietary chlorophyll derivatives. Nutr Res 81:19–37

Heimann K, Huerlimann R (2015) Microalgal classification: major classes and genera of commercial microalgal species. In: Handbook of marine microalgae. Academic Press, pp 25–41

Herrera A, Boussiba S, Napoleone V, Hohlberg A (1989) Recovery of C-phycocyanin from the Cyanobacterium *Spirulina maxima*. J Appl Phycol 1:325–331

Hounslow E, Noirel J, Gilmour DJ, Wright PC (2017) Lipid quantification techniques for screening oleaginous species of microalgae for biofuel production. Eur J Lipid Sci Technol 119(2):1500469

Ishika T, Moheimani NR, Bahri PA (2017) Sustainable saline microalgae co-cultivation for biofuel production: a critical review. Renew Sust Energ Rev 78:356–368

Iwamoto H (2004) Industrial production of microalgal cell-mass and secondary products—major industrial species *Chlorella*. In: Handbook of microalgal culture biotechnology and applied phycology. Blackwell Publishing Ltd., pp 255–263

Jafari Z, Bigham A, Sadeghi S, Dehdashti SM, Rabiee N, Abedivash A, Varma RS (2021) Nanotechnology-abetted astaxanthin formulations in multimodel therapeutic and biomedical applications. J Med Chem 65(1):2–36

Jayappriyan KR, Rajkumar R, Venkatakrishnan V, Nagaraj S, Rengasamy R (2013) In vitro anticancer activity of natural β-carotene from *Dunaliella salina* EU5891199 in PC-3 cells. Biomed Prev Nutr 3(2):99–105

Jensen GS, Wu X, Patterson KM, Barnes J, Carter SG, Scherwitz L et al (2008) Improved Bligh and Dyer extraction procedure. Lipid Technol 20:280–281

Jones J, Manning S, Montoya M, Keller K, Poenie M (2012) Extraction of algal lipids and their analysis by HPLC and mass spectrometry. J Am Oil Chem Soc 89:1371–1381

Julianti E, Singgih M, Mulyani LN (2019) Optimization of extraction method and characterization of phycocyanin pigment from *Spirulina platensis*. J Math Fundam Sci 51:168–176

Khalid W, Gill P, Arshad MS, Ali A, Ranjha MMAN, Mukhtar S, Maqbool Z (2022) Functional behavior of DHA and EPA in the formation of babies' brain at different stages of age and protection from different brain-related diseases. Int J Food Prop 25(1):1021–1044

Kim D, Hwang T, Oh Y, Han J (2014) Harvesting Chlorella sp. KR-1 using cross-flow electrofiltration. Algal Res 6:170–174

Kovaleski G, Kholany M, Dias LMS, Correia SFH, Ferreira RAS, Coutinho JAP, Ventura SPM (2022) Extraction and purification of phycobiliproteins from algae and their applications. Front Chem 10:1065355

Kumar V, Arora N, Nanda M, Pruthi V (2019) Different cell disruption and lipid extraction methods from microalgae for biodiesel production. In: Microalgae biotechnology for development of biofuel and wastewater treatment, pp 265–292

Kumar BR, Mathimani T, Sudhakar MP, Rajendran K, Nizami AS, Brindhadevi K, Pugazhendhi A (2021) A state-of-the-art review on the cultivation of algae for energy and other valuable products: application, challenges, and opportunities. Renew Sust Energ Rev 138:110649

Kupcinskas L, Lafolie P, Lignell Å, Kiudelis G, Jonaitis L, Adamonis K, Andersen LP, Wadström T (2008) Efficacy of the natural antioxidant astaxanthin in the treatment of functional dyspepsia in patients with or without *Helicobacter pylori* infection: a prospective, randomized, double-blind, and placebo-controlled study. Phytomedicine 15:391–399

Lam SM, Tian H, Shui G (2017) Lipidomics, en route to accurate quantitation. Biochim Biophys Acta (BBA)-Mol Cell Biol Lipids 1862(8):752–761

Lari Z, Ahmadzadeh H, Hosseini M (2019) Cell wall disruption: a critical upstream process for biofuel production. In: Advances in feedstock conversion technologies for alternative fuels and bioproducts. Woodhead Publishing, pp 21–35

Lee AK, Lewis DM, Ashman PJ (2012) Disruption of microalgal cells for the extraction of lipids for biofuels: processes and specific energy requirements. Biomass Bioenergy 46:89–101

Lee SY, Khoiroh I, Vo DVN, Senthil Kumar P, Show PL (2021) Techniques of lipid extraction from microalgae for biofuel production: a review. Environ Chem Lett 19:231–251

Legentil L, Paris F, Ballet C, Trouvelot S, Daire X, Vetvicka V, Ferrières V (2015) Molecular interactions of β-(1→3)-glucans with their receptors. Molecules 20(6):9745–9766

Leong HY, Chang C, Lim JW, Show PL, Lin D, Chang J (2019) Liquid biphasic systems for oil-rich algae bioproducts processing. Sustain For 11:4682

Levine RB, Pinnarat T, Savage PE (2010) Biodiesel production from wet algal biomass through in situ lipid hydrolysis and supercritical transesterification. Energy Fuel 24(9):5235–5243

Li X, Liu J, Chen G, Zhang J, Wang C, Liu B (2019) Extraction and purification of eicosapentaenoic acid and docosahexaenoic acid from microalgae: a critical review. Algal Res 43:101619

Lim KC, Yusoff FM, Shariff M, Kamarudin MS (2018) Astaxanthin as feed supplement in aquatic animals. Rev Aquac 10(3):738–773

Liu Y, Liu X, Cui Y, Yuan W (2022) Ultrasound for microalgal cell disruption and product extraction: a review. Ultrason Sonochem 87:106054

Ma M, Li Y, Chen J, Wang F, Yuan L, Li Y, Hu Q (2021) High-cell-density cultivation of the flagellate alga *Poterioochromonas malhamensis* for biomanufacturing the water-soluble β-1, 3-glucan with multiple biological activities. Bioresour Technol 337:125447

Mathiazakan P, Shing SY, Ying SS, Kek HK, Tang MSY, Show PL, Ling TC (2016) Pilot-scale aqueous two-phase floatation for direct recovery of lipase derived from *Burkholderia cepacia* strain ST8. Sep Purif Technol 171:206–213

Matyash V, Liebisch G, Kurzchalia TV, Shevchenko A, Schwudke D (2008) Lipid extraction by methyl-tert-butyl ether for high-throughput lipidomics. J Lipid Res 49(5):1137–1146

Mear H, Gillon P, Gifuni I, Lavenant L, Poidevin A, Couallier E (2023) Extraction of soluble proteins by bead milling from *Tetraselmis chui* in two different physiological states. Algal Res 74:103180

Mehariya S, Goswami RK, Karthikeysan OP, Verma P (2021) Microalgae for high-value products: a way towards green nutraceutical and pharmaceutical compounds. Chemosphere 280:130553

Meléndez-Martínez AJ, Mandić AI, Bantis F, Böhm V, Borge GIA, Brnčić M, Fikselová M (2022) A comprehensive review on carotenoids in foods and feeds: status quo, applications, patents, and research needs. Crit Rev Food Sci Nutr 62(8):1999–2049

Merlo S, Durany XG, Tonon AP, Rossi S (2021) Marine microalgae contribution to sustainable development. Water 13:1373

Meullemiestre A, Breil C, Abert-Vian M, Chemat F (2015) Modern techniques and solvents for the extraction of microbial oils. Springer International Publishing, Cham, pp 19–42

Mishra RR, Sharma AK (2016) Microwave–material interaction phenomena: heating mechanisms, challenges and opportunities in material processing. Compos A: Appl Sci Manuf 81:78–97

Mitra S, Rauf A, Tareq AM, Jahan S, Emran TB, Shahriar TG, Uddin MS (2021) Potential health benefits of carotenoid lutein: an updated review. Food Chem Toxicol 154:112328

Nakagawa K, Kiko T, Miyazawa T, Burdeos GC, Kimura F, Satoh A, Miyazawa T (2011) Antioxidant effect of astaxanthin on phospholipid peroxidation in human erythrocytes. Br J Nutr 105(11):1563–1571

Nishshanka GKSH, Liyanaarachchi VC, Premaratne M, Ariyadasa TU, Nimarshana PHV (2022) Sustainable cultivation of *Haematococcus pluvialis* and *Chromochloris zofingiensis* for the production of astaxanthin and co-products. Can J Chem Eng 100(10):2835–2849

Orr VC, Plechkova NV, Seddon KR, Rehmann L (2016) Disruption and wet extraction of the microalgae *Chlorella vulgaris* using room-temperature ionic liquids. ACS Sustain Chem Eng 4(2):591–600

Pagels F, Guedes AC, Amaro HM, Kijjoa A, Vasconcelos V (2019) Phycobiliproteins from cyanobacteria: chemistry and biotechnological applications. Biotechnol Adv 37(3):422–443

Patel A, Mikes F, Matsakas L (2018) An overview of current pretreatment methods used to improve lipid extraction from oleaginous microorganisms. Molecules 23(7):1562

Patel AK, Albarico FPJB, Perumal PK, Vadrale AP, Nian CT, Chau HTB, Senthilkumar B (2022) Algae as an emerging source of bioactive pigments. Bioresour Technol 351:126910

Pereira AG, Otero P, Echave J, Carreira-Casais A, Chamorro F, Collazo N, Prieto MA (2021) Xanthophylls from the sea: algae as source of bioactive carotenoids. Mar Drugs 19(4):188

Rachana CR, Lyju JV (2014) Three phase partitioning—a novel protein purification method. Int J Chem Technol Res 6:3467–3472

Rajput MS, Pandey K, Jain A, Gautam AS, Singh S, Kumar S, Tyagi RD (2022) Analysis of lipids. In: Biomass, biofuels, biochemicals. Elsevier, pp 69–88

Rakesh S, Dhar DW, Prasanna R, Saxena AK, Saha S, Shukla M, Sharma K (2015) Cell disruption methods for improving lipid extraction efficiency in unicellular microalgae. Eng Life Sci 15(4):443–447

Ranjith Kumar R, Hanumantha Rao P, Arumugam M (2015) Lipid extraction methods from microalgae: a comprehensive review. Front Energy Res 2:61

Refaat AA, El Sheltawy ST, Sadek KU (2008) Optimum reaction time, performance and exhaust emissions of biodiesel produced by microwave irradiation. Int J Environ Sci Technol 5:315–322

Rengasamy KR, Mahomoodally MF, Aumeeruddy MZ, Zengin G, Xiao J, Kim DH (2020) Bioactive compounds in seaweeds: an overview of their biological properties and safety. Food Chem Toxicol 135:111013

Saini RK, Prasad P, Sreedhar RV, Akhilender Naidu K, Shang X, Keum YS (2021) Omega− 3 polyunsaturated fatty acids (PUFAs): emerging plant and microbial sources, oxidative stability, bioavailability, and health benefits—a review. Antioxidants 10(10):1627

Salinas-Salazar C, Saul Garcia-Perez J, Chandra R, Castillo-Zacarias C, Iqbal HM, Parra-Saldívar R (2019) Methods for extraction of valuable products from microalgae biomass. In: Microalgae biotechnology for development of biofuel and wastewater treatment, pp 245–263

Sanjeewa KA, Herath KHINM, Kim YS, Jeon YJ, Kim SK (2023) Enzyme-assisted extraction of bioactive compounds from seaweeds and microalgae. TrAC Trends Anal Chem 117266:117266

Shen L, Pang S, Zhong M, Sun Y, Qayum A, Liu Y, Ren X (2023) A comprehensive review of ultrasonic assisted extraction (UAE) for bioactive components: principles, advantages, equipment, and combined technologies. Ultrason Sonochem 106646:106646

Sheng J, Vannela R, Rittmann BE (2011) Evaluation of methods to extract and quantify lipids from Synechocystis PCC 6803. Bioresour Technol 102(2):1697–1703

Show KY, Yan YG, Lee DJ (2021) Advances in drying and milling technologies for algae. In: Recent advances in micro and macroalgal processing: food and health perspectives. Wiley, pp 72–95

Shuba ES, Kifle D (2018) Microalgae to biofuels: 'Promising' alternative and renewable energy, review. Renew Sust Energ Rev 81:743–755

Silva LA, Kuhn KR, Moraes CC, Burket AV, Kalil SJ (2009) Experimental design as a tool for optimization of C-phycocyanin purification by precipitation from *Spirulina platensis*. J Braz Chem Soc 20(1):5–12

Sommer MC, Balazinski M, Rataj R, Wenske S, Kolb JF, Zocher K (2021) Assessment of phycocyanin extraction from Cyanidium caldarium by spark discharges, compared to freeze-thaw cycles, sonication, and pulsed electric fields. Microorganisms 9(7):1452

Spolaore P, Joannis-Cassan C, Duran E, Isambert A (2006) Commercial applications of microalgae. J Biosci Bioeng 101:87–96

Taher H, Al-Zuhair S, Al-Marzouqi AH, Haik Y, Farid M (2014) Enzymatic biodiesel production of microalgae lipids under supercritical carbon dioxide: process optimization and integration. Biochem Eng J 90:103–113

Tan X, Zhang D, Parajuli K, Upadhyay S, Jiang Y, Duan Z (2018) Comparison of four quantitative techniques for monitoring microalgae disruption by low-frequency ultrasound and acoustic energy efficiency. Environ Sci Technol 52(5):3295–3303

Tang Y, Zhang Y, Rosenberg JN, Sharif N, Betenbaugh MJ, Wang F (2016) Efficient lipid extraction and quantification of fatty acids from algal biomass using accelerated solvent extraction (ASE). RSC Adv 6(35):29127–29134

Thengodkar RRM, Sivakami S (2010) Degradation of Chlorpyrifos by an alkaline phosphatase from the Cyanobacterium *Spirulina platensis*. Biodegradation 21(5):637–644

Tounsi L, Ben Hlima H, Hentati F, Hentati O, Derbel H, Michaud P, Abdelkafi S (2023) Microalgae: a promising source of bioactive phycobiliproteins. Mar Drugs 21(8):440

Tran T, Lafarge C, Pradelles R, Perrier-Cornet J, Cayot N, Loupiac C (2019) Effect of high hydrostatic pressure on the structure of the soluble protein fraction in *Porphyridium cruentum* extracts. Innovative Food Sci Emerg Technol 58:102226

Vasistha S, Khanra A, Clifford M, Rai MP (2021) Current advances in microalgae harvesting and lipid extraction processes for improved biodiesel production: a review. Renew Sust Energ Rev 137:110498

Vetvicka V, Teplyakova TV, Shintyapina AB, Korolenko TA (2021) Effects of medicinal fungi-derived β-glucan on tumor progression. J Fungus 7(4):250

Wang J, Hu X, Chen J, Wang T, Huang X, Chen G (2022) The extraction of β-carotene from microalgae for testing their health benefits. Food Secur 11(4):502

Weerheim AM, Kolb AM, Sturk A, Nieuwland R (2002) Phospholipid composition of cell-derived microparticles determined by one-dimensional high-performance thin-layer chromatography. Anal Biochem 302(2):191–198

Xie Y, Xiong X, Chen S (2021) Challenges and potential in increasing lutein content in microalgae. Microorganisms 9(5):1068

Xu J, Zhao F, Su X (2021) Direct extraction of lipids from wet microalgae slurries by super-high hydrostatic pressure. Algal Res 58:102412

Yaakob Z, Ali E, Zainal A, Mohamad M, Takriff MS (2014) An overview: biomolecules from microalgae for animal feed and aquaculture. J Biol Res-Thessalon 21:1–10

Yoo G, Park WK, Kim CW, Choi YE, Yang JW (2012) Direct lipid extraction from wet *Chlamydomonas reinhardtii* biomass using osmotic shock. Bioresour Technol 123:717–722

Zhang C (2018) Biosynthesis of carotenoids and apocarotenoids by microorganisms and their industrial potential. In: Zepka LQ, Jacob-Lopes E, De Rosso VV (eds) Progress in carotenoid research. IntechOpen, London

Bioplastics and Biopolymers from Algae: Manufacturing and Applications

11

Sudarshan Sahu, Gursharan Singh, and Shailendra Kumar Arya

Contents

S. Sahu
Department of Biotechnology Engineering, University Institute of Engineering and Technology, Panjab University, Chandigarh, India

G. Singh
Lovely Professional University, Phagwara, Punjab, India

S. K. Arya (✉)
University Institute of Engineering and Technology (UIET), Panjab University, Chandigarh, India
e-mail: skarya@pu.ac.in

S. K. Arya et al. (eds.), *Value Added Products From Bioalgae Based Biorefineries: Opportunities and Challenges*,
https://doi.org/10.1007/978-981-97-1662-3_11

Abstract

The hunt for sustainable alternatives in the face of the growing global plastic issue has brought us to the intriguing field of bioplastics and biopolymers derived from algae. This chapter takes readers on a thorough exploration of the many dimensions of this emerging topic, delving into its possibilities, production methods, uses, and projected future. Exploring the world of algae, we reveal its various forms and traits, highlighting its potential as a feedstock for the manufacturing of bioplastics. The following sections explore the details of biopolymers that are present in algae, including proteins, polyhydroxyalkanoates (PHAs), cellulose, polysaccharides, and polysaccharides, revealing their special qualities and uses. This chapter then goes into production, including strategies for growing, extraction and purification, and sculpting algae bioplastics through mixing and modification, among other procedures. Optimal performance customization of bioplastics is highlighted, highlighting the fluid nature of material innovation. In the final section of the exploration, the future of algae-based bioplastics is explored, with an emphasis on innovative studies utilizing genetically modified algae, the identification of novel biopolymers, and state-of-the-art production techniques. This chapter discusses the remaining obstacles and presents a comprehensive plan for the broad use of products made from algae, putting a focus on consumer awareness, regulatory interventions, technological innovation, and cooperative efforts.

Keywords

Algae · Bioplastic · Biopolymers · Manufacturing

Abbreviations

°	Degree
3D	Three dimensions
BSP	Brown seaweed polysaccharides
CNF	Cellulose nanofibrils
FTIR	Fourier transform infrared spectroscopy
HPLC	High-performance liquid chromatography
HPSEC	High-performance size-exclusion chromatography
PHAs	Polyhydroxyalkanoates
PHB	Polyhydroxybutyrate
UATPP	Ultrasonic-assisted three-phase partitioning
α	Alpha
ε	Epsilon

11.1 Introduction

In the twenty-first century, the pervasive use of conventional plastics has led to an environmental crisis of unprecedented proportions. From clogging oceans to infiltrating terrestrial ecosystems, plastic pollution poses a severe threat to biodiversity and the health of our planet (Kumar et al. 2021). The urgency to address this crisis has spurred innovative research into sustainable alternatives, and one promising solution lies beneath the surface of our oceans—algae (Yong et al. 2022).

The magnitude of the global plastic crisis demands a radical shift in our approach to material production. Conventional plastics, derived predominantly from fossil fuels, have overwhelmed our ecosystems due to their non-biodegradable nature (Baranwal et al. 2022; Ibn-Mohammed et al. 2021). They persist for centuries, causing irreparable damage to marine life, wildlife, and human health (Dermawan et al. 2022; Gallo et al. 2020; Sahu et al. 2023a, b, c). As we stand at a critical juncture, the exploration of alternative materials becomes imperative for a sustainable future.

Algae, a varied collection of photosynthetic organisms, have the potential to transform the methodds by which we manufacture plastics (Manzi et al. 2022; Zanchetta et al. 2021). Algae, often overlooked in mainstream discussions, are emerging as a beacon of hope in the quest for eco-friendly alternatives. Their rapid growth, high photosynthetic efficiency, and ability to thrive in diverse environments make them an ideal candidate for sustainable bioplastics production (Agarwal et al. 2022; Samadhiya et al. 2022; Yashavanth et al. 2021). To understand the algal promise, we must first delve into the intricate world of algae. Algae encompass a vast array of organisms, ranging from microscopic phytoplankton to macroscopic seaweeds. Their adaptability to various environments, from freshwater ponds to harsh marine conditions, showcases the versatility that makes algae a compelling source for sustainable solutions (Schoenrock et al. 2021).

Exploring the diversity of algae reveals a rich tapestry of species, each with unique characteristics that can be harnessed for specific applications. Microalgae, for instance, are known for their rapid growth rates and high lipid content, making them a prime candidate for biofuel production (Sahu et al. 2023a, b, c). Macroalgae, on the other hand, boast strong cellulose fibers suitable for various industrial applications. The versatility of algae extends beyond their structural components, encompassing a myriad of compounds that can be extracted for bioplastic production. Within the cellular structure of algae lies a treasure trove of biopolymers—organic polymers synthesized by living organisms. These biopolymers exhibit remarkable properties and, importantly, are inherently biodegradable (Sahu et al. 2023a, b, c). PHAs, cellulose, polysaccharides, and proteins are among the key biopolymers found in algae, each offering distinct advantages for sustainable material development.

PHAs, derived from microbial fermentation of algae, represent a promising class of biopolymers with properties akin to traditional plastics (Costa et al. 2019). What sets PHAs apart is their ability to break down naturally in various environments, leaving no trace of harmful residues (Adeleye et al. 2020). This characteristic addresses the critical issue of plastic persistence and aligns with the principles of a circular, environmentally conscious economy. Cellulose, a ubiquitous biopolymer found in the cell walls of algae, provides another avenue for sustainable material development. The strong, flexible nature of cellulose fibers makes them suitable for a range of applications, from packaging materials to textiles (Koller 2018). Moreover, cellulose-based bioplastics have the advantage of being derived from renewable resources, further reducing the environmental footprint associated with traditional plastics (Behera et al. 2022).

Polysaccharides, complex sugar molecules abundant in algae, contribute to the development of bio-based materials with diverse applications. Alginate, a polysaccharide found in brown algae, is renowned for its ability to form gels, making it an ideal candidate for food packaging and biomedical applications (Sahu et al. 2023a, b, c). The versatility of polysaccharides positions them as valuable components in the quest for sustainable alternatives.

Algal proteins, often overlooked in the context of bioplastics, present yet another dimension to the potential of algae (Choi et al. 2022). These proteins can be utilized in the development of biodegradable films and coatings, offering a sustainable alternative to conventional plastic packaging materials (Asgher et al. 2020; Tyagi et al. 2021).

The algal promise in the realm of bioplastics and biopolymers holds immense potential to address the global plastic crisis. As we navigate the complexities of our environmental challenges, algae emerge as a beacon of hope, offering sustainable alternatives that align with the principles of circular economies and ecological balance. The exploration of algal bioplastics is not merely a scientific endeavor; it is a commitment to a future where material innovation coexists harmoniously with the well-being of our planet.

11.2 Manufacturing Bioplastics and Biopolymers from Algae: A Green Revolution Unveiled

In the pursuit of sustainable alternatives to traditional plastics, algae have emerged as green giants capable of transforming the landscape of material manufacturing (Althuri and Venkata Mohan 2022). The journey from cultivating these microscopic powerhouses to shaping them into versatile biopolymers involves a series of intricate processes, each contributing to the realization of an eco-friendly future (Sahu et al. 2023a, b, c). In this exploration, we dive into the cultivation methods, extraction and purification techniques, bioplastic manufacturing processes, and the tailoring of algal bioplastics for optimal performance.

11.2.1 Cultivation: Growing the Green Giants

The first step in harnessing the potential of algae for bioplastic production involves cultivating these green microorganisms on a scale that can meet the demands of a plastic-dependent world. Various methods are employed, each with its own set of advantages and challenges. Open ponds, a traditional method, involve growing algae in large, shallow ponds where sunlight and nutrients promote rapid growth (Kumar et al. 2015). Cost-effective, open ponds face challenges such as contamination and inefficient use of space. In contrast, closed photobioreactors provide a controlled environment, minimizing contamination risks and allowing for precise control of growth conditions (Assunção and Malcata 2020). However, the higher capital costs associated with photobioreactors remain a consideration.

Integrated systems represent a hybrid approach, combining the benefits of open ponds and closed photobioreactors. These systems aim to optimize efficiency by utilizing natural sunlight while mitigating contamination risks through controlled conditions (Liu et al. 2019; Yun et al. 2018). The choice of cultivation method depends on factors such as scalability, cost-effectiveness, and the specific requirements of the algae species being cultivated.

The optimization of biomass production is a key focus in algae cultivation. Researchers seek to enhance growth rates, lipid content, and overall biomass yield to ensure a sustainable and economically viable feedstock for bioplastics production.

11.2.2 From Biomass to Biopolymers: Extraction and Purification

Once algae biomass is harvested, the next crucial step involves extracting and purifying the biopolymers embedded within these microscopic organisms. Various extraction techniques are employed, depending on the type of biopolymer targeted. Figure 11.1 shows the various polymers extracted from brown algae.

PHAs, for instance, can be extracted through microbial fermentation or direct extraction from algae biomass. The efficiency and scalability of extraction processes play a pivotal role in determining the viability of PHAs as a feedstock for bioplastics. The pursuit of environmentally acceptable polymers to counteract worldwide pollution centers on biodegradable substitutes such as PHAs. The high expense of bacterial PHA production is mostly attributable to the organic substrates. By making use of photoautotrophic microalgae's capacity to flourish on light, CO_2, and nutrients, microalgal biomass utilization provides an affordable option. However, issues with cultivation, harvesting, biomass conversion, and molecular factors arise when producing PHA on a big scale from microalgal biomass (Tan et al. 2022). Traditional petroleum-based plastics, while convenient, contribute to fossil resource depletion and environmental degradation. PHAs offer a biodegradable alternative suitable for various applications. This study explored PHA production by the freshwater microalga *Scenedesmus* sp., not previously studied for this purpose. Using a Taguchi experimental design, factors like glucose, N, P, Fe, and salinity were investigated. Results identified phosphorus, glucose, and iron as

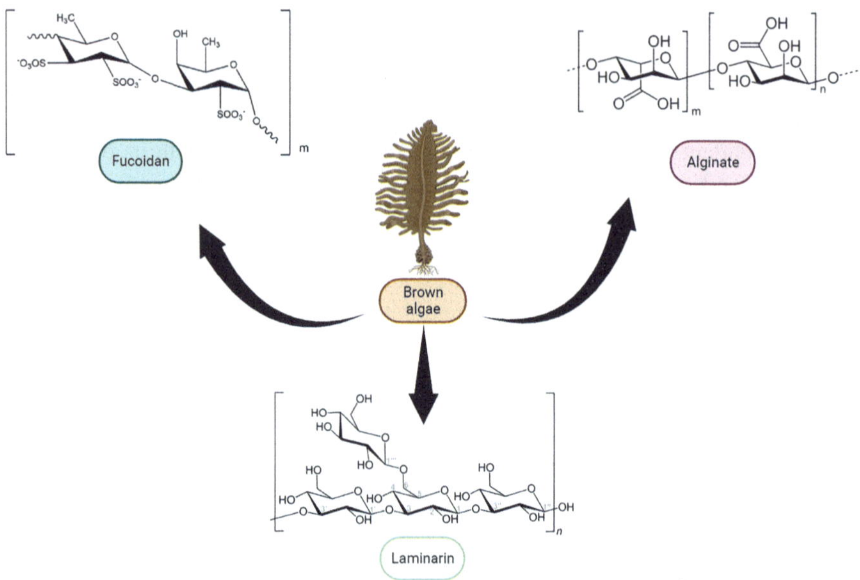

Fig. 11.1 Biopolymers extracted from brown algae

significant factors, showcasing *Scenedesmus* sp. as a promising species for commercial PHA production (García et al. 2021). *Laminaria japonica* biomass was explored as a carbon source for PHA production using three bacterial strains. Acid treatment increased reducing sugar production, reaching 6.1 g/L with H_2SO_4 and 5.9 g/L with HCl. Strains, including *Cupriavidus necator*, demonstrated PHA production up to 32% of dry cell weight. Batch and fed-batch cultures enhanced PHA content to 49%. *Paracoccus* sp. LL1 also produced carotenoids, with a fed-batch culture yielding 2.3 mg/L. The produced PHA polymers were identified as poly(3-hydroxybutyrate) or poly(3-hydroxybutyrate-*co*-3-hydroxyvalerate) (Muhammad et al. 2020).

Cellulose, a prevalent biopolymer in algae, requires specific methods such as chemical or enzymatic processes to isolate it from the complex matrix of algal cell walls. The challenge lies in achieving high purity levels while maintaining cost-effectiveness and minimizing environmental impact. Green algae, known for their high photosynthetic efficiency, serve as a sustainable source of biopolymers. This study focuses on extracting cellulose from *Ulva lactuca* algae by finely grinding them with solvents to enhance substance extraction. Optimal conditions yielded a 20,944% cellulose extraction, with hydrogel membranes formed from the extracted cellulose exhibiting swelling behavior in different concentrations of ethyl alcohol. The absorption process showed rapid initial absorption followed by a slower rate until saturation (Patrichi et al. 2023). Cellulose was successfully extracted from the green macroalgae *Ulva lactuca*, both collected from the Swedish west coast and cultivated indoors. The extracted cellulose was then used to produce lignin-free cellulose nanofibrils (CNF) without enzymatic pre-treatments. Characterization revealed semicrystalline CNF with a crystallinity index of 48%, while regenerated cellulose exhibited mostly amorphous characteristics. The cellulose fractions were thermally stable and free from inorganic substances, with optotracing confirming the cellulose content and revealing the presence of xylose in small amounts (Wahlström et al. 2020). Green nanotechnology, utilizing macroalgal biomass, offers a safer and environmentally friendly approach for medical applications. *Ulva lactuca* was used to extract cellulose, which was further processed into nanocellulose using sulfuric acid hydrolysis. The resulting nanocellulose exhibited antibacterial properties against both gram positive and gram negative bacteria, comparable to antibiotics. Minimal inhibitory concentration (MIC) was estimated. Nanocellulose also demonstrated efficacy against fungi, suggesting its potential as a valuable and sustainable medical material (El-Sheekh et al. 2023).

Polysaccharides like alginate, found in brown algae, are typically extracted using techniques that preserve their structural integrity. Alginate extraction involves breaking down the cell wall matrix and subsequently purifying the polysaccharide to meet the desired specifications for bioplastics production. Brown algae, consumed historically, are now recognized for their bioactive compounds, serving as functional foods with various health benefits. Predominantly composed of polysaccharides, including alginate, laminarin, and fucoidan, these compounds exhibit diverse structures contributing to immunomodulation, antibacterial, antioxidant, prebiotic, antihypertensive, antidiabetic, antitumor, and anticoagulant activities (Li et al. 2021). The intestinal barrier plays a crucial role in gut homeostasis, impacting overall health. This review focuses on brown seaweed polysaccharides (BSP), including laminarin, fucoidan, and alginate. BSP exhibit physicochemical properties influencing their positive effects on intestinal health. Resistant to upper gastrointestinal digestion, BSP undergo microbial fermentation, preserving intestinal integrity by mitigating oxidative stress, reducing tight junction dysfunction, and influencing gut microbiota composition. This comprehensive understanding suggests BSP as promising natural compounds for preventing and treating intestinal diseases effectively and safely (Yao et al. 2023). BSP are vital components in brown seaweed with diverse pharmaceutical and biomedical applications. Quality control is challenging due to their complex structure. This study used advanced techniques like HPLC, HPSEC-MALLS-Vis-RID, and FT-IR to analyze BSP from nine brown algae species. Results revealed β-D-glucans and β-1,3−1,4-glucan linkages, with varying molecular weights, sizes, and antioxidant activities. Understanding the structure-function relationship enhances BSP quality control and suggests their potential as antioxidant supplements (Chen et al. 2021).

Algal proteins can be extracted through various methods such as mechanical disruption or enzymatic hydrolysis. The choice of extraction method influences the yield and purity of proteins, which are crucial factors in determining the properties of protein-based bioplastics. Population growth and resource limitations necessitate alternative protein sources. Macroalgae and microalgae, considered under-exploited "crops," offer a promising solution, avoiding competition with traditional food crops. This review explores the characteristics of commonly consumed algae and their potential as protein sources, emphasizing protein quality, amino acid composition, and digestibility. Various protein extraction methods, including novel techniques, are discussed, along with potential applications in human nutrition, animal feed, and aquaculture (Bleakley and Hayes 2017). Green microalgae, rich in bioactive compounds and macronutrients, hold potential in food industries due to their high protein content and resilience. This study introduces ultrasonic-assisted three-phase partitioning (UATPP) as an efficient method for extracting proteins from *Chlorella vulgaris* FSP-E. UATPP outperformed conventional three-phase partitioning (TPP), achieving a 74.59% separation efficiency and 56.57% yield under optimized conditions. Scale-up studies validated UATPP's reliability, establishing it as a promising approach for microalgae downstream processing (Chia et al. 2019).

Efficiency and scalability are paramount in extraction and purification processes. Advances in technology aim to streamline these processes, making them not only environmentally friendly but also economically viable for large-scale production.

11.2.3 Shaping the Future: Bioplastic Manufacturing from Algae

With purified biopolymers in hand, the focus shifts to the manufacturing processes that transform these raw materials into usable bioplastics. Several approaches exist, each with its own technical and economic considerations. Direct extraction involves taking the purified biopolymers and processing them into bioplastics without significant chemical alteration. This method is advantageous in preserving the inherent properties of the biopolymers but may face challenges in achieving desired material characteristics.

Chemical conversion and polymerization represent a more involved approach, where extracted biopolymers undergo chemical transformations to create polymers with specific properties. This method offers greater control over the final product but may involve additional steps and resources. The choice between these manufacturing processes depends on factors such as the desired properties of the bioplastics, cost considerations, and the overall environmental impact. Striking a balance between technical feasibility and economic viability is crucial in developing scalable manufacturing solutions.

11.2.4 Blending and Modification: Tailoring Bioplastics for Performance

To expand the application range of algal bioplastics and enhance their performance, blending techniques come into play. Blending involves combining algal biopolymers with other biopolymers or synthetic additives to achieve a desired set of properties. The green macroscopic alga, *Ulva* sp., identified through morphological and molecular analyses, was utilized with a plant-based plasticizer to create a bioplastic with strong tensile properties and thermal stability. FTIR analysis confirmed the formation of a starch-based blend, combining starch and cellulose from *Ulva* sp.

and *P. ovata*. The blend, displaying biodegradability in a soil burial test, is composed of natural polysaccharides, making it environmentally friendly with suitable strength and stability (El Semary et al. 2022). The escalating demand for plastic products contributes to widespread plastic pollution, prompting a need for innovative, environmentally friendly alternatives. Biodegradable bioplastics, particularly those derived from microalgal species, offer sustainability, reduced carbon footprint, lower toxicity, and rapid degradability. Abundant and easily cultivated, microalgae can be processed for polysaccharides, serving as a promising feedstock for biopolymers (Arora et al. 2023).

Blending with other biopolymers allows for the creation of composite materials with improved mechanical strength, flexibility, or thermal stability. For example, combining algal cellulose with chitosan, a biopolymer derived from crustacean shells, can result in materials with enhanced strength and biodegradability. The rise of eco-friendly polymers is driven by the environmental impact of marine plastics and emerging pollutants like micro- and nanoplastics. This review explores algae-derived natural polysaccharides and polyesters, emphasizing their structure, biosynthetic mechanisms, production processes, and biodegradability. The review advocates for intensified research to enhance the cost-effectiveness of algal biorefineries, integrating biofuel, high-value products, and biopolymer production using waste and wastewater-grown algal biomass. This approach aligns with circular, net-zero carbon future economies and promotes sustainability (Aswathi Mohan et al. 2022). Synthetic additives may be introduced to modify specific properties of algal bioplastics, such as enhancing their resistance to UV radiation, improving processability, or imparting specific textures. The challenge lies in selecting additives that align with sustainability goals and do not compromise the eco-friendly nature of the bioplastics.

The tailoring of algal bioplastics through blending and modification is a dynamic area of research, with the goal of creating materials that not only rival traditional plastics in performance but also surpass them in environmental sustainability.

11.2.5 Paving the Green Path Forward

The journey from cultivating algae to manufacturing bioplastics represents a multifaceted endeavor that holds the potential to redefine our approach to material production. The green promise of algae, with their rapid growth, diverse biopolymers, and inherent sustainability, positions them as key players in addressing the global plastic crisis (Rosenboom et al. 2022). As research and innovation continue to propel the field forward, the integration of algae into the bioplastics landscape offers a glimpse into a future where materials are not only functional but also harmonious with the delicate balance of our planet's ecosystems. The cultivation, extraction, manufacturing, and modification processes outlined herein collectively contribute to paving the green path forward, ushering in a new era of sustainable material solutions.

11.3 Applications of Algae-Based Bioplastics and Biopolymers: Green Solutions for a Sustainable Future

In the pursuit of a sustainable and eco-friendly future, the applications of algae-based bioplastics and biopolymers stand as beacons of innovation, offering viable alternatives to traditional plastics. This chapter delves into the diverse and promising applications of algal bioplastics, ranging from packaging solutions to a multitude of fields, while also addressing the sustainability challenges and opportunities that accompany their large-scale production.

11.3.1 Packaging with a Purpose: Replacing Plastic Waste with Bioplastics

11.3.1.1 Food Packaging: A Greener Wrapper

Algal bioplastics have found a compelling niche in the world of food packaging, providing a sustainable alternative to conventional plastics that inundate landfills and oceans. These bioplastics, often derived from PHAs or algal cellulose, exhibit biodegradability without compromising on performance or shelf life. From wraps to containers, algal bioplastics are emerging as a key player in reducing the environmental impact of food packaging.

Plastic pollution, exacerbated by the COVID-19 pandemic, has spurred the search for sustainable alternatives like bioplastics. Microalgae, with rapid growth, minimal nutrient requirements, and wastewater remediation capability, are a promising source. Despite the current boom in algal bioplastics, challenges include cost competitiveness against petroleum-based plastics (Nanda and Bharadvaja 2022). Modern living heavily relies on petroleum-based plastics, causing environmental damage and depleting fossil fuels. Bioplastics, with properties akin to conventional plastics, offer a promising solution. Despite the potential of algae and cyanobacteria as bioplastic sources, limited research exists on strain selection and culture condition optimization. Genetic engineering of *Synechocystis* sp. under abiotic stress recently demonstrated an 81% increase in polyhydroxybutyrate (PHB) levels (Sreenikethanam et al. 2021).

Notably, the biodegradability of algal bioplastics aligns with the circular economy model, ensuring that packaging materials can return to the environment harmlessly. This contrasts starkly with conventional plastics that persist for centuries, causing ecological harm. The ability of algal bioplastics to decompose into harmless by-products signifies a crucial step toward mitigating the plastic crisis.

11.3.1.2 Beverage Bottles: Refreshing Sustainability

The beverage industry, a major contributor to plastic pollution, has seen a transformative shift with the introduction of algal bioplastics. These bioplastics, with their inherent strength and versatility, are well-suited for the production of bottles that not only rival traditional PET bottles but also contribute to reducing carbon footprints. Algal bioplastics in beverage bottles offer a sustainable choice without compromising on functionality, maintaining the integrity and safety of the packaged liquids while minimizing environmental impact (Chia et al. 2020; Nanda and Bharadvaja 2022).

The adoption of algal bioplastics in beverage packaging reflects a commitment to a sustainable future, encouraging responsible consumption practices and fostering awareness among consumers about the environmental impact of their choices.

11.3.1.3 Other Packaging Applications: Beyond the Basics

Beyond food and beverage packaging, algal bioplastics find applications in a wide array of packaging solutions. From cosmetic containers to electronics packaging, the versatility of these bioplastics allows for customization to meet specific industry needs. The integration of algal bioplastics in packaging materials extends to various sectors, promoting a circular economy where waste is minimized and environmental impact is reduced.

11.3.2 Beyond Packaging: Biopolymers for Diverse Applications

11.3.2.1 Textiles: Green Fibers Weaving a Sustainable Future

Algal biopolymers, particularly cellulose, play a crucial role in revolutionizing the textile industry. The demand for sustainable and eco-friendly textiles has driven the exploration of algal cellulose as a green alternative to traditional fibers.

The rapid pace of global industrialization necessitates a shift from conventional petroleum-based polymers. Microalgae-derived biopolymers offer a sustainable solution, presenting biodegradability and enhanced biomass productivity. This review proposes a novel, cost-effective, light-depleted wastewater cultivation approach to accelerate biopolymer production. It explores various microalgae strains for biopolymer synthesis, and extraction techniques, and addresses challenges in the circular bioeconomy (Khanra et al. 2022).

The unique properties of algal cellulose as shown in Fig. 11.2, including strength, flexibility, and biodegradability, position it as an ideal candidate for the development of environmentally friendly textiles. Algal-based fabrics cater to a growing market of environmentally conscious consumers, contributing to the reduction of synthetic fiber pollution.

11.3.2.2 Cosmetics: Natural Elegance

The cosmetic industry, known for its extensive use of packaging and single-use plastics, is turning to algal biopolymers to align with sustainable practices. Algal biopolymers, rich in proteins and polysaccharides, offer a natural and renewable source for cosmetic formulations. These biopolymers serve as alternatives to synthetic thickeners, emulsifiers, and film-forming agents commonly found in cosmetic products.

Microalgae are gaining attention in cosmetics for their potential in treating skin issues like aging and pigmentation disorders. While some algae species are used in

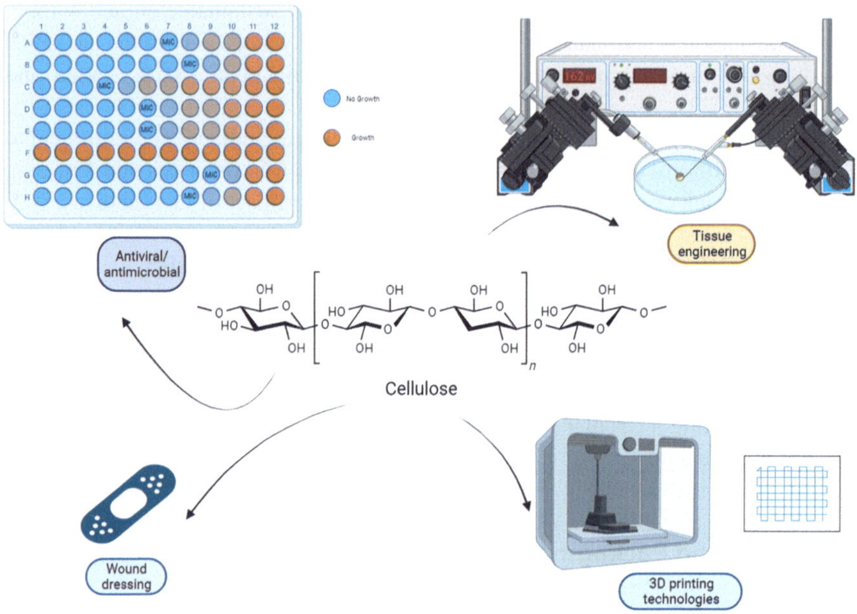

Fig. 11.2 Unique properties of cellulose fibers

cosmetic formulations for moisturizing and thickening, their full potential remains untapped as a primary active ingredient (Wang et al. 2015). Skin health is crucial, and algae, rich in bioactive compounds like vitamins and polyphenols, are recognized for their skin benefits with minimal cytotoxicity as shown in Fig. 11.3. Commercialized as skin care products, algae demonstrate efficacy in treating issues like rashes, pigmentation, and aging (El-Chaghaby and Rashad 2021).

The incorporation of algal biopolymers in cosmetics not only enhances the sustainability profile of the products but also appeals to consumers seeking natural and biodegradable alternatives. The potential of algal biopolymers to replace conventional cosmetic ingredients represents a step toward a more environmentally conscious beauty industry.

11.3.2.3 Medicine: Healing with Green Polymers

In the medical field, algal biopolymers demonstrate their versatility and biocompatibility, opening new avenues for sustainable medical applications. PHAs, in particular, have gained attention for their potential in drug delivery systems, tissue engineering, and medical implants. The biodegradability and non-toxic nature of PHAs make them suitable for medical applications, where minimizing environmental impact and ensuring patient safety are paramount.

The demand for eco-friendly alternatives to traditional plastics has led to the exploration of algae-based bioplastic production, offering biodegradability and cost-effectiveness. PHAs, naturally occurring in microbes, serve as a foundation for various biomolecules. This review outlines the global status, precise algae-based

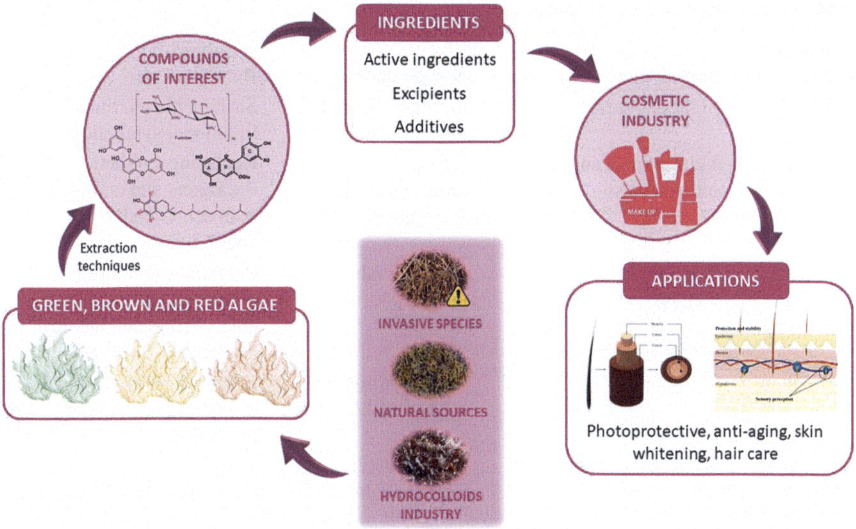

Fig. 11.3 Cosmetics production from various algae

PHA extraction methods, and bioplastic production. It also covers techniques for PHA valorization, biodegradability, and commercial applications, addressing environmental pollution concerns associated with conventional plastics (Mal et al. 2022).

Algal biopolymers offer a sustainable solution to the often-wasteful practices in the medical industry, contributing to the development of green and responsible healthcare technologies.

11.3.2.4 Agriculture: Nurturing Crops with Algae

In agriculture, algal biopolymers contribute to sustainable practices by enhancing soil health and promoting eco-friendly crop protection (Kumar et al. 2022). Alginate, a polysaccharide derived from brown algae, has been utilized as a soil conditioner and a carrier for agricultural inputs (Ma et al. 2022). The biodegradability of algal biopolymers ensures that these applications do not leave harmful residues in the soil, aligning with organic farming practices (Martínez-Cano et al. 2022).

The integration of algal biopolymers in agriculture highlights their potential not only in reducing plastic waste but also in fostering environmentally conscious farming methods that prioritize long-term soil health. Table 11.1 provides a concise synopsis of the numerous uses of biopolymers outside of the packaging industry, highlighting the particular biopolymers that are employed, their salient characteristics, and the advantages they provide in each field of application.

Table 11.1 Applications of biopolymers in various fields

Application area	Biopolymer	Key properties	Benefits
Textiles	Algal cellulose	Strength, flexibility, biodegradability	Sustainable alternative to traditional synthetic fibers
Cosmetics	Algal polysaccharides and proteins	Natural, biodegradable, versatile	Replacing synthetic thickeners and emulsifiers with eco-friendly alternatives
Medicine	Polyhydroxyalkanoates (PHAs)	Biodegradable, biocompatible	Potential use in drug delivery systems, tissue engineering, and medical implants
Agriculture	Alginate	Soil conditioning, biodegradable	Enhancing soil health and promoting eco-friendly crop protection
3D Printing	Algal biopolymers	Customizable, biodegradable	Sustainable alternative to traditional plastic filaments

11.3.3 Sustainability Challenges and Opportunities

While the applications of algal bioplastics and biopolymers hold immense promise, their large-scale production comes with inherent challenges that need to be addressed to ensure true sustainability. These challenges span economic, environmental, and social aspects, necessitating a holistic approach to overcome them.

11.3.3.1 Economic Challenges: Scaling up Responsibly

The economic viability of large-scale algal bioplastics production remains a challenge. The initial capital investment in cultivation systems, extraction processes, and manufacturing facilities can be substantial (Beckstrom et al. 2020; Bhattacharya and Goswami 2020; Shahid et al. 2020). To overcome this hurdle, research focuses on optimizing processes, reducing production costs, and exploring public-private partnerships to incentivize the adoption of algal bioplastics by industries.

11.3.3.2 Environmental Considerations: Balancing Act

While algal bioplastics boast biodegradability and reduced carbon footprints, the environmental impact of large-scale cultivation and processing must be carefully managed. Cultivating algae on a massive scale can potentially lead to eutrophication and other ecological imbalances (Glibert 2017, 2020). Sustainable practices, such as integrated cultivation systems and the use of non-arable land, must be emphasized to minimize environmental footprints (Jeswani et al. 2020; Serio et al. 2022).

Additionally, the life cycle analysis of algal bioplastics should consider the energy inputs, water usage, and overall ecological impact throughout the production process. Striking a balance between environmental benefits and potential drawbacks is essential for ensuring the overall sustainability of algal bioplastics.

11.3.3.3 Social Implications: Ethical Production and Fair Practices

As the demand for algal bioplastics grows, considerations of fair labor practices, ethical sourcing, and community involvement become crucial. Ensuring that the benefits of algal bioplastics extend beyond environmental gains to encompass social and economic benefits is essential for fostering a truly sustainable industry.

Opportunities lie in developing certifications and standards for ethical and sustainable algal bioplastics production. Collaborations with local communities and transparent supply chains can contribute to the social sustainability of the industry.

11.3.4 Future Sustainability Strategies

Addressing the challenges associated with algal bioplastics opens avenues for future sustainability strategies that can further enhance their positive impact. These strategies encompass technological innovations, policy interventions, and collaborative initiatives.

11.3.4.1 Technological Innovations: Efficiency and beyond

Advancements in cultivation techniques, extraction processes, and manufacturing technologies are critical for improving efficiency and reducing the costs associated with algal bioplastics production. Research into the genetic engineering of algae for enhanced biopolymer production, as well as the development of novel extraction methods, contributes to the technological evolution of the industry.

11.3.4.2 Policy Interventions: Incentivizing Sustainable Practices

Governmental policies play a pivotal role in shaping the landscape of sustainable materials. Incentives for industries adopting algal bioplastics, subsidies for research and development, and regulations that encourage eco-friendly alternatives contribute to the widespread adoption of algal bioplastics. Policy frameworks that prioritize sustainability and circular economy principles can accelerate the transition toward a more environmentally conscious materials industry.

11.3.4.3 Collaborative Initiatives: Uniting for Sustainability

Collaboration among stakeholders, including governments, industries, research institutions, and non-governmental organizations, is key to overcoming the multifaceted challenges associated with algal bioplastics. Knowledge-sharing, joint research projects, and global initiatives can foster a collaborative ecosystem that propels the sustainable materials industry forward.

The applications of algae-based bioplastics and biopolymers herald a new era in material science, where sustainability and functionality converge (Álvarez-Castillo et al. 2021). From replacing plastic waste in packaging to infiltrating diverse industries, algal bioplastics demonstrate their versatility and potential to reshape the way we produce and consume materials (Karimi Sani et al. 2023). While challenges exist, the commitment to addressing economic, environmental, and social considerations propels the industry toward a green horizon. Through ongoing research, technological innovation, and collaborative efforts, algal bioplastics are poised to play a pivotal role in mitigating the global plastic crisis and fostering a circular economy where materials contribute to the well-being of the planet rather than its detriment. The journey toward a sustainable future is dynamic, and as the applications of algal bioplastics continue to expand, so too does the potential for transformative change. With every application and innovation, we inch closer to a world where materials not only serve our needs but also contribute to the resilience and health of the ecosystems that sustain us. The green horizon beckons, and with algal bioplastics leading the way, we navigate toward a future where environmental harmony and human progress coexist.

11.4 The Future of Algae-Based Bioplastics and Biopolymers: Navigating Toward a Green Tomorrow

As we stand at the crossroads of environmental crisis and innovation, the future of algae-based bioplastics and biopolymers beckons as a beacon of hope. In this exploration, we delve into the research frontiers propelling this green revolution and conclude with perspectives on the potential, challenges, and a visionary road map for the widespread adoption of algae-based materials.

11.4.1 Research Frontiers: Innovation for a Greener Future

11.4.1.1 Genetically Engineered Algae: Unlocking Potential

Cutting-edge research in the field of algae-based bioplastics is charting new territories through the genetic engineering of algae. Scientists are exploring ways to enhance the natural capabilities of algae for biopolymer production. This includes modifying algae to produce higher yields of specific biopolymers, tailoring their properties for various applications, and even engineering algae to thrive in specific environments.

Genetic engineering holds the promise of optimizing the inherent potential of algae, accelerating the development of strains that are more efficient, resilient, and well-suited for large-scale bioplastics production. By leveraging the power of genetic tools, researchers aim to create algae that act as biofactories, churning out biopolymers with precision and efficiency.

11.4.1.2 New Biopolymer Discovery: Nature's Blueprints

The exploration of algae's biochemical diversity continues to yield exciting discoveries in the realm of new biopolymers. Beyond the well-known PHAs, cellulose, and polysaccharides, researchers are uncovering novel biopolymers with unique properties. These discoveries open doors to a broader palette of materials, each with its specific set of characteristics and applications.

The identification and characterization of new biopolymers contribute to the expansion of the material toolbox, offering researchers and industries more options to tailor materials to specific needs. From enhanced mechanical properties to novel functionalities, the discovery of new biopolymers adds layers to the versatility of algae-based materials.

11.4.1.3 Advanced Bioplastic Manufacturing Technologies: Precision and Sustainability

In parallel with advancements in cultivation and extraction, the manufacturing technologies for algae-based bioplastics are undergoing transformative developments. Researchers are exploring innovative methods that enhance precision, reduce waste, and improve overall efficiency in the production process.

Advanced bioplastic manufacturing technologies include techniques such as 3D printing with algal biopolymers, allowing for intricate and customized designs. Additionally, advancements in chemical processes and polymerization methods aim to streamline the transformation of extracted biopolymers into final bioplastic products. These technologies not only enhance the technical feasibility of algae-based bioplastics but also contribute to their economic viability and scalability.

11.5 A Sea of Change: Conclusion and Perspectives

11.5.1 Summarizing the Potential: A Green Resurgence

As we summarize the potential of algae-based bioplastics and biopolymers, it becomes evident that these materials have the capacity to usher in a green resurgence in the world of materials science. The applications span from sustainable packaging solutions to diverse industries such as textiles, cosmetics, medicine, agriculture, and 3D printing. The inherent biodegradability, coupled with the versatility of algal biopolymers, positions them as true contenders in the quest for sustainable alternatives to traditional plastics.

The environmental benefits of algae-based materials extend beyond their application domains. These materials have the potential to address the global plastic crisis by offering biodegradable solutions that mitigate the long-lasting impact of conventional plastics. Algae, with their rapid growth and diverse biochemical composition, embody the essence of a circular economy where materials are not only functional but also regenerative.

11.5.2 Remaining Hurdles: Navigating Challenges

While the future of algae-based bioplastics is promising, several hurdles remain on the path to widespread adoption. Economic challenges, including the initial investment in cultivation and processing infrastructure, demand innovative solutions and increased efficiency to reduce production costs. Environmental considerations, such as the ecological impact of large-scale algae cultivation, require careful management to ensure a net positive environmental outcome.

Social implications, including ethical production practices and fair labor considerations, must be integrated into the fabric of the algae-based bioplastics industry. Ensuring that the benefits of this emerging field extend beyond environmental gains to encompass social and economic benefits is crucial for the industry's long-term sustainability.

11.5.3 Vision for Widespread Adoption: A Holistic Approach

As we envision the widespread adoption of algae-based bioplastics and biopolymers, a holistic approach is paramount. This involves collaborative efforts among scientists, industry stakeholders, policymakers, and consumers. Technological innovations must continue to drive the efficiency and economic feasibility of algae-based bioplastics, while policies that incentivize sustainable practices can accelerate their integration into mainstream industries.

Education and awareness play a crucial role in fostering consumer demand for sustainable alternatives. By highlighting the environmental benefits and superior performance of algae-based materials, consumers can be empowered to make choices that contribute to a greener future. Additionally, certification and labeling standards can provide transparency, assuring consumers that products labeled as algae-based meet specific sustainability criteria.

11.6 Conclusion

In conclusion, the future of algae-based bioplastics and biopolymers is characterized by a convergence of innovation, sustainability, and a collective commitment to addressing the challenges of our time. The ongoing research frontiers in genetically engineered algae, new biopolymer discovery, and advanced manufacturing technologies underscore the dynamic nature of this field. As scientists push the boundaries of what is possible, the potential applications of algae-based materials continue to expand, offering solutions to some of the most pressing environmental issues.

As we navigate the sea of change toward a green horizon, the journey requires collaboration, ingenuity, and a shared vision for a sustainable future. Algae-based bioplastics represent more than just materials; they embody a paradigm shift toward responsible consumption and production. Through research, innovation, and collective action, we have the opportunity to steer the course toward a world where materials harmonize with the environment, ensuring a legacy of sustainability for generations to come. The green horizon beckons, and with algae-based bioplastics at the helm, we sail toward a future where the oceans are clear, and the earth thrives in balance.

References

Adeleye AT, Odoh CK, Enudi OC, Banjoko OO, Osiboye OO, Toluwalope Odediran E, Louis H (2020) Sustainable synthesis and applications of polyhydroxyalkanoates (PHAs) from biomass. Process Biochem 96:174–193. https://doi.org/10.1016/J.PROCBIO.2020.05.032

Agarwal P, Soni R, Kaur P, Madan A, Mishra R, Pandey J, Singh S, Singh G (2022) Cyanobacteria as a promising alternative for sustainable environment: synthesis of biofuel and biodegradable plastics. Front Microbiol 13:939347. https://doi.org/10.3389/FMICB.2022.939347/BIBTEX

Althuri A, Venkata Mohan S (2022) Emerging innovations for sustainable production of bioethanol and other mercantile products from circular economy perspective. Bioresour Technol 363:128013. https://doi.org/10.1016/J.BIORTECH.2022.128013

Álvarez-Castillo E, Felix M, Bengoechea C, Guerrero A (2021) Proteins from agri-food industrial biowastes or co-products and their applications as green materials. Foods 10(5):981. https://doi.org/10.3390/FOODS10050981

Arora Y, Sharma S, Sharma V (2023) Microalgae in bioplastic production: a comprehensive review. Arab J Sci Eng 48(6):7225–7241. https://doi.org/10.1007/S13369-023-07871-0/TABLES/1

Asgher M, Qamar SA, Bilal M, Iqbal HMN (2020) Bio-based active food packaging materials: sustainable alternative to conventional petrochemical-based packaging materials. Food Res Int 137:109625. https://doi.org/10.1016/J.FOODRES.2020.109625

Assunção J, Malcata FX (2020) Enclosed "non-conventional" photobioreactors for microalga production: a review. Algal Res 52:102107. https://doi.org/10.1016/J.ALGAL.2020.102107

Aswathi Mohan A, Robert Antony A, Greeshma K, Yun JH, Ramanan R, Kim HS (2022) Algal biopolymers as sustainable resources for a net-zero carbon bioeconomy. Bioresour Technol 344:126397. https://doi.org/10.1016/J.BIORTECH.2021.126397

Baranwal J, Barse B, Fais A, Delogu GL, Kumar A (2022) Biopolymer: a sustainable material for food and medical applications. *Polymers 14*(5):983. https://doi.org/10.3390/POLYM14050983

Beckstrom BD, Wilson MH, Crocker M, Quinn JC (2020) Bioplastic feedstock production from microalgae with fuel co-products: a techno-economic and life cycle impact assessment. Algal Res 46:101769. https://doi.org/10.1016/J.ALGAL.2019.101769

Behera S, Priyadarshanee M, Vandana, Das S (2022) Polyhydroxyalkanoates, the bioplastics of microbial origin: properties, biochemical synthesis, and their applications. Chemosphere 294:133723. https://doi.org/10.1016/J.CHEMOSPHERE.2022.133723

Bhattacharya M, Goswami S (2020) Microalgae—a green multi-product biorefinery for future industrial prospects. Biocatal Agric Biotechnol 25:101580. https://doi.org/10.1016/J.BCAB.2020.101580

Bleakley S, Hayes M (2017) Algal proteins: extraction, application, and challenges concerning production. Food Secur 6(5):1–34. https://doi.org/10.3390/FOODS6050033

Chen S, Sathuvan M, Zhang X, Zhang W, Tang S, Liu Y, Cheong KL (2021) Characterization of polysaccharides from different species of brown seaweed using saccharide mapping and chromatographic analysis. BMC Chem 15(1):1–11. https://doi.org/10.1186/S13065-020-00727-W/TABLES/5

Chia SR, Chew KW, Zaid HFM, Chu DT, Tao Y, Show PL (2019) Microalgal protein extraction from Chlorella vulgaris FSP-E using Triphasic partitioning technique with sonication. Front Bioeng Biotechnol 7:493723. https://doi.org/10.3389/FBIOE.2019.00396/BIBTEX

Chia WY, Ying Tang DY, Khoo KS, Kay Lup AN, Chew KW (2020) Nature's fight against plastic pollution: algae for plastic biodegradation and bioplastics production. Environ Sci Ecotechnol 4:100065. https://doi.org/10.1016/J.ESE.2020.100065

Choi HI, Sung YJ, Hong ME, Han J, Min BK, Sim SJ (2022) Reconsidering the potential of direct microalgal biomass utilization as end-products: a review. Renew Sust Energ Rev 155:111930. https://doi.org/10.1016/J.RSER.2021.111930

Costa SS, Miranda AL, de Morais MG, Costa JAV, Druzian JI (2019) Microalgae as source of polyhydroxyalkanoates (PHAs)—a review. Int J Biol Macromol 131:536–547. https://doi.org/10.1016/J.IJBIOMAC.2019.03.099

Dermawan D, Wang YF, You SJ, Jiang JJ, Hsieh YK (2022) Impact of climatic and non-climatic stressors on ocean life and human health: a review. Sci Total Environ 821:153387. https://doi.org/10.1016/J.SCITOTENV.2022.153387

El Semary N, Alsuhail M, Al Amer K, AlNaim A (2022) Applications of algae for environmental sustainability: novel bioplastic formulation method from marine green alga. Front Mar Sci 9:1047284. https://doi.org/10.3389/FMARS.2022.1047284/BIBTEX

El-Chaghaby G, Rashad S (2021) An overview of algae prospects in cosmeceuticals. J Egypt Women's Dermatol Soc 18(3):161–166. https://doi.org/10.4103/JEWD.JEWD_22_21

El-Sheekh MM, Yousuf WE, Kenawy ER, Mohamed TM (2023) Biosynthesis of cellulose from Ulva lactuca, manufacture of nanocellulose and its application as antimicrobial polymer. Sci Rep 13(1):1–12. https://doi.org/10.1038/s41598-023-37287-7

Gallo F, Fossi C, Weber R, Santillo D, Sousa J, Ingram I, Nadal A, Romano D (2020) Marine litter plastics and microplastics and their toxic chemicals components: the need for urgent preventive measures *. Analysis of Nanoplastics and Microplastics in Food, 159–179. https://doi.org/10.1201/9780429469596-10

García G, Sosa-Hernández JE, Rodas-Zuluaga LI, Castillo-Zacarías C, Iqbal H, Parra-Saldívar R (2021) Accumulation of PHA in the Microalgae scenedesmus sp. under nutrient-deficient conditions. Polymers 13(1):1–23. https://doi.org/10.3390/POLYM13010131

Glibert PM (2017) Eutrophication, harmful algae and biodiversity—challenging paradigms in a world of complex nutrient changes. Mar Pollut Bull 124(2):591–606. https://doi.org/10.1016/J.MARPOLBUL.2017.04.027

Glibert PM (2020) Harmful algae at the complex nexus of eutrophication and climate change. Harmful Algae 91:101583. https://doi.org/10.1016/J.HAL.2019.03.001

Ibn-Mohammed T, Mustapha KB, Godsell J, Adamu Z, Babatunde KA, Akintade DD, Acquaye A, Fujii H, Ndiaye MM, Yamoah FA, Koh SCL (2021) A critical analysis of the impacts of COVID-19 on the global economy and ecosystems and opportunities for circular economy strategies. Resour Conserv Recycl 164:105169. https://doi.org/10.1016/J.RESCONREC.2020.105169

Jeswani HK, Chilvers A, Azapagic A (2020) Environmental sustainability of biofuels: a review. Proc R Soc A 476:2243. https://doi.org/10.1098/RSPA.2020.0351

Karimi Sani I, Masoudpour-Behabadi M, Alizadeh Sani M, Motalebinejad H, Juma ASM, Asdagh A, Eghbaljoo H, Khodaei SM, Rhim JW, Mohammadi F (2023) Value-added utilization of fruit and vegetable processing by-products for the manufacture of biodegradable food packaging films. Food Chem 405:134964. https://doi.org/10.1016/J.FOODCHEM.2022.134964

Khanra A, Vasistha S, Rai MP, Cheah WY, Khoo KS, Chew KW, Chuah LF, Show PL (2022) Green bioprocessing and applications of microalgae-derived biopolymers as a renewable feedstock: circular bioeconomy approach. Environ Technol Innov 28:102872. https://doi.org/10.1016/J.ETI.2022.102872

Koller M (2018) A review on established and emerging fermentation schemes for microbial production of Polyhydroxyalkanoate (PHA) biopolyesters. *Fermentation 4*(2):30. https://doi.org/10.3390/FERMENTATION4020030

Kumar K, Mishra SK, Shrivastav A, Park MS, Yang JW (2015) Recent trends in the mass cultivation of algae in raceway ponds. Renew Sust Energ Rev 51:875–885. https://doi.org/10.1016/J.RSER.2015.06.033

Kumar R, Kumar N, Rajput VD, Mandzhieva S, Minkina T, Saharan BS, Kumar D, Sadh PK, Duhan JS (2022) Advances in biopolymeric Nanopesticides: a new eco-friendly/eco-protective perspective in precision agriculture. *Nanomaterials 12*(22):3964. https://doi.org/10.3390/NANO12223964

Kumar R, Verma A, Shome A, Sinha R, Sinha S, Jha PK, Kumar R, Kumar P, Shubham, Das S, Sharma P, Prasad PVV (2021) Impacts of plastic pollution on ecosystem services, sustain-

able development goals, and need to focus on circular economy and policy interventions. *Sustainability 13*(17):9963. https://doi.org/10.3390/SU13179963
Li Y, Zheng Y, Zhang Y, Yang Y, Wang P, Imre B, Wong ACY, Hsieh YSY, Wang D (2021) Brown algae carbohydrates: structures, pharmaceutical properties, and research challenges. Mar Drugs 19:11. https://doi.org/10.3390/MD19110620
Liu W, Chen Y, Wang J, Liu T (2019) Biomass productivity of Scenedesmus dimorphus (Chlorophyceae) was improved by using an open pond–photobioreactor hybrid system. Eur J Phycol 54(2):127–134. https://doi.org/10.1080/09670262.2018.1519601
Ma C, Song W, Yang J, Ren C, Du H, Tang T, Qin S, Liu Z, Cui H (2022) The role and mechanism of commercial macroalgae for soil conditioner and nutrient uptake catalyzer. Plant Growth Regul 97(3):455–476. https://doi.org/10.1007/S10725-022-00819-8
Mal N, Satpati GG, Raghunathan S, Davoodbasha MA (2022) Current strategies on algae-based biopolymer production and scale-up. Chemosphere 289:133178. https://doi.org/10.1016/J.CHEMOSPHERE.2021.133178
Manzi HP, Abou-Shanab RAI, Jeon BH, Wang J, Salama ES (2022) Algae: a frontline photosynthetic organism in the microplastic catastrophe. Trends Plant Sci 27(11):1159–1172. https://doi.org/10.1016/J.TPLANTS.2022.06.005
Martínez-Cano B, Mendoza-Meneses CJ, García-Trejo JF, Macías-Bobadilla G, Aguirre-Becerra H, Soto-Zarazúa GM, Feregrino-Pérez AA (2022) Review and perspectives of the use of alginate as a polymer matrix for microorganisms applied in agro-industry. *Molecules 27*(13):4248. https://doi.org/10.3390/MOLECULES27134248
Muhammad M, Aloui H, Khomlaem C, Hou CT, Kim BS (2020) Production of polyhydroxyalkanoates and carotenoids through cultivation of different bacterial strains using brown algae hydrolysate as a carbon source. Biocatal Agric Biotechnol 30:101852. https://doi.org/10.1016/J.BCAB.2020.101852
Nanda N, Bharadvaja N (2022) Algal bioplastics: current market trends and technical aspects. Clean Techn Environ Policy 24(9):2659. https://doi.org/10.1007/S10098-022-02353-7
Patrichi CAM, Cioroiu Tirpan DR, Aljanabi AAA, Trica B, Gifu IC, Dobre T (2023) Extraction of cellulose from ulva lactuca algae and its use for membrane synthesis. Polymers 15(24):4673. https://doi.org/10.3390/POLYM15244673
Rosenboom JG, Langer R, Traverso G (2022) Bioplastics for a circular economy. Nat Rev Mater 7(2):117–137. https://doi.org/10.1038/s41578-021-00407-8
Sahu S, Kaur A, Khatri M, Singh G, Arya SK (2023a) A review on cutinases enzyme in degradation of microplastics. J Environ Manag 347:119193. https://doi.org/10.1016/J.JENVMAN.2023.119193
Sahu S, Kaur A, Singh G, Kumar Arya S (2023b) Harnessing the potential of microalgae-bacteria interaction for eco-friendly wastewater treatment: a review on new strategies involving machine learning and artificial intelligence. J Environ Manag 346:119004. https://doi.org/10.1016/J.JENVMAN.2023.119004
Sahu S, Sharma S, Kaur A, Singh G, Khatri M, Arya SK (2023c) Algal carbohydrate polymers: catalytic innovations for sustainable development. Carbohydr Polym 327:121691. https://doi.org/10.1016/J.CARBPOL.2023.121691
Samadhiya K, Sangtani R, Nogueira R, Bala K (2022) Insightful advancement and opportunities for microbial bioplastic production. Front Microbiol 12:674864. https://doi.org/10.3389/FMICB.2021.674864/BIBTEX
Schoenrock KM, McHugh TA, Krueger-Hadfield SA (2021) Revisiting the 'bank of microscopic forms' in macroalgal-dominated ecosystems. J Phycol 57(1):14–29. https://doi.org/10.1111/JPY.13092
Serio F, Manuel R, Machado A, Fussy A, Papenbrock J (2022) An overview of soil and soilless cultivation techniques—chances, challenges and the neglected question of sustainability. *Plants 11*(9):1153. https://doi.org/10.3390/PLANTS11091153
Shahid A, Malik S, Khan AZ, Liu C-G, Mehmood MA (2020) Multiproduct algal biorefineries: challenges and opportunities:513–537. https://doi.org/10.1007/978-981-15-9593-6_20

Sreenikethanam A, Bajhaiya A, Sreenikethanam A, Bajhaiya A (2021) Algae based bio-plastics: future of green economy. Biorefineries – Selected Processes. https://doi.org/10.5772/INTECHOPEN.100981

Tan FHP, Nadir N, Sudesh K (2022) Microalgal biomass as feedstock for bacterial production of PHA: advances and future prospects. Front Bioeng Biotechnol 10. https://doi.org/10.3389/FBIOE.2022.879476

Tyagi P, Salem KS, Hubbe MA, Pal L (2021) Advances in barrier coatings and film technologies for achieving sustainable packaging of food products—a review. Trends Food Sci Technol 115:461–485. https://doi.org/10.1016/J.TIFS.2021.06.036

Wahlström N, Edlund U, Pavia H, Toth G, Jaworski A, Pell AJ, Choong FX, Shirani H, Nilsson KPR, Richter-Dahlfors A (2020) Cellulose from the green macroalgae Ulva lactuca: isolation, characterization, optotracing, and production of cellulose nanofibrils. Cellulose 27(7):3707–3725. https://doi.org/10.1007/S10570-020-03029-5/TABLES/5

Wang HMD, Chen CC, Huynh P, Chang JS (2015) Exploring the potential of using algae in cosmetics. Bioresour Technol 184:355–362. https://doi.org/10.1016/J.BIORTECH.2014.12.001

Yao W, Kong Q, You L, Zhong S, Hileuskaya K (2023) Polysaccharides from brown seaweed: physicochemical properties, absorption in the intestine, and beneficial effects on intestinal barrier. Food Front 4(4):1547–1560. https://doi.org/10.1002/FFT2.288

Yashavanth PR, Das M, Maiti SK (2021) Recent progress and challenges in cyanobacterial autotrophic production of polyhydroxybutyrate (PHB), a bioplastic. J Environ Chem Eng 9(4):105379. https://doi.org/10.1016/J.JECE.2021.105379

Yong WTL, Thien VY, Rupert R, Rodrigues KF (2022) Seaweed: a potential climate change solution. Renew Sust Energ Rev 159:112222. https://doi.org/10.1016/J.RSER.2022.112222

Yun JH, Cho DH, Lee S, Heo J, Tran QG, Chang YK, Kim HS (2018) Hybrid operation of photobioreactor and wastewater-fed open raceway ponds enhances the dominance of target algal species and algal biomass production. Algal Res 29:319–329. https://doi.org/10.1016/J.ALGAL.2017.11.037

Zanchetta E, Damergi E, Patel B, Borgmeyer T, Pick H, Pulgarin A, Ludwig C (2021) Algal cellulose, production and potential use in plastics: challenges and opportunities. Algal Res 56:102288. https://doi.org/10.1016/J.ALGAL.2021.102288

Bioactive Compounds from Algae: Extraction, Purification, and Applications

12

Uplabdhi Tyagi and Anil Kumar Sarma

Contents

U. Tyagi
University School of Chemical Technology, Guru Gobind Singh Indraprastha University, Dwarka, Delhi, India

A. K. Sarma (✉)
Chemical Conversion Division, Sardar Swaran Singh National Institute of Bio-Energy (An Autonomous Institute of MNRE, Government of India), Kapurthala, Punjab, India
e-mail: anil.sarma16@gov.in

S. K. Arya et al. (eds.), *Value Added Products From Bioalgae Based Biorefineries: Opportunities and Challenges*,
https://doi.org/10.1007/978-981-97-1662-3_12

Abstract

Algae are a diverse group of photosynthetic organisms that can be found in various aquatic environments, including oceans, rivers, lakes, and even in some terrestrial habitats. These organisms have gained significant attention in recent years due to their rich content of bioactive compounds, which are molecules that have positive effects on living organisms. These include fatty acids, phycobiliproteins, chlorophylls, carotenoids, and vitamins that can be widely used in pharmaceuticals, cosmetics, food additives, and ingredients. The natural bioactive compounds from microalgae are attractive and may have great possibilities for commercialization in the near future due to their potential therapeutic activities. These properties include antioxidant, antiviral, antibacterial, antifungal, anti-inflammatory, antitumor, and antimalarial effects. Multiple improvements in the purification and analytical tools to characterize these compounds have been reported in recent years, aiming to gain further insight into the complexity of different molecular structures of bioactive peptides and carbohydrates. However, despite the extensive use of algae-derived compounds and extracts in the food industry, there are still no FDA-approved anticancer, antioxidant, anti-inflammatory, and antidiabetic drugs. Technology transfer from the pre-clinical results to the clinical application of secondary metabolites extracted from marine algae is still in its infancy and not fully exploited and more clinical studies are needed to really evaluate the pharmaceutical efficacy algal compounds.

Keywords

Bioactive compounds · Microalgae · Purifications · Extraction strategy · Utilization

Abbreviations

(DHA)	Docosahexaenoic acid
ASE	Accelerated solvent extraction
BCA	Bicinchoninic acid (Bca)
EPA	Eicosapentaenoic acid
GC	Gas chromatography
HPLC	High-performance liquid chromatography
IEF	Isoelectric focusing
LC-MS/MS	Liquid chromatography-mass spectrometry
PLE	Pressurized liquid extraction
PSE	Pressurized solvent extraction

PUFAs	Polyunsaturated fatty acids
SDS-PAGE	Sodium dodecyl sulfate-polyacrylamide gel electrophoresis
SFE	Supercritical fluid extraction
SPE	Solid-phase extraction

12.1 Introduction

The potential of traditional food ingredients to produce nutraceuticals and functional foods has drawn more scientific interest. Proteins, carbs, phenols, vitamins, minerals, and other important macro- and micronutrients are all abundant in algae. Due to their nutritional value—for example, as a source of proteins rich in important amino acids or as a means of obtaining non-digestible carbohydrates from seaweed as a source of dietary fiber—these substances have historically been utilized by the food and animal feed industries. Algal composition is rich and varied; it is being investigated for its potential to yield functional elements including carbohydrates and bioactive peptides. There is no question that the identification and extraction of bioactive chemicals from natural sources remain relevant as a research topic in the modern era. One of the significant emerging sources of bioactive compounds is microalgae; they have a number of intriguing properties, including the large number of strains and species that may be screened to find intriguing bioactive components that may then be used in commercial products, the potential to separate them from marine environments or use bioreactors to grow biomass, the least amount of arable land needed for their production, as coastal regions are not always associated with microalgal farming, the potential to cultivate them in carefully regulated environments that may be adjusted and changed to promote the overproduction of specific target chemicals. Modern, ecologically friendly extraction methods and procedures are being preferred. It is a known fact that conventional extraction techniques have a low extraction efficiency and require large amounts of organic solvents. This makes it essential to develop innovative procedures that boost sustainability and efficiency without requiring the use of hazardous solvents.

12.2 Extraction of Bioactive Compounds from Microalgae

12.2.1 Extraction of Lipids from Microalgae

Extraction of lipids from microalgae is a crucial process in various industries, including biofuel production, nutraceuticals, and cosmetics. Microalgae are a diverse group of microscopic photosynthetic organisms that can accumulate a significant amount of lipids, which are valuable for their high energy content and potential health benefits. The first step involves selecting a microalgal strain that has the potential to accumulate high levels of lipids. Various species of microalgae, such as *Chlorella*, *Nannochloropsis*, and *Scenedesmus*, are commonly used due to their lipid-rich composition. The microalgae are then cultivated under controlled conditions, including light, temperature, and nutrient supply, to promote lipid production. Once the microalgae have reached a sufficient lipid content, they need to be harvested from the culture medium. This can be achieved through different methods such as centrifugation, flocculation, filtration, and flotation. The choice of method depends on factors such as the size of the microalgae, the lipid content, and the desired scale of production. Microalgal cells have a rigid cell wall that needs to be disrupted to release the intracellular lipids. Cell disruption methods include mechanical techniques (e.g., bead milling, ultrasonication), chemical methods (e.g., using solvents), and enzymatic methods. The goal is to break open the cells and release the lipid-rich content. After cell disruption, the lipid molecules are extracted from the cellular material. Solvent extraction is a commonly used method, where non-polar solvents (e.g., hexane, chloroform) are employed to dissolve lipids. This process takes advantage of the lipophilic nature of lipids, allowing them to be separated from other cellular components. However, solvent extraction can be resource-intensive and environmentally unfriendly. An important consideration in the lipid extraction process is the recovery of solvents used in the extraction step. Solvent recovery systems can be employed to minimize waste and reduce the environmental impact of the extraction process. The extracted lipid mixture may contain impurities such as pigments, proteins, and carbohydrates. Additional purification steps may be required to refine the lipid extract and improve its quality. Techniques like chromatography, crystallization, and winterization can be used for this purpose. Analytical methods such as gas chromatography (GC) and high-performance liquid chromatography (HPLC) are used to quantify and characterize the lipid composition of the extracted material. This information is valuable for process optimization and quality control.

12.2.1.1 Supercritical Fluid Extraction for Microalgal Lipid Extraction

Supercritical fluid extraction (SFE) is an advanced and efficient method for extracting lipids from microalgae. It utilizes a supercritical fluid, usually carbon dioxide (CO_2), as the solvent to extract lipids from the microalgal biomass as shown in Fig. 12.1. Supercritical fluids have properties that make them ideal for extraction processes, including their ability to behave as both a gas and a liquid, high diffusivity, low viscosity, and tunable solvating power. Carbon dioxide becomes supercritical at a specific combination of pressure and temperature, known as the critical point. At this point, CO_2 has properties of both a gas and a liquid, making it an excellent solvent for extracting lipids from microalgal cells. Supercritical CO_2 can penetrate the cell walls, dissolve lipids, and then be easily separated from the lipid-rich extract. The supercritical fluid extraction process involves the following steps:

- Preparation: The microalgal biomass is usually dried to remove excess water, as water can interfere with the supercritical CO_2 extraction process.
- Loading: The dried microalgal biomass is loaded into an extraction vessel.

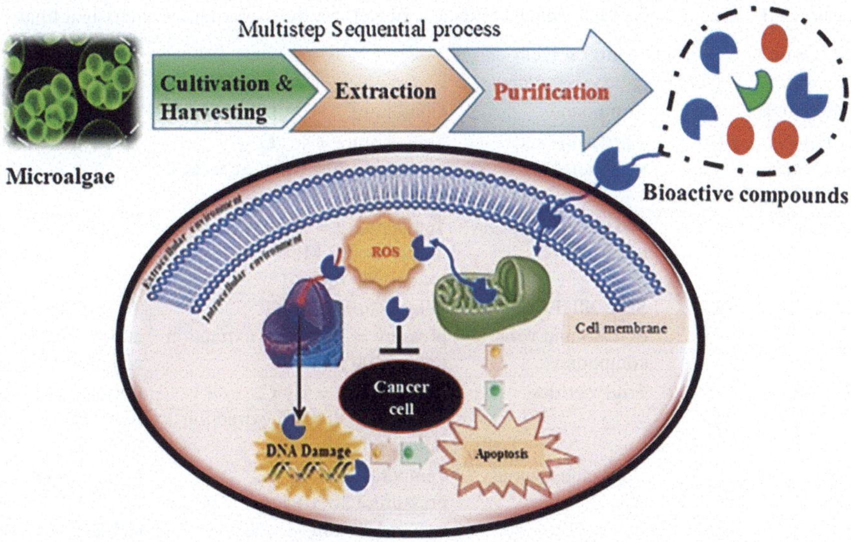

Fig. 12.1 Extracting bioactive compounds from microalgae using supercritical fluid extraction

- Pressurization and temperature control: The extraction vessel is pressurized with CO_2 to reach supercritical conditions. The temperature is carefully controlled to ensure optimal lipid solubility without causing degradation.
- Extraction: Supercritical CO_2 is passed through the biomass, allowing it to dissolve the lipids. The supercritical CO_2 becomes enriched with lipids as it interacts with the microalgal cells.
- Separation: The lipid-enriched supercritical CO_2 is then separated from the microalgal biomass using a pressure reduction or a temperature increase. As the CO_2 returns to its gaseous state, it leaves behind the extracted lipids.

Supercritical fluid extraction offers several advantages over traditional solvent extraction methods such as supercritical CO_2's solubility can be fine-tuned by adjusting pressure and temperature, allowing selective extraction of lipids while leaving other cellular components behind. Unlike traditional solvent extraction methods that may leave behind solvent residues, SFE with CO_2 is considered cleaner and leaves minimal residue in the extracted lipids. Further, SFE is widely applied for the recovery of products from natural matrices due to the properties of supercritical fluids which combine liquid- and gas-like behavior that favors the extraction of numerous compounds compared to conventional solvents in terms of quality and quantity. Furthermore, the selectivity of SFE can be easily tuned by changing the extraction conditions, that is, pressure and temperature, or by using a co-solvent. Supercritical CO_2 extraction is carried out at relatively mild temperatures, reducing the potential for thermal degradation of sensitive lipid molecules. CO_2 is non-toxic, non-flammable, and readily available, making it an environmentally friendly solvent choice compared to organic solvents. Table 12.1 summarizes the recent

Table 12.1 Recent applications related to the use of SFE for the extraction of microalgal lipids

SFE with sc-CO_2 as extraction solvent			
Microalgae	Compounds	Conditions	Yield and/or goal
Nannochloropsis oculata	Fatty acids and tocopherol	Temperature = 80 °C, pressure = 20.7 MPa, extraction time = 240 min	71%
Tetraselmis suecica	Fatty acids and tocopherol	Temperature = 40 °C, pressure = 62 MPa, extraction time = 240 min	69.9%
Dunaliella salina	SFA, MUFAs, PUFAS and volatile compounds	Temperature = 9.8 °C, pressure = 31.4 MPa, extraction time = 100 min	To maximized antioxidant activity
Isochrysis galbana	Triglycerides	Temperature = 51 °C, pressure = 30 MPa, extraction time = 60 min	Purified and rich-TG fraction
Crypthecodinium cohnii	Docosahexaenoic acid	Temperature = 50 °C, pressure = 30 MPa, extraction time = 180 min	Lipid extract with 72% DHA composition
Scenedesmus obliquus and Scenedesmus obtusiusculus	Triglycerides and PUFAs	Temperature = 20 °C, pressure = 10 MPa, extraction time = 540 min	Recovery of 92% of lipids with 59% of PUFAs
SFE with sc-CO_2 + co-solvent as extraction solvent			
Isochrysis galbana	Mid and highly polar lipids	Extraction solvent = Sc-CO_2/ ethanol, temperature = 50 °C, pressure = 7.38 MPa, extraction time = 60 min	Polar lipid-purified fraction
Scenedesmus obliquus and Chlorella protothecoides	ALA	Extraction solvent = Sc-CO_2/ ethanol, temperature = 60 °C, pressure = 30 MPa, extraction time = 90 min	Low ω6:ω3 ratio
–	Polar lipids	Extraction solvent = Sc-CO_2/ ethanol, temperature = 51.85 °C, pressure = 30 MPa, extraction time = 270 min	13.4/24.0% extraction yield
Cell wall pretreatment + SFE extraction			
Chlorella vulgaris	SFA, MUFAs, PUFAs	Pretreatment; extraction Solvent = MW-SFE; sc-CO_2, temperature = 40 °C, pressure = 28 MPa, extraction time = 540 min	Extract with 80% of FA
Phaeodactylum tricornutum	Triglycerides-rich fraction	Pretreatment; extraction Solvent = MW-SFE; sc-CO_2, temperature = 45 °C, pressure = 35 MPa, extraction time = 125 min	7% extraction yield of a purified TG fraction

(continued)

Table 12.1 (continued)

SFE with sc-CO_2 as extraction solvent			
Microalgae	Compounds	Conditions	Yield and/or goal
Schizochytrium limacinum	Docosahexaenoic acid US	Pretreatment; extraction Solvent = sc-CO_2 ethanol + urea purification, temperature = 45 °C, pressure = 35 MPa, extraction time = 60 min	60.4% DHA
Chlorella vulgaris	Neutral and polar lipids	Pretreatment; extraction Solvent = Bead milling-SFE; sc-CO_2/ethanol, temperature = 60 °C, pressure = 60 MPa, extraction time = 90 min	53% extraction yield
Nannochloropsis oculata	Triglycerides	Pretreatment; extraction Solvent = Air-flow-SFE; sc-CO_2, temperature = 60 °C, pressure = 40 MPa, extraction time = 15 min	Recovery of more than 90% of triglycerides

applications related to the use of SFE for the extraction of microalgal lipids. Supercritical fluids have better penetration and mass transfer properties compared to liquid solvents, resulting in higher extraction efficiency. SFE operates at lower temperatures compared to traditional extraction methods, leading to lower energy consumption and potentially less damage to thermally sensitive compounds (Agarwal et al. 2023). While supercritical fluid extraction offers numerous advantages, it is important to consider factors like equipment costs, pressure vessel safety, and the potential for co-extraction of unwanted compounds. Process optimization is essential to achieve the desired lipid yield and quality. Until 2022, more than 150 patents regarding SFE of microalgae have been recorded, concerning both pigment and lipid extraction, laboratory and large-scale application. Indicatively, approximately 43% of patents cover pigments, of which 84% referred to carotenoids and 13% to chlorophylls, while 29% of the total concerned extraction of lipids from microalgae.

12.2.1.2 Pressurized Liquid Extraction for Microalgal Lipid Extraction

Pressurized liquid extraction (PLE), also known as accelerated solvent extraction (ASE) or pressurized solvent extraction (PSE), is a modern extraction technique used to efficiently extract lipids and other compounds from solid samples like microalgal biomass. PLE involves the use of elevated temperatures and pressures to enhance the solubility of target compounds in the extraction solvent, thus increasing the extraction efficiency. In PLE, a solid sample (such as dried microalgal biomass) is placed in a sample cell or cartridge within an extraction instrument. The sample cell is then filled with a suitable solvent. The instrument uses controlled temperature and pressure to create conditions where the solvent behaves as both a liquid and a gas, effectively enhancing its ability to solubilize the target lipids. The PLE process involves several key steps:

- Sample loading: The dried microalgal biomass is loaded into the sample cell or cartridge. The sample is typically ground or homogenized to increase the surface area for efficient extraction.
- Solvent selection: A solvent with high affinity for lipids is chosen. Commonly used solvents include organic solvents like hexane, ethanol, or a mixture of solvents.
- Temperature and pressure control: The extraction instrument raises the temperature and pressure within the sample cell to conditions above the solvent's critical point. This promotes the solubility of lipids in the solvent.
- Extraction: The solvent is forced through the sample cell or cartridge containing the microalgal biomass. As the solvent passes through the sample, it dissolves the lipids and other target compounds. The elevated temperature and pressure accelerate the extraction process.
- Collection: The lipid-enriched solvent, now containing the extracted lipids, is collected and separated from the solid residue. This collected extract is then ready for further processing and analysis.

Pressurized liquid extraction offers several advantages for lipid extraction from microalgae such as the combination of elevated temperature and pressure results in rapid and efficient extraction, requiring shorter extraction times compared to traditional methods. Table 12.2 summarizes the recent applications related to the use of PLE for the extraction of microalgal lipids. PLE can achieve higher lipid yields due to enhanced solubility of lipids at elevated temperatures and pressures. PLE uses a lower volume of solvent compared to traditional extraction methods, making it a more environmentally friendly option. By adjusting the solvent composition and extraction conditions, PLE can selectively target specific lipid fractions (Ampofo and Abbey 2022). PLE instruments can be automated, ensuring consistent and reproducible extraction results across samples. PLE requires less manual intervention compared to some traditional extraction methods, saving time and effort. Further, Table 12.3 summarizes the economic feasibility of different extraction methods.

12.2.2 Extraction of Carbohydrates from Microalgae

The extraction of carbohydrates from microalgae is an important process for various applications, including biofuel production, food supplements, and the development of functional foods as shown in Fig. 12.2. Carbohydrates are essential compounds present in microalgae that can serve as a valuable source of energy and bioactive molecules. Choosing the right microalgal strain is crucial for optimizing carbohydrate yield. Some microalgae species naturally accumulate higher levels of

Table 12.2 Recent applications related to the use of PLE for the extraction of microalgal lipids

Microalgae	Compounds	Operating conditions	Yield and/or goal
Haematococcus pluvialis	Short-chain fatty acid	Extraction solvents = ethanol, temperature = 100 °C, pressure = 10.34 MPa, extraction time = 20 min	Antiviral activity
Dunaliella salina	Short-chain fatty acid	Extraction solvents = ethanol, temperature = 160 °C, pressure = 10.34 MPa, extraction time = 15 min	Antiviral activity
Chlorella vulgaris	Phytol, tocopherol and phytosterols	Extraction solvents = water, temperature = 150–200 °C, pressure = 10.34 MPa, extraction time = 20 min	Antimicrobial and antioxidant activity
Nannochloropsis oculata	Polyunsaturated fatty acid	Extraction solvents = ethanol, temperature = 60 °C, pressure = 10–12 MPa, extraction time = 48 min	36.4% extraction yield
Arthrospira platensis (Spirulina), Phormidium sp, Anabaena planctonica Stigeoclonium sp	Medium and polar fatty acid	Extraction solvents = limonene/ethanol, temperature = 200 °C, pressure = 20.7 MPa, extraction time = 15 min	Recovery of 70% of lipids
Haematococcus pluvialis	Short-chain fatty acid and tocopherol	Extraction solvents = water, temperature = 200 °C, pressure = 10.34 MPa, extraction time = 20 min	Antimicrobial and antioxidant activity

carbohydrates, while others may prioritize lipid production. The cultivation conditions, including light, temperature, and nutrient availability, can influence carbohydrate content in microalgae (Bhatnagar et al. 2023). Microalgal biomass needs to be harvested from the culture medium. Various methods such as centrifugation, filtration, and flocculation can be used based on the scale of production and the nature of the microalgae. Just like in lipid extraction, microalgal cells need to be disrupted to release the intracellular carbohydrates. Mechanical techniques (e.g., bead milling, ultrasonication) or enzymatic methods can be employed to break open the cells. The extracted microalgal biomass is then subjected to carbohydrate extraction. Several methods can be used for this purpose:

Hot Water Extraction: This method involves subjecting the microalgal biomass to high-temperature water to extract soluble carbohydrates. The resulting extract can be concentrated and further processed.

Enzymatic Hydrolysis: Enzymes can be used to break down complex carbohydrates into simpler sugars. This method can yield a mixture of mono- and oligosaccharides, which can be further separated and purified.

Table 12.3 Comparison of different extraction methods: cost and energy efficiency

Method	Lipid extraction efficiency	Efficiency rating	Cost involved	Energy requirement
Use of organic solvents such as chloroform/ methanol, hexane, and ether	Extraction efficiency depends on the species used, volume of the extractor, reaction time, sample volume, moisture content, types of lipids present, and in case of solvent-based methods, choice of the solvents, solvent ratios, etc.	Moderate	High due to the use of solvents. Re-use may help save some costs but again an energy-intensive process	Energy intensive
Pressurized solvent extraction		High	High because of cumulative costs incurred by use of solvent as well as use of pressurized nitrogen	Energy intensive
Isotonic extraction		Moderate-high	High cost of the solvents as the solvents used are synthetic "green" non-volatile solvents	Energy intensive
Supercritical Fluid extraction		High	High cost	Energy-intensive due to use of high pressure
Expeller press		Low-moderate	High cost	Energy intensive
Bead beating		Moderate	Cost-effective	Energy intensive. Reactor should be suitably designed to reduce energy inputs
Microwave		Very high	Initial investment and maintenance costs high	Energy demand is too high (also requires energy for cooling)

(continued)

Table 12.3 (continued)

Method	Lipid extraction efficiency	Efficiency rating	Cost involved	Energy requirement
Sonication method		High	Initial investment and maintenance costs high	Energy intensive (requires energy for both sonication and cooling)
Osmotic shock method		Moderate-high	Low-cost method	Less energy
Electroporation		Very high	Initial investment and maintenance costs high, but operates at comparatively lower costs	Less energy

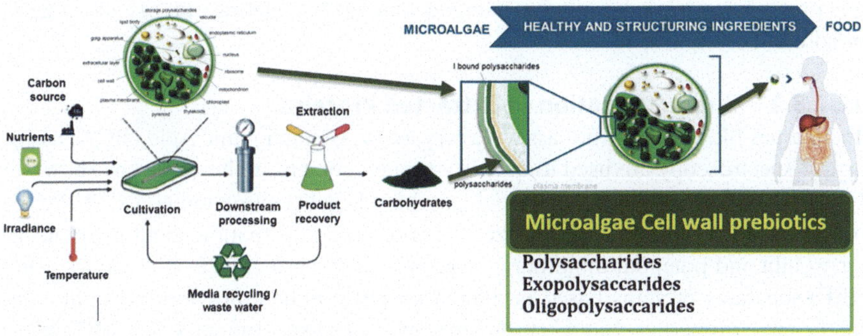

Fig. 12.2 Environmentally conscious and sustainable extraction of bioactive carbohydrates from microalgae

Acid Hydrolysis: Acidic conditions can be used to hydrolyze carbohydrates into their monomeric components. However, this method requires careful control to avoid degradation and loss of sugars.

After extraction, the carbohydrate-rich solution needs to be separated from cellular debris and impurities. Centrifugation, filtration, and precipitation are common techniques used for separation. Further purification steps may be needed to obtain a high-purity carbohydrate extract. Analytical techniques such as high-performance liquid chromatography (HPLC) and gas chromatography (GC) can be used to quantify and characterize the carbohydrate content of the extract (Chen et al. 2020). This information helps in optimizing the extraction process and assessing the quality of the carbohydrate extract.

12.2.3 Extraction and Characterization of Proteins from Microalgae

Extraction and characterization of proteins from microalgae are important processes with implications for various industries, including food, feed, nutraceuticals, and biotechnology. Microalgae are rich sources of proteins that can offer high nutritional value and bioactive properties.

12.2.3.1 Extraction of Proteins

Similar to other microalgal compounds, the first step is to disrupt the microalgal cells to release proteins. Mechanical methods like bead milling, ultrasonication, and high-pressure homogenization can be used. Enzymatic and chemical methods can also be considered. After cell disruption, the mixture is typically centrifuged or filtered to separate the cell debris and solid components from the liquid containing the protein extract (Cunha et al. 2023). The protein-rich liquid extract is subjected to protein solubilization using appropriate buffers or solvents. Adjusting pH and ionic strength can help enhance protein solubility. Proteins can be further concentrated and precipitated from the solution using techniques like salting out, pH adjustment, or addition of organic solvents.

12.2.3.2 Characterization of Extracted Proteins

Techniques like the Bradford assay, Lowry assay, Bicinchoninic Acid (BCA) assay, or UV spectroscopy are used to quantify protein content in the extract as shown in Fig. 12.3. Sodium dodecyl sulfate-polyacrylamide gel electrophoresis (SDS-PAGE) helps separate proteins based on size. This provides information about the molecular weight and potential multimeric structures of the proteins. Isoelectric focusing (IEF) separates proteins based on their isoelectric points, providing insights into their charge properties. Liquid chromatography-mass spectrometry (LC-MS/MS) is used for protein identification and structural characterization, providing information about amino acid sequences and post-translational modifications. Functional properties like solubility, emulsifying properties, and foaming properties can be determined (Hassan et al. 2022). These properties impact the potential applications of the extracted proteins. Quantifying the amino acid composition provides valuable information about the nutritional quality and potential health benefits of the proteins. Assays can be conducted to assess potential bioactive properties of the proteins, such as antioxidant activity, antimicrobial activity, and immunomodulatory effects.

12.2.4 Extraction of Minor Components from Microalgae

Within the chemical composition of microalgae, several minor components are present which may represent important targets from a bioactivity point of view. These components are considered as secondary metabolites and are produced by the cells in response to different growing conditions. For this reason, multiple approaches have been studied to achieve the over-production of certain compounds

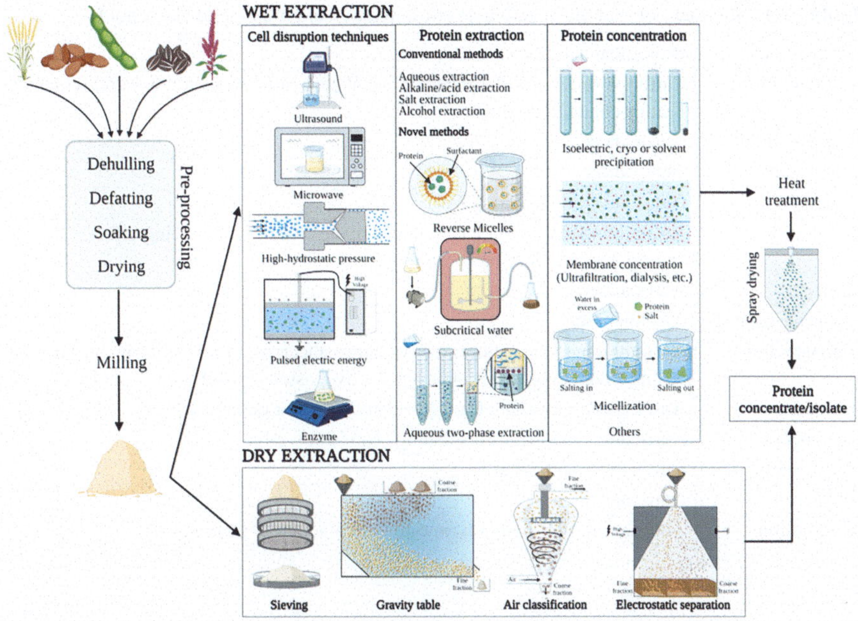

Fig. 12.3 Extraction of plant-based proteins from sustainable sources

included in this category, which is one of the strongest points to be considered for the large-scale cultivation of microalgae.

12.2.4.1 Carotenoids

Extraction of carotenoids from microalgae is a critical process due to the potential health benefits and various applications of these pigments. Table 12.4 summarizes the biological activities attributed to main microalgal carotenoids. Carotenoids are natural compounds responsible for the vibrant colors seen in many microalgae and are known for their antioxidant properties, provitamin A activity, and potential role in human health. Choosing a microalgal strain that naturally accumulates high levels of carotenoids is crucial. Factors such as light intensity, nutrient availability (especially nitrogen and carbon), and growth conditions can influence carotenoid content in microalgae. Microalgal biomass needs to be harvested from the culture medium. Various methods such as centrifugation, filtration, and flocculation can be used based on the scale of production and the nature of the microalgae. To extract carotenoids from microalgal cells, the cell walls need to be disrupted (Kaur et al. 2023). Mechanical methods like bead milling, ultrasonication, and high-pressure homogenization are commonly used. After cell disruption, carotenoids are extracted using appropriate solvents. Commonly used solvents include organic solvents like acetone, ethanol, and hexane, which have high affinity for carotenoids. Supercritical CO_2 extraction and pressurized liquid extraction (PLE) are also emerging methods for carotenoid extraction. Once extracted, the carotenoid-rich solvent is separated

Table 12.4 Summary of biological activities attributed to main microalgal carotenoids

Carotenoids	Biological activities
Astaxanthin	Cardioprotective, anti-cancer, anti-diabetes, anti-inflammatory, antioxidant, beauty-enhancing effect
Canthaxanthin	Anti-cancer
Capsanthin	Anti-cancer
α-Carotene	Anti-cancer, anti-diabetes
β-Carotene	Anti-angiogenic, anti-cancer, anti-diabetes, beauty-enhancing effect, age-related macular degeneration
Crocetin	Anti-cancer
β-Cryptoxanthin	Anti-diabetes
Fucoxanthin	Anti-angiogenic, cardioprotective, anti-cancer, anti-diabetes, anti-obesity, beauty-enhancing effect, neuroprotective, osteo-protective, weight loss
Lutein	Anti-cancer, anti-diabetes, age-related macular degeneration
Lycopene	Anti-cancer, anti-diabetes, beauty-enhancing effect
Neoxanthin	Anti-cancer, anti-obesity
Phytoene	Anti-cancer
Zeaxanthin	Anti-cancer, anti-diabetes, age-related macular degeneration

from cell debris and impurities. Additional purification steps such as column chromatography, flash chromatography, and solid-phase extraction may be used to isolate specific carotenoids. Several techniques are used to characterize extracted carotenoids such as HPLC which is commonly used to identify and quantify individual carotenoids within the extract. UV-Vis spectroscopy can provide information about the presence and concentration of carotenoids in the extract. Mass spectrometry can provide detailed structural information about carotenoids and their derivatives. TLC can be used to separate and visualize different carotenoid compounds in the extract.

12.2.4.2 Phenolic Compounds

Extraction of phenolic compounds from microalgae is a valuable process due to the potential health benefits and various applications of these bioactive compounds. Phenolic compounds are known for their antioxidant, anti-inflammatory, and potentially cardioprotective properties as shown in Fig. 12.4. Extracting phenolics from microalgae involves specific techniques to effectively release and capture these compounds. Selecting microalgal strains that naturally accumulate phenolic compounds is important. Environmental conditions, including light intensity, nutrient availability, and growth phase, can influence the content and type of phenolic compounds present in microalgae. Microalgal biomass needs to be harvested from the culture medium. Methods such as centrifugation, filtration, and flocculation are commonly used for this purpose. To release phenolic compounds from microalgal cells, the cell walls need to be disrupted. Mechanical methods like bead milling, ultrasonication, and high-pressure homogenization can be used (Liu et al. 2021). After cell disruption, phenolic compounds are extracted using appropriate solvents.

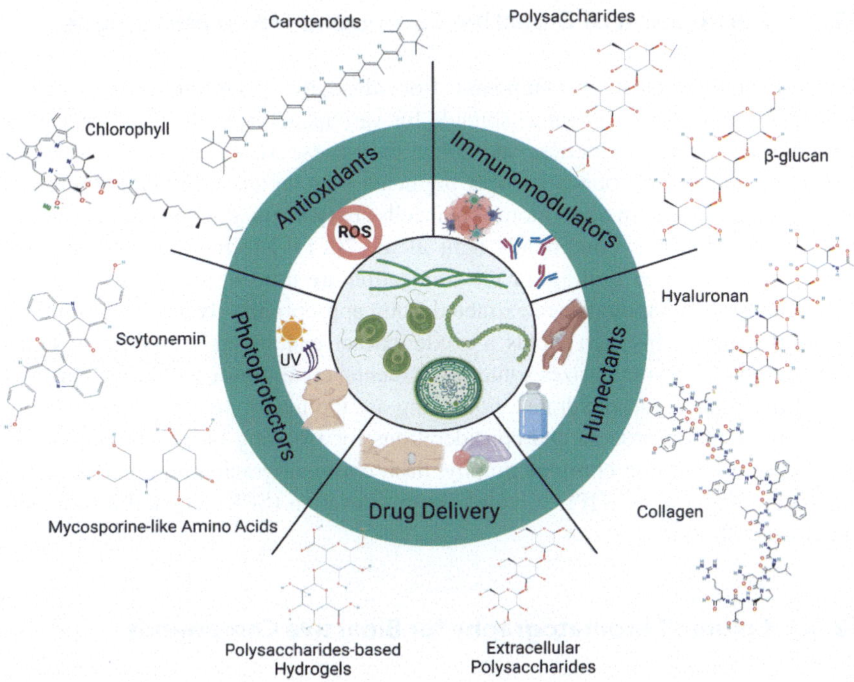

Fig. 12.4 Utilizations of bioactive compounds extracted from microalgae

Ethanol, methanol, and acetone are common solvents used for phenolic extraction. Aqueous solvents or mixtures of solvents can also be used, depending on the polarity of the target compounds. The phenolic-rich solvent needs to be separated from cell debris and impurities. Filtration, centrifugation, or decantation can be employed for this purpose. The extracted solution may be concentrated and further purified to isolate specific phenolic compounds. Techniques like rotary evaporation, vacuum distillation, and solid-phase extraction can be used. Extracted phenolic compounds from microalgae have various applications:

- Functional foods: Phenolic compounds can enhance the nutritional value of foods by providing antioxidant and potential health benefits.
- Pharmaceuticals: Phenolics are studied for their potential anti-inflammatory and cardioprotective effects.
- Cosmetics: Phenolic compounds are used in skincare products for their potential antioxidant and skin-soothing properties.
- Bioactive supplements: Extracted phenolics can be incorporated into dietary supplements for their health-promoting effects.

12.3 Purification of Bioactive Compounds from Microalgae

The purification of bioactive compounds from algae involves refining crude extracts to obtain high-purity compounds suitable for various applications in pharmaceuticals, nutraceuticals, cosmetics, and other industries. Algae are rich sources of diverse bioactive compounds such as pigments (carotenoids), lipids, polysaccharides, phenolics, proteins, and more. Algal cells are typically disrupted to release the intracellular bioactive compounds (Liu et al. 2021). Mechanical methods like homogenization, bead milling, or ultrasonication are used to break down the cell walls. Bioactive compounds are extracted using appropriate solvents based on their chemical nature. This step yields a crude extract containing a mixture of compounds. Table 12.5 summarizes some experimental conditions used for the purification of macroalgal carbohydrates. Following are the major steps that are essential for the purification process and considerations for isolating bioactive compounds from algae are column chromatography, flash chromatography, high-performance liquid chromatography (HPLC), solid-phase extraction (SPE), crystallization, and membrane filtration.

12.3.1 Column Chromatography for Bioactive Compounds

Column chromatography plays a significant role in the purification of bioactive compounds from microalgae. Its versatility, scalability, and ability to separate complex mixtures make it a valuable technique for isolating and obtaining high-purity bioactive compounds. Column chromatography allows for the separation of compounds based on their affinity for the stationary phase and their interaction with the mobile phase. This selectivity enables the isolation of specific bioactive compounds from complex mixtures present in microalgal extracts. Column chromatography can be adapted to purify a wide range of bioactive compounds, including pigments, lipids, polysaccharides, phenolics, proteins, and more. The choice of stationary phase and elution conditions can be adjusted to match the characteristics of the target compounds. Microalgal extracts may contain a mixture of compounds with varying polarities and molecular weights. Column chromatography allows for stepwise elution, enabling the separation of impurities and by-products from the desired bioactive compound, leading to enhanced purity (León-Vaz et al. 2023; Ong et al. 2021). Column chromatography is scalable, making it suitable for both laboratory-scale research and larger industrial applications. By adjusting column size and sample load, the technique can accommodate different sample volumes. Fractions collected during column chromatography can be analyzed using preliminary techniques like TLC to determine the presence of the desired bioactive compounds. The isolated fractions can then undergo more advanced analyses, such as HPLC and mass spectrometry, to confirm the identity and purity of the compounds. Many bioactive compounds from microalgae have potential applications in pharmaceuticals, nutraceuticals, cosmetics, and biotechnology. Column chromatography enables researchers to obtain sufficient quantities of these valuable compounds for further

Table 12.5 Experimental conditions used for the purification of macroalgal carbohydrates

Compounds of interest	Analytical method	Experimental conditions	Detector
Alginates	SEC	Shodex OHpak SB-806 HQ (8 × 300 mm) column. Eluted with 0.1 M NaOAc (pH 6). Flow rate 0.5 mL/min	Refractive index detector (RI)
Alginates	SEC	PSS Suprema (300 × 8 mm) column. Eluted with 0.01 M NaOH. Flow rate: 1 mL/min	Refractive index detector (RI)
Fucoidan	SEC	TSKgel GMPWXL (7.8 mm × 300 mm) with guard column TSK PWXL (6.0 mm × 4.0 mm). Eluted with HPLC-grade water. Flow rate: 1 mL/min	Evaporative light-scattering detector (ELSD)
Fucoidan	SEC	Sephacryl S-300 column (100 × 2.6 cm). Equilibrated and eluted with 0.2 M NaCl	Refractive index detector (RI)
Fucoidan	SEC	Shodex Asahipak GS-520 HQ and GS-620 HQ (7.5 mm × 300 mm) at 50 °C with elution by 0.5 M ammonium bicarbonate (0.8 mL/min)	Refractive index detector (RI)
Alginate	IEC	CarboPac PA1 (4 × 250 mm) column. Eluted using 0.1 M NaOH increasing up to 0.16 M NaOH with 0.19 M sodium acetate. Flow rate: 1.50 mL/min	Pulsed amperometric detector (PDA)
Carrageenan	IEC	Q-sepharosetm (EX 150 mm × 50 cm). Eluted with linear gradient 3 M NaCl. Flow rate: 1.50 mL/min	–
Fucoidan	IEC	DEAE-cellulose (50 × 5 cm, cl – form). Eluted stepwise with 0, 0.2, 0.4, 0.8 and 1.6 M NaCl, followed with 0.3 M NaOH	–
Laminarin and fucoidan	IEC	DEAE-cellulose (Fluca) column (3.0 × 14 cm). Eluted first with H2O and the fractions obtained further eluted with a linear gradient 0–2 M·NaCl	–

studies and applications. Purified bioactive compounds obtained from column chromatography can be used as starting materials for downstream processes such as chemical modification, formulation, and product development. Microalgal extracts often contain complex mixtures of compounds, including minor components that may have significant biological activities (Stirk and van Staden 2022). Column chromatography allows for the isolation of these minor compounds for detailed analysis and potential applications. Column chromatography's-controlled elution process minimizes the loss of target compounds while efficiently separating them from unwanted components.

12.3.2 Flash Chromatography for Bioactive Compounds

Flash chromatography is a powerful and efficient technique used for the purification of bioactive compounds from microalgae. It offers a rapid and versatile way to separate and isolate target compounds from complex mixtures. When applied to microalgae, flash chromatography becomes a crucial tool for obtaining high-purity bioactive compounds for various applications. Flash chromatography holds significant importance for the purification of bioactive compounds from microalgae, particularly due to its speed, simplicity, and versatility. This technique offers several advantages that make it suitable for isolating bioactive compounds efficiently. Flash chromatography is designed for rapid separations. It allows for quicker compound isolation compared to traditional column chromatography, making it ideal for handling larger sample volumes and reducing the time required for purification (Tzima et al. 2023). Flash chromatography can be easily scaled up to accommodate larger quantities of samples, making it suitable for both laboratory-scale research and preparative-scale purification of bioactive compounds from microalgae. Microalgal samples are often complex mixtures containing a range of bioactive compounds. Flash chromatography efficiently separates these compounds in a single run, simplifying the purification process. Flash chromatography supports a wide range of stationary phases and elution solvents, allowing researchers to tailor the purification conditions to the specific characteristics of the bioactive compounds in microalgae. Flash chromatography uses relatively less solvent compared to traditional column chromatography, making it more environmentally friendly and cost-effective in the long run. Setting up a flash chromatography system is relatively straightforward, requiring minimal equipment and expertise (Wang et al. 2022). This ease of setup makes it accessible to researchers and operators without extensive chromatographic experience. Flash chromatography provides good separation and elution control, resulting in the isolation of higher-purity fractions compared to other rapid purification techniques. Modern flash chromatography systems often include UV-Visible detectors that allow real-time monitoring of the elution process, enabling the collection of fractions containing the desired bioactive compounds. Flash chromatography's shorter run times and minimal exposure to stationary phases can help reduce the risk of compound degradation during purification, especially for sensitive bioactive compounds.

12.3.3 High-Performance Liquid Chromatography (HPLC) for Bioactive Compounds

High-performance liquid chromatography (HPLC) is a powerful analytical technique with immense significance in the field of bioactive compounds. It offers precise separation, quantification, and analysis of a wide range of compounds, making it an essential tool for researchers, scientists, and professionals working with bioactive molecules. HPLC provides exceptional separation efficiency due to its high-pressure capabilities and narrow-bore columns. It can resolve complex mixtures containing multiple bioactive compounds with different chemical properties. HPLC allows for sensitive detection and quantification of compounds at low concentrations. Coupling HPLC with various detectors such as UV-Visible, fluorescence, mass spectrometry, and refractive index detectors enhances the ability to identify and quantify bioactive compounds accurately (Cunha et al. 2023). HPLC is widely used for quantitative analysis of bioactive compounds in samples. By comparing peak areas or peak heights of known standards with those of sample peaks, the concentration of compounds can be determined. HPLC coupled with mass spectrometry (LC-MS) provides structural elucidation of bioactive compounds by measuring their mass-to-charge ratio and fragmentation patterns. This information is crucial for identifying and characterizing compounds. Many bioactive compounds exist as isomers with similar molecular weights but distinct structures. HPLC can separate these isomers based on their unique chemical properties, aiding in accurate identification and quantification. HPLC can separate chiral compounds into their enantiomers, allowing researchers to study their specific biological activities and interactions. HPLC is capable of analyzing complex mixtures in samples such as biological extracts, herbal formulations, and microbial metabolites, helping to identify and quantify multiple bioactive compounds simultaneously. HPLC is employed in pharmacokinetic studies to analyze the distribution, metabolism, and elimination of bioactive compounds within the body after administration. HPLC is used to analyze bioactive compounds in environmental samples, helping to assess pollution levels and the impact of pollutants on ecosystems. HPLC methods can be customized to suit the specific characteristics of the compounds being analyzed. Different columns, mobile phases, and detection techniques can be selected based on compound polarity, size, and other attributes (Liu et al. 2021). Modern HPLC systems allow for high-throughput analysis, enabling the rapid screening of large numbers of samples for the presence of specific bioactive compounds.

12.3.4 Solid-Phase Extraction (SPE) for Bioactive Compounds

Solid-phase extraction (SPE) is a widely used technique for the purification and concentration of bioactive compounds from complex mixtures. It offers several advantages over traditional liquid-liquid extraction methods, making it a valuable tool in various fields such as pharmaceuticals, environmental analysis, food and beverage, and natural product research. SPE involves the use of a solid sorbent (stationary phase) to selectively retain target compounds from a liquid sample. The stationary phase is typically packed into a cartridge, disk, or other formats. The sample is passed through the stationary phase, and the target compounds adsorb onto the sorbent while unwanted compounds are washed away. Afterward, the adsorbed compounds are eluted from the sorbent using a suitable solvent, resulting in their purification and concentration (Wang et al. 2022). SPE allows for selective extraction of specific bioactive compounds based on their chemical properties, such as polarity, size, and functional groups. This selectivity helps in removing unwanted matrix components that can interfere with analysis or applications. SPE concentrates target compounds by adsorbing them onto the sorbent, effectively reducing the volume of the sample and increasing compound concentration. For samples with limited volume, SPE allows researchers to enrich the target compounds while reducing the overall sample volume for subsequent analysis. Bioactive compounds are often present in complex matrices like biological fluids, environmental samples, or natural extracts. SPE removes interfering matrix components, enhancing the accuracy and sensitivity of downstream analyses. SPE acts as a sample cleanup step by eliminating contaminants that could interfere with sensitive analytical techniques like high-performance liquid chromatography (HPLC) or mass spectrometry (MS). SPE concentrates trace levels of bioactive compounds, allowing their detection and quantification even when present in low concentrations. SPE can be tailored to isolate both polar and nonpolar bioactive compounds by selecting appropriate sorbents and elution solvents. SPE procedures can be automated using dedicated instruments, enhancing reproducibility and reducing the risk of human error (Yang et al. 2023). SPE is suitable for various sample matrices, including aqueous, organic, and complex mixtures like biological fluids, plant extracts, and environmental samples. Different types of sorbents, cartridge sizes, and elution solvents can be chosen based on the target compounds and the sample matrix, allowing for customization of the extraction process.

12.3.5 Crystallization for Bioactive Compounds

Crystallization is a widely used technique for the purification of bioactive compounds. It involves the controlled formation of crystals from a solution containing the compound of interest. Crystallization is significant for purifying bioactive compounds due to its ability to separate compounds based on differences in solubility and crystal formation. Crystallization relies on the principle that compounds have varying solubilities in different solvents at different temperatures. By carefully manipulating solvent composition and temperature, the compound of interest can be induced to form crystals, while impurities remain in solution or form separate crystals. The crystals are then separated from the solution, washed, and dried to obtain a purified compound. Crystallization is effective at separating the compound of interest from impurities, resulting in a higher-purity final product. Impurities often have different solubilities and crystallization behaviors compared to the target compound. Crystallization can yield a relatively high amount of purified compound, making it suitable for preparative-scale purification. Crystallization is a relatively simple process that requires minimal equipment and is suitable for a wide range of compounds (Yang et al. 2022; Agarwal et al. 2023). Crystallization can selectively separate compounds with similar structures but slightly different properties, such as enantiomers. Crystallization effectively concentrates the compound of interest, allowing for easier handling and downstream applications. Crystallization can separate stereoisomers, providing access to pure enantiomers or diastereomers. Crystallization provides a straightforward way to obtain compounds with well-defined characteristics, aiding in quality control and standardization. Crystallization can be performed under mild conditions, making it suitable for isolating sensitive bioactive compounds that may degrade under harsher conditions.

12.3.6 Membrane Filtration for Bioactive Compounds

Membrane filtration is a separation technique that plays a crucial role in the purification of bioactive compounds from complex mixtures. Membrane filtration is a versatile technique used for the purification of bioactive compounds from complex mixtures. It involves the use of semi-permeable membranes to separate and concentrate target compounds based on their size, molecular weight, and other physical properties. Membrane filtration can be applied to various types of bioactive compounds, including proteins, peptides, enzymes, polysaccharides, and more.

The first step is to choose the right type of membrane based on the characteristics of the bioactive compounds and the desired separation objectives. Different types of membranes with varying pore sizes (microfiltration, ultrafiltration, nanofiltration, reverse osmosis) are available, each suitable for specific compound sizes and molecular weights (Ampofo and Abbey 2022; Zhou et al. 2023). The crude extract or solution containing the bioactive compounds is typically pre-processed to remove larger particles, debris, and contaminants. This step helps prevent membrane fouling and ensures efficient filtration. The sample is then passed through the selected membrane under controlled conditions. The membrane acts as a barrier, allowing smaller molecules to pass through while retaining larger molecules. The target bioactive compounds are either retained on the membrane's surface (in the case of dead-end filtration) or allowed to pass through the membrane and collected in the filtrate (in the case of cross-flow or tangential flow filtration). Membrane filtration not only purifies but also concentrates the bioactive compounds by removing solvent or water. Concentrated solutions are often preferred for further analysis, characterization, or downstream processing. One of the key benefits of membrane filtration is the removal of impurities such as contaminants, salts, and other undesired molecules that might interfere with subsequent analyses or applications. In cross-flow or tangential flow filtration, the filtrate containing the purified and concentrated bioactive compounds is collected separately from the retentate (the portion retained on the membrane). The collected fractions can be analyzed to ensure the presence and quality of the target compounds. The process parameters, such as pressure, flow rate, and membrane type, are optimized to achieve the desired separation efficiency and compound recovery. Membrane filtration can be scaled up for larger sample volumes and preparative applications. The purified and concentrated bioactive compounds obtained from membrane filtration can be used for various downstream applications, such as pharmaceutical formulation, nutraceutical production, cosmetics, and biotechnology (Liu et al. 2021). Regular maintenance and cleaning of the membrane are essential to prevent fouling, maintain membrane integrity, and ensure consistent performance over time.

Membrane filtration offers several advantages in terms of scalability, efficiency, and versatility, making it significant for the purification of bioactive compounds.

Membrane filtration relies on the concept of selectively allowing certain molecules to pass through a membrane while retaining larger molecules. The membrane acts as a barrier, with pores or channels of specific sizes that determine what can pass through and what is retained. Bioactive compounds can be separated from larger particles, contaminants, or solvents using this principle. Membrane filtration is effective for separating bioactive compounds based on their size and molecular weight. This is particularly useful for isolating proteins, peptides, enzymes, and other biomolecules. Membrane filtration is conducted under mild conditions, making it suitable for isolating sensitive bioactive compounds that may be prone to denaturation under harsher separation methods. Membrane filtration is scalable, making it suitable for both laboratory-scale research and larger-scale purification of bioactive compounds. Membrane filtration can be performed as a batch process or as a continuous process, allowing for efficient separation of compounds (Chen et al. 2020). Membrane filtration can be used to concentrate bioactive compounds by removing solvent or water. Concentrated solutions are often preferred for downstream processing and analysis. Different types of membranes with varying pore sizes can be used to achieve the desired selectivity for the target bioactive compounds. Membrane filtration can effectively remove larger particles, debris, and contaminants present in the sample, leading to a higher-purity final product. Unlike other separation techniques, membrane filtration does not involve phase changes (e.g., evaporation), which can lead to compound degradation.

12.4 Applications of Bioactive Compounds Produced from Microalgae

Bioactive compounds produced from microalgae have a wide range of applications across various industries due to their diverse biological activities and potential health benefits as shown in Fig. 12.5. These compounds are of growing interest for their potential as functional ingredients, nutraceuticals, pharmaceuticals, and more.

12.4.1 Pharmaceuticals Industry

Bioactive compounds produced from microalgae have gained significant attention in the pharmaceutical industry due to their potential therapeutic properties and diverse biological activities. Compounds with antimicrobial, anticancer, anti-inflammatory, and antioxidant properties are being studied for their therapeutic applications. These compounds offer a promising source for the development of new drugs and treatments. Microalgae-derived bioactive compounds, such as antimicrobial peptides, fatty acids, and secondary metabolites, have demonstrated antimicrobial activities against various pathogens, including bacteria, fungi, and viruses. These compounds can be explored as alternatives or adjuncts to conventional antibiotics. Certain microalgae-derived compounds have shown anticancer properties, including apoptosis induction, inhibition of tumor growth, and antiangiogenic

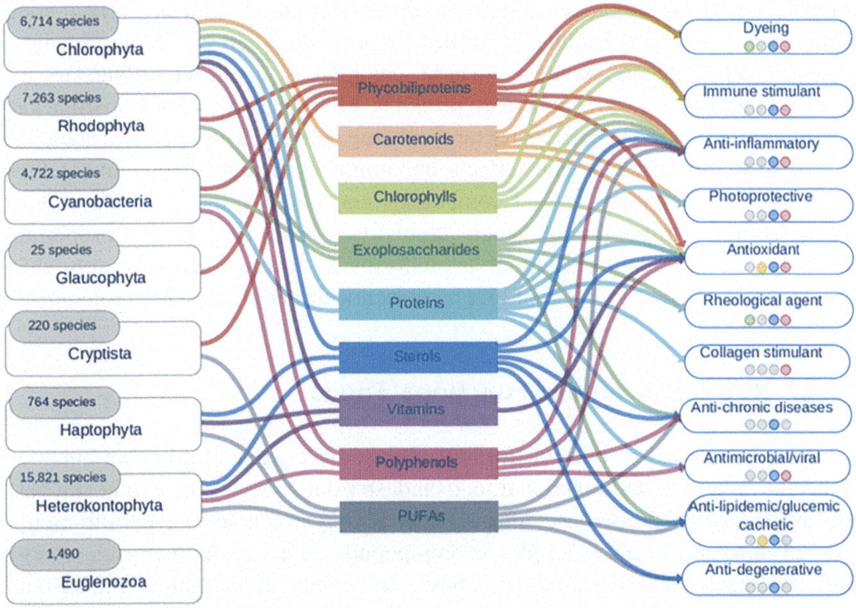

Fig. 12.5 Primary bioactive compounds found in microalgae, categorized and explored for potential applications as high-value products

effects (Yang et al. 2023). Compounds like polysaccharides, polyphenols, and carotenoids are being investigated for their potential in cancer therapy and prevention. Bioactive compounds from microalgae, such as polyunsaturated fatty acids (PUFAs) and phycobiliproteins, possess anti-inflammatory properties. They can be used to develop drugs for treating chronic inflammatory diseases like rheumatoid arthritis, inflammatory bowel disease, and asthma. Microalgae-derived antioxidants, including carotenoids, tocopherols, and polyphenols, can neutralize free radicals and protect cells from oxidative damage. These compounds are studied for their potential in preventing age-related diseases and improving overall health. Certain microalgae-derived compounds, such as omega-3 fatty acids and phycocyanin, exhibit neuroprotective effects by supporting brain health, reducing neuroinflammation, and enhancing cognitive function (Yang et al. 2022). These compounds have implications for neurodegenerative disorders like Alzheimer's and Parkinson's diseases. Omega-3 fatty acids, phytosterols, and polysaccharides from microalgae have cardiovascular benefits, including lowering cholesterol levels, reducing blood pressure, and promoting overall heart health. These compounds could contribute to the development of drugs for cardiovascular diseases. Microalgae-derived compounds can modulate the immune response by enhancing immune cell activity or suppressing excessive immune reactions. These compounds may find applications in autoimmune disease treatments and immunotherapy. Bioactive compounds from microalgae have demonstrated potential antiviral activities against various viruses, including HIV, herpes simplex virus, and influenza virus. These compounds can be explored for the development of antiviral drugs. Certain microalgae-derived compounds possess wound healing properties by promoting cell proliferation, angiogenesis, and collagen synthesis. These compounds could contribute to the development of advanced wound care products. Microalgae-based nanoparticles or liposomes can be used as carriers for targeted drug delivery. This approach enhances drug bioavailability, reduces side effects, and improves therapeutic outcomes (Zhou et al. 2023). Microalgae-derived compounds can improve the bioavailability of poorly soluble drugs through their emulsifying and solubilizing properties. Microalgae are also being explored as platforms for recombinant protein production, enabling the development of biopharmaceuticals and therapeutic proteins.

12.4.2 Nutraceuticals and Functional Foods

Bioactive compounds produced from microalgae have gained significant attention in the nutraceuticals and functional foods industry due to their potential health benefits and functional properties. Omega-3 fatty acids, carotenoids, phycobiliproteins, and polysaccharides are examples of compounds added to supplements, energy bars, beverages, and other products. These compounds offer a natural and sustainable source of valuable nutrients and bioactive molecules that can enhance the nutritional profile and functionality of various food products. Microalgae are a primary source of omega-3 fatty acids, including eicosapentaenoic acid (EPA) and docosahexaenoic acid (DHA) (Zhou et al. 2022). These essential fatty acids are associated with cardiovascular health, brain function, and anti-inflammatory effects. Microalgae-derived omega-3 s are used as supplements in various functional foods and nutraceutical products. Microalgae produce a range of antioxidants such as carotenoids (astaxanthin, beta-carotene), tocopherols (vitamin E), and polyphenols. These antioxidants can neutralize free radicals, protect cells from oxidative stress, and support overall health. They are added to products like beverages, snacks, and dietary supplements. Certain microalgae species are rich in high-quality proteins and essential amino acids. These proteins can be incorporated into protein-enriched foods, sports nutrition products, and vegetarian/vegan alternatives. Microalgae-derived carotenoids, such as astaxanthin and beta-carotene, are used as natural colorants and antioxidants in functional foods. They are added to beverages, dairy products, confectionery, and more. Phycobiliproteins are vibrant pigments found in certain microalgae, offering natural coloring options for foods and beverages. They are often used in products aiming to replace synthetic colorants. Certain microalgae contain polysaccharides and dietary fiber that can act as prebiotics, promoting the growth of beneficial gut microorganisms (Hassan et al. 2022; Tzima et al. 2023). These compounds contribute to digestive health and may be added to functional foods like yogurt and dietary supplements. Microalgae can be used to fortify foods with essential nutrients, such as vitamins, minerals, and trace elements. This is particularly beneficial for addressing nutrient deficiencies. Some microalgae-derived compounds, like beta-glucans, have cholesterol-lowering effects and are used in functional foods aimed at improving heart health. Bioactive compounds from microalgae, such as polysaccharides and peptides, can have immunomodulatory effects. They are incorporated into immune-supporting products like dietary supplements and beverages. Microalgae-derived ingredients are used in plant-based and vegan foods as alternatives to animal-derived nutrients, contributing to the nutritional adequacy of such diets. Microalgae-based ingredients can be incorporated into nutrient-dense snacks, energy bars, and functional beverages, enhancing their nutritional value. Microalgae-derived compounds, like certain pigments and lipids, can enhance the bioavailability of other nutrients, making them suitable for products aiming to improve nutrient absorption.

12.4.3 Cosmetics and Skincare

Bioactive compounds produced from microalgae have found a variety of applications in the cosmetics and skincare industry due to their unique properties and potential benefits for skin health and beauty. Microalgae-derived compounds are used in cosmetics and skincare products due to their antioxidant, anti-aging, and skin-soothing properties. Compounds like astaxanthin and polyunsaturated fatty acids are incorporated into creams, lotions, serums, and sunscreens (Zhuang et al. 2022). Microalgae-derived antioxidants, such as astaxanthin, beta-carotene, and tocopherols, protect the skin from oxidative stress caused by UV radiation, pollution, and other environmental factors. These antioxidants help prevent premature aging, reduce fine lines, and maintain skin health. Compounds like peptides, polysaccharides, and fatty acids from microalgae contribute to the development of anti-aging skincare products. They can stimulate collagen production, improve skin elasticity, and reduce the appearance of wrinkles. Certain microalgae-derived compounds have moisturizing and hydrating properties, making them suitable for creams, lotions, and serums designed to improve skin hydration and barrier function. Microalgae-derived compounds, including certain pigments and enzymes, are used in products designed to address hyperpigmentation, uneven skin tone, and dark spots, contributing to a brighter complexion. Microalgae-derived bioactive compounds with anti-inflammatory properties, such as polyphenols and phycocyanin, are used to soothe irritated skin and reduce redness. Microalgae-derived compounds like astaxanthin and mycosporine-like amino acids (MAAs) offer natural sun protection by absorbing UV radiation and reducing skin damage (Hassan et al. 2022). Certain microalgae-derived compounds can accelerate skin regeneration and wound healing by promoting cell proliferation and tissue repair. Bioactive compounds from microalgae are used in cleansing products to remove impurities and toxins from the skin, leaving it refreshed and revitalized. Phycobiliproteins and carotenoids from microalgae are used as natural colorants in cosmetics, providing vibrant and natural shades to makeup products. Microalgae-derived ingredients align with the growing demand for sustainable and environmentally friendly cosmetic ingredients, making them appealing to conscious consumers. Microalgae offer innovative opportunities for the development of novel cosmetic products with unique selling points and potential differentiators in the market. Microalgae-derived compounds are often used in natural and organic cosmetics due to their minimal processing and potential health benefits.

12.4.4 Food Colorants

Bioactive compounds produced from microalgae have found applications as natural food colorants due to their vibrant pigments and potential health benefits. Phycobiliproteins and carotenoids obtained from microalgae are used as natural food colorants, replacing synthetic dyes in various food products. They provide vibrant colors while also offering potential health benefits. These compounds offer a sustainable alternative to synthetic food dyes, which have raised concerns about their safety and environmental impact. Microalgae-derived colorants are used to add appealing colors to a wide range of food and beverage products. Carotenoids are natural pigments found in microalgae that contribute to the colors of fruits, vegetables, and seafood (Yang et al. 2022). Microalgae such as *Dunaliella salina* produce carotenoids like beta-carotene, which can be used to create yellow, orange, and red hues in various food products. Phycobiliproteins are water-soluble pigments present in certain microalgae, such as spirulina. They provide vivid blue and green colors, making them suitable for coloring dairy products, beverages, confectionery, and more. Astaxanthin, a carotenoid found in microalgae and some seafood, imparts a bright red color. It is used to color salmon, shrimp, and other seafood products. Astaxanthin's antioxidant properties also offer potential health benefits. MAAs are UV-absorbing compounds produced by microalgae as a defense against UV radiation. They can be used to create a range of colors, from yellow to pink, and are particularly suitable for beverages and gummy candies. Chlorophylls are responsible for the green color in plants and some microalgae. They can be used to create natural green shades in food and beverages. Microalgae-derived colorants can be used to create novelty and specialty products, such as blue pasta or green beverages, to capture consumer interest. The natural origin and potential health benefits of microalgae-derived colorants align with consumer preferences for clean label and health-conscious products (Stirk and van Staden 2022). Microalgae-derived colorants are often used in clean label formulations to replace synthetic colorants, offering a more natural and consumer-friendly ingredient list. Microalgae-derived colorants are used to enhance the appearance of various beverages, including juices, energy drinks, smoothies, and flavored waters. Food colorants from microalgae are used to add visually appealing colors to candies, cakes, pastries, and other baked goods. Microalgae-derived colorants are employed to create appealing colors in dairy products like yogurt, ice cream, and cheese, as well as in plant-based alternatives. Microalgae-derived colorants can be used to enhance the visual appeal of snack foods, including chips, crackers, and packaged snacks.

12.4.5 Aquaculture and Animal Feed

Bioactive compounds produced from microalgae have significant applications in the aquaculture and animal feed industries, offering nutritional benefits, health promotion, and improved growth in aquatic organisms and livestock. These compounds provide a sustainable and efficient way to enhance the quality of feed and promote the health and performance of animals. Microalgae are used as a source of high-quality nutrients for aquaculture and animal feed. They provide essential fatty acids, proteins, vitamins, and minerals that promote growth and health in fish, shellfish, and livestock. Microalgae are natural producers of omega-3 fatty acids, such as eicosapentaenoic acid (EPA) and docosahexaenoic acid (DHA) (Chen et al. 2020; Ong et al. 2021). These essential fatty acids are vital for the growth and health of aquatic organisms and livestock. Microalgae-derived omega-3 s are used to enrich feed and enhance the nutritional profile of farmed fish and animal products. Certain microalgae species are rich in high-quality proteins and essential amino acids. Microalgae-derived proteins can be incorporated into animal feed to improve protein content, amino acid balance, and overall nutrition. Microalgae-derived pigments, such as astaxanthin and canthaxanthin, are used in aquaculture feed to enhance the coloration of fish and crustaceans, making them more visually appealing to consumers. Bioactive compounds from microalgae, including polysaccharides and peptides, can have immunomodulatory effects. Incorporating these compounds into feed can enhance the immune response and disease resistance in aquatic organisms and animals. Microalgae-derived compounds can contribute to improved gut health and digestibility in animals, enhancing nutrient absorption and reducing the risk of digestive disorders. Microalgae are used as a source of nutrition for larval stages of aquatic organisms, providing essential nutrients for healthy growth and development (Cunha et al. 2023; Tzima et al. 2023). Microalgae-based feed ingredients are more environmentally sustainable compared to traditional feed sources like fish meal. Cultivating microalgae for feed production can reduce the strain on marine resources. Microalgae-derived bioactive compounds can enhance the growth rate, weight gain, and overall performance of farmed fish and animals. Certain microalgae-derived compounds possess antimicrobial and antiviral properties that can help prevent disease outbreaks in aquaculture and livestock farming. In aquaculture, using microalgae-derived feed can help reduce the fish-in-fish-out ratio by replacing fish meal with alternative protein sources. Microalgae can be cultivated as part of integrated aquaculture systems, providing both feed and nutrient recycling benefits.

12.4.6 Biofuels and Bioproducts

Bioactive compounds produced from microalgae have applications beyond traditional industries and are being explored for their potential in biofuels and bioproducts. Microalgae are known for their ability to accumulate lipids, carbohydrates, and other valuable compounds, making them a promising feedstock for the production of biofuels and various biochemicals. Microalgae are known for their high lipid content, which can be converted into biofuels like biodiesel and renewable diesel through processes such as transesterification. These biofuels can serve as alternatives to conventional fossil fuels, reducing greenhouse gas emissions and dependence on non-renewable resources (Zhou et al. 2022). Microalgae-derived compounds can be used as feedstock for fermentation and biotransformation processes to produce valuable chemicals, such as bioethanol, organic acids, and bioplastics. Microalgae-derived compounds, such as omega-3 fatty acids and carotenoids, can be extracted and incorporated into dietary supplements and nutraceutical products.

12.4.7 Environmental Remediation

Bioactive compounds produced from microalgae have shown potential applications in environmental remediation due to their unique properties and capabilities to address pollution and environmental challenges. These compounds can be utilized in various strategies to remove contaminants, improve water quality, and restore ecosystems. Microalgae-derived compounds can adsorb or absorb pollutants such as heavy metals, dyes, and organic compounds from water and soil (León-Vaz et al. 2023). The bioactive compounds in microalgae cell walls, such as polysaccharides, proteins, and lipids, contribute to this process.

12.4.8 Bioplastics and Biomaterials

Bioactive compounds produced from microalgae have shown promising applications in the development of bioplastics and biomaterials. These compounds offer a sustainable and eco-friendly alternative to conventional plastics and synthetic materials, contributing to reduced environmental impact and the advancement of green technologies. Microalgae-derived compounds can be used to produce biodegradable plastics, also known as bioplastics. These plastics break down naturally over time, reducing plastic waste and pollution. Bioactive compounds such as polysaccharides and proteins from microalgae can serve as building blocks for creating biodegradable polymer matrices. Polysaccharides extracted from microalgae, such as alginate and carrageenan, can be used to create edible films and coatings for food packaging. These materials are not only biodegradable but also provide barriers against moisture and gas. Microalgae-derived compounds, such as cellulose and chitin, can be incorporated into composite materials to enhance their mechanical properties. These materials can be used in various applications, including construction, automotive, and packaging (Agarwal et al. 2023; Zhuang et al. 2022). Microalgae-derived biomaterials can serve as scaffolds for tissue engineering applications. These materials provide a structure for cells to adhere to and grow, enabling the regeneration of tissues and organs. Microalgae-derived biomaterials can be used in wound dressings and medical implants due to their biocompatibility and potential for controlled drug release. These can be used to create biodegradable microbeads, replacing conventional plastic microbeads in personal care products like exfoliating scrubs. Microalgae-derived bioplastics can be used to produce packaging materials for a wide range of products, including food, beverages, and consumer goods. These bioplastics can be integrated into energy storage devices, such as batteries and supercapacitors, as environmentally friendly alternatives to conventional materials.

12.5 Conclusion and Future Prospects

The use of supercritical fluid extraction and pressurized liquid extraction techniques coupled with advanced analytical tools is widely extended for the extraction and characterization of bioactive compounds from microalgae. Although supercritical fluid extraction using carbon dioxide will continue in the future, plenty of new solvents could be introduced for their use in pressurized liquid extraction processes. Moreover, ionic liquids, and more importantly, deep eutectic solvents (DES) will surely open new possibilities for the extraction of bioactive natural molecules. Further, the significance of algae as sources of novel bioactive compounds goes beyond dietary supplements and extends to therapeutics. Recent improvements in the purification and characterization methodologies gradually increased the use of novel chromatographic techniques, that is, ultra-performance liquid chromatography (UPLC) and analytical techniques such as NMR spectroscopy (^{1}H and ^{13}C NMR) for qualitative and quantitative analysis and mass spectrometry (ESI and MALDI) to gain further insight into the complexity of different molecular structures of bioactive peptides and carbohydrates.

Nonetheless, microalgae still face a great deal of difficulty in competing with the well-established platforms of today, in part because recombinant proteins are not secreted into the media. As a result, the full potential of microalgae as cell factories remains unrealized, and future research should focus on taking advantage of their unique benefits, like their GRAS status, low cost of cultivating, and potential scalability. Subsequent research on microalgal development should concentrate on the following areas: (1) safe transgenic strain evaluations; (2) recombinant protein standards developed under Good Manufacturing Practices; (3) preclinical testing completed; and (4) finalized clinical trials.

Acknowledgement The authors would like to thank University School of Chemical Technology, Guru Gobind Singh Indraprastha University, Delhi, India and Chemical Conversion Division, Sardar Swaran Singh National Institute of Bio-Energy (An Autonomous Institute of MNRE, Government of India), Kapurthala, Punjab, India.

References

Agarwal A, Selvam A, Majood M, Agrawal O, Chakrabarti S, Mukherjee M (2023) Carbon nanosheets to unravel the production of bioactive compounds from microalgae: a robust approach in drug discovery. Drug Discov Today 28:103586

Ampofo J, Abbey L (2022) Microalgae: bioactive composition, health benefits, safety and prospects as potential high-value ingredients for the functional food industry. Food Secur 11(12):1744
Bhatnagar P, Gururani P, Joshi S, Singh YP, Vlaskin MS, Kumar V (2023) Enhancing the bio-prospects of microalgal-derived bioactive compounds in food industry: a review. Biomass Convers Biorefinery:1–17
Chen X, Han W, Wang G, Zhao X (2020) Application prospect of polysaccharides in the development of anti-novel coronavirus drugs and vaccines. Int J Biol Macromol 164:331–343
Cunha SA, Borges S, Baptista-Silva S, Ribeiro T, Oliveira-Silva P, Pintado M, Batista P (2023) Astaxanthin impact on brain: health potential and market perspective. Crit Rev Food Sci Nutr:1–24
Hassan S, Meenatchi R, Pachillu K, Bansal S, Brindangnanam P, Arockiaraj J, Selvin J (2022) Identification and characterization of the novel bioactive compounds from microalgae and cyanobacteria for pharmaceutical and nutraceutical applications. J Basic Microbiol 62(9):999–1029
Kaur M, Bhatia S, Gupta U, Decker E, Tak Y, Bali M, Bala S (2023) Microalgal bioactive metabolites as promising implements in nutraceuticals and pharmaceuticals: inspiring therapy for health benefits. Phytochem Rev:1–31
León-Vaz A, León R, Vigara J, Funk C (2023) Exploring Nordic microalgae as a potential novel source of antioxidant and bioactive compounds. New Biotechnol 73:1–8
Liu JY, Zeng LH, Ren ZH (2021) The application of spectroscopy technology in the monitoring of microalgae cells concentration. Appl Spectrosc Rev 56(3):171–192
Ong G, Kasi R, Subramaniam R (2021) A review on plant extracts as natural additives in coating applications. Prog Org Coat 151:106091
Stirk WA, van Staden J (2022) Bioprospecting for bioactive compounds in microalgae: antimicrobial compounds. Biotechnol Adv 59:107977
Tzima S, Georgiopoulou I, Louli V, Magoulas K (2023) Recent advances in supercritical CO2 extraction of pigments, lipids and bioactive compounds from microalgae. Molecules 28(3):1410
Wang J, Hu X, Chen J, Wang T, Huang X, Chen G (2022) The extraction of β-carotene from microalgae for testing their health benefits. Food Secur 11(4):502
Yang N, Zhang Q, Chen J, Wu S, Chen R, Yao L, Zhang Z (2023) Study on bioactive compounds of microalgae as antioxidants in a bibliometric analysis and visualization perspective. Front Plant Sci 14:1144326
Yang Y, Hassan SH, Awasthi MK, Gajendran B, Sharma M, Ji MK, Salama ES (2022) The recent progress on the bioactive compounds from algal biomass for human health applications. Food Biosci 51:102267
Zhou J, Wang M, Grimi N, Dar BN, Calvo-Lerma J, Barba FJ (2023) Research progress in microalgae nutrients: emerging extraction and purification technologies, digestive behavior, and potential effects on human gut. Crit Rev Food Sci Nutr:1–21
Zhou L, Li K, Duan X, Hill D, Barrow C, Dunshea F, Suleria H (2022) Bioactive compounds in microalgae and their potential health benefits. Food Biosci 49:101932
Zhuang D, He N, Khoo KS, Ng EP, Chew KW, Ling TC (2022) Application progress of bioactive compounds in microalgae on pharmaceutical and cosmetics. Chemosphere 291:132932

Nutraceuticals and Functional Foods from Algae: Formulation and Health Benefits

13

M. Devaprakash, R. Thirumalaivasan, N. Sivakumar, R. Shyam Kumar, and K. Ponmurugan

Contents

M. Devaprakash · R. Thirumalaivasan · N. Sivakumar (✉)
Department of Molecular Microbiology, School of Biotechnology, Madurai Kamaraj University, Madurai, Tamil Nadu, India
e-mail: siva.biotech@mkuniversity.ac.in

R. Shyam Kumar
Department of Biotechnology, Kamaraj College of Engineering and Technology, Virudhunagar, Tamil Nadu, India

K. Ponmurugan
Department of Botany and Microbiology, College of Science, King Saud University, Riyadh, Saudi Arabia

S. K. Arya et al. (eds.), *Value Added Products From Bioalgae Based Biorefineries: Opportunities and Challenges*,
https://doi.org/10.1007/978-981-97-1662-3_13

Abstract

Algae are a large and diverse group of autotrophic photosynthetic organisms in the aquatic environment that includes eukaryotic multicellular algae and unicellular prokaryotic cyanobacteria. Algae comprise proteins, amino acids, fatty acids (PUFA), vitamins, polysaccharides, pigments, and other secondary metabolites. Algae are considered as suitable vegetable-like food sources and are also used as supplements for healthy food. The presence of bioactive metabolites and beneficial compounds in some macro- and microalgae can be an active ingredient for various nutraceuticals and healthcare medicines. Algae also represent a valuable source of minerals, antioxidants, and natural colorants; hence, the whole algae biomass and the extracted compounds have used as active food and feed supplements for infant food and specialist food for some diseased patients. This chapter outlines the food, feed, nutrient, and nutraceutical uses of macro- and microalgae and their health benefits along with special function food formulations.

Keywords

Algae · Microalgae · Seaweeds · Nutraceuticals · Functional food · Feed formulations

Abbreviations

ACE	Angiotensin-converting enzyme
ALA	Alpha-linolenic acid
ARA	Arachidonic acid
CVD	Cardiovascular diseases
DHA	Docosahexaenoic
DW	Dry weight
EAAs	Essential aminoacids
EPA	Eicosapentaenoic
EPS	Exopolysaccharides
FAO	Food and Agriculture Organization
FAO	The Food and Agriculture Organization
FFA	Free fatty acid
GDP	Gross domestic product
GLA	Gamma-linolenic acid
H_2O_2	Hydrogen peroxide
HBV	Hepatitis B virus
HDL	High density lipoprotein
HPV	Papillomavirus
LA	Linoleic acid
LC-PUFAs	Long-chain polyunsaturated fatty acids

LDL	Long chain fatty acid
MGDGs	Mono galactosyl diacyl glycerols
MPTP	1-methyl-4-phenyl-1, 2, 3, 6-tetrahydropyridine
PUFA	Poly unsaturated fatty acids
ROS	Reactive oxygen species
sPS	Sulfated polysaccharides
TG	Triglyceride
USD	United States Dollar
UV	Ultra Violet
WHO	World Health Organization

13.1 Introduction

Algae are the most abundant photo-autotrophs that produce enormous biomass on the surface of the earth, especially in aquatic (marine and freshwater) and terrestrial (moisture surface) environments. They are the primary producers of the aquatic ecosystem and food for various aquatic animals. According to their size (microscopic or macroscopic), shape, photosynthetic pigments, nutritional properties, and reproduction, they are largely classified into ten different classes. Among them, Chlorophyta (green algae) Phaeophyceae (brown algae), Rhodophyta (red algae), and cyanophyceae (blue-green algae) are considered sources of food for various animals including humans. Macroalgae (seaweeds) are generally considered sea vegetables due to their exceptional nutritional and edible properties; it has been used as a direct food for humans since medieval times (Bassi et al. 2014; Levasseur et al. 2020; Singh et al. 2023; Mahadevan 2015). Since the dawn of human civilization, coastal peoples all over the world have consumed algae (Dillehay et al. 2008). Marine algae with notable varieties have a consolidated market in Asia and Europe, due to their health benefits and medicinal values which are raising awareness among consumers (Mouritsen et al. 2019a, b; Mendes et al. 2022). The overall market value of algae was around 20.94 billion USD in the year 2022 and is projected to reach 45.55 billion USD by the year 2030. The Asia-Pacific algae products market value was 1024.91 million USD in 2021, and it is expected to reach 1883.03 million USD by 2029 with a growth rate of 7.90% between 2022 and 2029. According to the Food and Agriculture Organization (FAO), algal-based food production has to increase at least 60% to meet the requirements of the growing global population (Alexandratos and Bruinsma 2012). Ireland, France, and Spain are the leading countries producing macroalgae, while Italy, Germany, and Spain are the leading microalgae producers. The countries include China, Indonesia, Japan, Philippines, South Korea, and North Korea accounting for approximately 97% of global seaweed production. Fresh, dried, and powdered forms of edible algae are used for various industrial preparations such as food, feed, cosmetics, and pharmaceuticals and also for biofuel. Figure 13.1 represents the food and biomedical applications of algae. Currently, algae are used as food ingredients to prepare functional food for various ailments (Araújo and Peteiro 2021; Ferdouse et al. 2018; Olawore and Nwinyi

Fig. 13.1 Food, feed, and biomedical applications of algae

2019), since algae are natural sources of protein, fibers, vitamins, minerals, and also some phyto-compounds. Hence, it could be used as a raw material for the preparation of special nutrient-enriched diet with pharmacological values (Gomez-Zavaglia et al. 2019). Moreover, algal lipids are rich in polyunsaturated fatty acids (PUFA), such as docosahexaenoic (DHA) and eicosapentaenoic (EPA) acids which have anti-cholesterol and cardioprotective functions (Heydarizadeh et al. 2013). Macroalgae and microalgae are rich in proteinogenic amino acids and short peptides which have strong antioxidative properties and antihypertensives, antiproliferative, and immunostimulatory functions. Polysaccharides such as alginate, carrageenan, and agar are generally used in food industries for emulsifying, stabilizing, and gelling agents to produce soft drinks and jellies. Value-added compounds such as *b*-carotene, lutein, lycopene, and astaxanthin are mainly used as a functional food as well as specialty food preparations. Algal pigments such as chlorophylls, phycobiliproteins, and carotenoids are used as natural food coloring agents for making cookies, pasta, bread, and beverages. Further, microalgae are a good source of nutrients for the food, feed, and cosmetics industry (Ariede et al. 2017). Accordingly, algae biomass and its compounds are used as food, dietary supplements, food additives, coloring agents, gelling agents, stabilizers and antioxidants, and alternatives to artificial chemical agents. This chapter deals with the applications of microalgae and macroalgae as functional food and nutraceuticals for various food and dietary formulations (Fig. 13.1).

13.2 Short History

The use of algae as food and medicine is not a recent past; it is a day-old process, when humans settled near the coastal and large water bodies, they started to consume algae as food. Based on the archaeological reports, humans were using seaweeds for food and medicine around 14,000 years ago (Ainis et al. 2019). People from Korea, China, Japan, Chile, and northeast Atlantic countries have been using seaweeds as food supplements and medicinal food since 500 BC (Abulafia 2019; Sultana et al. 2023). The oldest documents revealed the use of algae as food was around 1500 years ago. In China, Mongolia, and South America, a microalgae *Nostoc* sp. is used as traditional food. The recorded evidences revealed that the African peoples in the ninth century harvesting Spirulina from Lake Kossorom and sun-dried, to known as "Dihé," and served as a dish along with meat and vegetables (Mendes et al. 2022). In the fourteenth century, the Aztecs living near Lake Texcoco consumed microalgae *Arthrospira* sp. in the form of dry cake called "Tecuitlatl" (García et al. 2017). Due to the easily degradable nature of microalgae and seaweeds, they were not identified and documented in various archaeological surveys. However, their medicinal applications were traced back to the era of Hippocrates. Past 500 years since the late fourteenth century all over the globe humans consumed seaweeds and microalgae as raw food and also a dry pellet baked with other foods, soups and dressings. Seaweeds such as *Porphyra* sp., *Palmaria palmate*, *Ulva lactuca*, and *Chondracanthus teedei* were the traditional food for fishermen and beachgoers. Dry pellets of such seaweeds are served along with snacks, salads, bread dough, curds, and boiled fish in coastal restaurants in the Mediterranean (Mendes et al. 2022).

Through the long traditional uses and recent scientific validation on nutritional properties of micro- and macroalgae are considered an excellent source of protein, vitamin, amino acid, and mineral-rich natural food and nutraceuticals (antioxidative, antibacterial, antiviral, anti-helminthic, anticancer) (Wells et al. 2017). *Spirulina* is one of the best microalgae recognized as food, feed, and nutraceuticals since the 1960s. Cultivation of *Spirulina* and other microalgae for food, agriculture, and cosmetic industries provides a promising reputation to the cultivars and also gives GDP. In 1974, the United Nations World Food Conference acknowledged the merits of Spirulina and declared as "wonderful future food source" (Fernandes et al. 2023). In addition, alginate extracted from *Ascophyllum nodosum, Laminaria digitata* and *Laminaria hyperborean* and black carrageenan from *Chondrus crispus* and *Mastocarpus stellatus* have been dominating since the early twentieth century. Europe people regularly take seaweeds as a vegetable diet, specifically *Porphyra* sp., *Laminaria* sp., *Undaria pinnatifida*, *Eisenia bicyclis,* and *Hizikia fusiforme*, which are largely imported from Asia (Hernandez-Carmona et al. 2013; Araújo and Peteiro 2021). In recent years, Western countries, seaweeds have been mainly associated with industrial applications focused on the extraction of hydrocolloids for the food industry (Jönsson et al. 2020). Further, phycocoloid industries use seaweeds such as *Chondrus* sp. and *Mastocarpus* sp., *Gelidium corneum, Pterocladiella capillacea, Gracilariopsis longissima*, which are harvested from the natural environment as well as by manmade cultivation methods. At present, both the macro- and microalgae market have been expanded largely in addition to food industries; it is available in the form of powder, tablets, or capsules, and also special food formulations (Hosseinkhani et al. 2022).

13.3 Edible Macro- and Microalgal Diversity

Algae are prominent photosynthetic entities often referred to as the "lungs of the Earth," owing to their oxygen-producing capability contributing up to 80% of the oxygen present in the air we breathe. These organisms play a crucial role in the environment and showcase an ability to thrive in even the most challenging environments (Lewin and Andersen 2019). Based on the size, shape, and organization of algal cells, they can be broadly categorized into microalgae and macroalgae (Barzkar et al. 2019). The diversity within the macroalgae category is astonishing, with over 1200 distinct varieties thriving in both freshwater and marine habitats. This diversity contributes significantly to the overall ecological balance and health of aquatic ecosystems.

Macroalgae, as multicellular eukaryotic autotrophs, with their robust presence, play a crucial role in both freshwater and marine water environments. Their adaptability is noteworthy, allowing them to flourish in environments that might seem inhospitable (Lewin and Andersen 2019). Their intricate structure encompasses a broad spectrum of colors and forms, boasting pseudo roots and robust solid growth with the most substantial species reaching lengths exceeding 60 meters. Macroalgae are categorized based on the leaf color, structure, and physiology; they typically fall into three phyla: red algae, green algae, and brown algae. Green algae exhibit a vibrant array of pigments, including carotene, lutein, chlorophyll a, and chlorophyll b, with an additional production of orange-yellow pigments. Conversely, red algae are common inhabitants in warmer waters, rich with carotenoids, and chlorophylls 'a' and 'd', and also contain phycoerythrin pigments. Brown algae contain characteristic pigments like lithophane, chlorophylls 'a' and 'c', carotenoids, stored lipids and polysaccharides, as essential chemo components (Bhagea and Puchooa 2015; Marques and Américo-Pinheiro 2017) with taxonomical importance.

The most consumed species among macroalgae are brown (66.5%), red (33%), and green (5%) algae, respectively (Afonso et al. 2019). Seaweeds are marine plants used as a traditional food source in addition to medicinal as well as fertilizers. Seaweeds are considered a high nutritional food with low caloric value. Traditionally, people from China, Japan, and Korea have been used as phycocolloids (Paiva et al. 2018). However, seaweeds have been used as a food, nutraceutical, cosmetic component, and also a fertilizer in Asia. Seaweeds contain high levels of mono and polyunsaturated fatty acids with very low fat and high protein. Compared with vegetables, macroalgae are rich in vitamins A, C, E, and B complex, and all essential amino acids (Circuncisão et al. 2018). Seaweed polysaccharides are heteropolysaccharides rich in rhamnose, xylose, and uronic acids. *Undaria pinnatifida*, *Porphyra* spp., and *Caulerpa* spp. are primarily grown as direct human food. *Laminaria sp.* (kombu), *Undaria* sp. (wakame), *Hizikia fusiforme* (hiziki), *Cladosiphon okamuranus* (mozuku), Fucus vesiculosus, *Chondrus crispus, Ulva lactuca, Ulva* sp., *U. pinnatifida, Enteromorpha* sp., *Monostroma* sp., *Caulerpa* sp., *Codium* sp., *Caulerpa lentillifera, Sargassum* sp., *Monostromalatis simum, Grateloupia filicina, Enteromorpha prolifera, Ulva lactuca, Porphyra* spp. (laver), *Palmaria palmata, Pyropia tenera* and *Pyropia* spp. (nori) is an important source of food-grade macroalgae in various parts of the earth (Arufe et al. 2018; Imchen and Singh 2023). They can be used in the form of soup, salad, stew, and sushi. Commercially viable compounds such as alginate, carrageenan, agarose, and agar obtained from seaweeds are used in food industries due to their gelling properties, for example, fucoidan from brown algae, agar and carrageenan from red algae, sulfated ulvan from *Ulva ohnoi* (green algae), and phlorotannin from *Fucus vesiculosus* (Afonso et al. 2019). Figure 13.2 shows some representative macroalga used as food and feed.

In contrast, microalgae, unicellular organisms ranging from a few hundred to several thousand micrometers, exhibit extensive diversity in terrestrial, freshwater, and marine environments, thriving in conditions as diverse as deserts and Polar Regions (Varshney et al. 2015; Taubert et al. 2019; Naselli-Flores and Padisák 2023). It has been estimated that about 200,000–800,000 species in many different genera exist of which about 50,000 species are described. Microalgae stand out as versatile sources of various nutritional components, making them valuable for the production of food and food products (Koyande et al. 2019). Humans have been consuming microalgae since the 1950s, with certain strains recognized as generally

Fig. 13.2 Examples of some edible macroalgae. (**a**) *Ulva lactuca*, (**b**) *Gelidium corneum*, (**c**) *Sargassum* sp., (**d**) *Sargassum natans*, (**e**) *Chondrus crispus*, (**f**) *Pyropia yezoensis*

safe for consumption by the US Food and Drug Administration (Kusmayadi et al. 2020). The key nutrients present in microalgae, including lipids, proteins, pigments, and carbohydrates, necessitate meticulous cultivation conditions such as pH level, light intensity, nutrient supply, CO_2 availability, temperature, and mixing (Kusmayadi et al. 2020). *Chlorella, Dunaliella, Haematococcus, Schizochytrium*, and *Spirulina* are commonly used microalgae for human consumption, often dried powder containing abundant ω-3 fatty acids, protein, carbohydrates, pigments, and vitamins (Chacón-Lee and González-Mariño 2010). Marine microalgal species such as *Chlorella pyrenoidosa*, *Spirulina platensis*, *Desmosdesmus* sp., *Nannochloropsis oceanic, Arthrospira* sp., *Tsochrysis* sp., *Porphyridium* sp., *Pavlova* sp., *Phorphidium valderianum,* and *Tetraselmis chui* are considered as a source of protein and essential fatty acids (eicosapentaenoic acid), due to their promising health benefits (Costa et al. 2016; Madeira et al. 2017). Table 13.1 represents the list of microalgae used as food and feed, and Fig. 13.3 shows different microalgae under the microscope.

Table 13.1 Microalgae used as a food and feed

Name	Product type	Use	Reference
Chlorella sp	Biomass	Food supplement	Muller-Feuga (2000)
Chlorella vulgaris	Biomass	Food emulsions	Raymundo et al. (2005)
Chlorella sp	Biomass	Colored food emulsions	Gouveia et al. (2006)
Spirulina maxima, Diacronema vlkianum	Biomass	Vegetable gelled desserts	Gouveia et al. (2008)
Chlorella sp	Biomass	Novel functional ingredient in mixed gel systems	Batista et al. (2008)
Chlorella vulgaris	Biomass	Production of enzymatic protein hydrolysates.	Morris et al. (2008)
Chlorella sorokiniana	Biomass	Auto flocculation	Bazarnova et al. (2021)
Arthrospira platensis	Single-celled protein	Rich in vitamins, minerals, β-carotene, essential fatty acids, and antioxidants	Capelli and Cysewski (2010)
Chlorella vulgaris, Spirulina maxima	Biomass	Incorporation in pasta products	Fradique et al. (2010)
Chlorella vulgaris	Biomass	Proteins, amino acids, minerals, vitamins, and pigments	Prabakaran et al. (2019)
Chlorella vulgaris, Arthrospira platensis	Biomass	Prebiotic supplement	Beheshtipour et al. (2013)
Spirulina platensis, Chlorella vulgaris	Biomass	Enhance resistance against the tested induced diseases	Fadl et al. (2017)
Parietochloris incisa	Arachidonic acid	Nutritional supplement	Shanab et al. (2018)
Phaeodactylum sp.	PUFA's	Nutritional supplement	Kim et al. (2016)

(continued)

Table 13.1 (continued)

Name	Product type	Use	Reference
Dunaliella salina	Carotenoids	Natural food colorant, antioxidant	Rammuni et al. (2019)
Arthrospira platensis	Phycobiliproteins	Natural colorant for food and antioxidant	García et al. (2017)
Isochrysis galbana, *Diacronema vlkianum*	Biomass	Fortification of food products from grain	Bazarnova et al. (2021)
Cyanidioschyzon sp.	Phycocyanobilin	Food colorant	de Jesus Raposo et al. (2016); Dufossé (2023)

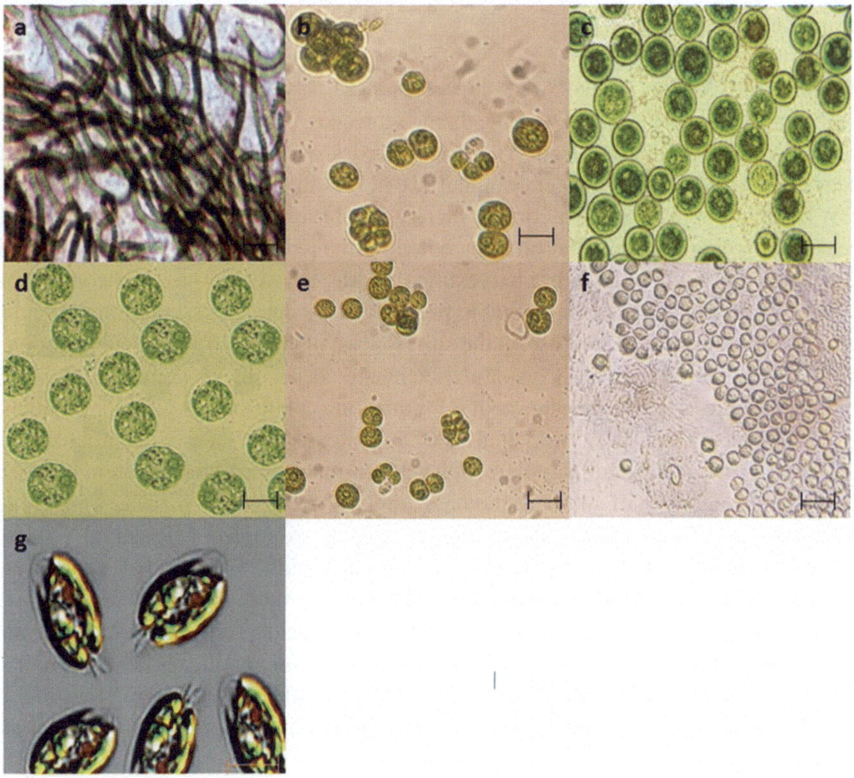

Fig. 13.3 Examples of some microalgae used for food and feed. (**a**) *Spirulina* sp., (**b**) *Schizochytrium* sp., (**c**) *Haematococcus pluvialis- Palmela,* vegetative cells at rest and without flagella, (**d**)*- Haematococcus pluvialis- a Zoospores, spherical, ellipsoidal, or pear-shaped vegetative cells, with two flagella,* (**e**) *Schizochytrium* sp., (**f**) *Nannochloropsis oceanica,* (**g**) *Tetraselmis chui* (Heterotrophic)

13.4 Nutrient Components in Algal

Algae are rich in proteins, carbohydrates, essential fatty acids and amino acids, minerals, and vitamins; they are good source of nutrition for various life-forms including human and animals. Both microalgae and macroalgae possess various forms of nutritive components in different proportions; however, selected group of algae has been recognized as a source of food as well as feed. Compared to animal-based foods, algae are also rich in protein and vitamins; hence, directly or indirectly algae can be considered as food, active ingredient, supplement, nutraceutical, and additive for the preparation of food for children, critical care patients and aged people (Imchen and Singh 2023; Show 2022). The following section describes the nutrient components of microalgae and macroalgae used for human food and animal feed.

Protein and Amino Acids Algae are considered as good vegetable protein rich with essential amino acids. Among them, seaweeds emerged as a viable protein source with levels ranging from 5% to 47% across species. Notably, green and red seaweeds exhibit higher protein content (10–47%) than brown seaweeds (5–24% DW). The amount of protein content therein algal biomass could vary in different seasons, for example, the total protein content of the red seaweed *Palmaria palmata* was 8% to 35%, under different harvesting periods. Seaweed proteins encompass a variety of amino acids, with a notable presence of lysine, an essential amino acid crucial for biological processes. As the shift toward plant-based diets gains traction, seaweed's potential to address lysine-deficient cereals becomes pivotal, presenting a unique avenue for the development of nutritious, plant-based protein products (Ramos et al. 2000; Dawczynski et al. 2007). Compared with seaweeds, microalgae are best choice for protein supplements. Certain microalgae boast protein content constituting nearly half of their biomass, ranging from 40% to 65%. *Spirulina*, *Chlorella*, and *Nannochloropsis* strains are notable examples. Microalgal protein serves as a valuable source of essential amino acids (EAAs) that cannot be synthesized independently in mammals. Microalgae, such as *Spirulina platensis*, provide easily digestible proteins, and their amino acid profiles encompass both EAAs and non-essential amino acids (NEAAs), offering significant health benefits (Da Silva Vaz et al. 2016; Buono et al. 2014; Barka and Blecker 2016). Tables 13.2 and 13.3 represent the examples of some macro- and microalgal protein content as well as EAAs.

Fatty Acids Algae are rich in saturated and unsaturated fatty acids. Certain brown seaweeds, such as *Alaria esculenta*, showcase richness in amino acids and polyunsaturated fatty acids (PUFAs) like eicosapentaenoic acid (EPA) and stearidonic acid. The omega-6: omega-3 ratio, a critical metric for health benefits, reveals favorable values, such as 1.93 for *Alaria esculenta* and 1.00 for *Saccharina latissima* (Afonso et al. 2020; Hayes 2015). Similar to macroalgae, microalgae are rich in essential fatty acids. Under stress conditions, they exhibit an extraordinary capacity to accumulate lipids, presenting a nutritionally valuable fatty acid profile,

Table 13.2 The protein content of selected edible algae on a dry weight basis

Source	Protein (%)	Reference
Microalgae		
Anabaena cylindrica	43–56	
Arthrospira fusiformis	62	
Arthrospira maxima	60–71	
Arthrospira platensis	55–70	
Botryococcus braunii	39–40	
Chlorella sp.	55–60	
Chlorella vulgaris	51–58	
Dunaliella salina	57	
Dunaliella tertiolecta	20	
Haematococcus pluvialis	48	
Isochrysis galbana	25–30	
Nannochloropsis spp.	25–30	
Pavlova salina	26	
Scenedesmus sp.	47–56	
Spirogyra sp.	15–20	
Synechococcus sp.	46–63	
Tetraselmis chui	30–47	
Macroalgae		
Padina sp.	3.48–9	
Sargassum sp.	13–15	
Ascophyllum sp.	3–20	
Fucus spp.	3–13	
Himanthalia elongata	6–11	
Laminaria digitata	8–15	
Saccharina japonica (kombu)	12	
Undaria pinnatifida (wakame)	11–24	Dawczynski et al. (2007)
Caulerpa lentillifera	19.38	
Cladophora rupestris	29.8	
Ulva intestinalis	10–18	
U. Lactuca (sea lettuce)	8.7–32.7	
Ulva rotundata	10.0	
Agarophyton vermiculophyllum	17.0%	Afonso et al. (2020)
Gracilaria spp.	7–13	
Gracilaria gracilis	31–45	
Palmaria palmata (dulse)	8–35	
Porphyra spp. (nori or purple laver)	33–50	
Porphyra tenera (nori)	33–47	
Porphyra umbilicalis (nori)	15–37	

Table 13.3 Comparison of essential amino acid composition (mg/g protein) of edible algae

Algae	His	Ile	Leu	Lys	Met + Cys	Phe + Tyr	Thr	Trp	Val	Reference
Macroalgae										
Undaria pinnatifida	68.2	50.8	78.5	69.6	NR	93.2	42.7	NR	35	
U. pinnatifida	21.6	47.3	89	58	73.1	110.4	53.6	NR	31.1	Cofrades et al. (2010)
Himanthalia elongata	20.2	43.6	79.3	60.9	68.8	113.7	54.8	NR	28.2	Cofrades et al. (2010)
Ascophyllum nodosum	49.8	48	58.8	14.3	11.2	46.6	62.4	5.8	54	
Ulva australis	8.6	25.9	52	30.1	NR	59.6	34.8	NR	39.1	
Ulva rigida	30.7	46.1	82	49.4	19.9	93.9	50.8	8.8	60.1	
Ulva lactuca	13.1	40	72.6	46.4	6.1	93.4	62	NR	70.1	
Porphyra umbilicalis	15.7	36.5	76.8	56.1	75.9	93	57.8	NR	12.3	Cofrades et al. (2010)
Porphyra palmata	18.5	65	81	107.8	34.7	93.3	47.4	NR	143.6	
Porphyra columbina	12.6	27.1	73.8	60.1	35.7	62.5	59.1	6.3	58.5	
Gracilaria changii	NR	42.3	53.4	48.7	16.1	63.5	49.3	NR	39.1	
Gracilaria vermiculophylla	10.7	54.9	84.5	54.4	12.9	90.8	58.2	4.0	64.1	
Microalgae										
Acutodesmus obliquus	14	36	85	41	33	92	59	20	60	Wang et al. (2020)
Arthrospira maxima	18	60	80	46	18	88	46	14	65	
Arthrospira platensis	22	45	98	71	39	157	46	12	78	
Botryococcus braunii	15	34	71	47	39	72	37	22	44	
Chaetoceros calcitrans	19	55	82	63	30	112	45	14	59	
Chlorella pyrenoidosa	16	62	34	81	61	51	35		52	
Chlorella sorokiniana	19	35	83	58	28	86	50	59	1.16	Wang et al. (2020)
Chlorella vulgaris	18	36	92	52	29	98	43	58	1.10	Wang et al. (2020)
Chlorella vulgaris	20	38	88	84	36	84	48	55	1.25	
Dunaliella tertiolecta	21	48	84	60	18	117	47	15	62	
Isochrysis galbana	21	48	87	62	27	108	52	13	62	

Nannochloropsis granulata (A)	23	56	110	85	51	104	54	28	71	
Nostoc sp.	20	37	95	65	38	140	53	10	72	
Pavlova lutheri	50	49	81	56	37	111	43	15	67	
Pavlova salina	15	44	90	62	20	92	52	9	61	
Scenedesmus obliquus	21	36	73	56	21	80	51	3	60	
Skeletonema costatum	16	52	83	57	26	109	51	13	63	
Spirulina platensis	22	67	98	48	34	106	62	3	71	
Tetraselmis chui	18	35	75	57	25	91	42	10	58	
Tetraselmis chui	16	34	73	56	52	77	40	23	48	

Amino acids are represented using the three-letter abbreviation code: His, histidine; Ile, isoleucine; Leu, leucine; Met, methionine; Cys, cysteine; Phe, phenylalanine; Tyr, Tyrosine; Thr, threonine; Trp, tryptophan; Val, valine notably, the synthesis of omega-3 and omega-6 fatty acids, and long-chain polyunsaturated fatty acids (LC-PUFAs), such as docosahexaenoic acid (DHA, 22:6, ω-3) and eicosapentaenoic acid (EPA, 20:5, ω-3), which have pivotal contributors to human and animal health. Microalgae emerge as a promising alternative, while compared to marine fish traditionally they serve as the primary source of DHA and EPA for human consumption (Richmond and Hu 2013).

Microalgae, particularly under heterotrophic cultivation, exhibit high productivity and can offer a superior, uncontaminated source of LC-PUFAs compared to fish oil. *Nannochloropsis* sp. and *Phaeodactylum tricornutum*, for instance, can contain up to 39% EPA, while *Thraustochytrium* sp. and *Schizochytrium limacinum* boast DHA concentrations ranging between 30 and 40% (Adarme-Vega et al. 2012). Beyond DHA and EPA, microalgae contribute other essential fatty acids, including linoleic acid (LA), arachidonic acid (ARA), gamma-linolenic acid (GLA), and alpha-linolenic acid (ALA) as shown in Table 13.4. These fatty acids play crucial roles in supporting the body's defense mechanisms, combating invasive microbes, and aiding in cell and tissue repair (Pulz and Gross 2004).

Polysaccharides Carbohydrates, comprising up to 60% of seaweed nutritional profiles, drive the end applications of seaweed species (Bauer et al. 2021). Algae contain high amounts of polysaccharides including structural polysaccharides (cell-wall), storage polysaccharides, and mucopolysaccharides. *Laminaria, Saccharina, Fucus, Ascophyllum, Undaria, Sargassum, Ulva, Chondrus, Porphyra, Gracilaria,* and *Palmaria are* some important macroalgal genera that produce 15 to 76% of polysaccharides of their total dry weight. The high amount of polysaccharides present in species such *Ascophyllum, Palmaria, Ulva, and Porphyra*, for example, green algae contain sulfated polysaccharides, sulfated galactans and xylans, red algae contain agars, xylans, carrageenans, water-soluble sulfated galactan, floridean starch (amylopectin-like glucan), as well as porphyran, and the brown algae contains alginic acid, fucoidan (sulfated fucose), laminarin (β-1, 3 glucan), and sargassan polysaccharides. Some important commercially valuable polysaccharide producing seaweeds are listed in the Table 13.5. In addition, seaweed species contain cellulose and hemicellulose content up to 2% to 10% and 9% of their dry weight, respectively. Lignin is found only in green seaweed species such as *Ulva* sp. up to 3% of their dry weight (Usman et al. 2017). These polysaccharides have nutritional and biomedical applications.

Moreover, microalgae are rich in carbohydrates, serving as both an energy reserve and structural components of cell walls. Carbohydrates can accumulate up to 50w/w of dry weight, showcasing the high photo conversion efficiency of microalgae (Markou et al. 2012; Raven and Beardall 2003). Cyanobacteria produces two different structural polysaccharides and storage cyanophycean granules (alpha-1, 4-linked glucose). Structural polysaccharides are generally denoted as exopolysaccharides (EPS) and are highly complex anionic heteropolymers containing six to ten different mono sugars (glucose, rhamnose, fucose, galactose, xylose, arabinose, and mannose) as shown in Fig. 13.4, and uronic acids (gluco- &galacto-) associated with certain functional moieties. These polysaccharides have biopotentials such as

Table 13.4 Fatty acid profiles of some macro- and microalgae

Fatty acids	[a]*Porphyra* sp.	[a]*Porphyra* sp	[a]*Undaria pinnatifida*	[a]*Laminaria* sp.	[a]*Hizikia fusiforme*	*Spirulina platensis*	*Dunaliella primolecta*	*Nannochloris* sp.
C12:0	0.02 ± 0.02	0.09 ± 0.06	n.d.	0.06 ± 0.03	n.d.	0.37 ± 0.05	–	–
C14:0	2.68 ± 2.17	4.07 ± 4.84	2.25 ± 0.07	2.88 ± 3.92	0.30 ± 0.09	1.74 ± 0.05	0.6 ± 0.2	13.3 ± 1.8
C15:0	0.34 ± 0.13	0.45 ± 0.23	0.21 ± 0.03	0.40 ± 0.21	0.17 ± 0.03	0.38 ± 0.03	n.d.	15.1–17.8
C16:0	30.8 ± 4.60	37.1 ± 17.74	13.5 ± 0.67	36.0 ± 10.04	26.8 ± 3.84b	24.5 ± 0.52	26.0 ± 3.2	18.0 ± 1.6
C17:0	0.19 ± 0.14	0.18 ± 0.07	0.20 ± 0.06	0.16 ± 0.06	0.04 ± 0.02	0.37 ± 0.04	2.5–4.5	–
C18:0	0.66 ± 0.02	1.89 ± 0.52	0.86 ± 0.06	1.49 ± 0.35	0.76 ± 0.31	5.32 ± 0.37	1.6 ± 0.40	1.0 ± 0.4
C20:0	0.21 ± 0.19	0.45 ± 0.63	0.39 ± 0.05	0.28 ± 0.49	0.04 ± 0.06	n.d.	n.d.	–
C22:0	0.41 ± 0.36	n.d	n.d.	n.d	. 0.01 ± 0.02	–	12.3±	4.5 ± 1.0
C24:0	0.17 ± 0.15	n.d.	n.d.	n.d.	n.d.	0.21 ± 0.02	1.2 ± 0.4	n.d.
C14:1n5	0.02 ± 0.02	0.03 ± 0.05	n.d.	n.d	. n.d	n.d.	n.d.	n.d.
C16:1n7	2.24 ± 1.89	1.82 ± 2.14	0.44 ± 0.02	1.71 ± 1.08	0.15 ± 0.07	5.53 ± 0.38	4.5 + 0.9	16.6 ± 1.4
C17:1n7	0.19 ± 0.03	0.20 ± 0.11	0.12 ± 0.02	0.13 ± 0.10	n.s.	0.49 ± 0.02	n.d.	
C18:1	7.16 ± 3.58	15.3 ± 11.43	5.95 ± 0.15	12.8 ± 8.27	7.68 ± 4.22	20.35 ± 0.07	16.3 ± 0.65	57.7 ± 04.4
C20:1n9	1.42 ± 2.29	1.52 ± 2.11	n.d.	1.55 ± 1.31	4.09 ± 2.13	1.34 ± 0.02	n.d.	n.d.
C22:1n11	2.96 ± 2.58	n.d.	n.d.	0.02 ± 0.04	n.d.	2.00 ± 0.27	n.d.	n.d.
C22:1n13	0.20 ± 0.25	0.96 ± 1.35	n.d.	0.96 ± 0.84	0.64 ± 0.32	–	n.d.	n.d.
C18:2n6	3.86 ± 1.47	7.06 ± 4.65	7.41 ± 0.47	5.48 ± 3.44	3.56 ± 1.45	13.21 ± 2.21	7.0 ± 0.25	10.8 ± 1.4
C20:2n6	0.51 ± 0.43	0.90 ± 0.71	0.12 ± 0.03	0.87 ± 0.31	0.97 ± 0.27	0.49 ± 0.03	n.d.	n.d.
$C18:3_{8t,10t,12c}$	0.81 ± 0.37	0.24 ± 0.17	0.34 ± 0.34	n.d.	0.56 ± 0.21	n.d.	n.d.	28.2 ± 1.1
C18:3n6	0.31 ± 0.01	2.04 ± 2.32	1.71 ± 0.13	1.60 ± 2.25	0.42 ± 0.06	7.12 ± 3.31	41.1 ± 3.8	n.d.
C18:3n3	5.66 ± 4.74	0.94 ± 1.15	11.2 ± 0.53	0.76 ± 0.93	0.41 ± 0.23	n.d.	n.d.	n.d.
C20:3n6	1.20 ± 1.28	0.97 ± 0.84	0.57 ± 0.04	1.17 ± 0.73	3.21 ± 0.56	0.39 ± 0.05	n.d.	n.d.
C20:3n3	0.15 ± 0.02	0.04 ± 0.05	0.14 ± 0.02	0.01 ± 0.02	0.09 ± 0.00	0.28 ± 0.08	n.d.	n.d.
C18:4n3	3.37 ± 2.72	1.51 ± 2.14	25.8 ± 1.55	1.24 ± 2.15	n.d.	0.93 ± 0.15	4.1 ± 0.4	6.1 ± 0.4

(continued)

Table 13.4 (continued)

Fatty acids	[a]*Porphyra* sp.	[a]*Porphyra* sp	[a]*Undaria pinnatifida*	[a]*Laminaria* sp.	[a]*Hizikia fusiforme*	*Spirulina platensis*	*Dunaliella primolecta*	*Nannochloris* sp.
C20:4n6	8.00 ± 4.69b	9.84 ± 1.60	13.3 ± 0.41	12.4 ± 1.61	5.30 ± 1.98	0.34 ± 0.03	n.d.	5.9 ± 0.4
C20:5n3	20.9 ± 26.54	10.4 ± 7.46	13.2 ± 0.66	16.2 ± 8.90	42.4 ± 11.88	1.79 ± 0.20	n.d.	n.d.
C22:5n3	0.05 ± 0.09	n.d.	n.d.	n.d	. 0.09 ± 0.04	1.37 ± 0.10	n.d.	n.d.
C22:6n3	n.d.	n.d.	n.d.	n.d.	n.d.	2.28 ± 0.96	n.d.	n.d.

Mean with ± SD (% of total FAME)
n.d. not detected
[a]Dawczynski et al. (2007)

Table 13.5 Polysaccharides of most common edible seaweeds

Class	Genus	Type of polysaccharide	Uses
Chlorophyta	*Monostroma*	Rhamnan sulfate	Human food
	Enteromorpha	Heteropolysaccharide	Human food
Rhodophyta	*Asparagopsis*	Complex sulfated galactans	Medical applications
	Gelidiella	Agar	Food and medical
	Gelidiopsis		
	Gracilaria		
	Pterocladia		
	Chondrus		
	Eucheuma	Carrageenan	Human food
	Kappaphycus		
	Gigartina		
	Hypnea		
	Iridaea		
	Palmaria	Sulfated polysaccharide	Human and animal feed
	Porphyra	Sulfated galactan	Human food
Phaeophyta	*Laminaria*	Alginates	Human food
	Undaria	Sulfated polysaccharide	Human food
	Cladosiphon	Sulfated l-fucose	Human food
	Saccharina	Sulfated polysaccharides	Human food

protection against dehydration, resistant to UV radiation, anticoagulants, antibacterial, antioxidants, anticancer, immunomodulation, and phagocytosis. Moreover, sulfated polysaccharides (sPS) and their derivatives also have as hypolipidemic or hypoglycemic, and drag-reducers activities.

Fiber Seaweeds are recognized for their high dietary fiber content, ranging widely from 25 to 75% of dry weight, mostly soluble. Noteworthy is the presence of laminarin, a beta-polymer of glucose, contributing to dietary fiber by passing through the upper digestive tract undigested (O'Sullivan et al. 2010).

Minerals and Vitamins Algae are rich in vitamins and minerals. Microalgae emerge as rich repositories of vitamins, including A, B1, B2, B6, B12, C, E, K, niacin, nicotinate, biotin, and folic acid. Strikingly, certain microalgae surpass traditional sources like spinach and baker's yeast in vitamin content. Abundant in minerals such as calcium, phosphorus, magnesium, potassium, sodium, zinc, iron, copper, and sulfur, microalgae contribute approximately 2.2% to 4.8% of the total dry weight (Hoseini et al. 2013; Fox and Zimba 2018). Similarly, seaweeds are also served as good sources of iron, zinc, and iodine, essential vitamins, including B1, B2, B12, C, beta-carotene, and E, offering potential health benefits such as immune system support and cardiovascular protection (Mouritsen et al. 2019a, b; Ganesan et al. 2019; Zhu et al. 2021). The specific mineral composition varies based on growth conditions and surrounding environments (Becker 2013).

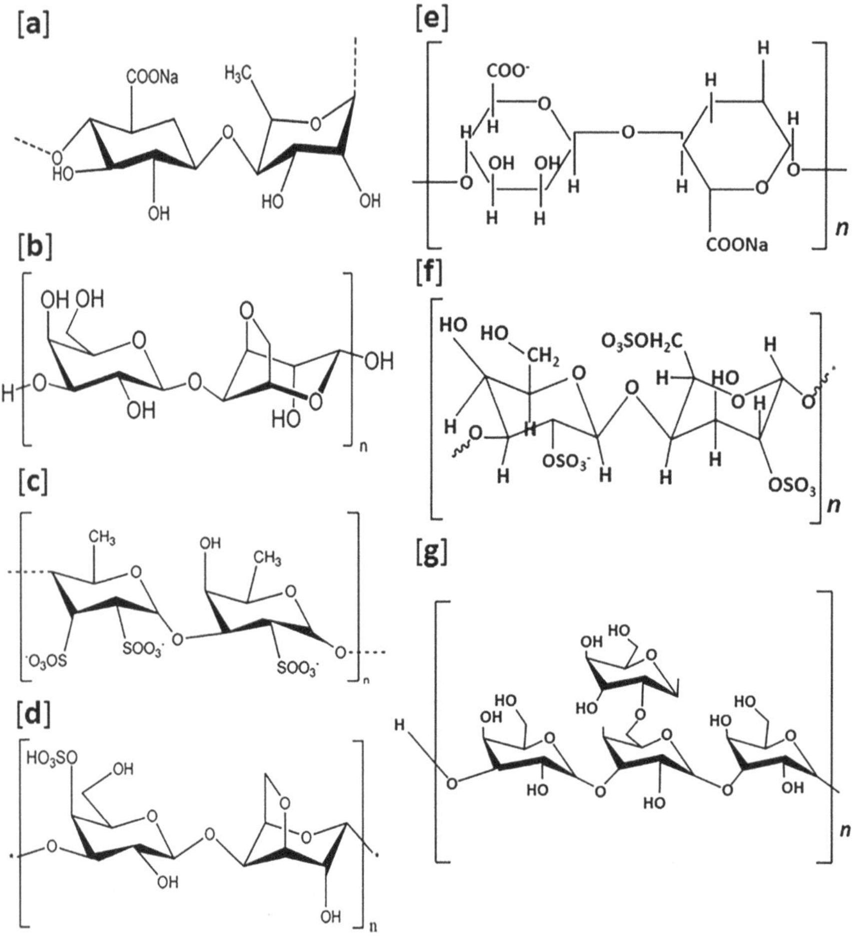

Fig. 13.4 Polysaccharides from edible macroalgae, used as food, feed and nutraceuticals. (**a**) Ulvan; (**b**) agar; (**c**) fucoidan; (**d**) carrageenan; (**e**) alginate; (**f**) Galactans; (**g**) Laminarin

Pigments The main natural pigments found in macroalgae are carotenoids (brown algae), chlorophylls (green algae) and phycobiliproteins (red algae). The vibrant colors of microalgae are attributed to pigments, essential for photosynthesis and offering health benefits. Carotenoids (fat-soluble), chlorophylls (fat-soluble), and phytobilin (water-soluble) are primary microalgal photosynthetic pigments (Fig. 13.5). These pigments possess immune-boosting, neuroprotective, antioxidant, cytotoxic, anticancer, anti-inflammatory, antihypertensive, and vitamin precursor qualities, addressing the growing demand for natural pigments in response to concerns about synthetic alternatives (Christaki et al. 2015). In essence, microalgae represent a biochemical marvel, offering a diverse array of nutrients that can redefine nutritional supplementation and functional food formulations, unlocking a wealth of health benefits for both humans and animals. Further, a rapid

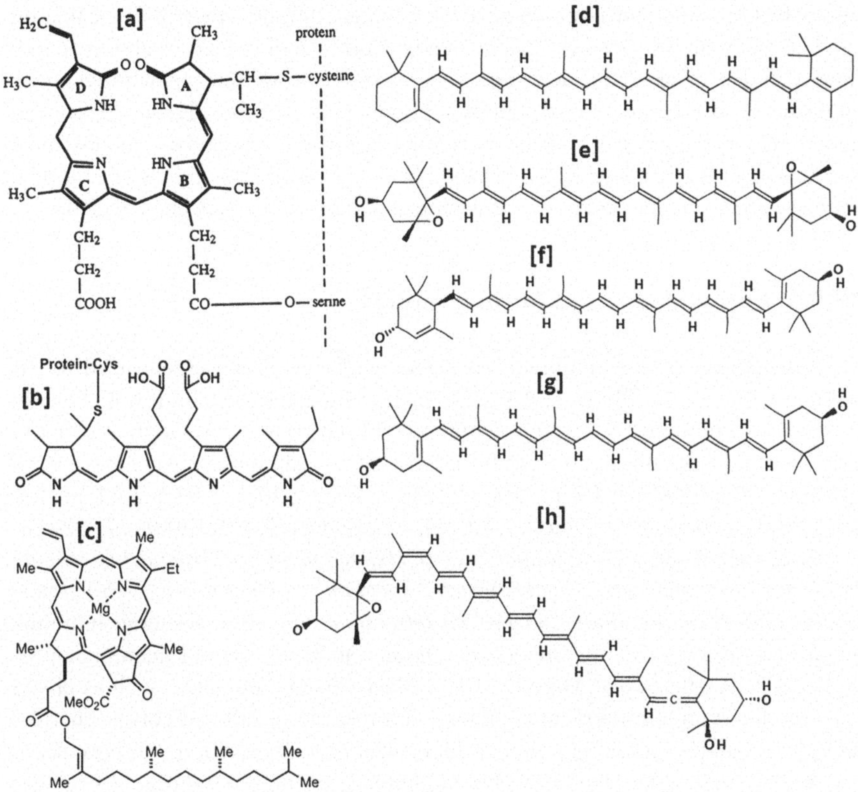

Fig. 13.5 Algal pigments as food colorants, antioxidants and nutraceuticals. (**a**) Phycoerythrin; (**b**) phycocyanobilin; (**c**) chlorophyll a; (**d**) β-carotene; (**e**) violaxanthin; (**f**) lutein; (**g**) zeaxanthin; (**h**) neoxanthin

growth rate of microalgae, their simple nutritional and cultivation strategies and biochemical's availability are making them as significant source for nutraceutical as well as therapeutic chemicals (Ramana et al. 2017).

13.5 Nutraceutical Potentials of Algae

A nutraceutical is defined as any substance that is a food or part of a food and provides health or medical benefits, including the prevention and treatment of disease (Daliu et al. 2019). The nutritive value of macro- and microalgae extends beyond basic nutrition, encompassing a wealth of bioactive compounds with significant potential in the treatment and prevention of diverse diseases (Guh et al. 2009) such as cancer, cardiovascular deficiencies, diabetes, Parkinson's disease, obesity-related disorders, and metabolic syndrome, as evidenced by Unnikrishnan et al. (2014), and Souza et al. (2017). Consequently, dietary patterns incorporating seaweed and microalgal biomass consumption have been linked, through epidemiological, clinical, and meta-analysis studies, to a reduced occurrence of numerous chronic conditions (Lordan et al. 2011; Rizwan et al. 2018). One of the noteworthy human applications of microalgal compounds is pigments β-carotenes, lycopene, astaxanthin, phycobilins, etc.), which have excellent preventive properties against heart diseases, neurological issues, eye-related conditions, and certain types of cancers, it could be a valuable resource for nutraceutical applications (Rizwan et al. 2018).

13.5.1 Antioxidants and Antimicrobials

The escalating threat of antibiotic resistance underscores the urgent need for the discovery of novel antibiotics. Algae are rich in bioactive metabolites and exhibit antioxidative properties as well as antimicrobial (Gram +ve and Gram −ve) activities. Several compounds extracted from both micro- and macroalgae have a broad spectrum of antimicrobial activities. Organic solvent extracts of seaweeds such as *Ulva fasciata, Bryopsis plumosa, Chaetomorpha antennina, Acrosiphonia orientalis, Sargassum wightii, Grateloupia filicina, Hypnea pannosa, Gracilaria corticata, Portieria hornemannii, Cheilosporum spectabile, Centroceras clavulatum, Chnoospora bicanaliculata,* and *Padina tetrastromatica* show bacterial inhibition against *Escherichia coli, Staphylococcus aureus,* and *S. pyogenes* (Shanmughapriya et al. 2008; Prarthana and Maruthi 2019). Macroalgal compounds, such as phycoerythrin (a red protein-pigment complex), carrageenan (a sulfated polysaccharide), and phlorotannins, are good antimicrobial substances. In addition, the presence of active ingredients like phenolics (anthraquinones, coumarins), flavanoids (anthraquinones, coumarins, rutin, quercetin, and kaempferol), phlorotannin (Fucol, phloroethol, Fucophlorethol, eckol, dieckol, fuhalol) and sterols of algal extracts showed broad range of antimicrobial activity against both Gram-positive and Gram-negative bacteria as illustrated in Table 13.6 (Negreanu-Pirjol et al. 2012; Wahidi et al. 2015; Rajauria 2018). In addition, compounds such as sulfated polysaccharides (e.g., fucoidan), sulfoglycolipids, carrageenan, and sesquiterpene hydroquinone found in algae have demonstrated antiviral properties.

Beyond direct antibacterial properties, microalgal compounds have exhibited antibiofilm activities, crucial in managing contagious diseases. *Streptococcus mutans* and *Lactobacillus* sp., are known contributors to dental plaque, face inhibition of biofilm formation and prevention of dental caries through extracts from *C. vulgaris* and *D. salina*. These findings underscore the potential of microalgae in contributing to the development of novel antibacterial agents in the face of growing antibiotic resistance concerns.

Microalgae are an excellent source of antioxidants. Microalgae can synthesize various biochemicals including β-carotenes, zeaxanthin, astaxanthin, chlorins, pheophytins, pyropheophytins neoxanthin, polyphenols, flavonoids, and sterols, which have strong antioxidant, anti-inflammatory as well as antimicrobial properties. They can scavenge free radicals and relieve oxidative stress. Pheophorbide b, pheophytin b, protoporphyrin methyl ester, and magnesium-chelated derivatives

Table 13.6 Antimicrobial properties of macro- and microalgae

Algal species	Metabolites	Pharmaceutical applications	Reference
Haematococcus pelvis	Short-chain fatty acids	Inhibits *Staphylococcus aureus*. and *Escherichia coli*	Santoyo et al. (2009)
Chlorococcum humicola	Phenolic compounds, β-carotene and Chlorophyll b	*Bacillus subtilis*, *Salmonella typhimurium*, *Vibrio cholera*, and *S. aureus*	Bhagavathy et al. (2011)
Cladosiphon okamuranus	Fucoidan	Inhibit the New Castle disease	Elizondo-Gonzalez et al. (2012)
Undaria pinnatifida *Cystoseira indica* *Ascophyllum nodosum*	Crude extract	Inhibitory activity against the DNA polymerase of the hepatitis B virus (HBV), influencing its replication	Wang et al. (2012)
Ulva reticulate *Caulerpa occidentalis* *Cladophora socialis* *Dictyota ciliolate* *Gracilaria dendroides*	Crude extract	*E. coli, Enterococcus faecalis, Pseudomonas aeruginosa,* and *S. aureus*	Al-Saif et al. (2014)
U. pinnatifida *Cystoseira indica* *Ascophyllum nodosum*	Crude extract	Inhibit virus adherence to host cells	Ahmadi et al. (2015)
Symphyocladia latiuscula	Crude extract	*Vibrio mimicus* and *Vibrio vulnificus*	Parsaeimehr and Lutzu (2016)
Chlorella vulgaris	Chlorellin	Anit-bacterial activity	Dineshkumar et al. (2017)
Grateloupia indica *Scinaiahatei* sp. *Gracilariacorticata* *Stoechospermum marginatum* *Cystoseira indica* *Caulerpa racemosa*	Sulfated polysaccharide	Anti -dengue viral activity	Pina-Pérez et al. (2017)
Coccomyxa sp.	Monogalactosyl diacylglycerol	Antiviral activities specifically on human Papillomavirus (HPV)	Hayashi et al. (2019)
Chlamydomonas reinhardtii	Terpenes	Anti-bacterial	Vishwakarma and Vavilala (2019)
Sargassum fusiforme	Fatty acids (PUFA)	Anti-bacterial	Albratty et al. (2021)
Saccharina japonica	Sulfated polysaccharide	Anti-viral	Liu et al. (2021)
Sargassum fusiforme	Polymannuroguluronate	Anti-viral	Dai et al. (2022)
Dictyota pfaffii	Diterpenes	Anti-viral	Cotas et al. (2022)

have been identified as potent antioxidant molecules, and their activities are attributed to the porphyrin ring's electron transfer capability (Lanfer-Marquez et al. 2005).

13.5.2 Algae-Based Pharmaceuticals

The market potential and economic benefit prospects of the algae pharmaceutical industry are huge. Global pharmaceutical markets are expanding annually, and by 2021, they could be valued at $1170 billion (Mishra et al. 2021). Algae-based nutraceutical and pharmaceutical markets anticipated that the products could find their way into the medical field. Polyunsaturated fatty acids, beta-carotene, astaxanthin, lutein, and phytobilin, for instance, have global market values expected to surpass $60 to $700 million annually (Markou and Nerantzis 2013). Therefore, additional promotion of the pharmacological activity and clinical assessment of algae products is necessary to encourage their commercialization.

The capacity of algae to contribute to the reduction in blood pressure, sugar, and fat levels is well-established. The presence of linolenic acid and its derivatives in algae plays a crucial role in decreasing blood viscosity and facilitating the interaction of vasoconstrictor substances with blood vessels. Studies have demonstrated that a mere 1% increase in linolenic acid concentration can lead to a noteworthy 5 mm Hg reduction in blood pressure (Djousse et al. 2005; Cardoso et al. 2015; Yimin 2018). Furthermore, the analysis by Zhong et al. (2020) revealed the potential of DHA algae oil to significantly decrease in blood pressure-lowering effects and control the heart rate.

Seaweed components such as alginate have been proven effective in lowering blood sugar levels in patients with type II diabetes (Hall et al. 2012; Sharifuddin et al. 2015). Additionally, porphyrin and peptides derived from *Porphyra yezoensis* have demonstrated the ability to lower both blood pressure and blood sugar levels in rats and rabbits (Qu et al. 2010). Phytocompounds like fucoidan presence in seaweeds prevent the absorption of fat, which directly contributes to the reduction in blood sugar and fat levels (Zheng et al. 2020). This effect is achieved through the activation of adenosine monophosphate-activated protein kinase, improving insulin sensitivity and restoring lipid homeostasis in mice with type II diabetes (Jeong et al. 2013; Derosa et al. 2019). The dietary intake of brown seaweed significantly decreases low-density lipoprotein, triglycerides and total cholesterol (Singh et al. 2021); for example, fucoidan supplementation in animal diet significantly reduces serum cholesterol, triglycerides, and LDL (Yokota et al. 2016). Ruqqia et al. (2020) resulted in low triglyceride levels in hyperlipidemic rats fed a high-fat diet supplemented with brown seaweeds *Sargassum binderi* and *Jolyna laminarioides*. Due to its low calorific value, increased daily intake of seaweed could reduce the level of cholesterol in dosage dependently (Sakai et al. 2009). In addition, seaweed contains soluble dietary fibers, potassium, and flavonoids which are inversely associated with the risk of coronary heart disease and mortality from stroke among Japanese men (Chichibu et al. 2021).

Microalgae strains have emerged as potential sources of neuroprotective compounds, playing a crucial role in the prevention or deceleration of neurological disorders. *Spirulina platensis*, in particular, has garnered attention in the literature for its reported neuroprotective properties against neurodegenerative diseases, including Parkinson's and Alzheimer's. Studies have documented the neuroprotective effects of *Spirulina maxima*, showcasing its ability to mitigate oxidative stress and counteract MPTP (1-methyl-4-phenyl-1, 2, 3, 6-tetrahydropyridine) neurotoxicity. These findings underscore the potential of microalgae-derived compounds, especially from Spirulina strains, in contributing to the development of neuroprotective products with applications in the management and prevention of various neurological disorders.

In addition, algal compound acts as a class of anti-cancer medications. For example, cryptophycin 1, a bioactive substance isolated from blue-green algae, has exhibited anti-cancer properties, marking a significant stride in the quest for effective cancer treatments (Kharkwal et al. 2012). In addition, certain pigments (β – carotene, Lutein, Zeaxanthin and Astaxanthin) and bioactive compounds isolated from cyanobacteria can induce apoptosis and activate the protein kinase-c family within cancer cells, showing promise in cancer therapy (Khalifa et al. 2019). These findings underscore the vast pharmacological potential of algae-derived compounds, offering new avenues for drug development in the battle against cancer. Table 13.7 represented certain nutraceutical potentials of algal active principles are given. As research in this field advances, it holds the promise of delivering innovative and sustainable solutions for cancer therapy.

13.5.3 Anticoagulants and Antithrombotic

Cardiovascular and cerebrovascular diseases have become the leading causes of death, as reported by the World Health Organization (WHO). Sulfated polysaccharides derived from algae exhibit noteworthy antithrombotic and anticoagulant properties (Déléris et al. 2016). Sulfated polysaccharides such as carrageenan, fucoidan, and ulvan are derived from marine algae mainly red algae, brown algae, and green algae, respectively. Fucoidan is a complex sulfated polysaccharide extracted from brown seaweeds *Sargassum duplicatum*, *Turbinaria decurrens,* and *Sargassum wightii*. Ulvan is obtained from the cell wall of green algae such as *Ulva* sp., *Caulerpa* sp., *Codium* sp., *Enteromorpha* sp., *Monostroma* sp., and *Gayralia* sp. Carrageenan is an extract from red seaweed, *Chondrus crispus*, and Agar is extracted from brown algae such as *Gracilaria* sp., *Gelidium* sp. and *Pterocladia* sp. These sulfate polysaccharides exhibit strong antithrombotic and anticoagulant properties (Pradhan et al. 2020; Bahari et al. 2021; Mohibbullah et al. 2023). In line with this, Ustyuzhanina et al. (2016) demonstrated that both the heparinoid Clexane (enoxaparin) and the native fucoidan from *S. latissima* exhibit comparable

Table 13.7 Algae-based medicinal compounds

Name of the algae	Compound	Nutraceutical use	Reference
Phaeodactylum tricornutum	Sulfated extracellular polysaccharide	Immunostimulant	Guzmán et al. (2003)
Porphyridium sp.	Sulfated polysaccharides	Inhibition of the migration of polymorphonuclear leukocytes	
Stephanodiscus sp.	Digalactosyl diacylglycerol	Enhanced cell differentiation	Xu et al. (2010)
Lyngbya majuscule	Lyngbic acid	Cytotoxic to colon cancer cells	
Nannochloropsiso culata	Docosapentaenoic acid	Inhibition of pro-inflammatory prostaglandin E2	
Spirulina platensis	Carotenoids, phycocyanin and polysaccharides	Enhances the immune response against microbial infections	El-Sheekh et al. (2010)
Ecklonia cava	Phlorotannins	Anti-hypertensive	Wijesinghe et al. (2011)
Oscillatoria sp.	Polyketide lactone derivative	Anti-inflammatory activity	Balunas et al. (2012)
Porphyridium cruentum	Polysaccharides	Lubricant, flocculant, anti-viral agent	de Jesus Raposo et al. (2016)
Calothrix sp.	Calothrixin B	Anti-proliferative activity against colon cancer cells	Hatae et al. (2014)
Isochrysis galbana	b-D-glucan sulfated exopolysaccharide	Cytotoxicity against lymphoma cells	Sadovskaya et al. (2014)
Fucus vesiculosus	Sulfated fucoidan	Anti-coagulant	Zhang et al. (2014)
Limnospira fusiformis	Biomass	Hepato-renal and gastroprotective activity	Dai et al. (2022)
Spirulina sp.	Biomass	Mitigate the toxic effects induced by lead	El-Tantawy (2016)
Aphanothece sacrum	Polysaccharide Sacran	Anti-inflammatory	Motoyama et al. (2016)
Spirulina platensis, *Chlorella vulgaris*	Biomass	Enhance resistance against the tested induced diseases	Fadl et al. (2017)
Microcystis aeruginosa	Mycosporine-like amino acids	Skin-care products	Sathasivam et al. (2019)
Spirulina sp	Biomass	Suppressing retinal damage	Okamoto et al. (2019)
Sprilunia sp., *Chlorella* sp.	Biomass	Mitigated the necrotic and degenerative changes	El-Habashi et al. (2019)

Alsidium corallinum	Fucoxanthin	Anti-inflammatory	Liu et al. (2021)
Sargassum fusiforme	Polysaccharides	Anti-hyperglycemic	Jia et al. (2020)
Ascophyllum nodosum	Phycocyanin	Anti-coagulant	Hachicha et al. (2022)
Ascophyllum nodosum	Fucoxanthin	Anti-obesity	Miyashita et al. (2020)
Sargassum fusiforme	Polysaccharides	Anti-aging	Zhang et al. (2020)
Scenedesmus rubescens	Carotenoids Phenolics Fatty acids	Anti-aging	Dai et al. (2022)
Arthrospira platensis	Bilirubin-like tetrapyrrolic	Anti-proliferative effects and antioxidant	Singh et al. (2022)

antithrombotic and anticoagulant properties. This similarity extends to the chemical transformation of branched xylofucans isolated from the brown algae *Punctariaplantaginea* into highly sulfated linear fucans, effectively inhibiting clot formation. Insights from in vitro studies and previously conducted human and animal trials suggest that *Ecklonia cava* stands out as a rich source of bioactive marine polyphenols, offering antihyperglycemic, antihyperlipidemic, anti-inflammatory, and antioxidant effects (Murray et al. 2018). Anionic fucose-containing polysaccharides fucoidans from the brown seaweeds are the source for the development of new anticoagulant drugs to treat cardiovascular and cerebrovascular diseases (Ustyuzhanina et al. 2016).

13.5.4 Anti-Aging and Anti-Allergic Properties

The ageing process is influenced by a multitude of physical and physiological factors, with the normal functioning of the kidneys and spleen significantly impacting human health. Seaweeds and microalgal bioactive substances play a crucial role in combating ageing by repairing DNA, enhancing immunity, neutralizing free radicals, regulating endocrine function, promoting healthy metabolism, and improving the functioning of the kidneys and spleen (Fujimura et al. 2002). Fucoidan alleviates the cells from both replicative and stress-induced senescence, mainly reduced senescence-associated beta-galactosidase (SA-β-Gal) activity, and uremic toxins (Lee et al. 2015, 2018). In addition, red algae *Pyropia yezoensis* derived pigment phycoerythrin also reduces the SA-β-Gal activity in hippocampal neuron cells (Oh et al. 2018). The use of the aqueous extract of *Fucus vesiculosus* results in increased expression of integrin molecules, leading to improved thickness and mechanical characteristics of human skin when applied directly (Tian et al. 2019). Sulfated polysaccharides found in seaweeds exhibit anti-inflammatory and antioxidant properties, offering potential therapeutic benefits in treating various disorders, including Alzheimer's (Ganesan et al. 2019; Bauer et al. 2021). Table 13.8 represents few examples of anti-ageing potentials of algae.

In the cosmetics industry, polysaccharides derived from seaweeds serve various functions, acting as suspending agents, rheology modifiers, or wound-healing agents (Wang et al. 2015). Carotenoids, powerful antioxidants abundant in seaweeds, exhibit anti-inflammatory and anti-aging properties. Astaxanthin, is a xanthophyll carotenoid present in macroalgae, has been demonstrated in multiple studies to mitigate oxidative stress and safeguard mitochondria from the cumulative damage caused by reactive oxygen species. Additionally, astaxanthin shows promising neuroprotective effects, suggesting its potential application in the treatment or prevention of neurodegenerative illnesses such as Parkinson's and Alzheimer's disease (Kidd 2011; Barbosa et al. 2014).

Seaweed phenolic compounds like curcumin, quercetin, flavonoids, and epigallocatechin gallate (EGCG) have demonstrated strong antiallergic properties (Fleurence and Gall 2016). Fucoidan derived from *U. pinnatifida* has been shown to reduce chemical and immunological reactions in animal models (Fleurence and

Table 13.8 Anti-aging effect of algae

Algae	Compound	Function	Reference
Chlorella vulgaris	Hot water extract	Hydrogen peroxide-induced DNA damage and telomere shortening of human fibroblasts	Makpol et al. (2009)
Porphyra haitanensis	Porphyran	Shinorine delays aging process	Rastogi et al. (2010)
Chlorella vulgaris	Hot water extract	Prevents replicative senescence in human diploid fibroblasts	Saberbaghi et al. (2013)
Sargassum sp.	Fucoxanthin	Increases lifespan of drosophila melanogaster and Caenorhabditis elegans	Lashmanova et al. (2015)
Porphyra haitanensis	Sulfated galactan	H_2O_2-induced premature senescence in WI-38 cells	Zhang et al. (2018)
Pyropia yezoensis	Phycoerythrin-derived tryptic peptide	Attenuates glutamate-induced ER stress and neuronal senescence in primary rat hippocampal neurons	Oh et al. (2018)

Gall 2016). Sulfate polysaccharide porphyran from *Porphyra tenera* and *Porphyra yezoensis* exhibits antiallergic properties. In an animal model with ear edema, administering 2% porphyran (from dried nori) inhibited disease progression (Ishihara et al. 2005). Porphyrans not only possess antiallergic properties but also exhibit anti-inflammatory effects, primarily attributed to their ability to scavenge reactive oxygen species (Fleurence and Gall 2016). Phenols from *E. arborea*, known for their antiallergic qualities, have been used in folk medicine since ancient times (Sugiura et al. 2007). Additionally, phycocyanins, carotenoids, polysaccharides, polyunsaturated fatty acids (PUFAs), and phenol tannins have all demonstrated antiallergic qualities. Table 13.9 illustrates the nutraceuticals and their application which are extracted from algae.

13.5.5 Algal Pigments as Therapeutic Leads

Pigments are rich in algae, which are non-toxic natural substances that unveil a spectrum of applications such as antioxidant, anti-inflammatory, immunoprophylactic, and antitumor activities among others (Patel et al. 2022). Algal pigments such as β-carotenes, lutein, fucoxanthin, β-cryptoxanthin, astaxanthin, violaxanthin, halocynthiaxanthin, zeaxanthin, canthaxanthin, peridinin, phycocyanin, and phycoerythrin which are largely produced from *Botryococcus braunii*, *Bracteacoccus* sp., *Chlorella* spp., *Dunaliella salina, Spirulina platensis, Scenedesmus* spp., *Haematococcus lacustris,* and *Scenedesmus* spp. (Patel et al. 2022). These pigments showcase their prowess in preventing acute and chronic coronary syndromes, atherosclerosis, rheumatoid arthritis, muscular dystrophy, cataracts, and neurological disorders, emphasizing their versatility in

Table 13.9 Algae-based nutraceuticals and their applications

Species	Nutraceutical compounds/ Functional group	Perceived health benefit	Reference
Undaria pinnatifida	Fucoxanthin	Antiangiogenic protective effects against retinol deficiency	Sangeetha et al. (2009)
Polyopes lancifolia	Bromophenol	α-Glucosidase inhibition	Kim et al. (2010)
Porphyra sp.	Phycoerythrin	Amelioration of diabetic complications	Yabuta et al. (2010)
Ecklonia cava	Dieckol Fucodiphloroethol G Phlorofucofuroeckol A	Inhibition of α-glucosidase and α-amylase	Lee et al. (2010)
Polyopes lancifolia	Bromophenol	Inhibition of α-glucosidases and inhibits absorption of maltase and sucrase	Kim et al. (2010)
Himanthalia elongata	Sulfated polysaccharide	Decreases cholesterol and cholic acid, inhibits hormone-sensitive lipase, acetyl CoA carboxylase, and fatty acid synthase	González-Torres et al. (2012)
Caulerpa lentillifera	Polyphenols and sterols	Decrease dipeptidyl peptidase-IV and α-glucosidase enzyme. Increases insulin secretion and glucose uptake in 3 T3-L1 adipocytes (10 μL CLE)	Sharma and Rhyu (2014)
Sargassum horneri	Monogalactosyldiacylglycerols (MGDGs), linoleyl [18:2(ω-6)]	Inhibitory effects on triglyceride (TG) accumulation and free fatty acid (FFA) levels in 3 T3-L1 adipocytes	Ma et al. (2014)
Sargassum horneri (akamoku)	Fucoxanthin (Fx)	Lowers HbA1c levels (2 mg/d Fx)	Ma et al. (2014)
Fucus vesiculosus	Polyphenol	Inhibition of α-glucosidase	Kim et al. (2014)
Porphyra dentate	Carotenoid and anthocyanin	Lower plasma glucose level and raised plasma insulin secretion	Chao et al. (2014)

(continued)

Table 13.9 (continued)

Species	Nutraceutical compounds/ Functional group	Perceived health benefit	Reference
Sargassum siliquosum, Sargassum polycystum	Fucoxanthin-rich fraction (FRF	Suppresses angiotensin-converting enzyme (ACE), α-amylase, and α-glucosidase	Nagappan et al. (2017)
Gelidiumamansii	Polyphenol	Decreases adipogenesis and suppresses adipogenic gene expression	Park et al. (2017)
Undaria pinnatifida	Fucoxanthin	Inhibits hepatic and adipocytic lipogenesis, modulating lipolysis in liver	Yoshinaga and Mitamura (2019)
Chaetomorpha antennina	Ascorbic acid, octadecadienoate, and fucosterol	Lowers serum cholesterol, triglycerides, ALT and AST	Unnikrishnan and Jayasri (2017)

maintaining overall health. Lutein, a prominent xanthophyll found in green microalgae, emerges as a guardian against age-related cataracts and macular degeneration by acting as a selective filter against harmful blue light. With its accumulation in the macula of the human retina, lutein not only shields the eyes from oxidative stress but also serves as a preventive measure against these vision-related disorders (Shao and Bock 2008; Granado-Lorencio et al. 2009). Fucoxanthin (fuco), through demonstrated anti-allergic activity in rodent mast cells, not only aids in allergic response modulation but also holds promise in reducing the risk of obesity (Sakai et al. 2009; Maeda et al. 2018). The hepatoprotective effects of microalgal pigments, including phycocyanin and phycoerythrin, are evident in vitro, safeguarding hepatocytes, while their neuroprotective effects extend to cultured rat cerebellar neurons, thanks to their antioxidant and anti-inflammatory activities (Sekar and Chandramohan 2008). Additionally, intriguingly, some pigments such as phycocyanin and allophycocyanin exhibit antifungal and antiviral properties (Wang et al. 2020).

Algal pigments have antiproliferative effects against breast and colon cancer cell lines by activation of apoptosis, DNA fragmentation, and arresting the cell cycle (Manochkumar et al. 2022). Photosynthetic pigments such as chlorophyll and carotenes can be used as natural photosensitizers, which generate reactive oxygen species (ROS) under specific light irradiation that can induce cell death in tumor cells (Xin et al. 2023). Carotenoids obtained from microalgae prevent cell proliferation (Zhong et al. 2020) and regulate ERK and cell cycle mechanism proteins to increase the expression of apoptotic protein p27. Phycocyanin extracted from *Spirulina* sp. induces pathological changes and DNA fragmentation in HeLa and MCF7 cell lines, and also upregulation of Fas and downregulation of Bcl2. Cucarin-A from *Lyngbya majuscule* exhibits antitumor activity due to the inhibition of tubulin polymerization selectively in diverse types of tumor cells such as breast, renal, and colon cell lines (Singh et al. 2021). The multifaceted benefits of microalgal pigments underscore their promising role in preventive and therapeutic interventions across a spectrum of health conditions.

13.6 Algae as a Functional Food

The concept of functional food has emerged based on the individual health of people and to reduce the high health costs. Food enriched with more active functional ingredients that provide additional health benefits beyond their nutritional value (Donato-Capel et al. 2014) are called functional foods. The Food and Agriculture Organization (FAO) of the United Nations defined "functional foods as foods that contain, in addition to nutrients, other components that may be beneficial to health". Different terminology is used, health food, novel food, bioactive food, medicinal food, therapeutic food, vitamin food, alternative medicine, and pharmafoods (Dominguez 2013). The main goal for the functional food targets is (a) to increase gastrointestinal functions along with healthy gut microflora, immune activity, nutrient bioavailability and transit time, (b) macro- and micronutrient metabolism and to reduce the pathogenic risk and cardiovascular disease, (c) good antioxidant, reduce toxicity and inflammation, (d) healthy and positive mood, behavior and mental health, and (e) fetal and infant diet along with simple digestion.

A large number of compounds such as polyunsaturated fatty acids (PUFA), polyphenols, pigments, and phytosterols obtained from various biological resources are used to prepare functional foods. Commonly, functional food ingredients are obtained from plants, fish, animals, and/or microbial by-products. In recent years, algal metabolites can be considered good sources of vitamins, minerals, protein and essential amino acids, polysaccharides, fatty acids, and pigments to prepare nutrients rich valuable functional food (Priyadarshani and Rath 2012) and have gained popularity as essential components of functional foods.

The algal biomass comprises various macromolecules and a smattering of minerals. Algal biomass, characterized by a rich biochemical makeup of carbohydrates, lipids, and proteins, serves as a valuable raw material for diverse bioproducts. Despite the challenges posed by cell walls, fibers, and carotene-protein complexes that hinder digestion, algae present a unique nutrient profile. The bioavailability of β-carotene, for example, is typically low in plant sources due to these hindrances, but algae overcome this through various processing methods (Ali et al. 2021; Dutta et al. 2022; Duan et al. 2023). Algae encompass a spectrum of pigments such as carotenoids, phycobilins, and chlorophylls. Carotenoids especially, β-carotene, lutein, and astaxanthin play a crucial role as antioxidants or provitamin A sources (Yusoff et al. 2020). In addition, phycobilins (phycocyanin and phycoerythrin) are found in cyanobacteria and the chloroplasts of red algae (Dufossé 2023). In essence, algae emerge as a treasure trove of bioactive compounds, offering a spectrum of health benefits and serving as a promising foundation for the development of functional foods that harness nature's nutrient bounty (Mouritsen et al. 2019a, b).

Algal oils are rich in long-chain polyunsaturated fatty acids (Makkar et al. 2016). Polyunsaturated fatty acids (PUFA), mainly those from omega-3 and omega-6 fatty acids, have high nutritional importance. Linoleic acid and α –linolenic acids present in food could reduce LDL-cholesterol, increase HDL-cholesterol and reduce cardiovascular diseases (CVD). PUFA endorses to increase of PGI 2, a vasodilating factor that can reduce the risk of CVD. Macroalgae and microalgae are rich in PUFA, specifically, linolenic acid (18:3 omega-3), linoleic acid (18:2 omega-6), eicosapentaenoic acid (20:5, n-3), and docosahexaenoic acid (22:6, n-3) are the most important which have strong biological functions. Microalgae such as *Dunaliella salina, Haematococcus pluvialis, Chlorella* sp., *and Arthrospira platensis* are rich in ω-3 fatty acids (Ibañez and Cifuentes 2013). Seaweeds and microalgae offer a wide range of bioactive compounds such as peptides, proteins, polysaccharides, fatty acids, vitamins, minerals, and sterols which show health benefits to the consumers (Afonso et al. 2020; Tanna et al. 2021). Scientific validation of algal metabolites such as polyphenols, carotenoids, and flavonoids showed high nutraceutical potentials like antioxidant, free radical scavenging, immunomodulating, anti-cancer and anti-proliferative effects (Abiusi et al. 2022; Terriente-Palacios and Castellari 2022). Further, sulfated polysaccharides and other hetero-polysaccharides promote beneficial microbiomes in the gut of animals to maintain intestinal health and gut homeostasis (Pratap et al. 2022). Microalgae such as *Isochrysis galbana*, *Nannochloropsis* sp., *Crypthecodinium cohnii*, and *Phaeodactylum tricornutum* are exceptional producers of PUFAs. These PUFAs have been associated with the prevention of cardiovascular, diabetic, and hypertension diseases, making them valuable in the nutraceutical, pharmaceutical, and therapeutic industries (Colussi et al. 2017). Moreover, *Isochrysis galbana* and *Diacronem avlkianum* (Haptophyta, Pavlovophyceae) biomass have been incorporated into macaroni to serve as a source of polyunsaturated fatty acids (PUFAs) (Fradique et al. 2010). Additionally, *Isochrysis galbana* PUFA-3 has been added to biscuits and gelled desserts, along with *Limnospira maximum* and *Haematococcus lacustris* (Sousa et al. 2008). Despite this, *Isochrysis galbana* remains a promising biomass for nutritional applications.

At present, several products including health drinks, ice cream, and protein powders are supplemented with algal ingredients. For example, to increase the consistency of the foods, emulsification and gel formation carrageenan and agar are used. *Chlorella* sp., *Spirulina* sp. and *D. salina* are suitable algal sources for making vegetable creams, biscuits, pasta, and fermented drinks (Boukid and Castellari 2021; Tork et al. 2022). The addition of whole algae or extracts provides improved nutritional properties and also the rheological quality of the food. A special food formulation prepared with protein-rich vegetable cream contains the active ingredients of *Chlorella vulgaris, Spirulina* sp., *Nannochloropsis oceanic,* and *Tetraselmis chui* (Boukid and Castellari 2021; Bazarnova et al. 2021). Further, algal pigments are considered natural coloring ingredients for making attractive foods with enriched nutritional properties. For example, chlorophylls, carotenoids, and phycobiliproteins obtained from *Spirulina sp*, *Chlorella* sp., and *Dunaliella* sp. are suitable food colours for making colourful cookies, bread creams, health drinks (Sun et al. 2018), and also for ornamental fish feed (Sathasivam et al. 2019).

Vitamin A deficiency and malnutrition can be treated with carotene-rich food supplements as an alternative to vitamin tablets. Research has consistently shown that biofortified crops offer a dependable source of carotenoids for animals. For example, introducing carotenoid-biofortified corn into the diet of chickens has proven effective in enhancing the accumulation of carotenoids in muscles, skin, and egg yolks. This dietary intervention has demonstrated its efficacy in increasing the vitamin A content of the liver, both in controlled laboratory conditions and real-world agricultural settings (Díaz-Gómez et al. 2017, Green and Fascetti 2016). The food ingredients rich with bioavailable carotenoids are desirable for functional food. The term "bioaccessibility" refers to the number of carotenoids that can be released from the food matrix and become accessible to the digestive system and are absorbed and rendered useful for normal physiological processes, metabolism, and storage (Thakur et al. 2020). Additionally, bioconversion, bioactivity, or bioefficacy focuses on the proportion of the absorbed provitamin A carotenoids that undergo biotransformation into vitamin A (Mondloch et al. 2015).

Carotenoids, serving as the bioactive substances in all algae, are essential photosynthetic pigments responsible for the red, yellow, and orange colors observed in these organisms (Langi et al. 2018; Darvin et al. 2022; Hossain et al. 2021). Among microalgae, *Spirulina platensis* pigments, particularly carotenoids and phycocyanin, have been extensively studied for human and animal nutrition their absorption and overall metabolism (Deng and Chow 2010; Liu et al. 2016, Prabakaran et al. 2020). This supplementation leads to increased levels of liver enzymes associated with the glutathione system, along with reduced levels of reactive oxygen species (ROS) and lipoperoxides. Astaxanthin from *H. lacustris* exhibited greater bioavailability and antioxidant effects compared to other carotenoids (Rao et al. 2013). Furthermore, lymphocytes exhibit less DNA damage compared to those on a normal diet. Spirulina has also shown efficacy in diminishing lipid peroxidation in the liver and kidneys, lowering blood glucose levels, and elevating glycogen levels. These effects highlight Spirulina's effectiveness in preventing oxidative stress and hyperglycemia associated with diabetes (Kim et al. 2010). Additionally, various improvements were observed in their lipid profile (Kanwugu et al. 2022). Categorizing dietary fibers based on their solubility, viscosity, and fermentability provides insights into how these fibers are processed by the body. Various examples of dietary fiber include polysaccharides, oligosaccharides, lignin, cellulose, hemicellulose, and related algal compounds. Dietary fiber contributes to physiological benefits in humans, promoting laxation and assisting in the reduction in blood glucose or cholesterol levels (Guillon and

Champ 2000). Furthermore, the presence of polysaccharides in dietary fiber brings additional advantages, such as antioxidant, anti-inflammatory, and cardioprotective properties, contributing to overall human health (Salanță et al. 2020).

Polysaccharides constitute a crucial component of algae. The cell walls of these algae are composed of cellulose, hemicellulose, pectin molecules, and glycoproteins, providing resistance to digestive enzymes (Yadav et al. 2021). The potential benefits of algal polysaccharides acting as prebiotics in the gastrointestinal system are realized through oral processing, stomach digestion, and intestinal absorption. Due to glycosidic linkages in polysaccharides, they resist the digestive enzymes, acids, and bile secretions of the host gut (Goodman 2010) considered a suitable prebiotic additive in food and feed formulations. Fermentable algal polysaccharides, such as galactose-6-sulfate and 3,6-anhydrogalactose, exhibit significant prebiotic effects by altering gut flora composition, promoting the growth of beneficial bacterial communities (Zheng et al. 2020). Through in vitro digestion with pepsin and pancreatic enzymes, dietary fiber from *Auxenochlorella pyrenoidosa* and *Arthrospira platensis* showed digestibility values of approximately 62.0% and 82%, respectively, and were rich in dietary fiber and carbohydrates (Demarco et al. 2022). Studies utilizing *Arthrospira platensis* (Cyanobacteria), *Haematococcus lacustris*, and *Botryococcus braunii* (Chlorophyta) biomasses explored the in-vivo bioavailability of carotenoids (−carotene, astaxanthin, and lutein) and monitored postprandial plasma, liver, and ocular responses in rats over 9 h.

WHO/FAO recognized microalgae as alternative protein sources. Proteins derived from microalgae, such as Chlorella and Spirulina, meet WHO/FAO recommendations due to their favorable essential amino acid composition (Matos 2019). Microalgae such as *Arthrospira platensis* and *Chlorella vulgaris* exhibit high-protein digestibility rates, establishing them as excellent sources of proteins and essential amino acids. However, the marine diatom *Phaeodactylum tricornutum* faces challenges with protein availability (Kumar et al. 2022). Consequently, a pre-treatment involving cell disruption is necessary to enhance protein content and digestibility (Demarco et al. 2022). On the other hand, macroalgae such as Ulva and Gracilaria demonstrated a high-protein diet with easy digestion. Notably, 90% of Ulva sp. protein and 100% of *Gracilaria* protein are digested, indicating their potential for human consumption and absorption (Demarco et al. 2022).

Microalgal biomass rich in carotenoids, by removing free radicals from live cells and reactivating enzymes like catalase, superoxide dismutase, and peroxidase, may prevent lipid peroxidation. Lipid peroxidation products generated during this process have been shown to have various biological actions in vivo, including controlling gene expression, signaling, activating receptors, and inducing adaptive responses. Regular intake of carotenoids by humans may thus serve as a protective measure against cellular damage to lipid-containing molecules under oxidative stress (Gammone et al. 2015). Additionally, a study investigated the bioavailability of phenolic compounds, lutein, and elements (zinc, iron, and magnesium) in synthetic digestive juices from commercially available *Chlorella vulgaris* tablets and powder. The results indicated high bioaccessibility of phenolic compounds and lutein in preparations containing *C. vulgaris* biomass after incubation in synthetic digestive juices. However, the bioavailability of elements (zinc, iron, and magnesium) was found to be minimal. The authors suggested that *C. vulgaris* biomass preparations in the form of powder and tablets might not be considered a suitable source of iron, magnesium, and zinc alone (Sikiru et al. 2019).

13.7 Commercial Formulations

13.7.1 Algae in Animal Feed

Approximately 30% of the current global algal production is marketed for animal feed, showcasing the profitability of algal biomass as an animal feed and aquaculture supplement. Algae present a promising alternative protein source for the animal nutrition industry, which is actively exploring new protein sources. Algal biomass is rich in pigments, polyunsaturated fatty acids, and other bioactive substances (Dineshbabu et al. 2019). Pre-treated algal biomass is used as feed supplements for cattle, sheep, and other ruminants. Early research on incorporating algae into the diets of pigs, poultry, and horses intended for human consumption demonstrated that algae are well-tolerated, easily digested, and assimilated. Several studies reported the use of whole algal biomass as functional feed supplements for various animals, including sheep, chickens, mice, and rats. In addition, microalgal compounds have been largely used in aquatic animal feeds such as rainbow trout, Atlantic salmon, Nile tilapia, and Pacific white prawns and shrimps, due to high-protein content, PUFA, and minerals (Gamboa-Delgado et al. 2019).

The potential of microalgae as a feed supplement lies in its composition, which is optimal for fish health and includes proteins, lipids, carbohydrates, polyphenols, flavonoids, carotenoids, and pigments comparable to feed ingredients commonly used in the aquafeed sector. Microalgae, such as *Chlorella, Chlamydomonas, Porphyridium, Isochrysis*, and *Nannochloropsis*, contain substantial amounts of methionine, a crucial amino acid, especially when compared to plant-based sources like *Chlorella* and *Chlamydomonas* (Nagappan et al. 2021; Yadav et al. 2021). The starch content in various microalgal species ranges from 7% to 45%, with some, like *Tetraselmis subcordiformis, C. rheinhardtii*, and *C. vulgaris*, exhibiting lower starch contents (30–49%) (Yao et al. 2018). A viable substitute for live or fresh feed in aquaculture is a microalgal paste, which can be found in dry, flocculated, micro-encapsulated, or cryopreserved forms. Algae pastes derived from microalgae serve as a protein alternative in aquaculture, providing nutrition to prawns, fish larvae, zooplankton, and bivalves (Hemaiswarya et al. 2011). Historically, the microalgae industry has primarily concentrated on food products, with a focus on species like *Spirulina* sp. and *Dunaliella salina* that are associated with cosmetics. Microalgae such as *Chlorella* sp., *Tetraselmis* sp., *Isochrysis* sp., *Synechococcus* sp. and *Phormidium* sp. are a source of life feed for shrimp aquaculture, specifically *Penaeus monodon* (Sivakumar et al. 2011) and also a live feed for larvae in fish hatcheries, but these businesses are typically small-scale. Various microalgae species, including *Spirulina* sp., *Dunaliella* sp., *Isochrysis* sp., and *Pavlova* sp., are utilized as larval feed in fish hatcheries, although cultivation is often not conducted on a large scale (Becker 2013; Ma and Hu 2023). Similarly, cyanobacteria such as *Phormidium tenue*, *Synechococcus elongates, and Spirulina* sp. are considered alternative protein supplements used for dietary formulations for shrimp (Sivakumar et al. 2014, 2018) and fish aquaculture (Ma and Hu 2023). Seaweeds such as *Padina* sp., *Sargassum* sp. and *Gracilaria* sp. are considered alternative feed supplements replacing fishmeal protein in the shrimp diet (Sivakumar and Sundararaman 2006). Algae require simple nutritional needs compared to insects and bacteria hold potential in the context of biorefineries for the production of fish feed. This concept allows for the co-production of valuable metabolites such as colors alongside fish diet, enhancing the sustainability and efficiency of the aquafeed sector (Ahmad et al. 2022).

Further, prebiotics play a beneficial role in fish health, similar to probiotics, by enhancing gut health. Algal cell wall polysaccharides and fibers act as prebiotics to promote the growth of beneficial bacteria, such as Bifidobacteria and Lactobacilli, in the intestine (Rezende et al. 2021). For example, beta-glucan from *Chlorella vulgaris* and homogalactan from *Gyrodinium* sp., are reported as fish prebiotics (Kumar et al. 2022). Algal biomass contains easily digestible carbohydrates like starch in varying amounts ranging from 7% to 45% (Mondal et al. 2021). Some species, such as *Tetraselmis subcordiformis*, *Chlamydomonas rheinhardtii*, *Spirulina platensis* and *Chlorella vulgaris*, have higher starch concentrations (30–49%) compared to other microalgae. The carbohydrate composition can influence the overall nutritional value of the feed, impacting the growth and weight gain of fish (Azaza et al. 2020; Tibbetts et al. 2020; Nagappan et al. 2021; Xiang et al. 2021). The addition of algal meal or components in the fish diet increases body weight, standard growth rate, and protein efficiency ratio in Nile tilapia, Atlantic salmon Pacific white prawns (*Litopenaeus vannamei*) (Nagappan et al. 2021), and also with enrichment of PUFA. This could have a positive effect on the consumers, through this kind of biofortification. Recent research has explored the potential of microalgae as a supplement for fingerlings and adult fish for mass feeding. Microalgae boast the highest net biomass productivity compared to any terrestrial plant or animal. As interest in the microalgae market expands, there is an increasing focus on exploring diverse applications beyond traditional uses, such as in aquaculture, where microalgae are being considered for broader applications in fish nutrition (Ahmad et al. 2022; Ma and Hu 2023).

13.7.2 Formulation of Health Food Supplements

Increasing consumer awareness regarding the health benefits and nutritional quality of the food the market value of functional food has steadily enlarged in the last two decades. The consumption of seaweed is deeply rooted in culinary traditions globally, with East Asia, particularly Japan, China, Korea, Taiwan, and Southeast Asia, being the most prominent region for its popularity. The consciousness about the consumption of carbohydrate and fat-rich foods is associated with several metabolic diseases including obesity, diabetes, and cardiovascular issues. Commonly, protein-rich foods derived from animals provide a significant amount of essential amino acids, dietary calories, vitamins, and some trace minerals (McLeod et al. 2018). Some of the essential amino acids and vitamins are often lacking in many plant-based foods. To ensure nutritious food security on a global scale, new kinds of formulation are required. Alternative to plant and animal-based protein and vitamin supplements, algae are potential and cheapest resources to compensate for all essential nutrients in functional foods (Gerber et al. 2016). For example, seaweed polysaccharides play a vital role in various cookery products, including frozen goods, ice cream, jam, and beverages. Commercially available seaweed-based foods include raw, unroasted seaweed, rolling crispy, grilled thin seaweed, and potato chips with seaweed (Mouritsen et al. 2019a, b).

Further, algae are a good source of protein and amino acids. In comparison to other alternative components like yeast and bacteria, microalgae provide a significant source of protein and fat. The well-balanced amino acid composition of microalgae eliminates the need for expensive amino acid supplements in the diet, addressing deficiencies like the lack of methionine in plant-derived substances (Nagappan et al. 2021). Among the various benefits, microalgae are rich in protein, antioxidants, fibers, and minerals that have become important nutritious food supplements in various industries such as human special food formulations, animal feeds, and beverages. There are several reasons behind the use of whole algae or part of the algal component as a special food supplement for functional foods, which may be due to (a) high protein content with essential amino acids; (b) rich in antioxidants, vitamins and minerals; (c) low calorific value with nutraceutical; (d) high productivity in simple culture technique with minimum requirements; and (e) environmental safeties with less metal contaminations (Navarro-López et al. 2020). Through various laboratory pre-experiments, different edible and highly nutritious algae are harvested from natural resources and cultured in controlled environmental conditions (Borowitzka and Vonshak 2017) and are used for commercial purposes.

Microalgae are rich sources of pigments with antioxidant properties, and specific types produce a variety of vitamins and immunostimulants in their cells, contributing to the health of consumers. Pigments like carotenoids and other microalgal pigments not only have the potential to enhance the coloration of shrimp (Sivakumar et al. 2018) and fish, making them more appealing in the market but also offer nutritional benefits. Certain microalgae are particularly noteworthy for their high concentrations of omega-3 fatty acids, including docosahexaenoic acid (DHA) and eicosapentaenoic acid (EPA), which are beneficial for both human and fish health (Liu et al. 2022). The organic compounds in microalgae include macronutrients, and the synthesis of high-value metabolites, such as pigments, using biotechnology, is anticipated to contribute significantly to earnings in the coming years (Camacho et al. 2019; Mehariya et al. 2021; Yadav et al. 2021; Ahmad et al. 2022).

13.8 Challenges in Algae Production and Commercial Food Preparation

13.8.1 Microalgae

Large-scale cultivation of microalgae biomass will positively contribute to the expansion of several industries such as food, feed, agriculture, cosmetics, bioethanol, and pharmacy. Several species of microalgae biomass have been produced through large-scale cultivation methods. The success of large-scale cultivation requires advanced methods with well-established culture facilities that will produce high-graded biomass for functional food industries. Microalgae are rich with protein, vitamins, PUFA, and minerals and are produced through modern photobioreactor (flat panel, tubular, stirred tank, and column bioreactors)-based culture systems. For large-scale commercial production, closed photobioreactors (PBRs) and open raceway ponds (OPRs) are in practice (Uyar and Kapucu 2015). A flat and sunny area with access to water and a gentle slope to assist drainage is the ideal site for an open pond, but a closed-water environment equipped with an artificial aeration and agitation system is required for a closed pond method of microalgae cultivation. Algae require high light intensity (2000–4000 lux) with12:12 or 16:8 h light-dark cycles, ideal growth temperature (20–30 °C), pH (7.0–9.0), CO2 at the concentration of 2–10%, and macro- and micronutrients. Each system has its own merits and demerits. Table 13.10 shows the advantages and disadvantages of different commercial microalgal cultivation. The following are some of the microalgal species *Chaetoceros* spp., *Dunaliella* sp., *Diacronema* sp., *Haematococcus* sp., *Nannochloropsis* spp., *Tetraselmis* sp., *Tisochrysis* spp., *Nostoc* sp., and *Spirulina* sp. (Novoveská et al. 2023) are cultivated in large scale for feed, food and pharmaceuticals.

In microalgae production, harvesting is the most important process after the mass production. Several methods including physical (centrifugation, filtration, gravity sedimentation, flotation, electrical, magnetic), chemical, and biological methods have been largely used depending on the type and cell density of microalgae. Chemical agents such as aluminum, chitosan, or polyacrylamide are used to flocculate or coagulate microalgae cells into aggregates or flocs, which can be easily separated from the water (Choy et al. 2018). However, biologically, natural substances or organisms selectively make aggregates of microalgae to form flocs or bio-flocs that can be easier to separate from water (Abdelfattah et al. 2023). Generally, majority of microalgal industries use physical methods such as filtration and sedimentation. However, microalgal industries face continuous challenges in cultivation methods such as high production costs, limited scale-up possibility, infrastructure, technical challenges, high energy requirements, contamination risks and water usage (Brown et al. 2018; Vieira de Mendonça et al. 2021; Yuan et al. 2023).

Despite the enormous array of nutritional and bioactive potentials, microalgae offer a source for food, feed and pharmaceutical industries. The major disadvantage of the use of microalgae biomass is not having suitable taste, order, and color, which

Table 13.10 Comparison of large-scale cultivation methods

Si. no.	System	Advantages	Disadvantages
1	Open pond	Cheapest method with simple maintenance, large surface area and natural sunlight for algae growth, simple scaling up with higher production rate	High risk for contamination, larger evaporation, environmental factors affect cell density
2	Flat panel photo bioreactor	Efficient light penetration, easy maintenance of pH and temperature, low energy requirement	Low biomass productivity, high biofilm risk, and risk for scale up
3	Tubular photo bioreactor	High biomass productivity, good light distribution, and easy scaling up	Limited temperature control and high energy consumption for circulation of culture media (fluid)
4	Stirred tank photo bioreactor	Efficient mixing of fluid and air, controlled pH and temperature	Limited light penetration, less biomass productivity and high energy consumption
5	Column photo bioreactor	High biomass productivity, efficient aeration, easy to control temperature and pH	Limited light penetration, difficult to scale up and high shear stress on cells

do not attract consumers (Yatipanthalawa and Martin 2021). However, in the last few decades, there are several microalgae-based functional foods that have been successfully commercialized due to their potential benefits on health. Consumer satisfaction is the main objective for the commercialization of microalgae in food. Several factors influence how microalgae are used to prepare functional as well as healthy food. Due to its unpleasant taste, microalgae change the sensory properties of the food. Consumers who previously tasted microalgae-enriched food are less likely to be dissatisfied with the taste of microalgae in food products (Moons et al. 2018). Further, scientific validation of the nutritional properties of microalgae gave knowledge to the people to like microalgae-enriched commercial food. In addition, in recent days, people have more health consciousness, hence the health-conscious individuals appear to be more willing to accept microalgae-enriched or supplemented food products. Microalgae-supplemented foods are generally called carrier foods which are a potential environmentally sustainable alternative vegetable protein source for humans (Grahl et al. 2018, 2020). Various health benefits of microalgae and the components of microalgae are discussed in the previous section. However, several limitations and challenges are there for the large-scale production, testing, manufacturing, and commercialization of microalgae-based foods. Further, research should be required for large-scale production of microalgal biomass rich in high protein and bioactive metabolites, and also to make awareness public for frequent uptake of microalgae food.

13.8.2 Macroalgae

Seaweed cultivation holds huge economic potential as well as a suitable resource for food, feed, agriculture and pharmaceutical industries. Seaweeds are commonly collected from wild or natural populations. In recent years, based on the regeneration capacities of the algal thallus and the environmental parameters either a single-step or a two-step or multistep farming could be in practice. Globally, seaweed such as *Eucheuma* spp., *Kappaphycus* spp., *Undaria* spp., *Porphyra* spp., *Caulerpa* spp., *Sargassum* spp., *Laminaria* spp., *Chondrus* spp., *Sarcothalia* spp., *Gigartina* spp., *Chondracanthus* spp. and *Gracilaria* spp. are produced through various large-scale cultivation methods (Prasad Behera et al. 2022). Seaweeds are cultivated either by an intensive culture system (artificial tanks or ponds) or a comprehensive open sea farming system (rafts or net). The commercially successful large-scale seaweed cultivation methods such as fixed off-bottom, raft methods, and net methods also have a few limitations depending on the space and other environmental factors. For example, single rafts, hanging longline and off-bottom cultivation of seaweeds are best in shallow waters systems. In India, the floating bamboo raft and rope-hanging longline systems are effective for *Gracillaria* spp. *Kappaphycus* spp. and *Sargassum* spp.. The following Table 13.11, represents the merits and demerits of various large-scale open sea cultivation of seaweeds.

The overabundance of nutrients that wash off from macroalgae farms may influence water quality, due to eutrophication. Sustainable and effective harvesting techniques are crucial to the surrounding ecosystem. The scalability of mass production is impacted by environmental factors and climatic changes. Large-scale processing factories and farming facilities are extremely costly and require high energy (Wood et al. 2017). Moreover, getting the permissions required for large-scale macroalgae farming and dealing with zoning laws is difficult and time-consuming. It is mandatory to create efficient and customized storage strategies to

Table 13.11 Advantages and disadvantages of seaweed cultivation methods

Si. no.	Cultivation method	Advantages	Disadvantages
1	Off-bottom monoline	Easy, farmers friendly and low-cost method	Heavy seedling loss; sensitive to wind, tides, storms, bad weathers, epiphytes, etc.
2	Floating raft	More economical and ecofriendly method	No uniform growth; sensitive to climatic and environmental changes
3	PVC pipe raft	Easy to handle	More expensive
4	Tube net	Uniform growth rate and lower risks	Non-economic
5	Multiple rafts longline	Large scale production, more economic	Risks for loss in case of bad weather
6	Cage systems	Sustain in the bad weather conditions	Highly expensive cages are required; increased biofouling risk

increase the shelf life of macroalgae products. To enlarge the demand, need to increase consumer acceptance (taste) for goods derived from macroalgae which require some culinary innovation (Blikra et al. 2021; Kostas et al. 2021). To improve macroalgae types for increased production, disease resistance, and environmental adaption, more biotechnology research and development are required.

13.9 Conclusion

Algae are a sustainable source of biologically active metabolites for human health and functional food supplements. Algal metabolites are natural substances used for the alleviation of lifestyle-related diseases such as diabetes, obesity, hypertension, cardiovascular diseases, and cancer. Seaweed polysaccharides are considered therapeutic nutraceuticals with multiple benefits to humans and animals. Microalgal exopolysaccharides and some algal polysaccharides are used as prebiotic components to support the growth of beneficial gut microflora, and underlying health benefits. Naturally, algae are rich in proteins with essential amino acids, which could be a suitable protein alternative to an animal-based protein diet and it could supplement all essential amino acids; hence, it is considered "food as pharma." Further, algae are low-calorific food that is rich with ω-3 fatty acids (EPA and DHA), carotenoids, vitamins, and minerals. The antioxidants present in the algae provide various health benefits such as anticancer, antiviral, and antibacterial activity. Incorporation of algae or algal-derived metabolites into our regular food or supplemented in functional food, that will replace the use of chemically synthesized antioxidants and antibiotics in drugs and cosmetics. Algal components are natural immune boosters, which rejuvenate our body tissues and system, also to enhance innate immunity against communicable diseases and various physiological disorders.

Large-scale cultivation provides enough supply of algae and algal-based nutraceuticals to the markets. Both macro- and micro-algae can be commercially produced by simple culture techniques without the costliest chemicals. Algae could be a suitable alternative protein supplement and coloring agent for aquaculture and other animal diets. Among the large number of algal diversity, limited numbers of algal resources have been recognized as food, feed, and nutraceuticals. Further scientific validation is required for unexplored algal species and their nutrients, which could be a desirable alternative food for human consumption.

Acknowledgement Authors thanks to RUSA-2.0, Madurai Kamaraj University to provide the internet and other facilities for preparation of this chapter.

Conflict of Interest No.

Authors Contribution Overall chapter design, introduction, historical prospects and conclusion by NSK; Algae as food & feed and Algal nutrients by MDP; Functional food and nutraceuticals by RTV; Formulations by RSK; and, overall correction and validation of chapter work by KPM.

References

Abdelfattah A, Ali SS, Ramadan H, El-Aswar EI, Eltawab R, Ho SH, Elsamahy T, Li S, El-Sheekh MM, Schagerl M, Kornaros M, Sun J (2023) Microalgaebased wastewater treatment: mechanisms, challenges, recent advances, and future prospects. Environ Sci Ecotechnol 13:100205

Abiusi F, Fernández PM, Canziani S, Janssen M, Wijffels RH, Barbosa M (2022) Mixotrophic cultivation of Galdieriasulphuraria for C-phycocyanin and protein production. Algal Res 61:102603

Abulafia D (2019) The Boundless Sea: A human history of the oceans. Penguin UK

Adarme-Vega TC, Lim DKY, Timmins M, Vernen F, Li Y, Schenk PM (2012) Microalgal biofactories: a promising approach towards sustainable omega-3 fatty acid production. Microb Cell Factories 11:96e105

Afonso C, Matos J, Guarda I, Gomes-Bispo A, Gomes R, Cardoso C, Gueifão S, Delgado I, Coelho I, Castanheira I (2020) Bioactive and nutritional potential of *Alaria esculenta* and *Saccharina latissima*. J Appl Phycol 33:501–513

Afonso NC, Catarino MD, Silva AM, Cardoso SM (2019) Brown macroalgae as valuable food ingredients. Antioxidants 8(9):365

Ahmad A, Hassan WS, Banat F (2022) An overview of microalgae biomass as a sustainable aquaculture feed ingredient: food security and circular economy. Bioengineered 13(4):9521–9547

Ahmadi A, ZorofchianMoghadamtousi S, Abubakar S, Zandi K (2015) Antiviral potential of algae polysaccharides isolated from marine sources: a review. Biomed Res Int 2015:825203

Ainis AF, Erlandson JM, Gill KM, Graham MH, Vellanoweth RL (2019) The potential use of seaweeds and marine plants by native peoples of Alta and Baja California. In: An archaeology of abundance: reevaluating the marginality of California's islands. University Press of Florida, Gainesville, pp 135–170

Albratty M, Alhazmi HA, Meraya AM, Najmi A, Alam MS, Rehman Z, Moni SS (2021) Spectral analysis and antibacterial activity of the bioactive principles of Sargassum tenerrimum J. Agardh collected from the Red Sea, Jazan, Kingdom of Saudi Arabia. Braz J Biol 2021(83):e249536

Alexandratos N, Bruinsma J (2012) World agriculture towards 2030/2050: the 2012 revision. ESA Working Paper No. 12-03-June 2012

Ali S, Peter AP, Chew KW, Munawaroh HSH, Show PL (2021) Resource recovery from industrial effluents through the cultivation of microalgae: A review. Bioresour Technol 337:125461

Al-Saif SSAL, Abdel-Raouf N, El-Wazanani HA, Aref IA (2014) Antibacterial substances from marine algae isolated from Jeddah coast of Red Sea, Saudi Arabia. Saudi J Biol Sci 21(1):57–64

Araújo R, Peteiro C (2021) Algae as food and food supplements in Europe. Centro Oceanográfico de Santander

Ariede MB, Candido TM, Jacome ALM, Velasco MVR, de Carvalho JCM, Baby AR (2017) Cosmetic attributes of algae-A review. Algal Res 25:483–487

Arufe S, Della Valle G, Chiron H, Chenlo F, Sineiro J, Moreira R (2018) Effect of brown seaweed powder on physical and textural properties of wheat bread. Eur Food Res Technol 244:1–10

Azaza MS, Saidi SA, Dhraief MN, El-Feki A (2020) Growth performance, nutrient digestibility, hematological parameters, and hepatic oxidative stress response in juvenile Nile tilapia, *Oreochromis niloticus*, fed carbohydrates of different complexities. Animals 10(10):1913

Bahari A, Moelants K, Wallecan J, Mangiante G, Mazoyer J, Hendrickx M, Grauwet T (2021) Understanding the effect of time, temperature and salts on carrageenan extraction from *Chondrus crispus*. Algal Res 58:102371

Balunas MJ, Grosso MF, Villa FA, Engene N, McPhail KL, Tidgewell K et al (2012) Coibacins A–D, antileishmanial marine cyanobacterial polyketides with intriguing biosynthetic origins. Org Lett 14(15):3878–3881

Barbosa M, Valentão P, Andrade PB (2014) Bioactive compounds from macroalgae in the new millennium: implications for neurodegenerative diseases. Mar Drugs 12:4934–4972

Barka A, Blecker C (2016) Microalgae as a potential source of single-cell proteins. A review. Base

Barzkar N, Jahromi ST, Poorsaheli HB, Vianello F (2019) Metabolites from marine microorganisms, micro, and macroalgae: immense scope for pharmacology. Mar Drugs 17:464

Bassi A, Saxena P, Aguirre AM (2014) Mixotrophic algae cultivation for energy production and other applications. Algal biorefineries: volume 1: cultivation of cells and products, 177–202

Batista AP, Gouveia L, Nunes MC, Franco JM, Raymundo A (2008) Microalgae biomass as a novel functional ingredient in mixed gel systems. In: Gums and Stabilisers for the food industry 14. The Royal Society of Chemistry, pp 487–494

Bauer S, Jin W, Zhang F, Linhardt RJ (2021) The application of seaweed polysaccharides and their derived products with potential for the treatment of Alzheimer's disease. Mar Drugs 19:89

Bazarnova J, Nilova L, Trukhina E, Bernavskaya M, Smyatskaya Y, Aktar T (2021) Use of microalgae biomass for fortification of food products from grain. Food Secur 10(12):3018

Becker EW (2013) Micro-algae for human and animal consumption. In: Richmond AE, Hu Q (eds) Handbook of Microalgal Culture: Applied Phycology and Biotechnology (Second), pp 461–503

Beheshtipour H, Mortazavian AM, Mohammadi R, Sohrabvandi S, Khosravi-Darani K (2013) Supplementation of Spirulina platensis and Chlorella vulgaris algae into probiotic fermented milks. Compr Rev Food Sci Food Saf 12(2):144–154

Bhagavathy S, Sumathi P, Bell IJS (2011) Green algae Chlorococcumhumicola-a new source of bioactive compounds with antimicrobial activity. Asian Pac J Trop Biomed 1(1):S1–S7

Bhagea R, Puchooa D (2015) Isolation and characterisation of microalgae from University of Mauritius farm for bioethanol production. Asian. J Microbiol Biotechnol Environ Sci 17:1065–1070

Blikra MJ, Altintzoglou T, Løvdal T, Rognså G, Skipnes D, Skåra T et al (2021) Seaweed products for the future: using current tools to develop a sustainable food industry. Trends Food Sci Technol 118:765–776

Borowitzka MA, Vonshak A (2017) Scaling up microalgal cultures to commercial scale. Eur J Phycol 52(4):407–418

Boukid F, Castellari M (2021) Food and beverages containing algae and derived ingredients launched in the market from 2015 to 2019: a front-of-pack labeling perspective with a special focus on Spain. Food Secur 10(1):173

Brown JJ, Das P, Al-Saidi M (2018) Sustainable agriculture in the Arabian/Persian gulf region utilizing marginal water resources: making the best of a bad situation. Sustain For 10:1364

Buono S, Langellotti AL, Martello A, Rinna F, Fogliano V (2014) Functional ingredients from microalgae. Food Funct 5(8):1669–1685

Camacho F, Macedo A, Malcata F (2019) Potential industrial applications and commercialization of microalgae in the functional food and feed industries: A short review. Mar Drugs 17(6):312

Capelli B, Cysewski GR (2010) Potential health benefits of spirulina microalgae* A review of the existing literature. Forum Nutr 9(2):19–26

Cardoso SM, Pereira OR, Seca AML, Pinto DCGA, Silva AMS (2015) Seaweeds as preventive agents for cardiovascular diseases: from nutrients to functional foods. Mar Drugs 13:6838–6865

Chacón-Lee TL, González-Mariño GE (2010) Microalgae for "healthy" foods—possibilities and challenges. Compr Rev Food Sci Food Saf 9(6):655–675

Chao PC, Hsu CC, Liu WH (2014) Renal protective effects of Porphyra dentate aqueous extract in diabetic mice. Biomedicine 4:1–5

Chichibu H, Yamagishi K, Kishida R, Maruyama K, Hayama-Terada M, Shimizu Y et al (2021) Seaweed intake and risk of cardiovascular disease: the circulatory risk in communities study (CIRCS). J Atheroscler Thromb 28(12):1298–1306

Choy SY, Prasad KMN, Wu TY, Raghunandan ME, Phang SM, Juan JC, Ramanan RN (2018) Starch-based flocculant outperformed aluminium sulfate hydrate and polyaluminium chloride through effective bridging for harvesting acicular microalga ankistrodesmus. Algal Res 29:343–353

Christaki E, Bonos E, Florou-Paneri P (2015) Innovative microalgae pigments as functional ingredients in nutrition. In: Handbook of marine microalgae. Academic Press, pp 233–243

Circuncisão AR, Catarino MD, Cardoso SM, Silva AM (2018) Minerals from macroalgae origin: health benefits and risks for consumers. Mar Drugs 16(11):400

Cofrades S, López-Lopez I, Bravo L et al (2010) Nutritional and antioxidant properties of different brown and red Spanish edible seaweeds. Food Sci Technol Int 16(5):361–370. https://doi.org/10.1177/1082013210367049

Colussi G, Catena C, Novello M, Bertin N, Sechi LA (2017) Impact of omega-3 polyunsaturated fatty acids on vascular function and blood pressure: relevance for cardiovascular outcomes. Nutr Metab Cardiovasc Dis 27(3):191–200

Costa DFA, Quigley SP, Isherwood P, McLennan SR, Poppi DP (2016) Supplementation of cattle fed tropical grasses with microalgae increases microbial protein production and average daily gain. J Anim Sci 94(5):2047–2058

Cotas J, Pacheco D, Gonçalves AMM, Pereira L (2022) Antiviral compounds from seaweeds: an overview. Sustainable global resources of seaweeds volume 2: food, pharmaceutical and health applications, 441–453

Da Silva Vaz B, Botelho Moreira J, Greque de Morais M, Vieira Costa J (2016) Microalgae as a new source of bioactive compounds in food supplements. Curr Opin Food Sci 7:73–77

Dai N, Wang Q, Xu B, Chen H (2022) Remarkable natural biological resource of algae for medical applications. Front Mar Sci 9:912924

Daliu P, Santini A, Novellino E (2019) From pharmaceuticals to nutraceuticals: bridging disease prevention and management. Expert Rev Clin Pharmacol 12(1):1–7

Darvin ME, Lademann J, von Hagen J, Lohan SB, Kolmar H, Meinke MC, Jung S (2022) Carotenoids in human skin in vivo: antioxidant and photo-protectant role against external and internal stressors. Antioxidants 11(8):1451

Dawczynski C, Schubert R, Jahreis G (2007) Amino acids, fatty acids, and dietary fibre in edible seaweed products. Food Chem 103:891–899

Déléris P, Nazih H, Bard J-M (2016) Seaweeds in human health. In: Fleurence J, Levine I (eds) Seaweed in health and disease prevention, 1st edn. Academic Press, London, pp 319–367

Demarco M, de Moraes JO, Matos ÂP, Derner RB, de Farias Neves F, Tribuzi G (2022) Digestibility, bioaccessibility and bioactivity of compounds from algae. Trends Food Sci Technol 121:114–128

Deng R, Chow TJ (2010) Hypolipidemic, antioxidant, and antiinflammatory activities of microalgae spirulina. Cardiovasc Ther 28(4):e33–e45

Derosa G, Pascuzzo MD, D'Angelo A, Maffioli P (2019) Ascophyllum nodosum, Fucus vesiculosus and chromium picolinate nutraceutical composition can help to treat type 2 diabetic patients. Diabetes Metab Syndr Obes: Targets and Therapy 12:1861–1865

Díaz-Gómez J, Moreno JA, Angulo E, Sandmann G, Zhu C, Ramos AJ et al (2017) High-carotenoid biofortified maize is an alternative to color additives in poultry feed. Anim Feed Sci Technol 231:38–46

Dillehay TD, Ramírez C, Pino M, Collins MB, Rossen J, Pino-Navarro JD (2008) Monte Verde: seaweed, food, medicine, and the peopling of South America. Science 320(5877):784–786

Dineshbabu G, Goswami G, Kumar R, Sinha A, Das D (2019) Microalgae–nutritious, sustainable aqua-and animal feed source. J Funct Foods 62:103545

Dineshkumar R, Narendran R, Jayasingam P, Sampathkumar P (2017) Cultivation and chemical composition of microalgae Chlorella vulgaris and its antibacterial activity against human pathogens. J Aquac Mar Biol 5(3):00119

Djousse L, Arnett DK, Pankow JS, Hopkins PN, Province MA, Ellison RC (2005) Dietary linolenic acid is associated with a lower prevalence of hypertension in the NHLBI family heart study. Hypertension 45:368–373

Dominguez H (ed) (2013) Functional ingredients from algae for foods and nutraceuticals. Elsevier

Donato-Capel L, Garcia-Rodenas CL, Pouteau E, Lehmann U, Srichuwong S, Erkner A et al (2014) Technological means to modulate food digestion and physiological response. In: Food structures, digestion and health. Academic Press, pp 389–422

Duan X, Xie C, Hill DR, Barrow CJ, Dunshea FR, Martin GJ, Suleria HA (2023) Bioaccessibility, bioavailability and bioactivities of carotenoids in microalgae: a review. Food Rev Intl:1–30

Dufossé L (2023) Natural colorants from cyanobacteria and algae. Handbook of Natural Colorants, 417–437

Dutta N, Mahmood F, Sarwar A, Sultan M, Ahmad M, Khan MU, Omar MM (2022) Algae biomass conversion technologies. In: Handbook of research on algae as a sustainable solution for food, energy, and the environment. IGI Global, pp 524–546

El-Habashi N, Fadl SE, Farag HF, Gad DM, Elsadany AY, El Gohary MS (2019) Effect of using spirulina and chlorella as feed additives for elevating immunity status of Nile tilapia experimentally infected with Aeromonas hydrophila. Aquac Res 50(10):2769–2781

Elizondo-Gonzalez R, Cruz-Suarez LE, Ricque-Marie D, Mendoza-Gamboa E, Rodriguez-Padilla C, Trejo-Avila LM (2012) In vitro characterization of the antiviral activity of fucoidan from Cladosiphonokamuranus against Newcastle disease virus. Virol J 9:1–9

El-Sheekh MM, Mahmoud YG, Abo-Shady AM, Hamza W (2010) Efficacy of Rhodotorulaglutinis and Spirulina platensis carotenoids in immunopotentiation of mice infected with Candida albicans SC5314 and Pseudomonas aeruginosa 35. Folia Microbiol 55:61–67

El-Tantawy WH (2016) Antioxidant effects of spirulina supplement against lead acetate-induced hepatic injury in rats. J Tradit Complement Med 6(4):327–331

Fadl SE, ElGohary MS, Elsadany AY, Gad DM, Hanaa FF, El-Habashi NM (2017) Contribution of microalgae-enriched fodder for the Nile tilapia to growth and resistance to infection with Aeromonas hydrophila. Algal Res 27:82–88

Ferdouse F, Holdt SL, Smith R, Murúa P, Yang Z (2018) The global status of seaweed production, trade and utilization. Globefish Res Programme 124:I

Fernandes R, Campos J, Serra M, Fidalgo J, Almeida H, Casas A et al (2023) Exploring the benefits of phycocyanin: from spirulina cultivation to its widespread applications. Pharmaceuticals 16(4):592

Fleurence J, Gall EA (2016) Antiallergic properties. In: Fleurence J, Levine I (eds) Seaweed in health and disease prevention, 1st edn. Academic Press, London, UK, pp 389–420

Fox JM, Zimba PV (2018) Minerals and trace elements in microalgae. Microalgae in Health and Disease Prevention, 177–193

Fradique M, Batista AP, Nunes MC, Gouveia L, Bandarra NM, Raymundo A (2010) Incorporation of Chlorella vulgaris and Spirulina maxima biomass in pasta products. Part 1: preparation and evaluation. J Sci Food Agric 90(10):1656–1664

Fujimura T, Tsukahara K, Moriwaki S, Kitahara T, Sano T, Takema Y (2002) Treatment of human skin with an extract of Fucus vesiculosus changes its thickness and mechanical properties. J Cosmet Sci 53:1–9

Gamboa-Delgado J, Morales-Navarro YI, Nieto-López MG, Villarreal-Cavazos DA, Cruz-Suárez LE (2019) Assimilation of dietary nitrogen supplied by fish meal and microalgal biomass from Spirulina (*Arthrospira platensis*) and *Nannochloropsisoculata* in shrimp *Litopenaeusvannamei* fed compound diets. J Appl Phycol 31:2379–2389

Gammone MA, Riccioni G, D'Orazio N (2015) Marine carotenoids against oxidative stress: effects on human health. Mar Drugs 13(10):6226–6246

Ganesan AR, Tiwari U, Rajauria G (2019) Seaweed nutraceuticals and their therapeutic role in disease prevention. Food Sci Human Wellness 8:252–263

García JL, De Vicente M, Galán B (2017) Microalgae, old sustainable food and fashion nutraceuticals. Microb Biotechnol 10(5):1017

Gerber LN, Tester JW, Beal CM, Huntley ME, Sills DL (2016) Target cultivation and financing parameters for sustainable production of fuel and feed from microalgae. Environ Sci Technol 50(7):3333–3341

Gomez-Zavaglia A, Prieto Lage MA, Jimenez-Lopez C, Mejuto JC, Simal-Gandara J (2019) The potential of seaweeds as a source of functional ingredients of prebiotic and antioxidant value. Antioxidants 8(9):406

González-Torres L, Churruca I, Schultz Moreira AR, Bastida S, Benedí J, Portillo MP, Sánchez-Muniz FJ (2012) Effects of restructured pork containing Himanthaliaelongata on adipose tissue lipogenic and lipolytic enzyme expression of normo-and hypercholesterolemic rats. J Nutrigenet Nutrigenomics 5(3):158–167

Goodman BE (2010) Insights into digestion and absorption of major nutrients in humans. Adv Physiol Educ 34(2):44–53

Gouveia L, Batista AP, Raymundo A, Bandarra N (2008) Spirulina maxima and Diacronemavlkianum microalgae in vegetable gelled desserts. Nutr Food Sci 38(5):492–501

Gouveia L, Raymundo A, Batista AP, Sousa I, Empis J (2006) Chlorella vulgaris and *Haematococcuspluvialis* biomass as colouring and antioxidant in food emulsions. Eur Food Res Technol 222:362–367

Grahl S, Strack M, Mensching A, Mörlein D (2020) Alternative protein sources in Western diets: food product development and consumer acceptance of spirulina-filled pasta. Food Qual Prefer 84:103933

Grahl S, Strack M, Weinrich R, Mörlein D (2018) Consumer-oriented product development: the conceptualization of novel food products based on spirulina (*Arthrospira platensis*) and resulting consumer expectations. J Food Qual 1919482

Granado-Lorencio F, Herrero-Barbudo C, Acién-Fernández G, Molina-Grima E, Fernández-Sevilla JM, Pérez-Sacristán B, Blanco-Navarro I (2009) In vitro bioaccesibility of lutein and zeaxanthin from the microalgae Scenedesmus almeriensis. Food Chem 114(2):747–752

Green AS, Fascetti AJ (2016) Meeting the vitamin A requirement: the efficacy and importance of β-carotene in animal species. Sci World J 2016

Guh DP, Zhang W, Bansback N, Amarsi Z, Birmingham CL, Anis AH (2009) The incidence of co-morbidities related to obesity and overweight: a systematic review and meta-analysis. BMC Public Health 9:88

Guillon F, Champ M (2000) Structural and physical properties of dietary fibres, and consequences of processing on human physiology. Food Res Int 33(3–4):233–245

Guzmán S, Gato A, Lamela M, Freire-Garabal M, Calleja JM (2003) Anti-inflammatory and immunomodulatory activities of polysaccharide from Chlorella stigmatophora and Phaeodactylumtricornutum. Phytother Res 17(6):665–670

Hachicha R, Elleuch F, Ben Hlima H, Dubessay P, de Baynast H, Delattre C, Pierre G, Hachicha R, Abdelkafi S, Michaud P, et al (2022) Biomolecules from microalgae and cyanobacteria: applications and market survey. Appl Sci 12(4):1924. https://doi.org/10.3390/app12041924

Hall AC, Fairclough AC, Mahadevan K, Paxman JR (2012) Ascophyllum nodosum enriched bread reduces subsequent energy intake with no effect on post-prandial glucose and cholesterol in healthy, overweight males. A pilot study. Appetite 58:379–438

Hatae N, Satoh R, Chiba H, Osaki T, Nishiyama T, Ishikura M et al (2014) N-substituted calothrixin B derivatives inhibited the proliferation of HL-60 promyelocytic leukemia cells. Med Chem Res 23:4956–4961

Hayashi K, Lee JB, Atsumi K, Kanazashi M, Shibayama T, Okamoto K et al (2019) In vitro and in vivo anti-herpes simplex virus activity of monogalactosyl diacylglyceride from Coccomyxa sp. KJ (IPOD FERM BP-22254), a green microalga. *PLoS One 14*(7):e0219305

Hayes M (2015) Chapter 14—seaweeds: A nutraceutical and health food. In: Tiwari BK, Troy DJ (eds) Seaweed sustainability. Academic Press, San Diego, CA, USA, pp 365–387

Hemaiswarya S, Raja R, Ravi Kumar R, Ganesan V, Anbazhagan C (2011) Microalgae: a sustainable feed source for aquaculture. World J Microbiol Biotechnol 27:1737–1746

Hernandez-Carmona G, Freile-Pelegrín Y, Hernández-Garibay E (2013) Conventional and alternative technologies for the extraction of algal polysaccharides. In: Functional ingredients from algae for foods and nutraceuticals. Woodhead Publishing, pp 475–516

Heydarizadeh P, Poirier I, Loizeau D, Ulmann L, Mimouni V, Schoefs B, Bertrand M (2013) Plastids of marine phytoplankton produce bioactive pigments and lipids. Mar Drugs 11(9):3425–3471

Hoseini SM, Khosravi-Darani K, Mozafari MR (2013) Nutritional and medical applications of spirulina microalgae. Mini-Rev Med Chem 13(8):1231–1237

Hossain A, Raza A, Maitra S, Asaduzzaman M, Islam MR, Hossain MJ, … Aftab T (2021) Strigolactones: a novel carotenoid-derived phytohormone–biosynthesis, transporters, signalling, and mechanisms in abiotic stress. Plant Growth Regulators: Signalling Under Stress Conditions, 275–303

Hosseinkhani N, McCauley JI, Ralph PJ (2022) Key challenges for the commercial expansion of ingredients from algae into human food products. Algal Res 64:102696

Ibañez E, Cifuentes A (2013) Benefits of using algae as natural sources of functional ingredients. J Sci Food Agric 93(4):703–709

Imchen T, Singh KS (2023) Marine algae colorants: antioxidant, anti-diabetic properties and applications in food industry. Algal Res 69:102898

Ishihara K, Oyamada C, Matsushima R, Murata M, Muraoka T (2005) Inhibitory effect of porphyran, prepared from dried "nori", on contact hypersensitivity in mice. Biosci Biotechnol Biochem 69:1824–1830

Jeong YT, Kim YD, Jung YM, Park DC, Lee DS, Ku SK, Li X, Lu Y, Chao GH, Kim KJ et al (2013) Low molecular weight fucoidan improves endoplasmic reticulum stress-reduced insulin sensitivity through AMP-activated protein kinase activation in L6 myotubes and restores lipid homeostasis in a mouse model of type 2 diabetes. Mol Pharmacol 84:147–157

de Jesus Raposo MF, De Morais AMMB, De Morais RMSC (2016) Emergent sources of prebiotics: seaweeds and microalgae. Mar Drugs 14(2):27

Jia RB, Li ZR, Wu J, Ou ZR, Sun B, Lin L, Zhao M (2020) Antidiabetic effects and underlying mechanisms of anti-digestive dietary polysaccharides from Sargassum fusiforme in rats. Food Funct 11(8):7023–7036

Jönsson M, Allahgholi L, Sardari RR, Hreggviðsson GO, Nordberg Karlsson E (2020) Extraction and modification of macroalgal polysaccharides for current and next-generation applications. Molecules 25(4):930

Kanwugu ON, Glukhareva TV, Danilova IG, Kovaleva EG (2022) Natural antioxidants in diabetes treatment and management: prospects of astaxanthin. Crit Rev Food Sci Nutr 62(18):5005–5028

Khalifa SA, Elias N, Farag MA, Chen L, Saeed A, Hegazy MEF et al (2019) Marine natural products: A source of novel anticancer drugs. Mar Drugs 17(9):491

Kharkwal H, Joshi DD, Panthari PREETI, Pant MK, Kharkwal AC (2012) Algae as future drugs. Asian J Pharm Clin Res 5:1–4

Kidd P (2011) Astaxanthin, cell membrane nutrient with diverse clinical benefits and anti-aging potential. Altern Med Rev 16:355–364

Kim JH, Kim SM, Cha KH, Mok IK, Koo SY, Pan CH, Lee JK (2016) Evaluation of the anti-obesity effect of the microalga Phaeodactylumtricornutum. Appl Biol Chem 59(2):283–290

Kim KT, Rioux LE, Turgeon SL (2014) Alpha-amylase and alpha-glucosidase inhibition is differentially modulated by fucoidan obtained from Fucus vesiculosus and Ascophyllum nodosum. Phytochemistry 98:27–33

Kim KY, Nguyen TH, Kurihara H, Kim SM (2010) α-Glucosidase inhibitory activity of bromophenol purified from the red alga Polyopeslancifolia. J Food Sci 75(5):H145–H150

Kostas ET, Adams JM, Ruiz HA, Durán-Jiménez G, Lye GJ (2021) Macroalgal biorefinery concepts for the circular bioeconomy: A review on biotechnological developments and future perspectives. Renew Sust Energ Rev 151:111553

Koyande AK, Chew KW, Rambabu K, Tao Y, Chu D-T, Show P-L (2019) Microalgae: A potential alternative to health supplementation for humans. Food Sci Human Wellness 8:16–24

Kumar R, Hegde AS, Sharma K, Parmar P, Srivatsan V (2022) Microalgae as a sustainable source of edible proteins and bioactive peptides–current trends and future prospects. Food Res Int 157:111338

Kusmayadi A, Suyono EA, Nagarajan D, Chang J, Yen H (2020) Application of computational fluid dynamics (CFD) on the raceway design for the cultivation of microalgae: a review. J Ind Microbiol Biotechnol 47:373e382

Lanfer-Marquez UM, Barros RM, Sinnecker P (2005) Antioxidant activity of chlorophylls and their derivatives. Food Res Int 38(8–9):885–891

Langi P, Kiokias S, Varzakas T, Proestos C (2018) Carotenoids: from plants to food and feed industries. Microbial carotenoids: Methods and protocols, 57–71

Lashmanova E, Proshkina E, Zhikrivetskaya S, Shevchenko O, Marusich E, Leonov S et al (2015) Fucoxanthin increases lifespan of Drosophila melanogaster and Caenorhabditis elegans. Pharmacol Res 100:228–241

Lee JH, Lee SH, Choi SH, Asahara T, Kwon SM (2015) The sulfated polysaccharide fucoidan rescues senescence of endothelial colony-forming cells for ischemic repair. Stem Cells 33(6):1939–1951

Lee JH, Yun CW, Hur J, Lee SH (2018) Fucoidan rescues p-cresol-induced cellular senescence in mesenchymal stem cells via FAK-Akt-TWIST axis. Mar Drugs 16(4):121

Lee SH, Ko SC, Kang SM, Cha SH, Ahn GN, Um BH, Jeon YJ (2010) Antioxidative effect of Ecklonia cava dried by far infrared radiation drying. Food Sci Biotechnol 19:129–135

Levasseur W, Perré P, Pozzobon V (2020) A review of high value-added molecules production by microalgae in light of the classification. Biotechnol Adv 41:107545

Lewin RA, Andersen RA (2019) Algae. Encyclopedia Britannica. Algae Available: https://www.britannica.com/science/algae

Liu J, Obaidi I, Nagar S, Scalabrino G, Sheridan H (2021) The antiviral potential of algal-derived macromolecules. Curr Res Biotechnol 3:120–134

Liu Q, Huang Y, Zhang R, Cai T, Cai Y (2016) Medical application of Spirulina platensis derived C-phycocyanin. Evid Based Complement Alternat Med 2016(2016):7803846

Liu Y, Ren X, Fan C, Wu W, Zhang W, Wang Y (2022) Health benefits, food applications, and sustainability of microalgae-derived N-3 PUFA. Food Secur 11(13):1883

Lordan S, Ross RP, Stanton C (2011) Marine bioactives as functional food ingredients: potential to reduce the incidence of chronic diseases. Mar Drugs 9:1056–1100

Ma AC, Chen Z, Wang T, Song N, Yan Q, Fang YC et al (2014) Isolation of the molecular species of monogalactosyldiacylglycerols from brown edible seaweed Sargassum horneri and their inhibitory effects on triglyceride accumulation in 3T3-L1 adipocytes. J Agric Food Chem 62(46):11157–11162

Ma M, Hu Q (2023) Microalgae as feed sources and feed additives for sustainable aquaculture: prospects and challenges. Rev Aquac 2023:1–18

Madeira MS, Cardoso C, Lopes PA, Coelho D, Afonso C, Bandarra NM, Prates JAM (2017) Microalgae as feed ingredients for livestock production and meat quality: A review. Livest Sci 205(June):111–121

Maeda Y, Yoshino T, Matsunaga T, Matsumoto M, Tanaka T (2018) Marine microalgae for production of biofuels and chemicals. Curr Opin Biotechnol 50:111–120

Mahadevan K (2015) Seaweeds: a sustainable food source. In: Seaweed sustainability. Academic Press, pp 347–364

Makkar HP, Tran G, Heuzé V, Giger-Reverdin S, Lessire M, Lebas F, Ankers P (2016) Seaweeds for livestock diets: A review. Anim Feed Sci Technol 212:1–17

Makpol S, Yaacob N, Zainuddin A, Yusof Y, Ngah W (2009) Chlorella vulgaris modulates hydrogen peroxide-induced DNA damage and telomere shortening of human fibroblasts derived from different aged individuals. Afr J Tradit Complement Altern Med 6:4

Manochkumar J, Doss CGP, Efferth T, Ramamoorthy S (2022) Tumor preventive properties of selected marine pigments against colon and breast cancer. Algal Res 61:102594

Markou G, Angelidaki I, Georgakakis D (2012) Microalgal carbohydrates: an overview of the factors influencing carbohydrates production, and of main bioconversion technologies for production of biofuels. Appl Microbiol Biotechnol 96:631–645

Markou G, Nerantzis E (2013) Microalgae for high-value compounds and biofuels production: a review with focus on cultivation under stress conditions. Biotechnol Adv 31(8):1532–1542

Marques SM, Américo-Pinheiro JHP (2017) Algas comobioindicadores da qualidade da água. Rev Científica ANAP Brasil 10:19

Matos ÂP (2019) Microalgae as a potential source of proteins. In: Proteins: sustainable source, processing and applications. Academic Press, pp 63–96

McLeod C, Kiermeier A, Stewart I, Tan J, Turnbull A, Madigan T (2018) Paralytic shellfish toxins in Australian southern rock lobster (Jasusedwardsii): acute human exposure from consumption of hepatopancreas. Hum Ecol Risk Assess Int J 24(7):1872–1886

Mehariya S, Goswami RK, Karthikeysan OP, Verma P (2021) Microalgae for high-value products: A way towards green nutraceutical and pharmaceutical compounds. Chemosphere 280:130553

Mendes MC, Navalho S, Ferreira A, Paulino C, Figueiredo D, Silva D et al (2022) Algae as food in Europe: an overview of species diversity and their application. Food Secur 11(13):1871
Mishra N, Gupta E, Singh P, Prasad R (2021) Application of microalgae metabolites in food and pharmaceutical industry. In: Preparation of Phytopharmaceuticals for the Management of Disorders. Academic Press, pp 391–408
Miyashita K, Beppu F, Hosokawa M, Liu X, Wang S (2020) Nutraceutical characteristics of the brown seaweed carotenoid fucoxanthin. Arch Biochem Biophys 686:108364
Mohibbullah M, Talha MA, Baten MA, Newaz AW, Choi JS (2023) Yield optimization, physicochemical characterizations, and antioxidant properties of food grade agar from Gracilariatenuistipitata of Cox's Bazar coast, Bangladesh. Food Sci Nutr 11:2852
Mondal NK, Bhaumik R, Sen K, Debnath P (2021) Adsorption of fluoride in aqueous solutions using saline water algae (Rhodophyta sp.): an insight into isotherm, kinetics, thermodynamics and optimization studies. *Modeling Earth Systems and Environment*, 1–15
Mondloch S, Gannon BM, Davis CR, Chileshe J, Kaliwile C, Masi C et al (2015) High provitamin A carotenoid serum concentrations, elevated retinyl esters, and saturated retinol-binding protein in Zambian preschool children are consistent with the presence of high liver vitamin A stores. Am J Clin Nutr 102(2):497–504
Moons I, Barbarossa C, De Pelsmacker P (2018) The determinants of the adoption intention of eco-friendly functional food in different market segments. Ecol Econ 151(C):151–161
Morris HJ, Almarales A, Carrillo O, Bermúdez RC (2008) Utilisation of Chlorellavulgaris cell biomass for the production of enzymatic protein hydrolysates. Bioresour Technol 99(16):7723–7729
Motoyama K, Tanida Y, Hata K, Hayashi T, Hashim IIA, Higashi T et al (2016) Anti-inflammatory effects of novel polysaccharide sacran extracted from cyanobacterium Aphanothece sacrum in various inflammatory animal models. Biol Pharm Bull 39(7):1172–1178
Mouritsen OG, Duelund L, Petersen MA, Hartmann AL, Frøst MB (2019b) Umami taste, free amino acid composition, and volatile compounds of brown seaweeds. J Appl Phycol 31:1213–1232
Mouritsen OG, Rhatigan P, Pérez-Lloréns JL (2019a) The rise of seaweed gastronomy: phycogastronomy. Bot Mar 62(3):195–209
Muller-Feuga A (2000) The role of microalgae in aquaculture: situation and trends. J Appl Phycol 12(3–5):527–534
Murray M, Dordevic AL, Ryan L, Bonham MP (2018) An emerging trend in functional foods for the prevention of cardiovascular disease and diabetes: marine algal polyphenols. Crit Rev Food Sci Nutr 58:1342–1358
Nagappan H, Pee PP, Kee SHY, Ow JT, Yan SW, Chew LY, Kong KW (2017) Malaysian brown seaweeds Sargassum siliquosum and Sargassum polycystum: low density lipoprotein (LDL) oxidation, angiotensin converting enzyme (ACE), α-amylase, and α-glucosidase inhibition activities. Food Res Int 99:950–958
Nagappan S, Das P, AbdulQuadir M, Thaher M, Khan S et al (2021) Potential of microalgae as a sustainable feed ingredient for aquaculture. J Biotechnol 341:1–20
Naselli-Flores L, Padisák J (2023) Ecosystem services provided by marine and freshwater phytoplankton. Hydrobiologia 850(12–13):2691–2706
Navarro-López E, Ruíz-Nieto A, Ferreira A, Acién FG, Gouveia L (2020) Biostimulant potential of Scenedesmus obliquus grown in brewery wastewater. Molecules 25(3):664
Negreanu-Pirjol T, Negreanu-Pirjol B, Sirbu R, Paraschiva GM, Meghea A (2012) Comparative studies regarding the antioxidative activity of some therapeutic marine algae species along the Romanian Black Sea coast. J Environ Prot Ecol 13:1744–1750
Novoveská L, Nielsen SL, Eroldoğan OT, Haznedaroglu BZ, Rinkevich B, Fazi S, Robbens J, Vasquez M, Einarsson H (2023) Overview and challenges of large-scale cultivation of photosynthetic microalgae and cyanobacteria. Mar Drugs 21(8):445. https://doi.org/10.3390/md21080445. PMID: 37623726; PMCID: PMC10455696
O'Sullivan L, Murphy B, McLoughlin P, Duggan P, Lawlor PG, Hughes H, Gardiner GE (2010) Prebiotics from marine macroalgae for human and animal health applications. Mar Drugs 8:2038–2064

Oh JH, Kim EY, Nam TJ (2018) Phycoerythrin-derived tryptic peptide of a red alga pyropiayezoensis attenuates glutamate-induced ER stress and neuronal senescence in primary rat hippocampal neurons. Mol Nutr Food Res 62(8):1700469

Okamoto T, Kawashima H, Osada H, Toda E, Homma K, Nagai N et al (2019) Dietary spirulina supplementation protects visual function from photostress by suppressing retinal neurodegeneration in mice. Transl Vis Sci Technol 8(6):20–20

Olawore YA, Nwinyi OC (2019) Algae based biofuel and co-products-A review. Int J Adv Sci Res Rev 4:74–88

Paiva L, Lima E, Neto AI, Baptista J (2018) Seasonal variability of the biochemical composition and antioxidant properties of Fucus spiralis at two Azorean Islands. Mar Drugs 16(8):248

Park MH, Kang JH, Kim HJ, Han JS (2017) Gelidiumamansii ethanol extract suppresses fat accumulation by down-regulating adipogenic transcription factors in ob/ob mice model. Food Sci Biotechnol 26:207–212

Parsaeimehr A, Lutzu GA (2016) Algae as a novel source of antimicrobial compounds: current and future perspectives. In: Kon K, Rai M (eds) Antibiotic resistance, 1st edn, pp 377–413

Patel AK, Albarico FPJB, Perumal PK, Vadrale AP, Nian CT, Chau HTB et al (2022) Algae as an emerging source of bioactive pigments. Bioresour Technol 351:126910

Pina-Pérez MC, Rivas A, Martínez A, Rodrigo D (2017) Antimicrobial potential of macro and microalgae against pathogenic and spoilage microorganisms in food. Food Chem 235:34–44

Prabakaran G, Moovendhan M, Arumugam A, Matharasi A, Dineshkumar R, Sampathkumar P (2019) Evaluation of chemical composition and in vitro antiinflammatory effect of marine microalgae Chlorella vulgaris. Waste and Biomass Valorization 10:3263–3270

Prabakaran G, Sampathkumar P, Kavisri M, Moovendhan M (2020) Extraction and characterization of phycocyanin from Spirulina platensis and evaluation of its anticancer, antidiabetic and antiinflammatory effect. Int J Biol Macromol 153:256–263

Pradhan B, Patra S, Nayak R, Behera C, Dash SR, Nayak S et al (2020) Multifunctional role of fucoidan, sulfated polysaccharides in human health and disease: A journey under the sea in pursuit of potent therapeutic agents. Int J Biol Macromol 164:4263–4278

Prarthana J, Maruthi KR (2019) Fresh water algae as a potential source of bioactive compounds for aquaculture and significance of solvent system in extraction of antimicrobials. Asian J Sci Res 12:18–28

Prasad Behera D, Vadodariya V, Veeragurunathan V, Sigamani S, Moovendhan M, Srinivasan R, Kolandhasamy P, Nivrutti Ingle K (2022) Seaweeds cultivation methods and their role in climate mitigation and environmental cleanup. Total Environ Res Themes 93-40:100016

Pratap K, Majzoub ME, Taki AC, Hernandez SM, Magnusson M, Glasson CR et al (2022) The algal polysaccharide ulvan and carotenoid astaxanthin both positively modulate gut microbiota in mice. Food Secur 11(4):565

Priyadarshani I, Rath B (2012) Commercial and industrial applications of micro algae–A review. J Algal Biomass Util 3(4):89–100

Pulz O, Gross W (2004) Valuable products from biotechnology of microalgae. Appl Microbiol Biotechnol 65:635e648

Qu W, Ma H, Pan Z, Luo L, Wang Z, He R (2010) Preparation and antihypertensive activity of peptides from Porphyrayezoensis. Food Chem 123:14–20

Rajauria G (2018) Optimization and validation of reverse phase HPLC method forqualitative and quantitative assessment of polyphenols in seaweed. J Pharmaceut Biomed Anal 148:230–237

Ramana K, Xavier J, Sharma R (2017) Recent trends in pharmaceutical biotechnology. Pharm Biotechnol Curr Res 1(1):5

Rammuni MN, Ariyadasa TU, Nimarshana PHV, Attalage RA (2019) Comparative assessment on the extraction of carotenoids from microalgal sources: Astaxanthin from H. pluvialis and β-carotene from D. salina. Food Chem 277:128–134

Ramos MV, Monteiro ACO, Moreira RA, Carvalho ADFAFU (2000) Amino acid composition of some Brazilian seaweed species. J Food Biochem 24:33–39

Rao AR, Baskaran V, Sarada R, Ravishankar GA (2013) In vivo bioavailability and antioxidant activity of carotenoids from microalgal biomass—A repeated dose study. Food Res Int 54(1):711–717

Rastogi RP, Richa, Sinha RP, Singh SP, Häder DP (2010) Photoprotective compounds from marine organisms. J Ind Microbiol Biotechnol 37(6):537–558

Raven JA, Beardall J (2003) Carbohydrate metabolism and respiration in algae. In: Photosynthesis in algae. Springer Netherlands, Dordrecht, pp 205–224

Raymundo A, Gouveia L, Batista AP, Empis J, Sousa I (2005) Fat mimetic capacity of Chlorella vulgaris biomass in oil-in-water food emulsions stabilized by pea protein. Food Res Int 38(8–9):961–965

Rezende ESV, Lima GC, Naves MMV (2021) Dietary fibers as beneficial microbiota modulators: A proposed classification by prebiotic categories. Nutrition 89:111217

Richmond A, Hu Q (eds) (2013) Handbook of microalgal culture. John Wiley & Sons, Limited

Rizwan M, Mujtaba G, Memon SA, Lee K, Rashid N (2018) Exploring the potential of microalgae for new biotechnology applications and beyond: A review. Renew Sust Energ Rev 92:394–404

Ruqqia NS, Taj D, Sarwar G, Sultana V, Ara J, Haque SE (2020) Hypolipidemic potential of sterol containing fractions of Jolyna laminarioides: A brown alga. Pak J Pharm Sci 33(1):169–174

Saberbaghi T, Abbasian F, Mohd Yusof YA, Makpol S (2013) Modulation of cell cycle profile by Chlorella vulgaris prevents replicative senescence of human diploid fibroblasts. Evid-Based Complement Alternat Med 2013

Sadovskaya I, Souissi A, Souissi S, Grard T, Lencel P, Greene CM et al (2014) Chemical structure and biological activity of a highly branched (1→ 3, 1→ 6)-β-d-glucan from Isochrysisgalbana. Carbohydr Polym 111:139–148

Sakai H, Katayama H, Oguma K, Ohgaki S (2009) Kinetics of Microcystis aeruginosa growth and intracellular microcystins release after UV irradiation. Environ Sci Technol 43(3):896–901

Salanţă LC, Uifălean A, Iuga CA, Tofană M, Cropotova J, Pop OL, … González CV (2020) Valuable food molecules with potential benefits for human health. The Health Benefits of Foods-Current Knowledge and Further Development, 1–45

Sangeetha RK, Bhaskar N, Baskaran V (2009) Comparative effects of β-carotene and fucoxanthin on retinol deficiency induced oxidative stress in rats. Mol Cell Biochem 331:59–67

Santoyo S, Rodríguez-Meizoso I, Cifuentes A, Jaime L, Reina GGB, Señorans FJ, Ibáñez E (2009) Green processes based on the extraction with pressurized fluids to obtain potent antimicrobials from Haematococcuspluvialis microalgae. LWT Food Sci Technol 42(7):1213–1218

Sathasivam R, Radhakrishnan R, Hashem A, Abd-Allah EF (2019) Microalgae metabolites: A rich source for food and medicine. Saudi J Biol Sci 26(4):709–722

Sekar S, Chandramohan M (2008) Phycobiliproteins as a commodity: trends in applied research, patents and commercialization. J Appl Phycol 20:113–136

Shanab SM, Hafez RM, Fouad AS (2018) A review on algae and plants as potential source of arachidonic acid. J Adv Res 11:3–13

Shanmughapriya S, Manilal A, Sujith S, Selvin J, Kiran GS, Natarajaseenivasan K (2008) Antimicrobial activity of seaweeds extracts against multiresistant pathogens. Ann Microbiol 58:535–541

Shao N, Bock R (2008) A codon-optimized luciferase from Gaussia princeps facilitates the in vivo monitoring of gene expression in the model alga Chlamydomonas reinhardtii. Curr Genet 53:381–388

Sharifuddin Y, Chin Y-X, Lim P-E, Phang S-M (2015) Potential bioactive compounds from seaweed for diabetes management. Mar Drugs 13:5447–5491

Sharma BR, Rhyu DY (2014) Anti-diabetic effects of Caulerpa lentillifera: stimulation of insulin secretion in pancreatic β-cells and enhancement of glucose uptake in adipocytes. Asian Pac J Trop Biomed 4(7):575–580

Show PL (2022) Global market and economic analysis of microalgae technology: status and perspectives. Bioresour Technol 357:127329

Sikiru AB, Arangasamy A, Alemede IC, Guvvala PR, Egena SSA, Ippala JR, Bhatta R (2019) Chlorella vulgaris supplementation effects on performances, oxidative stress and antioxidant genes expression in liver and ovaries of New Zealand white rabbits. Heliyon 5(9):e02470

Singh KB, Kaushalendra, Rajan JP (2022) Therapeutical and nutraceutical roles of cyanobacterial Tetrapyrrole chromophore: recent advances and future implications. Front Microbiol 13:932459

Singh S, Kumar V, Datta S, Singh S, Dhanjal DS, Kaul N et al (2023) Marine algae as Green agriculture. In: Green chemistry in agriculture and food production. CRC Press, pp 99–110

Singh U, Singh AK, Asthana RK (2021) Bioactive molecules from microalgae and constraints in commercialization. Nat Bioact Compd:143–164

Sivakumar N, Sundararaman M (2006) The effect of different seaweeds as dietary supplements on growth and survival of black tiger prawn peneous monodon. Seaweed Res 28(1–2):175–182

Sivakumar N, Sundararaman M, Selvakumar G (2011) Efficacy of micro algae and cyanobacteria as a live feed for juveniles of shrimp *Penaeus monodon*. Afr J Biotechnol 10(55):11594–11599

Sivakumar N, Sundararaman M, Selvakumar G (2014) Efficacy of dietary supplementation of cyanobacterial meal on growth performance of black tiger shrimp Penaeus monodon (fabricius). Cibtech J Zool 3(1):61–69

Sivakumar N, Sundararaman M, Selvakumar G (2018) Evaluation of growth performance of *Penaeus monodon* (Fabricius) fed diet with partial replacement of fishmeal by Spirulina platensis (Sp) meal. Indian J Anim Res 52(12):1721–1726

Sousa I, Gouveia L, Batista AP, Raymundo A, Bandarra NM (2008) Microalgae in novel food products. Food Chem Res Dev:75–112

Souza RB, Frota AF, Sousa RS, Cezario NA, Santos TB, Souza LMF, da Cunha RMS (2017) Neuroprotective effects of sulphated agaran from marine alga Gracilaria cornea in rat 6-hydroxydopamine Parkinson's disease model: Behavioural, neurochemical and transcriptional alterations. Basic Clin Pharmacol Toxicol 120:159–170

Sugiura Y, Matusda K, Yamada Y (2007) Anti-allergic phlorotannins from the edible alga, Eisenia arborea. Food Sci Technol Res 13:54–60

Sultana N, Mohammad M, Lisa SA, Saha B (2023) Effects of dietary inclusion of seaweed in regular commercial fish feed on proximate composition, minerals and fatty acid profiling of Nile tilapia Oreochromis niloticus (Linnaeus, 1758). Aquat Food Stud 2:AFS112

Sun XM, Ren LJ, Zhao QY, Ji XJ, Huang H (2018) Microalgae for the production of lipid and carotenoids: a review with focus on stress regulation and adaptation. Biotechnol Biofuels 11:272

Tanna B, Choudhary B, Mishra A, Chauhan OP, Patel MK, Shokralla S et al (2021) Antioxidant, scavenging, reducing, and anti-proliferative activities of selected tropical brown seaweeds confirm the nutraceutical potential of Spatoglossum asperum. Food Secur 10(10):2482

Taubert A, Jakob T, Wilhelm C (2019) Glycolate from microalgae: an efficient carbon source for biotechnological applications. Plant Biotechnol J 17:1538

Terriente-Palacios C, Castellari M (2022) Levels of taurine, hypotaurine and homotaurine, and amino acids profiles in selected commercial seaweeds, microalgae, and algae-enriched food products. Food Chem 368:130770

Thakur N, Raigond P, Singh Y, Mishra T, Singh B, Lal MK, Dutt S (2020) Recent updates on bioaccessibility of phytonutrients. Trends Food Sci Technol 97:366–380

Tian T, Chang H, He K, Ni Y, Li C, Hou M, Chen L, Xu Z, Chen B, Ji M (2019) Fucoidan from seaweed Fucus vesiculosus inhibits 2, 4-dinitrochlorobenzene-induced atopic dermatitis. Int Immunopharmacol 75:105823

Tibbetts SM, Scaife MA, Armenta RE (2020) Apparent digestibility of proximate nutrients, energy and fatty acids in nutritionally-balanced diets with partial or complete replacement of dietary fish oil with microbial oil from a novel Schizochytrium sp.(T18) by juvenile Atlantic salmon (Salmo salar L.). Aquaculture 520:735003

Tork MB, Vazifedoost M, Hesarinejad MA, Didar Z, Zenoozian MS (2022) Fabrication of Dragee containing Spirulina platensis microalgae to enrich corn snack and evaluate its sensorial, physicochemical and nutritional properties. Foods 11:13

Unnikrishnan MK, Veerapur V, Nayak Y, Paul P, Mathew G (2014) Antidiabetic, antihyperlipidemic and antioxidant effects of the flavonoids. In: Watson R, Preedy V, Zibadi S (eds) Polyphenols in human death and disease, 1st edn. Elsevier, Amsterdam, The Netherlands, pp 143–161
Unnikrishnan PS, Jayasri MA (2017) Antidiabetic studies of Chaetomorphaantennina extract using experimental models. J Appl Phycol 29:1047–1056
Usman A, Khalid S, Usman A, Hussain Z, Wang Y (2017) Algal polysaccharides, novel application, and outlook. In: Algae based polymers, blends, and composites. Elsevier, pp 115–153
Ustyuzhanina NE, Bilan MI, Gerbst AG, Ushakova NA, Tsvetkova EA, Dmitrenok AS, Usov AI, Nifantiev NE (2016) Anticoagulant and antithrombotic activities of modified xylofucansulfate from the brown alga Punctariaplantaginea. Carbohydr Polym 136:826–833
Uyar B, Kapucu N (2015) Passive temperature control of an outdoor photobioreactor by phase change materials. J Chem Technol Biotechnol 90:915–920. https://doi.org/10.1002/JCTB.4398
Varshney P, Mikulic P, Vonshak A, Beardall J, Wangikar PP (2015) Bioresour Technol 184:363
Vieira de Mendonça H, Assemany P, Abreu M, Couto E, Maciel AM, Duarte RL, Barbosa dos Santos MG, Reis A (2021) Microalgae in a global world: new solutions for old problems? Renew. Energy 165:842–862
Vishwakarma J, Vavilala SL (2019) Evaluating the antibacterial and antibiofilm potential of sulphated polysaccharides extracted from green algae Chlamydomonas reinhardtii. J Appl Microbiol 127(4):1004–1017
Wahidi ME, Amraoui BE, Amraoui ME, Bamhaoud T (2015) Screening of antimicrobial activity of macroalgae extracts from the Moroccan Atlantic coast. Ann Pharm Fr 73:190–196
Wang HMD, Chen CC, Huynh P, Chang JS (2015) Exploring the potential of using algae in cosmetics. Bioresour Technol 184:355–362
Wang K, Pramod SN, Pavase TR, Ahmed I, Lin H, Liu L et al (2020) An overview on marine anti-allergic active substances for alleviating food-induced allergy. Crit Rev Food Sci Nutr 60(15):2549–2563
Wang W, Wang SX, Guan H-S (2012) The antiviral activities and mechanisms of marine polysaccharides: an overview. Mar Drugs 10:2795–2816
Wells ML, Potin P, Craigie JS, Raven JA, Merchant SS, Helliwell KE et al (2017) Algae as nutritional and functional food sources: revisiting our understanding. J Appl Phycol 29:949–982
Wijesinghe WAJP, Ko SC, Jeon YJ (2011) Effect of phlorotannins isolated from Ecklonia cava on angiotensin I-converting enzyme (ACE) inhibitory activity. Nutr Res Pract 5(2):93–100
Wood D, Capuzzo E, Kirby D, Mooney-McAuley K, Kerrison P (2017) UK macroalgae aquaculture: what are the key environmental and licensing considerations? Mar Policy 83:29–39
Xiang Q, Wei X, Yang Z, Xie T, Zhang Y, Li D et al (2021) Acclimation to a broad range of nitrate strength on a euryhaline marine microalga Tetraselmissubcordiformis for photosynthetic nitrate removal and high-quality biomass production. Sci Total Environ 781:146687
Xin Z, Zhang M, Cui H, Ding X, Zhang T, Wu L et al (2023) Algae: A robust living material against cancer. Int J Nanomedicine 18:5243–5264
Xu J, Chen D, Yan X, Chen J, Zhou C (2010) Global characterization of the photosynthetic glycerolipids from a marine diatom Stephanodiscus sp. by ultra performance liquid chromatography coupled with electrospray ionization-quadrupole-time of flight mass spectrometry. Anal Chim Acta 663(1):60–68
Yabuta Y, Fujimura H, Kwak CS, Enomoto T, Watanabe F (2010) Antioxidant activity of the phycoerythrobilin compound formed from a dried Korean purple laver (Porphyra sp.) during in vitro digestion. Food Sci Technol Res 16(4):347–352
Yadav DK, Singh A, Agrawal V, Yadav N (2021) Algal biomass: A natural resource of high-value biomolecules. Bioprospecting of Plant Biodiversity for Industrial Molecules, 303–334
Yao C, Jiang J, Cao X, Liu Y, Xue S, Zhang Y (2018) Phosphorus enhances photosynthetic storage starch production in a green microalga (Chlorophyta) Tetraselmissubcordiformis in nitrogen starvation conditions. J Agric Food Chem 66(41):10777–10787
Yatipanthalawa B and Martin G, (2021) Chapter 3—conventional and novel approaches to extract food ingredients and nutraceuticals from microalgae. In Cultured microalgae for the food industry, edited. Tomás Lafarga, Gabriel Acién, Academic Press, pp 73-96

Yimin Q (2018) Health benefits of bioactive seaweed substances. In: Quin Y (ed) Bioactive seaweeds for food applications, 1st edn. Academic Press, London, UK, pp 180–199

Yokota T, Nomura K, Nagashima M, Kamimura N (2016) Fucoidan alleviates high-fat diet-induced dyslipidemia and atherosclerosis in ApoEshl mice deficient in apolipoprotein E expression. J Nutr Biochem 32:46–54

Yoshinaga K, Mitamura R (2019) Effects of Undaria pinnatifida (Wakame) on postprandial glycemia and insulin levels in humans: a randomized crossover trial. Plant Foods Hum Nutr 74(4):461–467

Yuan S, Lei W, Liu Q, Liu R, Liu J, Fu J, Han Y (2023) Distribution and environmental impact of microalgae production potential under the carbon-neutral target. Energy 263:125584

Yusoff FM, Banerjee S, Nagao N, Imaizumi Y, Shariff M, Toda T (2020) Use of microalgae pigments in aquaculture. Pigments from microalgae handbook, 471–513

Zhang R, Zhang X, Tang Y, Mao J (2020) Composition, isolation, purification and biological activities of Sargassum fusiforme polysaccharides: A review. Carbohydr Polym 228:115381

Zhang Z, Till S, Jiang C, Knappe S, Reutterer S, Scheiflinger F et al (2014) Structure-activity relationship of the pro-and anticoagulant effects of Fucus vesiculosus fucoidan. Thromb Haemost 112(03):429–437

Zhang Z, Wang X, Su H, Pan Y, Han J, Zhang T, Mao G (2018) Effect of sulfated galactan from Porphyrahaitanensis on H2O2-induced premature senescence in WI-38 cells. Int J Biol Macromol 106:1235–1239

Zheng LX, Chen XQ, Cheong KL (2020) Current trends in marine algae polysaccharides: the digestive tract, microbial catabolism, and prebiotic potential. Int J Biol Macromol 151:344–354

Zhong R, Wan X, Wang D, Zhao C, Liu D, Gao L, Wang M, Wu C, Nabavid SM, Daglia M et al (2020) Polysaccharides from marine Enteromorpha: structure and function. Trends Food Sci Technol 99:11–20

Zhu X, Healy LE, Sevindik O, Sun D-W, Selli S, Kelebek H, Tiwari BK (2021) Impacts of novel blanching treatments combined with commercial drying methods on the physicochemical properties of Irish brown seaweed Alaria esculenta. Food Chem 369:130949

Biofertilizers and Soil Amendments from Algae: Production and Agronomic Benefits

14

Punam Vishwakarma, Vaniksha Pal, Jagriti Nagar, and Rohit Rai

Contents

Abstract

Due to changing climatic conditions and intense agricultural practices, the agricultural lands are on the edge of losing their fertility. The rising demand for food has intensified the use and production of synthetic fertilizers, which are resulting in environmental pollution and severe health issues. Microorganisms are gaining

P. Vishwakarma · R. Rai (✉)
Faculty of Applied Medical Sciences, Lovely Professional University, Phagwara, India
e-mail: rohit.25274@lpu.co.in

V. Pal · J. Nagar
Department of Microbiology, Lovely Professional University, Phagwara, India

S. K. Arya et al. (eds.), *Value Added Products From Bioalgae Based Biorefineries: Opportunities and Challenges*,
https://doi.org/10.1007/978-981-97-1662-3_14

attention as environmentally friendly alternatives to synthetic/chemical fertilizers. The biopesticides, biostimulants, and biofertilizers provide higher crop yield and productivity and aid in nutrient replenishment, soil amendment, and conditioning. Algae are a promising source for bio-based organic fertilizers due to their biostimulant and biofertilizer properties. To increase the sustainability of agriculture, microalgae are seeking the attention of agrochemical industries. The commercial production of biofertilizers can be enhanced by developing new technologies that are cost-effective, and economically & environmentally feasible. The integrated approach to produce algae-based biofertilizers through wastewater treatment supports the development of a circular bioeconomy. The mechanism of algae-based biofertilizers should be studied using molecular tools such as genomics, metagenomics, and proteomics, which help in validating the benefit of biofertilizers. Advanced technologies such as machine learning and artificial intelligence are promising tools to predict the potential of algae, optimize production, and enhance biofertilizers' quality and functionality. This chapter majorly focuses on algae-based biofertilizers, their production, and their contribution towards the sustainable development of modern agriculture.

Keywords

Sustainable agriculture · Biofertilizers · Soil amendments · Microalgae · Agronomic benefits

14.1 Introduction

The success of agronomy, which focuses on soil fertility and health, forms the foundation of organic farming systems. It requires nutrition from a fertile environment to support diverse and dynamic biotic organisms that help the soil resist environmental composition (Wijesinghe et al. 2023). Additionally, the concentration of heavy metals in the environment has increased in recent years due to human actions such as factories, energy, fuel generation, sludge treatment, and agriculture. These pollutants are highly toxic and not degraded by certain microorganisms (bacteria, fungi, etc.) and thus remain in the environment unequally (Garbowski et al. 2023). As a result, their concentrations exceed those in the soil and the recovery rate of most cropping systems is less than 50%. Furthermore, a major amount of applied nutrients is lost through volatilization, denitrification, leaching, and soil erosion. So, the world is in great need of cost-effective, ecofriendly biofertilizers, and soil amendments for agronomy to increase soil production and agronomical benefits.

A sustainable agronomical practice includes biofertilizers to upsurge the nutrient content of the soil, resulting in higher soil productivity (Suleiman et al. 2020). Biofertilizers are live microorganisms that increase the texture and quality (in terms of biological aspects) of soils and, encourage plant growth. Plants require necessary nutrients to thrive, and deficiency in this nutrient can be fulfilled by applying enough fertilizers and amendments. Meanwhile, long-term use of synthetic fertilizers has resulted in the contamination of soil, which would eventually continue to damage ecosystem. Moreover, algae found in almost all ecological niches belong to the most diverse group on the earth with potential agricultural applications as biofertilizers and soil amendments for fertility and productivity (Cao et al. 2023). Soil algae are moderate microorganisms that originate in the soil and are photosynthetic by nature. They also stay under the soil surface, and soil is a considerable habitation for algal diversity and evolution. It can help the soil to maximize its carbon-nitrogen content and aeration (Jassey et al. 2022). Furthermore, algae addition to other living organisms that are found in different soil types and states could indicate a healthy soil environment. The algal growth also weakens soil erosion by maintaining water flow, soil improvement, soil fertility, crust formation, bio-controlling, and agronomical wastewater management (Javeed et al. 2023).

On the other hand, these microscopic organisms contribute to biomass and organic matter enrichment in the soil after releasing organic compounds improving soil structure, moisture, and overall fertility. Additionally, algal-based metabolites, i.e., plant growth hormones, polysaccharides, and so on directly have agronomic benefits. Some algae (microalgae and macroalgae) are biocontrol agents (biopesticides) due to their ability to effectively resist the growth of plant pathogens (fungi, bacteria, and nematodes) (da Rosa et al. 2023). Algae can also enhance soil characteristics (heterogeneity, porosity, and humidity). Some algal species can excrete carbohydrates that form a mesh, which binds particles in soil on their surface and improves soil quality (Zhou et al. 2023). This ability plays a significant role in pH and thermal regulation, as well as protection against erosion. In turn, plants release exudates (e.g., sugars, amino acids, organic acids, and other compounds) that serve as a nutrient source for the algae. These algae form colonies around the plant roots, producing a biofilm that can help protect the roots from plant pathogens (Kholssi et al. 2022).

This chapter provides an efficient way of ecofriendly algal-based biofertilizers and soil amendments for agronomic benefits applied to increase soil production via different types of algae (micro and macroalgae). It can be used to increase soil fertility and remove various harmful chemicals from the soil.

14.2 Algae as a Source of Biofertilizers and Soil Amendments

The use of algae as biofertilizers is a viable practice that enhances the nutrient content of the soil, resulting in higher crop yields (Suleiman et al. 2020). These photosynthetic microorganisms live a few inches below the soil surface and help develop its nutrient content, such as carbon and nitrogen, which are essential for plant growth (Raheem et al. 2018). Moreover, the addition of algae to different soil types, along with other living organisms, indicates the health of the soil. Algae also play a vital role in preventing soil erosion by managing water flow. There are two types of algae used for biofertilizers and soil amendments: (a) Microalgae, and (b) Macroalgae.

14.2.1 Microalgae

Microalgae, such as eukaryotic green algae, are photosynthetic microorganisms that have many benefits in diverse fields like animal feed, healthcare, and fuel production. In the current agricultural scenario, these microorganisms can be used to enhance soil nutrition. Microalgae come in different forms, such as unicellular, multicellular, and filamentous, with more than 200,000 different species. To produce useful products like food, chemicals, animal feeds, fuel, and biofertilizers, these algae undergo various processes, including mass cultivation, biomass recovery, and downstream operations (Balasubramaniam et al. 2021). Moreover, these algae also produce growth hormones, polysaccharides, antibacterial, and other useful compounds to improve the quality of the soil (Ronga et al. 2019). They directly interact in the conversion of carbon dioxide from the atmosphere into organic algal biomass through the photosynthesis process. These algae are responsible for maintaining the carbon content in the soil by absorbing carbon dioxide. Additionally, other algae like *Nostoc, Anabaena*, and *Arthospira* are used to enhance plant growth, which strengthens the plants and results in higher yields and profits for farmers.

14.2.2 Macroalgae

Macroalgae, such as seaweed, have many practical applications, including use as animal feed, in skincare products, as soil fertilizer, in biofuel production, and for waste treatment (Khan et al. 2009). These algae also generate phytochemical compounds like phycocyanins, sterols, vitamins, tocopherol, polyunsaturated fatty acids, phycobilins, terpenoids, carotenoids, xanthophylls, chlorophylls, and polysaccharides. Although seaweeds are highly valuable, they are not yet recognized as a potential resource (Osório et al. 2020). On the other hand, Sargassum, a type of brown macroalgae found in tropical waters, has high levels of phenols, carotenoids, and antioxidants. It also contains fucoxanthin, which is well known for its anticancer properties (Silva et al. 2019). Sargassum is mostly found in deep water and on marine reefs and can be collected on the seashore. In some places, it is consumed as food and used as a fertilizer. Near seashore areas, these seaweed fertilizer products are used to enhance crop growth by improving germination, root growth, and nutrient absorption.

14.2.3 Algae as a Biofertilizers and Soil Amendment

Algae, as biofertilizers, are alternatives to synthetic fertilizers as they are considered to be cost-effective options in agriculture (Chatterjee et al. 2017). The algal pathway provides valuable byproducts that can be used as biofertilizers, which enhance soil health. This is a promising bioresource due to its technological, commercial, and environmental advantages. These algae are also adapted to extreme environmental conditions and can survive under conditions with low nutrient content, such as carbon and nitrogen levels, and limited water. Moreover, when algae decompose, they release all the macronutrients and micronutrients into the soil. However, agricultural production mainly depends on the nature and type of soil (Mahapatra et al. 2018). By growing different varieties of crops and using various nutrient measures, more agricultural land can be saved. Thus, it is advised to improve the use of natural bioresources, which can enhance the nutrient content.

Biofertilizers play a crucial role in mineral breakdown and the fertilization process. These fertilizers are ecofriendly in nature and can replace the chemical fertilizers used nowadays. In simple terms, the microbes containing live cells are directly applied to the plant and soil. They rapidly interact with the plant roots and help the plant grow better by converting the minerals into needed nutrients through various processes such as mineralization, nitrogen fixation, and rock phosphate dissolution (Priya et al. 2014). Algae play a vital role in soil amendment. They are used as biofertilizers, biopesticides, plant stimulants, and ground stabilizers. They are produced from live or dead seaweed biomass. This algal agent can be directly applied to the seed or to the leaf surface as soil amendments. These practices are used to improve the soil quality in terms of its chemical function.

14.3 Production of Biofertilizers and Soil Amendments from Algae

Algae-based fertilizers have been an interest of different stakeholders like farmers, ecologists, bioexperts, and business entities. These are highly important as they can create unique plant hormones that can be found only within microorganism communities or between species. These hormones help plants to grow better and improve the quality by providing nutrients and maximum yield of crops (Nur-Hafiza et al. 2023). To improve the yield of biomass production for industrial purposes, the current researches focus on multiple strategies under stressful conditions for robust development of microalgal strain. Some of the approaches that are adopted to enhance the bioproduction are omics approach and genetic engineering methods (Muhammad et al. 2020). Adaptive laboratory evolution (ALE) is a method used for the construction of microalgal strains based on phenotypic functions. This method was extensively applied for the examination of *Chlorella* sp. AE10. The strain produced 2.9 times (3.7 g/L) higher yield of biomass with 10% CO_2 in comparison to the wild type after 31 cycles (Li et al. 2015). The foremost step to produce algae-based biofertilizers is the selection of microalgae strains. Some of the strains that are under focus of researchers for commercial interests are *Haematococcus, Spirulina, Dunaliella*, and *Chlorella* (Mącik et al. 2020).

14.3.1 Nutrient Contents

As a stimulant, seaweed sap may contain a mannitol compound that increases the nutrient content and the uptake of mannitol at leaf level has been proposed by Shah

& his coworkers (Shah et al. 2013). In a study on wheat crops, when the crops were sprayed with plant sap containing *Kappaphycus alvarezii* and *Gracilaria edulis* seaweeds, the protein content increased by 15.6% and 13.1%, respectively. Another study reported that using seaweed extract (Biozyme) on wheat plants significantly increased the protein content due to root growth and absorption of essential nutrients like nitrogen, phosphorus, and sulfur, which results in protein synthesis (Singh and Chandel 2005). In potato plants, when seaweed is directly added to the soil, the phosphorous content increases because of the alginic acid content present in seaweeds (Khan et al. 2009). This acid mainly consists of Al and Fe, which help to form compounds that make phosphates more available (Illera-Vives et al. 2017).

In soil, seaweeds provide more nutrients compared to animal dung. The seaweed extracts also help grapevines to absorb more nutrients, making roots grow stronger, and causing cell membranes in leaves and stomata to open more. Many times, these seaweed extracts also act as hormones in plant processes like cell respiration, photosynthesis, and various chemical reactions (Turan and Köse 2004).

14.3.2 Cultivation Methods

Algae have the ability to increase their growth rate. They efficiently use the photosynthetic process to produce useful byproducts stored within their cell membranes, which can be used in various industries (Randrianarison and Ashraf 2017). Additionally, algae cultivation does not require fertile land, freshwater, weed control substances, or insecticides compared to other crops (Khan et al. 2018). Moreover, cultivation can even be done using wastewater, such as sewage water. Apart from water treatment, algae cultivation methods help in reducing carbon dioxide via photosynthesis. For the cultivation of microalgae, major requirements include natural or artificial sunlight, nutrients (nitrogen, phosphorus, and carbon), and specific culture conditions (temperature and pH) (Menegazzo and Fonseca 2019). Various cultivation methods depend on different factors like the selection of the microalgae strain, environmental conditions, and their target applications. Classically, there are two cultivation systems: open raceway ponds and closed photobioreactors (Tan et al. 2018).

Open raceway ponds are shallow basins made of concrete, about 0.3–0.5 m deep, that ensure a high growth rate. They have a paddle wheel installed, which allows the continuous flow of culture under a circular channel (Chisti 2007). This method reduces the requirement of nutrients and water by improving light exposure, aeration, and proper mixing of the culture. It is easy, scalable, and low-cost, but some main challenges include the vulnerability of the microalgae yield to contamination, environmental fluctuations (temperature, pH, and seasonal variation), low utilization of CO2 due to evaporation, and the requirement of large land areas (Brennan and Owende 2010).

A photobioreactor is a closed system that allows controlled cultivation under certain environmental conditions like temperature, pH, nutrients, and light intensity. Light is used as a source of energy in this method for photobiological reactions.

Compared to the raceway pond method, this reactor produces more microalgae biomass and requires less land for cultivation (Tan et al. 2018). The method helps in enhancing the productivity, consistency, and quality of the microalgae biomass. It also lowers the chance of predation and contamination. However, economically, the bioreactor method is more expensive compared to the raceway pond (Rawat et al. 2011).

There is an integrated method called a hybrid system for cultivation, which combines different methods such as open raceway and photobioreactors. The benefits of each method used in the integrated system help in optimizing the quality, productivity, and sustainability of microalgae. However, the integrated system requires complex operation and maintenance (Zeng et al. 2011).

14.3.3 Harvesting Techniques

After cultivation, the algal biomass is separated from the culture medium to obtain concentrated biomass. There are various techniques used for harvesting, which vary to optimize the efficiency, quality, and cost of the algal biomass depending on the type of the culture. The harvesting of microalgae biomass is the most expensive process due to the small size of microalgae cells and high water content in the culture system (Mobin and Alam 2017). Some of the harvesting techniques that are used commonly are membrane filtration, centrifugation, sedimentation, flotation, and chemical and bioflocculation. However, none of the above can serve as a multipurpose technique that can be used to deal with both cost and energy consumption. Upon harvesting, drying and storage of the biomass become important as the high moisture content of the harvested biomass may damage its quality at room temperature within no time. Different methods used for drying and storage of biomass are microwave, oven, solar, spray, cross-flow air drying, freeze, and incinerator (Muhammad et al. 2020).

In the centrifugation method, the algal biomass is separated from the culture media. It is a quick and effective method, but the need for maintenance and energy input makes it costly. Moreover, this method can damage internal cells, leading to nutrient loss. In the flotation method, a microbubble diffuser is used to float algal cells to the water's surface without the need for chemicals. It is one of the cheapest methods due to its low operational cost, simple process, and high yield of biomass (Muhammad et al. 2020). The biomass aggregation method involves the addition of cationic polymers and multivalent cations to the mixture. This causes algal cells to bind together, forming larger particles called flocs. The process neutralizes the surface charge of the cells. Filtration involves using a membrane with different pore sizes to extract algal biomass from the culture media.

In summary, these methods are crucial for efficiently harvesting algal biomass, each with its own advantages and considerations. The choice of method or combination depends on the specific characteristics of the algae and the intended use of the harvested biomass.

14.3.4 Processing & Formulation

As the human population increases, the need for more food escalates, necessitating more fertilizers for sustainable farming. Biofertilizers play a vital role in this context, providing formulations that contain living or latent microorganisms supported on a carrier material. The production of these biofertilizers involves several important steps. First, the appropriate microorganism, such as bacteria, algae, fungi, or actinomycetes, is selected. Then, a carrier for the microorganisms is chosen, which could be peat, wheat bran, talc, or vermiculite (Kumar et al. 2018).

Additional substances, such as additives and sticking materials, can also be used to enhance the biofertilizer. Depending on these factors, there are four main types of formulations:

1. **Carrier-Based Formulations**: These use materials like peat, wheat bran, talc, or vermiculite to carry the microorganisms.
2. **Granular Formulations**: Instead of a powdery mix, peat prills are used to create granules, thereby avoiding certain issues.
3. **Liquid Formulations**: These are liquid mixtures, making them easier to handle and more effective.
4. **Cell Immobilization**: This is a newer method that encapsulates the microorganisms, making them more stable and effective.

All of these steps are crucial for producing high-quality biofertilizers. The choice of method depends on the specific requirements and intended use of the biofertilizer.

14.3.5 Quality Control and Standardization

Industrial processes for producing microalgae biomass mainly depend on the choice of species. The species are selected primarily based on their chemical properties and their sensitivity to environmental conditions. Factors such as the cost of land, energy availability, water type, nutrients, seasonal conditions, and the desired target molecules also play a vital role in selecting the appropriate algal strain and technology for each stage of the industrial process (Garcia-Gonzalez and Sommerfeld 2016).

Microalgae produce various molecules through photosynthesis, but different species have different abilities to accumulate these molecules depending on growth conditions. Therefore, the method needs to be adjusted based on the particular algal strain relevant to the desired end product. The content is mostly influenced by operational parameters, which include the type of culture system and the composition of the growth medium (Pulz and Gross 2004).

It is crucial to ensure the quality of the biomass for any biorefinery. Quality control involves monitoring the physical, biochemical, and microbiological properties throughout the culture process. This is especially important when producing food-grade microalgae biomass. On-site labs perform these analyses, which should be validated by occasional analysis from certified labs. Adopting a quality management approach, including hazard analyses and critical control points practices, is essential to meet regulatory standards for nutritional aspects.

Quality criteria for microalgae products are typically defined by the consumers. Process assurance and quality control can be improved by choosing the right culture system and controlling environmental conditions. These factors impact the physiological integrity and biochemical metabolism of the cells. High levels of chlorophyll breakdown, leading to the formation of chlorophyllides, pheophorbides, pheophytins, indicate low quality of biomass and cellular health and damage (Serra and Costa 2015).

14.4 Agronomic Benefits and Applications

Algae is used as a pure form or crude compost to enhance crop productivity by increasing seed germination at various levels such as improving soil texture, stimulating growth, and stress tolerance. In the era of biofertilizers, climatic factors (temperature, humidity, etc.) support the degradation of organic material in organic fertilizers; thus, plants release them in the soil after the non-requirement of these nutrients. Hence, the nutritional content of organic fertilizers is quite low, and only a limited amount of it is available for plants. It is more difficult to supply nutrients only using organic/inorganic fertilizers. Further, fertilizers are vital for plant maturation and also not easily available for small-scaling crops or farmers (Ferreira et al. 2023). The use of algae in agronomy offers a wide range of benefits, not only for the environment but also for soil health. Figure 14.1 represents the applications of algal-based agronomy. In general, algae exhibit growth properties through three different modes, namely biofertilizers, biostimulants, and biopesticides. Also, the earliest algae (microalgae) are investigated as the most primitive microorganisms as an origin of the higher phytohormones-producing organisms (Asimakis et al. 2022).

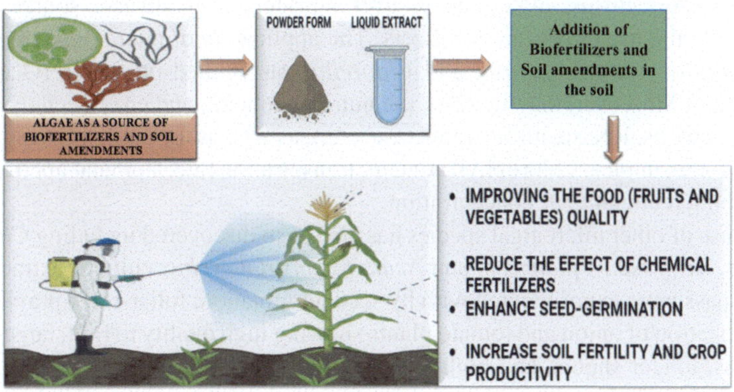

Fig. 14.1 The applications of algae used as a biofertilizer and soil amendments

Accordingly, a better understanding of the evolution, metabolic pathways, and regulatory networks of this algae internal system will seek attention to using phylogenetically diverse microalgal resources as substrates for plant growth hormone production and their agricultural uses. The modifications of plant growth hormones or phytohormone systems in higher plants by the introduction, increase, or manipulation of important agronomic traits such as tolerance to extreme conditions (salinity, drought, flooding, pH, and temperature) in crop areas. Importantly, it is considered a significant way to successfully lead to the "Green Revolution" (Abd-Elsalam et al. 2023). Furthermore, the external use of phytohormones can enhance the biomass of cultivated microalgal species, which can modulate plant growth and development. Furthermore, the direct applications of microalgae with other microorganisms, i.e., cyanobacteria are limited in several ways such as high inoculant rates and frequent watering. Moreover, their experimental applications (biofertilizers) are prominent for the modulation of the crops (Braun and Colla 2023).

Most literature showed microalgae as a biostimulant (microorganism helps to promote natural processes in plants). The most crucial application was by root and studies that applied seed soaking and germination were performed in the laboratory, which are usually the first experiments to study the biostimulant or biofertilizer potential of algae (Gupta et al. 2023). Two studies have proposed the effects of microalgae when used via foliar and root in greenhouse experiments. When comparing foliar spray to soil application, the microalgae provide faster nutrient uptake and remove nutrient deficiencies (Mutum et al. 2023). Some of the studies suggested that the microalgae-based biofertilizers and compounds are limited (polysaccharides and phytohormones). This novel strategy is relatively one of the most innovative agronomic trials as it is safe for the environment and promotes agricultural feasibility. Since the natural substances in algae are relatively lower than in organic fertilizers, the application (foliar) could be a more suitable method, when plant responses to these mentioned nutrients (N, P, and K) supplied via foliar sprays are normally more beneficial than when applied in the soil. The algal species are also well understood for their improving effect on soil via fixing atmospheric nitrogen or increasing soil's physical and chemical properties. Furthermore, as photosynthetic microorganisms, algae can recover nutrients from diverse sources, thus, minimizing the use of organic fertilizers. The application of microalgae (monocultures), *Microcystis aeruginosa,* and *Chlorella* was studied on maize (Occhipinti et al. 2023). Moreover, they increase the nutrient content, and enzyme activity, i.e., dehydrogenases, RNase, nitrate reductase, etc.), as well as the entire photosynthesis process for example—chlorophyll content, transpiration rate, stomatal aperture, and lowered intercellular CO_2 concentration.

The use of other microalgal species has also been discovered including *Chlorella vulgaris*, *Arthrospira platensis*, and *Acutodesmus dimorphus* culture, extracts, and dry biomass were experimented with by seed germination, foliar-root spraying, and biofertilization of onion and tomato plants showing high quality to root, largest stem thickness, longer shoots, and higher chlorophyll content (Ferreira et al. 2023). In general, microalgal treatment improved seed growth with a significant enhancement of soluble biomolecules (carbohydrates, protein, and total free amino acids) and mineral compositions (Na, K, P, Ca, Fe, Mg, and Mn), which results in increased market price (Parmar et al. 2023). The application of microalgal biomass into the soil improves the pure and dry weight of seedlings as well as their pigment quality (utilizing 2 and 3 g of dry microalgae per kg of soil) (Ramakrishnan et al. 2023).

Eukaryotic green algae are the main cause of enriching nutrient availability and improve the fertility and quality of soils by the production of compounds, plant growth hormones (abscisic acid, cytokinin, gibberellins, and ethylene), and other metabolites that can enhance plant growth. Together with small amounts of rhamnose, fucose, fructose, and methyl sugars microalgal carbohydrates can serve as biostimulants for improving nutrient availability, growth performance, and physiological state. Additionally, with the assistance of algal biofertilizers and amendments, the abiotic stress tolerance of crops can be enhanced through their mechanistic actions, including enhanced root growth, metal chelation in soils, increased forage of reactive oxygen species, Rubisco synthesis, and photosynthesis of the crop plants of algal biofertilizers and amendments (Kaur et al. 2022).

Algae (macroalgae) have a variety of applications (biofertilizers, soil amendments, pharmaceuticals, biomass treatment, animal feed, and fodder). Table 14.1 summarizes the major applications of algal-based biofertilizers and soil amendments. These seaweeds also secrete some biologically active compounds (pigments, carbohydrates, vitamins, and fats) (Ma et al. 2022). On a worldwide basis, seaweed is still considered an underestimated resource for biofertilizers. Some fast-growing macroalgae like *Sargassum* with high levels of these pigments and biomolecules, with the fucoxanthin (a well-known anti-cancer agent), considered it a feasible source for various applications (Deivasigamani and Sharmila 2022). The macroalgae, *Gracilaria* (Rhodophyta) can generate possible amounts of commercially useful dry mass and is essential for commercial microbiological preparations (Mouedden et al. 2024). Additionally, seaweed biofertilizer products have been created to increase seed germination, broad-area penetration, nutrient supply, and crop outcome. Additionally, drifted seaweeds, Jepara seaweeds such as *Gracilaria verrucosa*, and *Sargassum* sp. have long been used as soil amendments and were massed (in Central Java), cleared properly to contaminating soil. Also, seaweeds are a source of seaweed (*Sargassum* and *Gracilaria* powder) powder that provides nutrition and is used to increase the soil's organic material (Ammar et al. 2022).

Thus, novel ecofriendly technologies include the applications of biofertilizers and soil amendments that will be part of sustainable agronomy. Due to its enhanced nutritional value, biofertilizers improve soil fertility and increase plant productivity. It is possible to procure biofertilizers and soil amendments from a variety of microorganisms such as algae (*Chlorella vulgaris, Sargassum, Gracilaria, Haematococcus pluvialis* (Sani et al. 2022), *Chlamydomonas reinhardtii* (Muthukumaran et al. 2022), *Dunaliella salina* (Hussain et al. 2021), and *Acutodesmus dimorphus*). In addition to their biodegradable nature, the microorganisms, and their residues can safely improve soil fertility without aggregating contaminants, which preserves the strength of the environment and ecosystems, so this supports the principles of green revolution or sustainable development.

Table 14.1 Summary of algae (microalgae and macroalgae) as a source of biofertilizers and soil amendments for the increase in soil fertility, productivity, and agronomic applications

S.no.	Algae	Biofertilizer/ soil amendment type	Application on soil productivity and agronomic benefits	References
1.	*Chlorella vulgaris*	Solid biomass to the soil	Increase yield, germination of seed, plant height, and biochemical and mineral content	Das et al. (2019)
2.	*Scenedesmus dimorphus*	Dry biomass diluted in cyanobacteria and urea	Increase in plant height (40%)	Jochum et al. (2018)
3.	*Chlorococcum*	Solid biomass to the soil	Increase in plant biomass (33%), and increase of NPK (15%, 122%, and 23%)	Renuka et al. (2016)
4.	*Acutodesmus dimorphus*	Solid biomass to the soil, seed and leaf spraying diluted in water	Increase in flower buds (390%) and shoot length (33%)	Garcia-Gonzalez and Sommerfeld (2016)
5.	*Chlorella sorokiniana*	Algae with medium; filtered with bacteria only; filtered without bacteria and algae (liquid)	Increase soil carbon content (0.4%), increase bacterial biomass in the soil (up to 25%)	Marks et al. (2019)
6.	*Scenedesmus quadricauda*	Suspension	Increase in plant height (15%), stem diameter (5%), number of leaves (24%), number of buds (50%)	Lv et al. (2020)
7.	*Desmodesmus*	Supernatant after centrifugation	Increase soil mineral content and growth in plant height	Jo et al. (2020)
8.	*Heterocypris* sp.	Supernatant after centrifugation	Increase soil mineral content and growth in plant height	Jo et al. (2020)
9.	*Amoeboaphelidim*	Supernatant after centrifugation	Increase soil mineral content and growth in plant height	Jo et al. (2020)
10.	*Pseudodifugia*	Supernatant after centrifugation	Increase soil mineral content and growth in plant height	Jo et al. (2020)
11.	*Chlorella protothecoides*	Suspensions and aqueous extracts (liquid)	Increases in germination index (177%)	Kour et al. (2020)
12.	*Scenedesmus obliquus*	Irrigation with medium and algae (liquid)	Increase in the germination (100%)	Kumar et al. (2022)

S.no.	Algae	Biofertilizer/ soil amendment type	Application on soil productivity and agronomic benefits	References
13.	*Chlorella minutissima*	Solid biomass to the soil	Increase of N and P content (75% and 5%)	Tayoh (2020)
14.	*Monoraphidium*	Solid biomass to the soil	Reduction of nitrate leaching	Jimenez et al. (2020)
15.	*Dunaliella salina*	Crude polysaccharide extract	Increase plant growth; improve the nodes number	Rachidi et al. (2020)
16.	*Porphyridium* sp.	Crude polysaccharide extract	Increase carotenoid, chlorophyll, proteins, and nitrate reductase, NAD-glutamate dehydrogenase activities in the soil.	Rachidi et al. (2020)
17.	*Spirogyra*	Consortium	Improve grain yield; increase the availability of macronutrients; enhance microbial activity; increase crop productivity and yield.	Renuka et al. (2016)
18.	*Tetraselmis* sp.	Powder	Improve nutrient intake and plant growth	Saadaoui et al. (2019)
19.	*Ulva lactuca*	Powder	Accelerates growth and reduces the effect of high salt levels	Hashem et al. (2019)
20.	*Cystoseira* sp.	Powder	Reduce harmful effects of salinity and enhance the growth and productivity	Hashem et al. (2019)
21.	*Gelidium crinale*	Powder	Stimulate growth hormones, improve salt stress tolerance	Hashem et al. (2019)
22.	*Halimeda microloba*	Powder	Increases growth and yield	Muniswami et al. (2023)
23.	*Sargassum* sp.	Powder	Increase length and weight of pods	Muniswami et al. (2023)
24.	*Turbinaria ornata*	Powder	Increase the number of flowers, pods, length, and weight of pods	Muniswami et al. (2023)
25.	*Caulerpa racemosa*	Extract (aqueous)	Increase in biochemical factors in plants and increase overall growth	Uthirapandi et al. (2018)

14.5 Challenges and Limitations

Some of the serious problems such as excess utilization of chemical fertilizers, soil degradation, and biodiversity loss are challenging the sustainable development of the modern agriculture. The mineral fertilizers are being replaced by the organic ones due to their dependency on nonrenewable energy and environmental issues caused due to its accumulation (Kurniawati et al. 2023). The algae-based biofertilizers and soil amendments are encouraged as they serve as an alternate or greener solution to various agronomic problems. It helps in leading biowaste problems and aids in enhancing soil health conditions. There are some of the major problems or challenges that need to be addressed before implementing the use of bio-based fertilizers in agriculture. The bio-based fertilizers face obstacles in getting social acceptance due to their high production cost, challenges in transportation, nutrient concentration, and policy framework (Garbowski et al. 2023; Osorio-Reyes et al. 2023). Nevertheless, it is a long-term strategy to tackle global environmental health issues. One of the limitations to be considered is the high production cost of algal biomass with relatively lower productivity and extraction of the desired metabolites. Some of the studies state that the harvesting and dewatering itself cost about 20–30% and 3–15% of the total production cost, respectively (Vázquez-Romero et al. 2022). A recent approach has been proposed which involves an integration of biofertilizer production and wastewater treatment. The biofertilizers derived from wastewater introduce the accumulated contaminants and heavy metals into the agricultural soil which causes the risk to human health and the growth of crops (Gonçalves 2021). The more attention is required to develop technologies or processes that include environmental or health risk assessments because the algae-based biofertilizers have contamination of pathogens and heavy metals (Rawat et al. 2011). The integrated approach is technically and economically challenging due to the variability in the nutrient composition of the fertilizers derived from wastewater. The variability in the content and ratio of the biofertilizer affect its potential in promoting crop growth (Rawat et al. 2011). The incomplete understanding of the effect and mechanism of microalgae extracts (metabolites) on soil and plants; and their interaction with the environment prevents the use of microalgae in agriculture (Gonçalves 2021). To further encourage the utilization and production of algae biofertilizers, aforementioned challenges need to be addressed.

14.6 Future Prospects and Research Directions

The bio-based fertilizers have the potency to serve as an alternative source of energy to support the paradigm of circular economy. Algae-based biofertilizers play an important role in crop production by improving fertility and quality of soil and promoting plant growth. The nutrient uptake in crops can be facilitated by reducing the nutrient-related losses and toxicity by reducing the bulk form of fertilizers to nanoforms through various approaches such as mechano-physical, chemical, and biological. The biologically (microbially) derived nanonutrient synthesis is simple and ecofriendly in comparison to other approaches (Garg et al. 2023). Some of the examples of microorganisms used for the nanonutrient synthesis are bacteria (*Klebsiella aerogenes, Desulfovibrio desulfuricans,* and *Clostridium thermoaceticum*), Cyanobacteria (*Anabaena, Spirulina, Nostoc,* and *Oscillatoria*), green algae (*Chlamydomonas, Chlorella,* and *Chlorococcum*), fungus (*Fusarium oxysporum, Aspergillus* spp.). The addition of nanomaterials to the biofertilizers offers an effective and ecofriendly replacement to boost quality and productivity of crops and develops resistance to reduce the environmental stress (Mahapatra et al. 2022). After extraction of the compounds (metabolites), the remaining of the algae biomass serves as a very good nutrient source (contains essential macro and microelements) that can be used for some other applications such as feedstock, biofertilizers, and protein source (Muhammad et al. 2020).

The advantages of algae biotechnology bring in advanced technologies to improve soil nutrients. An integrated technique has emerged in recent years based on algae biofertilizer and wastewater remediation, which helps in providing ecofriendly agriculture (Zou et al. 2021). The production of biofertilizers in combination with algae-based wastewater remediation has certain advantages such as availability of wastewater at a very low cost, which can help in reducing the cost (selling price) of the algae biofertilizers; the waste nutrients get converted to biomass (value added), which provides economic benefits; the nutrients recovered through the growth of algae reduces the pollution of wastewater (Acién Fernández et al. 2018). This promotes resource recycling and reduces the generation of waste thus contributing to the implementation of a circular economy. This approach reduces the cost and also minimizes the amount of greenhouse gas generated with the classic cultivation of microalgae (Álvarez-González et al. 2022). However, this emerging technology is not yet used in agriculture due to some challenges such as overpriced algae-biofertilizers, ineffective production of algae and value-added products, and algal biomass contamination. The development of a cost-effective method can help to reduce labor costs, and improve the overall efficiency of the harvesting and dewatering process and productivity of algal biomass (Javeed et al. 2023a; Malik et al. 2023). The use of integrated approach for the production of algae-based biofertilizers is economically and environmentally feasible in the future (Muhammad et al. 2020).

To improve the sustainability of agriculture, still several points need to be addressed such as physiological, biochemical, molecular, and agronomic studies are required for a better understanding of how and when algal extracts are used (mechanism), changes created by biofertilizers in crop production will help agrochemical industries and farmers (Ronga et al. 2019). Some of the methods that can be used for the crops in modern agriculture such as artificial intelligence techniques, Big Data, IoT applications, cloud computing, and machine learning to determine the best microorganism used for extracting the biofertilizers, and to study the diversity of plants/crops in various regions. A campaign or awareness can be conducted to clarify and understand the importance of biofertilizers over synthetic ones. The farmers can be supported to initiate an idea of entrepreneurship to commercially produce biofertilizers (Ammar et al. 2023). The algae biofertilizers have enormous market potential with their growing awareness. The development of cost-effective technology for the production and field surveys can also be addressed to overcome the challenge of commercialization.

14.7 Conclusion

The research on algae-based biofertilizers and their application has exceedingly increased in the recent years. The interest in biofertilizers has been increased due to their potential significance in enriching soil nutrients and quality, increasing plant growth and tolerance to biotic and abiotic stress. The increase in soil degradation, global climatic changes, water scarcity, and emerging crises requires better solutions or innovations to protect and enhance the production of crops throughout the world. The availability of limited resources has raised the demand for food, which makes microalgae a suitable alternative to enhance the crop (agricultural) production and deliver environmental benefits to the agrochemical industries and farmers (Ferreira et al. 2023). Some of the studies state that microalgae serve as a viable biological alternative to the chemical fertilizers by improving the productivity of crops, providing resistance to the adverse conditions and it also enhances the nutritional quality of fruits and leaves. Microalgae also help in promoting plant growth by producing growth hormones, antimicrobial compounds, and various metabolites (Guo et al. 2020). The biofertilizers produced from algae support the sustainable development of the agriculture by maintaining the integrity of the ecosystem and the environment as it increases the fertility of the soil due to its biodegradable nature. The algae biofertilizers have enormous market potential with its growing awareness. The development of cost-effective technology for the production, aforementioned challenges and field surveys can be addressed to overcome the challenges associated with the commercialization and production of biofertilizers. In order to enhance the commercial production of biofertilizers, the researches should majorly focus on the mechanism of biofertilizers and their effect on the plant growth, soil structure, and biological systems. The economical and feasible strategies should be studied in depth and adopted for some other environmental benefits (Muhammad et al. 2020). The mechanism of the biofertilizers can be elucidated using advanced molecular tools such as genomics, metagenomics, and proteomics, which help in validating the efficacy of biofertilizers and their applications (Renuka et al. 2018).

Acknowledgments RR and PV acknowledge the Science and Engineering Research Board (SERB), Department of Science and Technology, and Government of India for their financial assistance under the Start-up Research Grant scheme (SRG/2021/001243). RR also acknowledges the financial support received from the Research and Development Cell, Lovely Professional University, India under the scheme of Matching Grant Seed Money (LPU/DRD/SEED/SAC/090922/25274).

Conflict of Interest Authors declare no conflict of interest.

References

Abd-Elsalam, Kamel A., Mousa A. Alghuthaymi, and Salah M. Abdel-Momen (2023). "Eco-Safety and Future Trends."

Acién Fernández FG, Gómez-Serrano C, Fernández-Sevilla JM (2018) Recovery of nutrients from wastewaters using microalgae. Front Sustain Food Syst 2:59

Álvarez-González A, Uggetti E, Serrano L, Gorchs G, Ferrer I, Díez-Montero R (2022) Can microalgae grown in wastewater reduce the use of inorganic fertilizers? J Environ Manag 323:116224

Ammar EE, Aioub AA, Elesawy AE, Karkour AM, Mouhamed MS, Amer AA, El-Shershaby NA (2022) Algae as bio-fertilizers: between current situation and future prospective. Saudi J Biol Sci 29:3083–3096

Ammar EE, Rady HA, Khattab AM, Amer MH, Mohamed SA, Elodamy NI, Al-Farga A, Aioub AA (2023) A comprehensive overview of eco-friendly bio-fertilizers extracted from living organisms. Environ Sci Pollut Res 30:113119–113137

Asimakis E, Shehata AA, Eisenreich W, Acheuk F, Lasram S, Basiouni S, Emekci M, Ntougias S, Taner G, May-Simera H (2022) Algae and their metabolites as potential bio-pesticides. Microorganisms 10:307

Balasubramaniam V, Gunasegavan RD-N, Mustar S, Lee JC, Mohd Noh MF (2021) Isolation of industrial important bioactive compounds from microalgae. Molecules 26:943

Braun JC, Colla LM (2023) Use of microalgae for the development of biofertilizers and biostimulants. Bioenergy Res 16:289–310

Brennan L, Owende P (2010) Biofuels from microalgae—a review of technologies for production, processing, and extractions of biofuels and co-products. Renew Sust Energ Rev 14:557–577

Cao TN-D, Mukhtar H, Le L-T, Tran DP-H, Ngo MTT, Nguyen T-B, Bui X-T (2023) Roles of microalgae-based biofertilizer in sustainability of green agriculture and food-water-energy security nexus. Sci Total Environ 870:161927

Chatterjee A, Singh S, Agrawal C, Yadav S, Rai R, Rai L (2017) Role of algae as a biofertilizer, algal green chemistry. Elsevier, pp 189–200

Chisti Y (2007) Biodiesel from microalgae. Biotechnol Adv 25:294–306

Das P, Quadir M, Thaher M, Alghasal G, Aljabri H (2019) Microalgal nutrients recycling from the primary effluent of municipal wastewater and use of the produced biomass as bio-fertilizer. Int J Environ Sci Technol 16:3355–3364

Deivasigamani, B., and R. Sharmila (2022). "A Review on L-Asparaginase: Important Sources and Its Applications." Marine Biochemistry, 147–170.

Ferreira A, Bastos CR, Marques-dos-Santos C, Acién-Fernandez FG, Gouveia L (2023) Algaeculture for agriculture: from past to future. Front Agron 5:1064041

Garbowski T, Bar-Michalczyk D, Charazińska S, Grabowska-Polanowska B, Kowalczyk A, Lochyński P (2023) An overview of natural soil amendments in agriculture. Soil Tillage Res 225:105462

Garcia-Gonzalez J, Sommerfeld M (2016) Biofertilizer and biostimulant properties of the microalga Acutodesmus dimorphus. J Appl Phycol 28:1051–1061

Garg D, Sridhar K, Stephen Inbaraj B, Chawla P, Tripathi M, Sharma M (2023) Nano-biofertilizer formulations for agriculture: a systematic review on recent advances and prospective applications. Bioengineering 10:1010

Gonçalves AL (2021) The use of microalgae and cyanobacteria in the improvement of agricultural practices: a review on their biofertilising, biostimulating and biopesticide roles. Appl Sci 11:871

Guo, Suolian, et al. (2020) "Microalgae as biofertilizer in modern agriculture." Microalgae biotechnology for food, health and high value products, 397–411.

Gupta S, Bhattacharyya P, Kulkarni MG, Doležal K (2023) Growth regulators and biostimulants: upcoming opportunities. Front Plant Sci 14:1209499

Hashem H, Mansour H, El-Khawas S, Hassanein R (2019) The potentiality of marine macro-algae as bio-fertilizers to improve the productivity and salt stress tolerance of canola (Brassica napus L.) plants. Agronomy 9:146
Hussain F, Shah SZ, Ahmad H, Abubshait SA, Abubshait HA, Laref A, Manikandan A, Kusuma HS, Iqbal M (2021) Microalgae an ecofriendly and sustainable wastewater treatment option: biomass application in biofuel and bio-fertilizer production. A review. Renew Sust Energ Rev 137:110603
Illera-Vives M, Seoane Labandeira S, Iglesias Loureiro L, López-Mosquera M (2017) Agronomic assessment of a compost consisting of seaweed and fish waste as an organic fertilizer for organic potato crops. J Appl Phycol 29:1663–1671
Jassey VE, Walcker R, Kardol P, Geisen S, Heger T, Lamentowicz M, Hamard S, Lara E (2022) Contribution of soil algae to the global carbon cycle. New Phytol 234:64–76
Javeed HMR, Ali M, Kanwal N, Ahmad I, Jamal A, Qamar R, Zakir A, Andaleeb H, Jabeen R, Radicetti E (2023a) Nitrogen mineralization in texturally contrasting soils subjected to different organic amendments under semi-arid climates. Land 12:989
Javeed, Hafiz Muhammad Rashad, et al. (2023) "Potential Applications of Algae-based Biofertilizer." Biofertilizers for Sustainable Soil Management. CRC Press 185–196.
Jimenez R, Markou G, Tayibi S, Barakat A, Chapsal C, Monlau F (2020) Production of microalgal slow-release fertilizer by valorizing liquid agricultural digestate: growth experiments with tomatoes. Appl Sci 10:3890
Jo S-W, Do J-M, Na H, Hong JW, Kim I-S, Yoon H-S (2020) Assessment of biomass potentials of microalgal communities in open pond raceways using mass cultivation. PeerJ 8:e9418
Jochum M, Moncayo LP, Jo Y-K (2018) Microalgal cultivation for biofertilization in rice plants using a vertical semi-closed airlift photobioreactor. PLoS One 13:e0203456
Kaur T, Kour D, Yadav AN (2022) Trends of agricultural microbiology for sustainable crops production and economy: an introduction, trends of applied microbiology for sustainable economy. Elsevier, pp 1–44
Khan MI, Shin JH, Kim JD (2018) The promising future of microalgae: current status, challenges, and optimization of a sustainable and renewable industry for biofuels, feed, and other products. Microb Cell Factories 17:1–21
Khan W, Rayirath UP, Subramanian S, Jithesh MN, Rayorath P, Hodges DM, Critchley AT, Craigie JS, Norrie J, Prithiviraj B (2009) Seaweed extracts as biostimulants of plant growth and development. J Plant Growth Regul 28:386–399
Kholssi R, Lougraimzi H, Grina F, Lorentz JF, Silva I, Castaño-Sánchez O, Marks EA (2022) Green agriculture: a review of the application of micro-and macroalgae and their impact on crop production on soil quality. J Soil Sci Plant Nutr 22:4627–4641
Kour D, Rana KL, Yadav AN, Yadav N, Kumar M, Kumar V, Vyas P, Dhaliwal HS, Saxena AK (2020) Microbial biofertilizers: bioresources and eco-friendly technologies for agricultural and environmental sustainability. Biocatal Agric Biotechnol 23:101487
Kumar MS, Reddy GC, Phogat M, Korav S (2018) Role of bio-fertilizers towards sustainable agricultural development: a review. J Pharmacogn Phytochem 7:1915–1921
Kumar S, Islam R, Akash PB, Khan MHR, Proshad R, Karmoker J, MacFarlane GR (2022) Lead (Pb) contamination in agricultural products and human health risk assessment in Bangladesh. Water Air Soil Pollut 233:257
Kurniawati A, Toth G, Ylivainio K, Toth Z (2023) Opportunities and challenges of bio-based fertilizers utilization for improving soil health. Org Agric:1–16
Li D, Wang L, Zhao Q, Wei W, Sun Y (2015) Improving high carbon dioxide tolerance and carbon dioxide fixation capability of chlorella sp. by adaptive laboratory evolution. Bioresour Technol 185:269–275
Lv J, Liu S, Feng J, Liu Q, Guo J, Wang L, Jiao X, Xie S (2020) Effects of microalgal biomass as biofertilizer on the growth of cucumber and microbial communities in the cucumber rhizosphere. Turk J Bot 44:167–177

Ma C, Song W, Yang J, Ren C, Du H, Tang T, Qin S, Liu Z, Cui H (2022) The role and mechanism of commercial macroalgae for soil conditioner and nutrient uptake catalyzer. Plant Growth Regul 97:455–476

Mącik M, Gryta A, Frąc M (2020) Biofertilizers in agriculture: an overview on concepts, strategies and effects on soil microorganisms. Adv Agron 162:31–87

Mahapatra DM, Chanakya H, Joshi N, Ramachandra T, Murthy G (2018) Algae-based biofertilizers: a biorefinery approach. Microorganisms for green revolution: Volume 2: Microbes for Sustainable Agro-ecosystem, 177–196

Mahapatra DM, Satapathy KC, Panda B (2022) Biofertilizers and nanofertilizers for sustainable agriculture: Phycoprospects and challenges. Sci Total Environ 803:149990

Malik S, Kishore S, Dhasmana A, Kumari P, Mitra T, Chaudhary V, Kumari R, Bora J, Ranjan A, Minkina T (2023) A perspective review on microbial fuel cells in treatment and product recovery from wastewater. Water 15:316

Marks EA, Montero O, Rad C (2019) The biostimulating effects of viable microalgal cells applied to a calcareous soil: increases in bacterial biomass, phosphorus scavenging, and precipitation of carbonates. Sci Total Environ 692:784–790

Menegazzo ML, Fonseca GG (2019) Biomass recovery and lipid extraction processes for microalgae biofuels production: a review. Renew Sust Energ Rev 107:87–107

Mobin S, Alam F (2017) Some promising microalgal species for commercial applications: a review. Energy Procedia 110:510–517

Mouedden R, Abdellaoui S, El Madani F, El Ouamari N, Slimani D, Kasmi K, Taibi M, Zahir I, Chaabane K (2024) Gracilaria gracilis: a review of ecological knowledge, chemical composition, cultivation, and applications. Ecol Eng Environ Technol 25:275–286

Muhammad, Gul, et al. (2020) "Microalgae biomass production: an overview of dynamic operational methods." Microalgae biotechnology for food, health and high value products 415–432.

Muniswami DM, Chinnadurai S, Sachin M, Jithin H, Ajithkumar K, Narayanan GS, Gopinath B, Rasheeq AA, Preethi BA, Dineshkumar R (2023) Comparative study of biofertilizer/biostimulant from seaweeds and seagrass in Abelmoschus esculentus crop. Biomass Convers Biorefinery 13:11005–11022

Muthukumaran P, Arvind J, Kamaraj M, Manikandan A (2022) Algal-sourced biostimulants and biofertilizer for sustainable agriculture and soil enrichment: algae for fertilizers and soil conditioners, algal biorefineries and the circular bioeconomy. CRC Press, pp 211–235

Mutum L, Solomon W, Janda T, Molnár Z (2023) Time of application and cultivar influence on the effectiveness of microalgae biomass upon winter wheat (Triticum aestivum L.). Cereal Res Commun:1–9

Nur-Hafiza A, Rosazlin A, Wan-Rasidah K, Ajeng A, Mohamad-Fakhri I, Nur-Saadah A (2023) Variation in composition of organic marine deposits in sediments and their influence on the growth of RHIZOPHORA SPP. IN tanjung piai mangrove forest. J Trop For Sci 35:82–92

Occhipinti PS, Russo N, Foti P, Zingale IM, Pino A, Romeo FV, Randazzo CL, Caggia C (2023) Current challenges of microalgae applications: exploiting the potential of non-conventional microalgae species. J Sci Food Agric

Osório C, Machado S, Peixoto J, Bessada S, Pimentel FB, Alves CR, Oliveira MBP (2020) Pigments content (chlorophylls, fucoxanthin and phycobiliproteins) of different commercial dried algae. Separations 7:33

Osorio-Reyes JG, Valenzuela-Amaro HM, Pizaña-Aranda JJP, Ramírez-Gamboa D, Meléndez-Sánchez ER, López-Arellanes ME, Castañeda-Antonio MD, Coronado-Apodaca KG, Gomes Araújo R, Sosa-Hernández JE (2023) Microalgae-based biotechnology as alternative biofertilizers for soil enhancement and carbon footprint reduction: advantages and implications. Mar Drugs 21:93

Parmar P, Kumar R, Neha Y, Srivatsan V (2023) Microalgae as next generation plant growth additives: functions, applications, challenges and circular bioeconomy based solutions. Front Plant Sci 14:1073546

Priya M, Gurung N, Mukherjee K, Bose S (2014) Microalgae in removal of heavy metal and organic pollutants from soil, Microbial biodegradation and bioremediation. Elsevier, pp 519–537

Pulz O, Gross W (2004) Valuable products from biotechnology of microalgae. Appl Microbiol Biotechnol 65:635–648

Rachidi F, Benhima R, Sbabou L, El Arroussi H (2020) Microalgae polysaccharides bio-stimulating effect on tomato plants: growth and metabolic distribution. Biotechnol Rep 25:e00426

Raheem A, Prinsen P, Vuppaladadiyam AK, Zhao M, Luque R (2018) A review on sustainable microalgae based biofuel and bioenergy production: recent developments. J Clean Prod 181:42–59

Ramakrishnan B, Raju MN, Venkateswarlu K, Megharaj M (2023) Potential of microalgae and cyanobacteria in improving soil health and agricultural productivity—a critical view. Environ Sci Adv 2:586

Randrianarison G, Ashraf MA (2017) Microalgae: a potential plant for energy production. Geol Ecol Lands 1:104–120

Rawat I, Kumar RR, Mutanda T, Bux F (2011) Dual role of microalgae: phycoremediation of domestic wastewater and biomass production for sustainable biofuels production. Appl Energy 88:3411–3424

Renuka N, Guldhe A, Prasanna R, Singh P, Bux F (2018) Microalgae as multi-functional options in modern agriculture: current trends, prospects and challenges. Biotechnol Adv 36:1255–1273

Renuka N, Prasanna R, Sood A, Ahluwalia AS, Bansal R, Babu S, Singh R, Shivay YS, Nain L (2016) Exploring the efficacy of wastewater-grown microalgal biomass as a biofertilizer for wheat. Environ Sci Pollut Res 23:6608–6620

Ronga D, Biazzi E, Parati K, Carminati D, Carminati E, Tava A (2019) Microalgal biostimulants and biofertilisers in crop productions. Agronomy 9:192

da Rosa MDH, Alves CJ, dos Santos FN, de Souza AO, Zavareze EDR, Pinto E, Noseda MD, Ramos D, de Pereira CMP (2023) Macroalgae and microalgae biomass as feedstock for products applied to bioenergy and food industry: a brief review. Energies 16:1820

Saadaoui I, Sedky R, Rasheed R, Bounnit T, Almahmoud A, Elshekh A, Dalgamouni T, al Jmal K, Das P, Al Jabri H (2019) Assessment of the algae-based biofertilizer influence on date palm (Phoenix dactylifera L.) cultivation. J Appl Phycol 31:457–463

Sani I, Kurnia YW, Sinthya HC, Anthony R, Situmorang EC, Utomo C, Liwang T (2022) Exploring the potency of microalgae-based biofertilizer and its impact on oil palm seedlings growth. AGRIVITA, J Agric Sci 44:139–151

Serra, A., and L. Costa. (2015) "Process and product control in an industrial scale microalgae production plant." Conference Proceedings.

Shah (2013) Seaweed sap as an alternative liquid fertilizer for yield and quality improvement of wheat." Journal of plant Nutrition 36.2, 192–200

Silva LD, Bahcevandziev K, Pereira L (2019) Production of bio-fertilizer from Ascophyllum nodosum and Sargassum muticum (Phaeophyceae). J Oceanol Limnol 37:918–927

Singh P, Chandel A (2005) Effect of Biozyme on yield and quality of wheat (Triticum aestivum). Ind J Agron 50:58–60

Suleiman AKA, Lourenço KS, Clark C, Luz RL, da Silva GHR, Vet LE, Cantarella H, Fernandes TV, Kuramae EE (2020) From toilet to agriculture: fertilization with microalgal biomass from wastewater impacts the soil and rhizosphere active microbiomes, greenhouse gas emissions and plant growth. Resour Conserv Recycl 161:104924

Tan XB, Lam MK, Uemura Y, Lim JW, Wong CY, Lee KT (2018) Cultivation of microalgae for biodiesel production: a review on upstream and downstream processing. Chin J Chem Eng 26:17–30

Tayoh, Lekeanju Nguatem (2020) "Destruction of soil health and risk of food contamination by application of chemical fertilizer." Ecological and practical applications for sustainable agriculture 53–64.

Turan M, Köse C (2004) Seaweed extracts improve copper uptake of grapevine. Acta Agric Scand Section B Soil Plant Sci 54:213–220

Uthirapandi V, Suriya S, Boomibalagan P, Eswaran S, Ramya SS, Vijayanand N, Kathiresan D (2018) Biofertilizing potential of seaweed liquid extracts of marine macro algae on growth and biochemical parameters of Ocimum sanctum. J Pharmacogn Phytochem 7:3528–3532

Vázquez-Romero B, Perales JA, Pereira H, Barbosa M, Ruiz J (2022) Techno-economic assessment of microalgae production, harvesting and drying for food, feed, cosmetics, and agriculture. Sci Total Environ 837:155742

Wijesinghe J, Botheju SM, Nallaperuma B, Kanuwana N, (2023) Organic farming: the influence on soil health. One Health: Human, Animal, and Environment Triad, 185–197

Zeng G, Zhang J, Chen Y, Yu Z, Yu M, Li H, Liu Z, Chen M, Lu L, Hu C (2011) Relative contributions of archaea and bacteria to microbial ammonia oxidation differ under different conditions during agricultural waste composting. Bioresour Technol 102:9026–9032

Zhou L, Liu W, Duan H, Dong H, Li J, Zhang S, Zhang J, Ding S, Xu T, Guo B (2023) Improved effects of combined application of nitrogen-fixing bacteria Azotobacter beijerinckii and microalgae Chlorella pyrenoidosa on wheat growth and saline-alkali soil quality. Chemosphere 313:137409

Zou Y, Zeng Q, Li H, Liu H, Lu Q (2021) Emerging technologies of algae-based wastewater remediation for bio-fertilizer production: a promising pathway to sustainable agriculture. J Chem Technol Biotechnol 96:551–563

Algae-Based Biorefineries for Animal Feed and Aquaculture Applications

15

Sandeep Kaur and Rattandeep Singh

Contents

Abstract

There are innumerable species of microalgae and their approximate number is between 30,000 and 1,000,000. Algae species especially microalgae have

S. Kaur (✉)
School of Allied Medical Sciences, Lovely Professional University, Phagwara, Punjab, India
e-mail: sandeep.28639@lpu.co.in

R. Singh
School of Bioengineering and Biosciences, Lovely Professional University, Phagwara, Punjab, India

S. K. Arya et al. (eds.), *Value Added Products From Bioalgae Based Biorefineries: Opportunities and Challenges*,
https://doi.org/10.1007/978-981-97-1662-3_15

distinctive metabolism, biomass diversity and their ability to tolerate different water environments makes them potential candidates for commercial applications. Microalgae are unicellular photosynthetic microorganisms with a broad spectrum of applications in numerous sectors. They are essential to primary production. They have high photosynthetic and metabolic capacities and can produce a variety of valuable metabolites, such as lipids, carbohydrates, pigments, and proteins. There are three different ways by which microalgae can be cultivated that are photoautotrophic, heterotrophic, and mixotrophic. The major mode by which they are cultivated mainly is photoautotrophic and they can generate energy directly from light, CO_2, and water, typically through photosynthesis, producing organic compounds. Microalgae have a lot of potential for use in the feed business as a feed additive because of their high nutritional and functional value, which can improve animal performance. The usage of microalgae incorporation in the feeds of animals and aquaculture has beneficial effects like they produce naturally immunostimulant and antimicrobial compounds.

Keywords

Microalgae · Aquaculture · Animal feed · Biomass

Abbreviations

AA	Archidonic acid
ALA	Alpha-linolenic acid
DHA	Docosahexaenoic acid
EAA	Essential amino acids
EPA	Eicosatetraenoic acid
GRAS	Generally recognized as safe
LA	Linoleic acid
PUFA	Polyunsaturated fatty acids

15.1 Introduction

Microalgae-based biorefineries could help solve the global energy, environment, and food crises. The exponential need for natural ingredients in aquaculture, nutraceuticals, and animal feeds makes them useful as natural sources of bioactive chemicals. The FDA (FAO 2019) lists only a few species as GRAS. Ancient people ate microalgae. They may grow in lakes, ponds, rivers, oceans, and wastewater because of their photosynthetic nature. Eukaryotes and prokaryotic cyanobacteria (blue-green algae) thrive in watery settings and range in size from tiny cells to huge seaweeds. Microalgae can be seen microscopically while macroalgae can be seen by the naked eye (Ketzer et al. 2018). Microalgae are becoming a renewable supply and alternative to conventional components in aquaculture and animal feed. They can survive temperature, salinity, and pH stress and develop in reservoirs and alongside other organisms (Graham et al. 2009). The many bioactive nutrients like algae pigments can benefit aquaculture and animal feeding. Chlorophylls, carotenoids, and phycobilins are the primary types (Khan et al. 2015). Microalgae pigments are useful in treating cancer, neurological problems, and ocular illnesses. Controlling growing conditions allows microalgae to create biomass rich in lipids, proteins, carbohydrates, vitamins, mineral salts, and carotenoids (Ketzer et al. 2018; Shahid et al. 2020). Few decades ago, it became clear that the world's fast-growing population will likely lack protein-rich diets. Microalgae production began. Antibiotics and vitamins make microalgae a valuable food source for people and animals (Cuellar-Bermudez et al. 2015). Thus, microalgae characterization can be a new study topic to evaluate their biological and therapeutic uses. In this chapter, we evaluate algal feed options that can meet the needs of aquatic organisms, humans, and animals (Jensen 2001).

15.2 Microalgae Cultivation

Algae are a broad group of aquatic photosynthetic organisms that are divided into two categories: microalgae, which are unicellular and filamentous, and macroalgae, which include seaweed. According to Rosenberg et al., microalgae can thrive in environments that are rich in oxygen, light, carbon dioxide, and other nutrients needed for growth. They can grow not only photosynthetically but can also be heterotrophic and mixotrophic. They can be prevalent in freshwater or marine environments, and even in harsh environments such as hot springs, deserts, and cold lands. Microalgae when compared with terrestrial plants exhibit greater photon efficiency in conversion, retention of carbon dioxide, and biomass output. Microalgae have higher photon conversion efficiency, CO_2 sequestration, and enhanced biomass yield than the terrestrial plants (Gouveia 2011; Malcata 2011). Approximately 20 and 50% of total biomass is made up of photosynthetic products, which can accumulate in different forms, such as cell components or substances for storage (Chisti 2008). Although nitrogen and phosphorus make up about 10–20% of algae biomass, they are essential for algal growth (Benemann 2013). Other nutrients that are needed for growth include trace elements, macronutrients like Na, Mg, Ca, and K, and micronutrients like Mo, Mn, B, Co, Fe, and Zn. Thus, microalgae can be nurtured by the application of organic wastewater from the food and agricultural sectors. The requirements of various microalgal species for propagation can differ. Algal cells traverse several phases as they grow, spanning lag, exponential, declining, and death. Employing bioreactors is an essential component of the algal culture process. This needs to be made with the species and cultural intent in mind. The species, targeted added-value compounds, and intended use all influence the microalgae cultivation system. For example, Spirulina has been considered by the World Health Organization as one of the most protruding superfoods (Chacón-Lee and González-Mariño 2010). The potential of microalgae as a novel source of functional suppers has been explored by firms as well as scientists due to its promising ability and increasing popularity in the global population (Seyidoglu et al. 2017). The market for microalgae benefits from the recent rise in attention to healthy living and nutrition, with consumers being urged to eat "superfoods," which are extremely nutrient-dense, improve overall health, and assist in warding off various kinds of chronic ailments (Griffiths et al. 2016). The choice between using closed or open systems is crucial for growing microalgae. In addition to being inexpensive and simple to set up and run, open systems also require little maintenance. However, these systems are more prone to contamination and difficult to handle and high cell densities are not reached due to self-shading effects. Closed photobioreactors (PBRs) or open or covered raceway ponds (HRPs) can be used for autotrophic microalgae culture. The closed contain an extensive variety of shapes and sizes with various configurations such as horizontal, inclined, helical, vertical, rotating, underwater, floating, and many more. Individual PBRs cannot be scaled significantly above approximately 100 m^2 due to gas exchange constraints (O_2 removal and CO_2 supply to the algae), but single open raceway ponds can be well over one hectare (>10,000 m^2) in size (Formighieri et al. 2012). Thousands of independent PBR undistracted production ponds will be required to produce commodities on hundreds of hectares for commercial-scale systems, compared to a few score multi-hectare raceway ponds (HRPs). Therefore, the PBRs would be limited to a small portion of the process's total cost. Algae can be cultivated in open ponds (high-rate ponds) on a massive scale.

15.3 Compounds Produced by Microalgae

The tiny photosynthetic organisms known as microalgae convert the energy from light and carbon dioxide in the air into lipids, proteins, carbohydrates, and microelements such minerals, vitamins, polyphenols, flavonoids, and carotenoids as shown in (Barkia et al. 2019; Tibbetts et al. 2015). The production of animal feed makes use of a variety of microalgae species, including *Arthrospira platensis*, *Haematococcus pluvialis*, *Tetraselmis chui*, and *Nannochloropsis granulate*. Disparities in growing conditions, such as temperature, geographical location, and sunshine availability, cause microalgae to display remarkable dynamism and variety across strains. Not only do microalgae include recognized metabolites, but they also provide a wealth of undiscovered compounds with interesting properties and potential uses. Lipoproteins, sterols, and alkaloids are among these substances (Tan 2007). One of the many health benefits of these chemicals is that they strengthen the immune system, which means that antibiotics used in cattle, poultry, and fish farming may be less necessary in the future (Tibbetts 2018). Carotenoids primary functions include acting as potent antioxidants and converting light energy to chlorophyll. *Cladophora* species (Chlorophyta) are rich in carotenoids, including lutein, beta-carotene, and zeaxanthin, as well as chlorophylls (a and b) (Tibbetts et al. 2015). Microalgae are more productive than any other terrestrial plant or animal in terms of net biomass. Microalgae can be grown with wastewater or even seawater, unlike land-based plants, which require the fertile ground to flourish.

The primary factor supporting the microalgae's potential is that it contains the proper ratio of protein, fat, and carbohydrates to protect fish health. Algal biorefineries are particularly high in protein and fat when compared to other substitute ingredients such as bacteria and yeast. Additionally, the well-balanced amino acid composition of microalgae negates the need for expensive dietary supplements of amino acids. For example, methionine, which is frequently deficient in plant-based components, is abundant in microalgae including *Chlorella*, *Chlamydomonas*, *Porphyridium*, *Isochrysis*, and *Nannochloropsis* and carbohydrate is also an essential ingredient of the feed. For instance, starch, an easily digested carbohydrate, can make up anywhere from 7% to 45% of microalgal species. In comparison to other microalgae, species belonging to *Tetraselmis subcordiformis*, *Chlamydomonas reinhardtii*, and *Chlorella vulgaris* have a greater starch concentration (30–49%). The content of fiber—a complex carbohydrate in microalgae ranges from 5% to 18%. Some species of microalgae contain sufficient protein that it comprises nearly half of their biomass. The protein content of the majority of Spirulina strains, a small number of Chlorella strains, and Nannochloropsis strains range from 40 to 65% (De Morais et al. 2015). Food is the only source of essential amino acids (EAA), as humans are not able to synthesize all amino acids. Algal protein contains almost all of the essential amino acids (EAA); therefore, microalgae tablets or capsules are a useful and simple dietary supplement. Large amounts of lipids, which vary among species and fluctuate by a number of parameters, are also produced and accumulated by microalgae. Algal cells mostly contain lipids in the form of glycerol, esterified carbohydrates, and various fatty acid types with 12–22 carbon atoms. These fatty acids are used in medicine and nutrition.

15.3.1 Protein

The identification of various and innovative alternative protein sources is crucial for ensuring food security (Becker 2003). A variety of protein sources have been consumed as both food and feed throughout the past decade. Nevertheless, the excessive utilization of these sources has resulted in a growing recognition of the need to discover innovative solutions to address the risk of protein scarcity (Becker 2003). Microalgae possess great potential as a highly promising alternative. Microalgae, in particular some strains, have the potential to be a sustainable source of feed additives due to their high protein production. An example of a cyanobacterium with a high protein content is *Arthrospira*. For example, the essential amino acid profile of microalgae is quite similar to that of soybean and other high-quality crop proteins (Hemaiswarya et al. 2011). According to FAO/WHO, microalgae like *Chlorella and Arthrospira* (Spirulina) are potential microorganisms to get sustainable proteins that people can eat. This is due to the fact that, like more traditional protein sources like soybeans and eggs, they contain necessary amino acids (FAO 2019). Muscle growth in chicken breast and thighs, together with improved meat quality and quantity, can be achieved by feeding stocks that include enhanced levels of lysine and methionine. Furthermore, microalgae produce bioactive peptides that have anti-inflammatory, anti-cancer, anticoagulant, antihypertensive, and immune-stimulating properties (Samarakoon and Jeon 2012).

15.3.2 Carbohydrates

Carbohydrates, found in microalgae, are crucial to their nutritional and therapeutic qualities. One soluble fiber found in *chlorella sp.*, beta-1-3-glucan, is an antioxidant and an important factor in lowering blood cholesterol levels (De Morais et al. 2015). Another unicellular red alga with promising economic potential for carbohydrate production is *Porphyridium cruentum*. As an alternative to carrageenan, it creates a sulfated galactan exopolysaccharide, which finds usage in the meat and dairy industries (Pierre et al. 2015). Mannose, glucose, xylose, and rhamnose are the most common monosaccharides that can be produced by microalgal polysaccharides. Among the numerous carbohydrates contained in green microalgae, glucose accounts for 40–80% of the total (Markou et al. 2012). But diatoms like *Phaeodactylum tricornutum* show much higher mannose concentrations, up to 40% of the total sugars.

15.3.3 Lipids

There are a number of microalgae species that have long been considered an excellent food source for lipids. Under some strains and culture conditions, microalgae can generate lipids that can make up as much as half of their dry weight. The lipid concentration may go above this proportion in certain instances (Chacón-Lee and González-Mariño 2010). The polyunsaturated fatty acids alpha-linolenic acid (ALA), arachidonic acid (AA), linoleic acid (LA), and docosahexaenoic acid (DHA) might improve the long-chain fatty acid composition of microalgae. Because neither people nor animals are able to manufacture omega fatty acids internally, getting them through food is essential. In addition to their many biological roles, DHA and EPA improve mental health, reduce the risk of cardiovascular diseases like arrhythmia, stroke, rheumatoid arthritis, and hypertension, and act as antioxidants and anti-inflammatory agents (López et al. 2019). New studies show that infertile men whose diets are rich in omega-3 fatty acids have significantly improved sperm motility and DHA concentration in sperm transport fluid (Udayan et al. 2017).

15.4 Microalgae in Aquaculture

Aquaculture, which includes the raising, breeding, and harvesting of freshwater and marine species of fish as well as aquatic plants in all aquatic environments, is another essential component of agriculture. In 2008, the worldwide market for aquatic edible plants and farmed fish was estimated to be worth US $106 billion. In 2022, the world's aquaculture output reached unprecedented heights. Aqua feed usage increased sixfold from 8 to 48 million tons (Tacon 2020) in response to the growing demand, which was driven by strong production profitability. Limiting the consumption of fish-based oils while continuing to generate finished products with high concentrations of long chains of omega-3 fatty acids, which are good for consumer health, is the major challenge in fish nutrition. The markets for omega-3 oils have undergone extensive growth in the human nutraceuticals and animal feed industries. Open ocean fishing has increased the number of fish consumed by humans and other organisms. High-quality protein, vitamins, minerals, carotenoids, and long-chain polyunsaturated fatty acids (PUFAs)—particularly those of the n-3 and n-6 series, such as arachidonic (AA), docosahexaenoic (DHA), and eicosapentaenoic (EPA)—can all be found in substantial quantities in algae. In contrast to live biomass, dried microalgae are less required in fisheries. The concentrated form of microalgae and the subsequent necessity to store them in freezers or extremely cold rooms drive up production costs (Camacho-Rodríguez et al. 2016). The bioactive elements of biomass may degrade if it is stored for an extended period of time.

15.4.1 Microalgae for Various Marine Species

A wide range of microalgal species, including *Phormidium valderianum*, have proven to be highly nutritious and non-toxic feed for aquaculture. In a similar vein, diatom *Thalassiosira pseudonana* is given to a wide variety of mollusks, including rock scallops and the Pacific oyster Crassostrea gigas. Two extremely high-protein microalgae species, *Hypnea cervicornis* and *Cryptonemia crenulata*, were studied using shrimp diets. The addition of algae to fish meal led to an enormous rise in shrimp growth rate. The long-chain polyunsaturated fatty acids (PUFAs) in microalgal biomass are oxidable because they are unsaturated. However, oxidation of the highly unsaturated fish oil can be prevented by adding natural antioxidants to the oil. Studies have shown that incorporating DHA and AA in the diet of shrimp and marine fish raised immunological indicators such as phenol oxidase enzyme activity, antioxidant activity, total hemocyte count, and bactericidal activity. In recent times, it has been reported that the addition of 0.5% Spirulina meal along with 14% of Peruvian fishmeal in a complete diet for shrimp (Litopenaeus *vannamei juveniles*) has proved as a nutritionally efficient feeding attractant. Feed derived from microalgae might be given either by itself or in addition to conventional feed. Numerous investigations on the physiological profile and toxicological properties of microalgae have supported their use as feed additives. Incorporating microalgae feed in aquaculture also increased other biological processes and fish phenotypic traits like color. Vital nutrients like proteins, polysaccharides, and omega-3 fatty acids like EPA and DHA are genetically improved into prokaryotic species like cyanobacteria and eukaryotic microalgae. Additionally, they contain vitamins A, B1, B2, B6, B12, C, E, niacin, nicotinate, biotin, folic acid, and pantothenic acid, minerals like calcium, magnesium, and potassium, and pigments like phycobiliproteins, carotenoids, and chlorophylls (Gouveia 2011). The biomass of microalgae not only provides essential nutrients to aquatic organisms but also has immune-stimulating and disease-resistant qualities. Several species of microalgae belonging to the genus Isochrysis and Pavlova are considered important strains for use as aquaculture feed including *Nannochloropsis*, *Arthrospira*, *Chlorella*, *Dunaliella*, *Haematococcus*, and *Schizochytrium* (Madeira et al. 2017; Yaakob et al. 2015). Microalgae feed is also used in fish farming, mollusks, crustacean, and zooplankton hatchery operations.

15.4.2 Zooplankton

The primary food supply for the small animals kept in farmhouses is zooplanktons, which are raised in special hatcheries. There is a special area in these zooplankton hatcheries where microalgae are cultured for the purpose of feeding the zooplankton. Microalgae produce nutrients that include fatty acids, protein, vitamins, and pigments. Fish eat the zooplanktons that are grown in ponds or tanks once they are harvested and fed to them. Since curing tends to denature some proteins and other key elements, dried feed is less nutrient-dense than live prey. Zooplankton can be cultivated using microalgae culture due to a technique termed as green water aquaculture. When using microalgae as a feed, a few conditions must be adhered to: (1) sufficient number of cells to be ingested; (2) easily digestible cell walls; (3) absence of hazardous compounds; (4) high-value end products that shouldn't change throughout the culture; and (5) simple cultivation. In fish production, the amount of unsaturated fatty acid plays a crucial role without its dependency on fish oil. Fish feed derived from microalgae has been gaining popularity and will likely take over the market in the coming years. Microalgae feed is used by several marine fish species, including Senegalese sole, Atlantic cod, halibut, striped mullet, seabream, seabass, and turbot (Tredici et al. 2009). The use of *H. pluvialis* algal meal in salmonid feed recently obtained approval from the US Food and Drug Administration (US FDA). Similar to this, ornamental fish coloration is significant from a commercial perspective since it raises the fish's selling price. The inclusion of microalgal food improved the ornamental goldfish's (*Carassius auratus*) coloration. Similar to this, algal supplementation is also necessary for other aquatic species such as lobster, shrimp, seabream, and koi carp in order to increase the color of their muscles (Gouveia 2011). Microalgae have an important role in aquaculture due to a number of excellent nutritional qualities. They can be cultivated to produce rotifers, which are then given to fish and crustaceans, or they can be used as the primary food supply for shrimp or bivalve mollusk larvae. Microalgae are used as live feed in aquaculture for a variety of reasons, including the ability to adjust pH, improve and stabilize water quality, and release bioactive chemicals that have been shown to have antibacterial and immune-stimulating qualities. Spirulina sp. and Chlorella sp. are the two most common microalgae utilized in aquaculture systems, with sizes ranging from 5 μm to 50 μm. Studies have shown that the second-largest farmed fish in the world, the Nile tilapia (*Oreochromis niloticus*), can be fed as a mixture of two microalgal species: *Schizochytrium sp.* and *Nannochloropsis oculata*. Because of their appropriate cell size (3–50 μm) and high nutritional content, microalgae are an essential source of food for fish, shrimp, and bivalves, especially in their larval and juvenile stages. Furthermore, hatcheries that raise sea cucumbers and crabs frequently employ microalgae.

15.4.3 Bivalves

Algal biorefineries are considered as the essential food source for fish, shrimp, and bivalves due to their suitable cell size range of 3–50 μm and high nutritional value, especially during their larval and juvenile stages. There are 12 bivalves in the Phylum Mollusca, and each one has two symmetric and asymmetric shells. Additionally, microalgae species are often used in hatcheries raising crabs and sea cucumbers. The life cycle of bivalves comprises six phases, ranging from conception to adulthood. Bivalves begin eating microalgae at the second stage of the D-larval, and most of them use microalgae as their main food source for the duration of their developmental stages. However, other materials have also been found as a substitute food source for bivalves like bacteria, yeast, and dried microalgae but they have been found nutritionally insufficient to substitute microalgae. The several strains of microalgae that are primarily utilized in the manufacture of bivalves are *Phaeodactylum tricornutum*, *Isochrysis galbana*, *Thalassiosira pseudonana*, *Chaetoceros calcitrans*, and *Pavlova lutheri*.

15.4.4 Shrimp

The shrimp has a complex life cycle. It consists of six stages including egg comprising, nauplius, zoea, mysis, post-larva, juvenile, and broodstock. The shrimp larvae generally depend upon internal nutrients at the time of the nauplius stage. Shrimp larvae begin feeding on food after reaching the zoea phase, mostly filtering microalgae. The larvae can thereafter survive on microalgae during the mysis and post-larval phases, although they do have a little preference for zooplankton like rotifer, artemia, and copepods. The most commonly utilized microalgae species in shrimp *production include the diatoms* (e.g., *Skeletonema costatum*, *P. tricornutum*, *Conticribra weissflogii*, and *Chaetoceros gracilis*), chlorophytes (e.g., *Tetraselmis chuii* and *Chlorella* sp.), and chrysophytes (e.g., *Isochrysis* sp.).

15.5 Utilizing Microalgae as a Source of Animal Feed

Microalgae may one day provide an ecologically sound and nutritionally beneficial substitute for conventional animal and aquatic feed production methods (Hosseini et al. 2019). Numerous animals, such as chickens, pigs, sheep, and mink, can benefit from these bacterial feed supplements, which can be mass-produced through the use of photobioreactors and open ponds (Dineshbabu et al. 2019). By adjusting the growing conditions, one can improve the feed's quality by increasing the content of beneficial components including carotenoids, essential amino acids, and omega fatty acids (Slegers et al. 2013). So, the nutritious value of meat, eggs, and milk can be enhanced by adding fortified microalgae to the diet of animals. In turn, this has the potential to provide a number of health benefits to people, such as antioxidant, antiviral, and anti-cancer effects when consumed.

15.5.1 Utilizing Microalgae as a Dietary Supplement to Enhance Egg Production

There is an immense amount of literature on dietary modification of polyunsaturated fatty acids (PUFAs). Because of this, high-PUFA egg yolks that are customized to meet individual dietary needs have emerged (Abd El-Baky et al. 2009). Eggs can be nutritionally improved for human consumption by adding microalgae, which are alternative sources of essential omega-3 fatty acids (ϋ-3 fatty acids) (González-Esquerra and Leeson 2001). In addition to being essential for brain development and maintenance and cardiovascular disease prevention, a number of omega-3 fatty acids also have potent anti-inflammatory properties. The presence of these ϋ-3 fatty acids in the egg yolk was caused by the addition *of Nannochloropsis gaditana*, a plant that provides long-chaeicosapentaenoic acid (EPA) and docosahexaenoic acid (DHA), to the diets of laying hens (Bourre 2005). Various factors, including the chicken's age and breed, and its digestion of microalgae, impact the amount of ϋ-3 fatty acids in its eggs (Bruneel et al. 2013).

In contrast to eggs made from regular feed, the concentration of beneficial fatty acids was higher in eggs from hens fed feedstocks mixed with algae and supplemented with ϋ-3 fatty acids (Feng et al. 2020). These lengthy fatty acids are predominantly contained in the yolk's phospholipids. Eggs' ϋ-3 fatty acid composition underwent significant changes when small quantities of microalgal biomass were added to feedstocks during feed trials. Also, when chickens were fed a diet with 2.4% algae (Fredriksson et al. 2006) and (Herber and Van Elswyk 1996) demonstrated that the rate of DHA absorption from microalgae to eggs was 42.6%. Chickens given a diet rich in algae laid eggs with six times the amount of DHA found in the control group's eggs (Moran et al. 2019). Eggs with low EPA and high DHA levels are most often produced by feedstocks, including *Nannochloropsis oculata*, that contain a lot of eicosapentaenoic acid (EPA) and very little docosahexaenoic acid (DHA). This could mean that biological systems preferentially absorb DHA over EPA, or that fatty-acid chain elongation converts EPA to DHA. Hens that were given algal feedstocks such as *Porphyridium sp.* ate 10% less food compared to the control groups. Additionally, the egg yolks from these hens had lower cholesterol levels and lower ratios of ω-6: ω-3 fatty acids (Ginzberg et al. 2000). The combination of 2.5% dried-fermented *Schizochytrium sp.* with flax seeds resulted in omega-rich feedstocks that produced 150 mg of omega-3 fatty acids per egg (Barclay et al. 1998). Seedstocks with very high amounts of ω-3 fatty acids do not have any negative impact on the bird's performance or health. However, it is important to mention that feedstock rich in omega-3 can lead to a reduction in the tocopherols required for the appropriate synthesis of egg yolk and oxidative stability (Cherian et al. 1996). According to reports, microalgae in broiler feed at levels ranging from 1.5% to 10% has positive effects. However, if the inclusion exceeds this range, it might have a negative impact on the growth and nutritional value of eggs. An experiment was carried out by (Fredriksson et al. 2006) to assess the effects of feeding hens a diet that contained 20% *Nannochloropsis oculata.* The levels of lutein and zeaxanthin in the eggs were found to be approximately 1.3 mg per egg after 4 weeks of supplementing the hens' diet with microalgae. Egg yolks with greater carotenoid content tend to be opaque and have a reddish-orange hue (Kassis et al. 2010). Egg size and nutrition conversion can be greatly improved with the help of both natural and synthetic carotenoids, such as chlorophylls. However, the research showed that the oxidative stability of yolk lipids was significantly improved when the naturally occurring lutein in Chlorella-supplemented feedstocks was absorbed into the system more efficiently (Englmaierová et al. 2013). Diets high in carotenoids also enhanced eggshell thickness and other desirable physical characteristics.

15.5.2 Utilizing Microalgae to Enhance the Quality of Meat

Due to the fact that neither people nor cattle are able to produce these fatty acids on their own, they are considered essential. Polyunsaturated fatty acids (PUFAs) like ALA, EPA, and DHA have abundantly documented health benefits (Connor 2000). So, it has been suggested that eating foods high in ӧ-3 PUFAs can have significant health benefits, and these compounds also have a lot of potential economically (Wallace et al. 2000).

A considerable amount of polyunsaturated fatty acids (PUFAs) is bio-hydrogenated in the rumen if the feedstocks are not coated. Linoleic acid (LA, C18:2 n-6) and linoleic acid (ALA, C18:3 n-3) are the primary fatty acids consumed by animals that ruminate through their diet of cereals and pasture (Scott et al. 1971). Ruminant meat does not contain a lot of polyunsaturated fatty acids (PUFAs) though (Wachira et al. 2002). Rumen biohydrogenation occurs in the presence of around 71–94% linoleic acid (LA) and 84–100% alpha-linolenic acid (ALA) prior to excretion (Shingfield et al. 2008). Meat that is higher in polyunsaturated fatty acids (PUFAs), such as eicosapentaenoic acid (EPA) and docosahexaenoic acid (DHA), has been enhanced using a variety of methods. Incorporating fish, marine microalgae, and bacteria that resemble algae into animals' typical diets is one of these techniques. The results of these bioassays showed that the algae-based feed supplement significantly increased the amounts of EPA and DHA in the animals" flesh. The meat quality of Suffolk and Soya lambs had been enhanced when fed with linseed and fish oils (Elmore et al. 2000). The ALA content was doubled, while the EPA and DHA levels were quadrupled, to accomplish this.

Research has shown that pigs whose diets include algae rich in polyunsaturated fatty acids (PUFAs)—for example, Chlorella sp., and *Arthrospira platensis*—consume pork with a balanced lipid profile (Lum et al. 2013). Microalgae, and more especially Schizochytrium sp., were found to enhance the amounts of omega-3 fatty acids in the flesh of growing pigs, according to a recent study. The inclusion rates used in the study were 5% and 7%. The pigs' skeletal muscle, phenotypic features, and functional abilities were all altered as a result of this nutritional alteration (Kalbe et al. 2019). There was no negative impact on young pigs when microalgae were included in their diets at a maximum level of 33%, the highest proportion recorded so far, according to an independent study. However, no one reported gaining weight (Arney et al. 2015) The addition of *Arthrospira* to conventional chicken feed (at concentrations of 4% or 8%), according to research by (Toyomizu et al. 2001), had no effect on the growth performance of broilers. It did, however, cause the chickens' skin, fat, liver, and muscles to have a yellowish hue. The orange hue raises the meat's monetary worth in the marketplace. Incorporating a 1% solution of fresh liquid algae into chicken feed improves the immune system, growth rate, and generation of Lactobacillus bacteria in the gut microbiome of broiler chickens (Kang et al. 2013). In the end, the rendered biomass of *Arthrospira* and *Chlorella*, obtained through biofuel production, improved the flavor and texture of chicken (Wiaętkiewicz et al. 2015).

15.5.3 Utilizing Microalgae to Enhance Milk Production and Improve Its Quality

Research on microalgae supplementation to increase milk's beneficial fatty acid concentration is expanding (Lamminen et al. 2019). But whether microalgal metabolites affect lactation and nutritional transmission to milk depends largely on the metabolic capacity of the animal and whether it has a ruminant or non-ruminant digestive tract. In ruminants, the rumen bacteria and the small intestine enzymes have a role in the digestion and absorption of dietary fatty acids. Intestinal and microbial enzymes convert the unsaturated fats into the more absorbable short-chain saturated fats. During this procedure, the molecules that will eventually become a part of the animal's tissue undergo a change in composition. Fatty acids consumed by nonruminant animals can be absorbed through the small intestine and incorporated into tissues without any changes.

A cow's milk quality is influenced by the type and number of fatty acids it consumes, according to research. Thus, avoiding biohydrogenation in the rumen is of utmost importance. Microalgae have recently shown promise as a viable substitute for soya bean meal in the protein feed for dairy cows (Santos et al. 2008). Hence, ruminant animals should be fed coated microalgae biomass to preserve the feedstock's nutritional properties. The small intestine will be able to absorb ϋ-3 fatty acids more effectively, and then they can be carried to the mammary glands (Stamey et al. 2012).

15.6 Challenges and Future Perspectives

The microalgae have restricted applications in aquaculture because of their high price during production. Microalgae have the capability to cover up the production cost and distribution cost (Becker 2003; Sarker et al. 2016). In microalgae, it is hard to remove moisture and pelletize, improper drying might change their both physical and nutritional qualities, thus lowering their effectiveness as feed. Challenges might arise from marine microalgae species utilized as food for fish because of their sluggish digestion and substantial salt accumulation. In experimental diets, algae can only meet 10–15% of the required amount of protein without impairing growth or food intake. The microalgae contain low protein and high carbohydrate composition because of their strong cell walls which prevent fats and proteins from being digested (Skrede et al. 2011). Microalgal biomass possesses a high amount of trace elements and toxic substances, making it unsuitable to be used as aquafeed. It is also important to investigate the influence of surroundings and stress on the chemistry of microalgae. For example, nitrogen deficit may result in an accumulation of PUFA and carbohydrates, which fish may ingest.

In spite of providing microalgal biomass at an appropriate cost, microalgal and aquafeed providers must deal with significant heterogeneity in basic structure, digestion, and developing situations. Aquaculture needs a varied spectrum of microalgal organisms. Microalgal varieties with superior dietary or developmental capabilities might be more effective. This implies that the outer layer of cells needs to be broken away in order to make algal components accessible to the enzymes that break down food (Berge et al. 2013). Nevertheless, more processing steps may lead to higher expenses. To promote the health of fish, microalgal products must be screened for hazardous substances before distribution. Along with the nutrition quantity and digestion, the processing demands of chemical feed manufacturing lines have to be examined.

References

Abd El-Baky HH, El Baz FK, El-Baroty GS (2009) Enhancement of antioxidant production in Spirulina platensis under oxidative stress. Acta Physiol Plant 31(3):623–631. https://doi.org/10.1007/s11738-009-0273-8

Arney B, Liu W, Forster IP, McKinley RS, Pearce CM (2015) Feasibility of dietary substitution of live microalgae with spray-dried Schizochytrium sp. or Spirulina in the hatchery culture of juveniles of the Pacific geoduck clam (Panopea generosa). Aquaculture 444:117–133. https://doi.org/10.1016/j.aquaculture.2015.02.014

Barclay W, Abril R, Abril P, Weaver C, Ashford A (1998) Production of docosahexaenoic acid from microalgae and its benefits for use in animal feeds. World Rev Nutr Diet 83:61–76. https://doi.org/10.1159/000059671

Barkia I, Saari N, Manning SR (2019) Microalgae for high-value products towards human health and nutrition. Mar Drugs 17(5):304. https://doi.org/10.3390/md17050304

Becker W (2003) Microalgae in human and animal nutrition. In: Handbook of Microalgal Culture. Wiley, Hoboken. https://doi.org/10.1002/9780470995280.ch18

Benemann J (2013) Microalgae for biofuels and animal feeds. Energies 6(11):5869–5886. https://doi.org/10.3390/en6115869

Berge GM, Hatlen B, Odom JM, Ruyter B (2013) Physical treatment of high EPA Yarrowia lipolytica biomass increases the availability of n-3 highly unsaturated fatty acids when fed to Atlantic salmon. Aquac Nutr 19(1):110–121. https://doi.org/10.1111/anu.12092

Bourre JM (2005) Where to find omega-3 fatty acids and how feeding animals with diet enriched in omega-3 fatty acids to increase nutritional value of derived products for human: what is actually useful? J Nutr Health Aging 9(4):232–242

Bruneel C, Lemahieu C, Fraeye I, Ryckebosch E, Muylaert K, Buyse J, Foubert I (2013) Impact of microalgal feed supplementation on omega-3 fatty acid enrichment of hen eggs. J Funct Foods 5(2):897–904. https://doi.org/10.1016/j.jff.2013.01.039

Camacho-Rodríguez J, Cerón-García MC, Macías-Sánchez MD, Fernández-Sevilla JM, López-Rosales L, Molina-Grima E (2016) Long-term preservation of concentrated Nannochloropsis gaditana cultures for use in aquaculture. Journal of Applied Phycology 28(1):299–312. https://doi.org/10.1007/s10811-015-0572-y

Chacón-Lee TL, González-Mariño GE (2010) Microalgae for "healthy" foods-possibilities and challenges. Compr Rev Food Sci Food Saf 9(6):655–675. https://doi.org/10.1111/j.1541-4337.2010.00132.x

Cherian G, Wolfe FW, Sim JS (1996) Dietary oils with added tocopherols: effects on egg or tissue tocopherols, fatty acids, and oxidative stability. Poult Sci 75(3):423. https://doi.org/10.3382/ps.0750423

Chisti Y (2008) Biodiesel from microalgae beats bioethanol. Trends Biotechnol 26(3):126–131. https://doi.org/10.1016/j.tibtech.2007.12.002

Connor WE (2000) Importance of n-3 fatty acids in health and disease. Am J Clin Nutr 71(1):171–175. https://doi.org/10.1093/ajcn/71.1.171s

Cuellar-Bermudez SP, Aguilar-Hernandez I, Cardenas-Chavez DL, Ornelas-Soto N, Romero-Ogawa MA, Parra-Saldivar R (2015) Extraction and purification of high-value metabolites from microalgae: essential lipids, astaxanthin and phycobiliproteins. Microb Biotechnol 8(2):12167. https://doi.org/10.1111/1751-7915.12167

De Morais MG, Vaz BDS, De Morais EG, Costa JAV (2015) Biologically active metabolites synthesized by microalgae. Biomed Res Int 2015:835761. https://doi.org/10.1155/2015/835761

Dineshbabu G, Goswami G, Kumar R, Sinha A, Das D (2019) Microalgae–nutritious, sustainable aqua- and animal feed source. J Funct Foods 62:103545. https://doi.org/10.1016/j.jff.2019.103545

Elmore JS, Mottram DS, Enser M, Wood JD (2000) The effects of diet and breed on the volatile compounds of cooked lamb. Meat Sci 55(2):149–159. https://doi.org/10.1016/S0309-1740(99)00137-0

Englmaierová M, Skřivan M, Bubancová I (2013) A comparison of lutein, spray-dried Chlorella, and synthetic carotenoids effects on yolk colour, oxidative stability, and reproductive performance of laying hens. Czeh J Anim Sci 58(9):412–419. https://doi.org/10.17221/6941-cjas

FAO (2019) The state of world fisheries and aquaculture 2018: meeting the sustainable development goals. FAO, Rome

Feng J, Long S, Zhang HJ, Wu SG, Qi GH, Wang J (2020) Comparative effects of dietary microalgae oil and fish oil on fatty acid composition and sensory quality of table eggs. Poult Sci 99(3):1734–1743. https://doi.org/10.1016/j.psj.2019.11.005

Formighieri C, Franck F, Bassi R (2012) Regulation of the pigment optical density of an algal cell: filling the gap between photosynthetic productivity in the laboratory and in mass culture. J Biotechnol 162(1):115–123. https://doi.org/10.1016/j.jbiotec.2012.02.021

Fredriksson S, Elwinger K, Pickova J (2006) Fatty acid and carotenoid composition of egg yolk as an effect of microalgae addition to feed formula for laying hens. Food Chem 99(3):530–537. https://doi.org/10.1016/j.foodchem.2005.08.018

Ginzberg A, Cohen M, Sod-Moriah UA, Shany S, Rosenshtrauch A, Arad S (2000) Chickens fed with biomass of the red microalga Porphyridium sp. have reduced blood cholesterol level and modified fatty acid composition in egg yolk. J Appl Phycol 12(3-5, 325):–330. https://doi.org/10.1023/a:1008102622276

González-Esquerra R, Leeson S (2001) Alternatives for enrichment of eggs and chicken meat with omega-3 fatty acids. Can J Anim Sci 81(3):295–305. https://doi.org/10.4141/A00-092
Gouveia L (2011) Microalgae as a feedstock for biofuels. In: Microalgae as a feedstock for biofuels. Springer, Cham. https://doi.org/10.1007/978-3-642-17997-6
Graham LE, Graham JM, Wilcox LW (2009) Algae, 2nd edn. Benjamin Cummings, San Francisco
Griffiths M, Harrison STL, Smit M, Maharajh D (2016) Major commercial products from micro- and macroalgae. In: Green energy and technology. Springer, Cham. https://doi.org/10.1007/978-3-319-12334-9_14
Hemaiswarya S, Raja R, Kumar RR, Ganesan V, Anbazhagan C (2011) Microalgae: a sustainable feed source for aquaculture. World J Microbiol Biotechnol 27(8):1737–1746. https://doi.org/10.1007/s11274-010-0632-z
Herber SM, Van Elswyk ME (1996) Dietary marine algae promotes efficient deposition of n-3 fatty acids for the production of enriched shell eggs. Poult Sci 75(12):1501–1507. https://doi.org/10.3382/ps.0751501
Hosseini B, Nourmohamadi M, Hajipour S, Taghizadeh M, Asemi Z, Keshavarz SA, Jafarnejad S (2019) The effect of omega-3 fatty acids, EPA, and/or DHA on male infertility: a systematic review and meta-analysis. J Dietary Supplements 16(2):1431753. https://doi.org/10.1080/19390211.2018.1431753
Jensen GS (2001) Blue-green algae as an immuno-enhancer and biomodulator. JANA 3:4
Kalbe C, Priepke A, Nürnberg G, Dannenberger D (2019) Effects of long-term microalgae supplementation on muscle microstructure, meat quality and fatty acid composition in growing pigs. J Anim Physiol Anim Nutr 103(2):13037. https://doi.org/10.1111/jpn.13037
Kang HK, Salim HM, Akter N, Kim DW, Kim JH, Bang HT, Kim MJ, Na JC, Hwangbo J, Choi HC, Suh OS (2013) Effect of various forms of dietary Chlorella supplementation on growth performance, immune characteristics, and intestinal microflora population of broiler chickens. J Appl Poult Res 22(1):100–108. https://doi.org/10.3382/japr.2012-00622
Kassis N, Drake SR, Beamer SK, Matak KE, Jaczynski J (2010) Development of nutraceutical egg products with omega-3-rich oils. LWT 43(5):777–783. https://doi.org/10.1016/j.lwt.2009.12.014
Ketzer F, Skarka J, Rösch C (2018) Critical review of microalgae LCA studies for bioenergy production. Bioenergy Res 11(1):95–105. https://doi.org/10.1007/s12155-017-9880-1
Khan N, Mobin, Abbas ZK (2015) Variation in photosynthetic pigments, antioxidant enzymes and osmolyte accumulation in seaweeds of red sea. Int J Plant Biol Res 2015:3
Lamminen M, Halmemies-Beauchet-Filleau A, Kokkonen T, Jaakkola S, Vanhatalo A (2019) Different microalgae species as a substitutive protein feed for soya bean meal in grass silage based dairy cow diets. Anim Feed Sci Technol 247:112–126. https://doi.org/10.1016/j.anifeedsci.2018.11.005
López G, Yate C, Ramos FA, Cala MP, Restrepo S, Baena S (2019) Production of polyunsaturated fatty acids and lipids from autotrophic, mixotrophic and heterotrophic cultivation of Galdieria sp. strain USBA-GBX-832. Sci Rep 9(1):10791. https://doi.org/10.1038/s41598-019-46645-3
Lum KK, Kim J, Lei XG (2013) Dual potential of microalgae as a sustainable biofuel feedstock and animal feed. J Anim Sci Biotechnol 4(1):53. http://www.jasbsci.com/content/4/1/53
Madeira MS, Cardoso C, Lopes PA, Coelho D, Afonso C, Bandarra NM, Prates JAM (2017) Microalgae as feed ingredients for livestock production and meat quality: A review. Livestock Science 205:111–121. https://doi.org/10.1016/J.LIVSCI.2017.09.020
Malcata FX (2011) Microalgae and biofuels: a promising partnership? Trends Biotechnol 29(11):542–549. https://doi.org/10.1016/j.tibtech.2011.05.005
Markou G, Angelidaki I, Georgakakis D (2012) Microalgal carbohydrates: an overview of the factors influencing carbohydrates production, and of main bioconversion technologies for production of biofuels. Appl Microbiol Biotechnol 96(3):631–645. https://doi.org/10.1007/s00253-012-4398-0
Moran CA, Morlacchini M, Keegan JD, Fusconi G (2019) Increasing the omega-3 content of hen's eggs through dietary supplementation with Aurantiochytrium limacinum microalgae: effect of inclusion rate on the temporal pattern of docosahexaenoic acid enrichment, efficiency of transfer, and egg characteristics. J Appl Poult Res 28(2):75. https://doi.org/10.3382/japr/pfy075

Pierre G, Delattre C, Laroche C, Michaud P (2015) Galactans and its applications. In: Polysaccharides: bioactivity and biotechnology. Springer, Cham. https://doi.org/10.1007/978-3-319-16298-0_69
Samarakoon K, Jeon YJ (2012) Bio-functionalities of proteins derived from marine algae - a review. Food Res Int 48(2):948–960. https://doi.org/10.1016/j.foodres.2012.03.013
Santos JEP, Bilby TR, Thatcher WW, Staples CR, Silvestre FT (2008) Long chain fatty acids of diet as factors influencing reproduction in cattle. Reprod Domest Anim 43(2):23–30. https://doi.org/10.1111/j.1439-0531.2008.01139.x
Sarker PK, Gamble MM, Kelson S, Kapuscinski AR (2016) Nile tilapia (Oreochromis niloticus) show high digestibility of lipid and fatty acids from marine Schizochytrium sp. and of protein and essential amino acids from freshwater Spirulina sp. feed ingredients. Aquac Nutr 22(1):12230. https://doi.org/10.1111/anu.12230
Scott TW, Cook LJ, Mills SC (1971) Protection of dietary polyunsaturated fatty acids against microbial hydrogenation in ruminants. J Am Oil Chem Soc 48(7):358–364. https://doi.org/10.1007/BF02890762
Seyidoglu N, Inan S, Aydin C (2017) A prominent superfood: spirulina platensis. In: Superfood and functional food - the development of superfoods and their roles as medicine. IntechOpen, London. https://doi.org/10.5772/66118
Shahid A, Malik S, Zhu H, Xu J, Nawaz MZ, Nawaz S, Asraful Alam M, Mehmood MA (2020) Cultivating microalgae in wastewater for biomass production, pollutant removal, and atmospheric carbon mitigation; a review. Sci Total Environ 704:135303. https://doi.org/10.1016/j.scitotenv.2019.135303
Shingfield KJ, Ahvenjärvi S, Toivonen V, Vanhatalo A, Huhtanen P, Griinari JM (2008) Effect of incremental levels of sunflower-seed oil in the diet on ruminal lipid metabolism in lactating cows. Br J Nutr 99(5):971–983. https://doi.org/10.1017/S0007114507853323
Skrede A, Mydland LT, Ahlstrem O, Reitan KI, Gislered HR, Overland M (2011) Evaluation of microalgae as sources of digestible nutrients for monogastric animals. J Anim Feed Sci 20(1):2011. https://doi.org/10.22358/jafs/66164/2011
Slegers PM, van Beveren PJM, Wijffels RH, Van Straten G, Van Boxtel AJB (2013) Scenario analysis of large scale algae production in tubular photobioreactors. Appl Energy 105:68. https://doi.org/10.1016/j.apenergy.2012.12.068
Stamey JA, Shepherd DM, de Veth MJ, Corl BA (2012) Use of algae or algal oil rich in n-3 fatty acids as a feed supplement for dairy cattle. J Dairy Sci 95(9):5412. https://doi.org/10.3168/jds.2012-5412
Tacon AGJ (2020) Trends in Global Aquaculture and Aquafeed Production: 2000–2017. Reviews in Fisheries Science and Aquaculture 28(1):43–56. https://doi.org/10.1080/23308249.2019.1649634/ASSET//CMS/ASSET/4E33B014-662E-4555-84BE-4FD2E53F9DA4/23308249.2019.1649634.FP.PNG
Tan LT (2007) Bioactive natural products from marine cyanobacteria for drug discovery. Phytochemistry 68(7):12. https://doi.org/10.1016/j.phytochem.2007.01.012
Tibbetts SM (2018) The potential for 'next-generation', microalgae-based feed ingredients for salmonid aquaculture in context of the blue revolution. In: Microalgal biotechnology. IntechOpen, London. https://doi.org/10.5772/intechopen.73551
Tibbetts SM, Milley JE, Lall SP (2015) Chemical composition and nutritional properties of freshwater and marine microalgal biomass cultured in photobioreactors. J Appl Phycol 27(3):1109–1119. https://doi.org/10.1007/s10811-014-0428-x
Toyomizu M, Sato K, Taroda H, Kato T, Akiba Y (2001) Effects of dietary Spirulina on meat colour in muscle of broiler chickens. Br Poult Sci 42(2):197–202. https://doi.org/10.1080/00071660120048447
Tredici MR, Biondi N, Ponis E, Rodolfi L, Chini Zittelli G (2009) Advances in microalgal culture for aquaculture feed and other uses. New Technologies in Aquaculture: Improving Production Efficiency, Quality and Environmental Management 610–676. https://doi.org/10.1533/9781845696474.3.610

Udayan A, Arumugam M, Pandey A (2017) Nutraceuticals from algae and cyanobacteria. In: Algal green chemistry: recent progress in biotechnology. Elsevier, Amsterdam. https://doi.org/10.1016/B978-0-444-63784-0.00004-7

Wachira AM, Sinclair LA, Wilkinson RG, Enser M, Wood JD, Fisher AV (2002) Effects of dietary fat source and breed on the carcass composition, n-3 polyunsaturated fatty acid and conjugated linoleic acid content of sheep meat and adipose tissue. Br J Nutr 88(6):697–709. https://doi.org/10.1079/bjn2002727

Wallace JMW, Mccabe AJ, Robson PJ, Keogh MK, Murray CA, Kelly PM, Márquez-Ruiz G, McGlynn H, Gilmore WS, Strain JJ (2000) Bioavailability of n-3 polyunsaturated fatty acids (PUFA) in foods enriched with microencapsulated fish oil. Ann Nutr Metab 44(4):157–162. https://doi.org/10.1159/000012839

Wiaętkiewicz S, Arczewska-Włosek A, Józefiak D (2015) Application of microalgae biomass in poultry nutrition. Worlds Poult Sci J 71(4):663–672. https://doi.org/10.1017/S0043933915002457

Yaakob Z, Ali E, Zainal A, Mohamad M, Takriff M (2015) 2016 Biomolecules From Microalgae for Animals.Pdf. Journal of Biological Research-Thessaloniki 21(1):1–10. http://jbiolres.biomedcentral.com/articles/10.1186/2241-5793-21-6

Promising Microalgae for Nutraceutical and Food Applications: Solution for Global Food Problems

16

Naziya Tarannum, Shailesh Kumar Patidar, and Nivedita Chaudhary

Contents

Abstract

Microalgae offer a sustainable solution to food security and nutrition by producing a varied collection of high-value bioactive compounds, such as, single-cell proteins, polyunsaturated fatty acids, glycolipids, phycobiliproteins, and pigments. These compounds, which comprise both primary and secondary metabolites (carotenoids, PUFAs like EPA and DHA, polypeptides, and proteins), present a lucrative opportunity for biorefinery optimization, allowing for the extraction of multiple valuable products from the same algal biomass. This chap-

N. Tarannum · S. K. Patidar (✉) · N. Chaudhary
Department of Environmental Science, School of Earth Sciences, Central University of Rajasthan, Ajmer, Rajasthan, India

S. K. Arya et al. (eds.), *Value Added Products From Bioalgae Based Biorefineries: Opportunities and Challenges*, https://doi.org/10.1007/978-981-97-1662-3_16

ter dives into the world of high-value microalgal compounds, exploring factors influencing their production and their potential as nutraceuticals for human health.

Keywords

Microalgae · Nutraceutical · PUFAs · Antioxidants · Carotenoids

Abbreviations

ALA	α-Linolenic acid
CFTRI	Centre Food Technology Research Institute
CSIR	Council of Scientific and Industrial
DHA	Docosahexaenoic acid
DPA	Docosapentaenoic acid
EAA	Essential amino acid
EPA	Eicosapentaenoic acid
LA	Linoleic acid
PUFA	Polyunsaturated fatty acids
RDA	Required daily allowance
ROS	Reactive oxygen species
SDA	Stearidonic acid
WHO	World Health Organization

16.1 Introduction

Microalgae are photosynthetic microorganisms classified as prokaryotes or eukaryotes. They can be unicellular, multicellular, filamentous, siphonaceous, and colonial (Pereira et al. 2020). Microalgae, which constitute about 200,000 species (Norton et al. 1996), account for 40% of global photosynthesis (Saide et al. 2021). While microalgae share similar photosynthetic processes to terrestrial plants, their efficiency in converting sunlight into biomass through photoautotrophy is significantly higher. This advantage stems from their simpler cellular structure, aquatic habitat, and easy access to CO_2 and nutrients (Habib and Ahsan 2008). Microalgae are considered a varied collection of cryptogamic plants that include 13 main phyla and numerous smaller subgroups that are yet to be understood (Reynolds 2006). Cyanobacteria are oxygenic photosynthetic prokaryotes with an extensive range of morphological, ecological, physiological, and biochemical features among several microalgae phyla. Cryptophyta have uni- and multicellular, filamentous, siphonous, and mostly found in freshwater, brackish, and marine system; Chlorophyta are filamentous, siphonous, multicellular, and primarily aquatic algae; dinophyta are single-celled, with two distinct flagella, and most of them survive in fresh, brackish, and sea waters (Begum et al. 2016).

Microalgae hold immense potential for biopigment production due to (1) their unique ability to flourish on non-arable land, alleviating the pressure on arable land resources (Pathak et al. 2018); (2) they can also be cultivated under open and closed systems and comparatively robust under varied seasonal conditions and harvesting can be managed through various methods (Varshney et al. 2015). Therefore, due to the brief life span and primary structural and cellular organization, microalgae are more accessible to cultivate, handle, and control according to the physicochemical requirements (Saini et al. 2020). Further, they can grow in diverse habitats and even thrive in harsh environments with variations in temperatures, pH, and salinity. Compared to marine plants and macro-organisms, their development is very easy, with rapid rate of growth after bacteria. These are also ecologically sound and environmentally sustainable due to their unmatched ecological services. Microalgae are surprisingly low-maintenance. They thrive with just a small amount of growth and development nutrients, either brackish/marine water (or even freshwater), and some sunshine or low-energy light (Patidar and Mishra 2017; Saha and Murray 2018). Popular commercially cultivated species like *Chlorella vulgaris*, *Arthrospira maxima*, *Arthrospira platensis,* and *Haematococcus pluvialis* are all photoautotrophic, meaning they use CO_2 as their carbon source, which can be provided through gassing or adding soluble carbonates and bicarbonates to the water (Kratzer and Murkovic 2021). Interestingly, some microalgae, like *Crypthecodinium cohnii* and *Schizochytrium* sp., can skip the sunshine altogether and grow heterotrophically using organic carbon sources (Karnaouri et al. 2020).

Microalgae occupy a preeminent position in the production of diverse pigments, including chlorophylls (a, b, c), astaxanthin, xanthophylls, phycobiliproteins, and β-carotene. These pigments exhibit a remarkable spectrum of applications, encompassing synthetic food coloring, cosmetic formulations, nutraceutical supplements, and pharmaceutical preparations. Their synthesis is demonstrably influenced by environmental factors such as salinity, pH, nutrient accessibility, temperature, and intensity of light (Patidar et al. 2014; Patidar and Mishra 2017). β carotene, zeaxanthin, lutein, lycopene, chlorophyll, and phycobiliproteins are just a few of the pigments that may be synthesized under optimum environmental and culture conditions with improved productivity. Some pigments (carotenoids) are required for microalgae development because they guard against reactive oxygen species (ROS) and intense light (Saini et al. 2020). Microalgae produce polyunsaturated fatty acids (PUFA) with numerous health benefits and applications; for instance, it is a significant source of omega-3 PUFA (dietary), abundant in seafood. The focus of recent microalgae research has shifted towards fatty acids with major health benefits, particularly eicosapentaenoic acid (EPA) and docosahexaenoic acid (DHA). Additionally, less explored PUFAs like docosapentaenoic acid (DPA) and stearidonic acid (SDA) are gaining attention due to their potential to alleviate disease-causing physiological processes. PUFAs are also being ameliorated and investigated for the expression of gene and cytokine, neurodevelopment, eicosanoid synthesis, and gut bacteria control, causing a substantial impact on human health (Robertson et al. 2013). Microalgae-derived protein sources and micronutrient-rich supplements could significantly improve dietary diversity and combat malnutrition in resource-limited countries.

It will also address challenges such as climate change and, perhaps, the food requirements of the world's growing population. Climate change poses a serious concern to the existence of many plant and animal species, and people. Microalgae's potential to revolutionize biofuels is rooted in their remarkable capabilities. These microalgae are ideal for commercial cultivation in optimized or suitable media to produce optimum biomass and lipids. They may thrive and multiply cell numbers even under stressed environmental conditions, such as in saline or marine waterways and various types of wastewater, and can be grown in photo-bioreactors or salt pans utilizing barren lands (where crops cannot be grown easily). All they need to flourish is sunlight, nutrients, and CO_2 (Chisti 2013; Saifullah et al. 2014). Microalgae-based biological carbon capture and sequestration have been regarded as one of the world's most essential and successful carbon sequestration approaches. As part of a climate change mitigation approach, these microbes are also being considered as a potential carbon stock (Greene et al. 2016).

Microalgae offer a powerful solution for sustainable development. They act as natural CO_2 capturers, converting it into biomass. This biomass, rich in carbon, serves as a valuable resource for bioenergy production through lipid extraction. Beyond biofuels, microalgae hold immense potential in creating various high-value products. Their versatility extends beyond energy, contributing to food security by combating malnutrition and treating wastewater. Moreover, cultivating microalgae creates jobs and boosts economic growth. This makes them especially appealing in metropolitan areas, where reducing pollution and minimizing waste are top priorities. Microalgae have ability to fix carbon about 10–50 times greater than other plants, can harvest 100% of their biomass, can grow on salt and wastewater, and not require fertile soil for growth and development (Chisti 2007). Therefore, they are not competing with production of food with land utilization resources. Their biomass may then be used in biofuels, nutrition for humans, pharmaceuticals, cosmetics, animal feeds, and fertilizers (Merlo et al. 2021). This chapter provides insight into extracted compounds from microalgae and explores their applicability primarily as nutraceuticals.

16.2 Microalgae: Nutraceuticals/Food

Microalgae are being hailed as a revolutionary natural source of bioactive compounds and vitamins for nutraceutical and food/dietary applications, with promising pharmaceutical potential (Saha et al. 2015; Del Mondo et al. 2020). Dubbed a "treasure trove" for their diverse health benefits, these microscopic marvels produce a vast array of compounds with numerous therapeutic properties. From anti-inflammatory and antioxidant to anti-cancer and antimicrobial, microalgae offer a natural arsenal beside a wide range of diseases, including diabetes, tuberculosis, epilepsy, hypertension, and osteoporosis (Lauritano and Ianora 2016; Lauritano et al. 2016, 2018; Brillatz et al. 2018; Martínez Andrade et al. 2018; Giordano et al. 2018; Elagoz et al. 2020). Their antiviral, antifungal, and immunomodulatory activities further highlight their potential, even extending to areas like hepatic fibrosis inhibition and enzyme regulation (Riccio and Lauritano 2020). Humans have been consuming microalgae for long period of time and adding with microalgae for stress alleviating on very source-intensive terrestrial crops (Koyande et al. 2019). Several beneficial chemicals have been identified and isolated from marine microalgae, including *marennine* (Pouvreau et al. 2008), sulfated polysaccharides (Raposo et al. 2013), carotenoids (fucoxanthin, astaxanthin, β-carotene) (Peng et al. 2011; Gammone et al. 2015), polyphenols (Goiris et al. 2012), and omega-3 fatty acids (Haimeur et al. 2012). The biological activity of several of these metabolites is established as a substantial antioxidant against inflammation, cancer, and viruses (Barkia et al. 2019). Also, microalgae and its derived products offer much potential as dietary supplements to prevent and manage physiological therapy anomalies and provide sustainable natural resources which are replaceable to synthesized dietary supplements (Beetul et al. 2016). *Chlorella vulgaris* the first commercially cultivated microalgae achieved commercial viability in Japan during the 1960s (Mobin and Alam 2017). Figure 16.1 illustrates the versatility of microalgae across various fields, including their nutraceutical value.

Compared to traditional dietary sources, microalgae comprise vitamins (A, E, B1) and folic acid in high concentration. According to Koyande et al. (2019), *Dunaliella tertiolecta* was shown to synthesize cobalamin, riboflavin, tocopherol, and β-carotene. *Tetraselmis suecica* having high concentration of thiamin, nicotinic acid, Vitamin B_5 (thenic acid), pyridoxine, and ascorbic acid (Fabregas and Herrero 1990). *Chlorella* spp. produce biotin in abundance (Watanabe et al. 2002), and around 9% to 8% of *Chlorella* sp. have a high vitamin B12 (Islam et al. 2017).

16.2.1 Challenges Associated with Incorporating Microalgae into Everyday Diets, Such as Taste, Processing Methods, and Potential Side Effects

In today's demanding environment with a large working population, maintaining a balanced diet is crucial for managing stress and supporting the body's internal processes. Microalgae products emerge as a promising alternative to traditional food sources, boasting a wealth of nutrients, vitamins, and minerals. Diverse species like *Spirulina, Chlorella, Dunaliella, Haematococcus*, and *Schizochytrium* find their way into various food applications. *Chlorella,* for instance, enriches yogurts, cheeses, and even lends a vibrant color to cookies. *Arthrospira* shines in snacks, pastas, and even spaghetti, gaining widespread acceptance. Their versatility extends further, adding a healthy twist to candies, gums, snacks, pastries, noodles, cereal flour, and even beverages like wine (Caporgno and Mathys 2018). While the high purity of algae-biomass makes it attractive for feed products and the food-processing industry, extracting its valuable assets like lipids and minerals remains a cost-prohibitive challenge. Existing techniques such as mechanical, chemical, and enzymatic methods often come with high operational costs (Musa et al. 2019). Additionally, the presence of certain pigments can negatively impact the taste, color, and flavor of microalgae-based food supplements. Moreover, some microalgae due

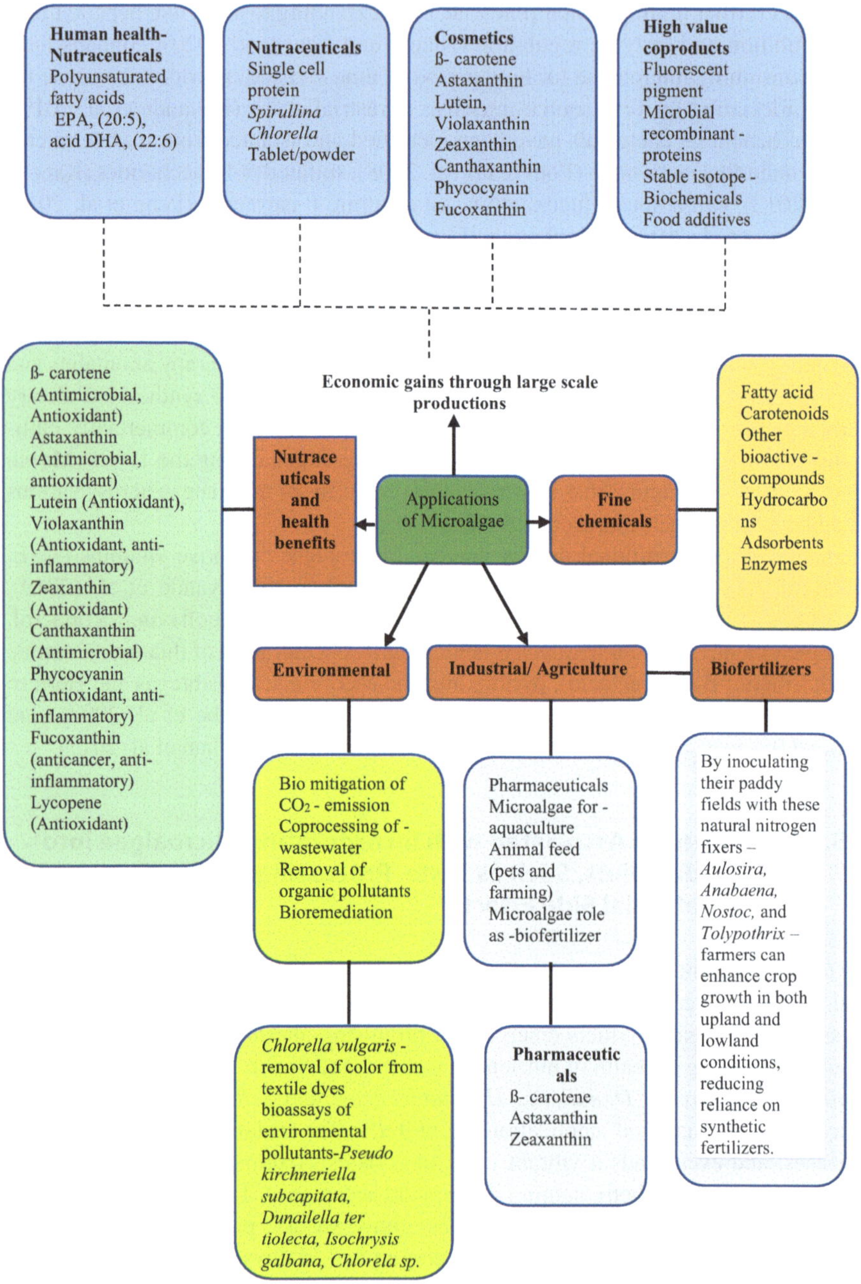

Fig. 16.1 Flowchart depicting functions of microalgae in several fields and microalgal nutraceuticals (Spolaore et al. 2006; Begum et al. 2016; Jha et al. 2017)

to the contamination of harmful cyanobacteria produce harmful toxins like microcystin, anatoxin, and saxitoxin, which have been linked to health problems like gout, kidney stones, and even renal failure in severe cases (Ampofo and Abbey 2022). For instance, *Chlorella* has been associated with renal failure, while excessive consumption of *Arthrospira* can lead to diarrhea, nausea, and vomiting. These challenges highlight the need for further research and development in efficient and safe extraction methods to unlock the full potential of algae-biomass.

16.3 Role of Microalgae in Human Health

In today's fast-paced world, people in many industrialized countries tend to make unhealthy food choices due to busy schedules. These choices can have serious health consequences, for instance, high blood pressure, obesity, diabetes, and heart-related disease. To counteract these risks and promote overall well-being, a healthy lifestyle must prioritize a balanced diet rich in essential nutrients like vitamins, antioxidants, and PUFAs. Microalgae are a treasure trove of valuable bioactive compounds. These include: Pigments: β-carotene, chlorophyll, lutein, phycocyanin, phycoerythrin, astaxanthin, and allophycocyanin (responsible for vibrant colors and potential health benefits). Antioxidants: ascorbate, peroxidase, superoxide dismutase, glutathione, and ergothioneine (protect cells from damage). Other beneficial compounds: polyphenols, α-tocopherols, and PUFAs (important for various bodily functions). Microalgae are also abundant in vitamins (A, B-complex, C and E) and some minerals (K, Fe, Ca, Mg, and I) (Becker 2004). As a result, microalgae began to gain popularity as a source of health benefits depicted in Fig. 16.1. Saha et al. (2015) reported several cyanobacteria and microalgae, for instance, *Chlorella vulgaris, Arthrospira (Spirulina) platensis, Arthrospira (Spirulina) maxima, Haematococcus pluvialis, Aphanizomenon flos-aquae, Nannochloropsis oculata,* and *Dunaliella salina,* can be economically cultivated for a variety of applications having commercial importance as nutraceuticals and food applications are listed in Table 16.1.

Spirulina is known as a "superfood" (Farag et al. 2016) among microalgae because of its high vitamin contents (A, B-complex, E, K) and macromolecular compositions of proteins and other biologically active substances (Farag et al. 2016; Jung et al. 2019). In terms of β-carotene and vitamin B12, 1 g of commercial *Spirulina platensis* powder provides up to half of the required daily allowance (RDA) (Vonshak and Guy 1992). Among the health benefits of high concentration in thiamine, folic acid, riboflavin, and biotin, *Chlorella pyrenoidosa* powder decreased proteinuria, anemia, and edema in pregnancy (Nakano et al. 2010; Del Mondo et al. 2020). For an average person's RDA of vitamin B1, B2, and B3, a spoonful of dried *Spirulina* sp. contains about 4 g of protein and 1 g of fat (containing PUFAs; omega-3 and -6 fatty acids). It also contains copper and iron, which account for almost 21% and 11% of RDA, respectively. Additional minerals such as Mn, Mg, and K have also been found in moderate levels (Koyande et al. 2019). According to Puyfoulhoux et al. (2001), *Spirulina* sp. is an abundant source of Fe

Table 16.1 Microalgae-based nutraceuticals: sources, benefits, and companies

Name of the microalgae	Growth type	Year of cultivation	Cultivation conditions	Targeted end products	Isolation/extraction strategies	Application	Industry/company	References
Chlorella vulgaris *Chlorella* sp.	Photoautotrophic	1951	Open raceway pond, tubular PBR, flat-plate photobioreactor	Biomass, pigments	Microwave-supported extraction	Health and food supplement	Nikken Sohonsha Corp. (Japan); Chlorella manufacturing and Co. (Taiwan); Klötze (Germany); Ocean Nutrition (Canada)	Spolaore et al. (2006), Pulz and Gross (2004), Saha and Murray (2018)
Crypthecodinium cohnii	Heterotrophic, non-photosynthetic	1999	Large stainless steel fermentor	DHA	Microwave	Infant formula	Martek (USA)	Mendes et al. (2009), Saha and Murray (2018)
Dunaliella salina	Photoautotrophic (two phase cultivation)	1980	Paddle wheel stimulated raceway ponds, unstirred open pond, lagoons, photobioreactors	β-carotene	Microwave and solvent extraction	Health and diet supplements, bio-colorant	Earthrise Nutritionals and Cyanotech (USA); Nature Beta Technologies Cognis (Israel); Betadene and Nature Beta Technologies Cognis (Australia)	Pulz and Gross (2004), Dufossé et al. (2005), Saha and Murray (2018)
Euglena gracilis				Biomass	Flocculation method	Health supplement food	Euglena (Japan)	Sathasivam et al. (2019)
Haematococcus pluvialis	Photoautotrophic (two phase growing), Mixotrophic	2000	Open raceway pond, bubble column, tubular bounded outdoor PBR, and airlift photobioreactors	Astaxanthin	Chemical and mechanical treatments by solvents, pressurized abstraction, ultrasounds, and microwaves	Food complemented, bio-colorant	Cyanotech Corporation (USA); Aquasearch Algatechnologies (Israel)	Pulz and Gross (2004), Molino et al. (2018a, b), Denery et al. (2004), Haque et al. (2016), Olaizola (2000), Saha and Murray (2018)

Nannochloropsis oculata	Photoautotrophic, mixotrophic	1997	Raceway pond, helical-tubular photobioreactor	Omega-3 PUFA	Solvent extraction; microwave	Omega-3 source	Cleanalgae SL (Spain); Astaxa (Germany)	Kagan et al. (2013, 2015), Saha and Murray (2018)
Odontella aurita	Photoautotrophic, heterotrophic or mixotrophic	1996	Open outdoor ponds, small tanks, cylindrical glass columns and flat-plate photobioreactors	Fatty acids		Health supplement	InnovalG (France)	Pulz and Gross (2004), Saha and Murray (2018)
Porphyridium spp.	Photoautotrophic	1996	Tubular PBR	Polysaccharides	Ultra-filtration	Food additives, nutrition	InnovalG (France)	Pulz and Gross (2004), Li et al. (2019), Marcati et al. (2014), Saha and Murray (2018)
Schizochytrium sp.	Heterotrophic	1999	Large stainless steel fermentor	DHA	Cell hydrolysis	Dietary supplement; health food	OmegaTech (USA)	Spolaore et al. (2006), Saha and Murray (2018)
Spirulina platensis	Photoautotrophic	1970	Open raceway pond, tanks, earthen	Algae protein	Enzymatic hydrolysis, physical processes, chemical abstraction, pulsed electric field, ultrasound-supported extraction, and microwave-assisted extraction	Health supplement, health food, infant food formula	Parry Nutraceuticals (India); Japan Spirulina Co., Ltd. (Japan); Siam Alga Co., Ltd. (Thailand); Saxony-Anhalt (Germany)	Spolaore et al. (2006), Bioeconomy (2020), Pulz and Gross (2004), Saha and Murray (2018)

(continued)

Table 16.1 (continued)

Spirulina platensis	Photoautotrophic	1970	Open raceway pond, tanks, earthen	Vitamin 12	Water extraction procedure	Diet supplement	Myanmar Spirulina Factory (Myanmar)	Pulz and Gross (2004), Kumudha et al. (2015), Saha and Murray (2018)
Spirulina platensis	Photoautotrophic	1970	Open raceway pond, tanks, earthen	Phycocyanin	Microwave-supported extraction	Bio-colorant, diet supplement	Yunnan Green A Biological Project Co., Ltd. (China) Panmol/Madaus (Austria)	Pulz and Gross (2004), Saha and Murray (2018)
Ulkenia sp.	Heterotrophic			DHA	Solvent extraction and microwave	Diet supplement	Nutrinova (Germany)	Spolaore et al. (2006)

for human intake. Another commercial species is rich in an important amino acid, proteins, antioxidants, vitamins B-complex, ascorbic acid, and minerals in *Chlorella* sp. (K, Na, Mg, and Fe). According to the WHO, *Spirulina* spp. should be included in the diet of NASA astronauts because it is an excellent and compact meal for space flight. Even when taken in smaller quantities, it provides a broad spectrum of nutrients (Khan et al. 2005). *Chlorella* sp. has been proposed as the preeminent option for a vegan source of vitamin B12, and it has been utilized in dietary supplements (Balasubramaniam et al. 2021). *Chlorella* sp. purchasing have increased as people inclining towards a healthy, nutritionally balanced diet. Globally, *Chlorella* sp. market was worth USD138 million in 2016 and was anticipated to be worth USD 164 million by the end of the current year 2021 (GCMI 2018).

16.4 Bioactive Compounds of Microalgae Origin

Presently, most microalgae and cyanobacteria production are intended for high-value applications due to pigments, proteins, PUFAs, vitamins, polysaccharides, and minerals. They are utilized in the formulation of natural products, which function in the food, pharmaceutical industries, and cosmetics (López and Soto 2020). Table 16.2 displays the bioactive compounds extracted from microalgae and their functions. Therefore, biochemicals produced from microalgae are employed in culinary and medical studies.

16.4.1 Microalgal Pigments

Pigments, which are highly visible light-absorbing chemicals and have fluorescent properties found in microalgae, have a potential market due to multiple applications. Major natural sources of pigments are vegetables, fruits, flowers, and photosynthetic microorganisms (microalgae and cyanobacteria) (Nakazawa et al. 2009). Pigments with complex conjugated ring structures can absorb much light in the visible spectrum. Furthermore, the conjugated double bonds allow for resonance, resulting in energy level changes, which is critical for energy transduction during photosynthesis. Based on their chemical backbones, algal pigments may be divided into tetrapyrrole pigments and isoprene. Tetrapyrrole rings make up chlorophyll and PBPs, while isoprene units make carotenoids (Chakdar and Pabbi 2017). Microalgal photosynthetic pigments comprise three main groups: chlorophylls (a, b, and c), lipid-soluble molecules, and water-soluble phycobilins. Carotenoids (Chlorophyll c) astaxanthin, and β-carotene from microalgae, are also widely used in food, supplements, functional components, and cosmetic preparations due to colored isoprenoid structured lipophilic pigments. *Haematococcus pluvialis* has far more potent antioxidant characteristics than β-carotene and vitamin E; also, it is a well-known microalga that is the abundant natural resource of the industrially produced carotenoid (astaxanthin) (Saha and Murray 2018). Pigments, with their potent antioxidant properties, act as a therapeutic shield against oxidative and free-radical

Table 16.2 Role of bioactive compounds

Bioactive compounds/stress metabolites	Name of the microalgae	Applications	References
Methanolic extracts of exopolysaccharides	*Arthrospira platensis*	Antioxidant	Raposo et al. (2013), López and Soto (2020)
Linear alkadienes, triene	*Botryococcus braunii*	Phycoremediation	Gani et al. (2017)
Zeaxanthin (b), violaxanthin	*Chlorella ellipsoidea*	For healthiness and cosmetics as UV protection, antioxidant and antibiofilm	López and Soto (2020)
Eicosapentaenoic acid (EPA)	*Chlorella minutissima*	Food supplements	Neumann et al. (2018)
Lutein, canthaxanthin, zeaxanthin	*Chlorella protothecoides*	Healthiness and cosmetic as UV protection	de Jesus Raposo et al. (2015), Sun et al. (2015)
Lutein, sulfated polysaccharide	*Chlorella pyrenoidosa*	For health and cosmetics as UV protection, antioxidant	Sun et al. (2015), Plaza et al. (2009)
Carotenoids, sulfated polysaccharides, sterols, fatty acids, PUFAs, chlorophyll	*Chlorella* sp.	Moisturizing and thickener and deodorants, antimicrobial, antibiofilm	Ibañez and Cifuentes (2013), Tang and Suter (2011), Jain et al. (2005)
Canthaxanthin, peptide, astaxanthin, oleic acid, diethyl phthalate (c), trimethyl (4-tertbutyl phenoxy) silane (d), *chlorella vulgaris* extracts	*Chlorella vulgaris*	Antioxidant, antibiofilm, antimicrobial, anti-ageing	Gayatri et al. (2019), Xin et al. (2017)
Astaxanthin	*Chlorella zofingiensis*	Health related and cosmetics as UV protection, antioxidants	Xin et al. (2017)
β-carotene, *trans*-betacarotene, *cis*-betacarotene, oleic acid, linolenic (e) acid, β-Cryptoxanthin palmitic acid (f), and glucosyltransferases (GTF) (g)	*Dunaliella salina*	Health and cosmetics as UV protection. Anti-inflammatory, antibacterial, and antibiofilm	Tang and Suter (2011), Bhattacharjee (2016)
Diacylglycerols	*Dunaliella* sp.	Acylation stimulating protein	Liu et al. (2014)
Astaxanthin, lutein, zeaxanthin, canthaxanthin, lutein, β-carotene, oleic acid	*Haematococcus pluvialis*	Required for health and cosmetic as UV protection	Molino et al. (2018a, b)
Boromycin, cryptophycin	*Nostoc linckia*/ *Nostoc spongiaeforme*	Antitumor compounds	Bajpai et al. (2018)

(continued)

Table 16.2 (continued)

Bioactive compounds/stress metabolites	Name of the microalgae	Applications	References
Oscillatoria sp. extract	*Oscillatoria* sp.	Antimicrobial, antioxidant, antibiofilm	Xin et al. (2017)
Diacylglycerols	*Spirulina fusiformis*	Acylation stimulating protein	Xin et al. (2017)
Phycocyanin, tocopherols, phenolic acids, neophytadiene, phytol, PUFAs (n-3) fatty acids, linolenic acid, palmitoleic acid, oleic acid	*Spirulina platensis*	Food, health and cosmetics, antibiofilm, antimicrobial	LewisOscar et al. (2017)
Polysaccharides	*Spirulina* sp.	For dietary and use in cosmetics manufacturing	Mourelle et al. (2017)

damage in humans and various animal species (Sathasivam et al. 2019). The worldwide carotenoids usage market was expected to grow to \$1.53 billion by 2021, although most of the microalgae established carotenoids manufacturing is based on chemical raw materials (Kobylewski and Jacobson 2010). Commercially pigments and fatty acids are valuable microalgal products with high market demand (Minhas et al. 2016). Carotene was produced around 1200 tons in 2010 (Enzing et al. 2014); derived from *Dunaliella salina*, it grew in open ponds with elevated salinity and intense light (Coesel et al. 2008). *Haematococcus pluvialis*, a freshwater green alga, produces another economically significant carotenoid, astaxanthin (Shah et al. 2016).

Astaxanthin is made in lower quantities than β-carotene (Panis and Carreon 2016). However, due to increased manufacturing expenses and the deficient of a prevailing market for astaxanthin as a nutraceutical for humans, its commercialization has been slower (Borowitzka 2013). *Dunaliella salina* can generate about 13% dry weight containing β-carotene; *Haematococcus pluvialis*, which generates up to 7% of the dry weight, contains astaxanthin; Canthaxanthin was obtained in abundance in *Coelastrella striolata*, accounting for about 5% of the dry weight. Different species also include the production of lutein, fucoxanthin, zeaxanthin, violaxanthin, and echinenone (Raposo et al. 2015), as shown in Table 16.3. β-carotene is a tetraterpenoid with a fundamental assembly of conjugated double bonds that stand in for 2 -ionone rings and 40 carbon atoms. At both the ionone end groups, 3-hydroxilation and 4-ketolation are used to make astaxanthin from β-carotene by catalyzing these processes β-carotene hydroxylase and ketolase, respectively. Higher plants are capable of hydroxylation, while ketolation is only seen in limited bacteria, fungus, and unicellular green algae (Saide et al. 2021). Orange-colored natural xanthophyll derivative of β-carotene is violaxanthin's exemption of four marginal groups: two-epoxy and two hydroxy. The polar groups are linked to the -ionone ring on two edges of respective molecule. Lutein is a yellow-colored xanthophyll pigment with a 40-carbon long-chain structure (Low et al. 2020). The chemical structures of these

Table 16.3 Microalgae-pigments and its applications

Pigments	Molecular Weight (g/mol)	Microalgae (strains)	Applications	References
Astaxanthin ($C_{40}H_{52}O_4$)	596.84	*Chlorella vulgaris, Ankistrodesmus braunii, Chlamydomonas nivalis, Chondria striolata, Haematococcus lacustris, Haematococcus pluvialis, Monoraphidium* sp., *Scenedesmus obliquus, Chlorococcus* spp.	Malignance defense, infection, diabetes, neuro-degenerative and optical disorders, reduced alveolar wall inflammation and enzyme myeloperoxidase activity, UV protectant, anti-oxidants, induce immune system, anti-aging characteristics, pharmaceutical, nutraceutical, cosmetic	Chakdar and Pabbi (2017), Cai et al. (2019), Capelli et al. (2019), Chang et al. (2010), Galasso et al. (2019), Park et al. (2010), Wu et al. (2014), Ambati et al. (2014), Talukdar et al. (2020)
Canthaxanthin ($C_{40}H_{52}O_2$)	564.82	*Chlamydocapsa* sp., *Chlorococcum* sp., *Chlamydomonas nivalis, Chlorella vulgaris, Chlorella zofingiensis, Neospongiococcum* sp., *Scenedesmus* spp.	Enhance vitamin E, anti-oxidative, anti-inflammatory, and neuroprotective characteristics, used as an additional to poultry feedstuff, preclude some blood-associated diseases	Chakdar and Pabbi (2017), Sathasivam et al. (2019), Kiran and Venkata Mohan (2021)
Chlorophylls ($C_{55}H_{72}N_4O_5Mg$) ($C_{55}H_{70}N_4O_6Mg$) ($C_{35}H_{30}N_4O_5Mg$) ($C_{54}H_{70}N_4O_6Mg$)	Chl a 892 Chl b 906 Chl c 610 Chl d 894	Found in all photosynthetic organism *Chlorella* sp., green algae	Consumed in the food manufacturing as a natural constituent of pigment, antioxidant property, cytotoxic regarding tumoral cells, immune stimulator	Singh et al. (2015), Khanra et al. (2018), Koller et al. (2014), Mishra et al. (2011), Odjadjare et al. (2017)

(continued)

Table 16.3 (continued)

Pigments	Molecular Weight (g/mol)	Microalgae (strains)	Applications	References
Fucoxanthin ($C_{42}H_{58}O_6$) and Zeaxanthin ($C_{40}H_{56}O_2$)	658.91 568.87	*Chlamydomonas nivalis, Chaetoceros gracilis, Dunaliella salina, Isochrysis* sp., *Ochromonas* sp., *Odontella aurita, Phaeodactylum tricornutum, Prymnesium parvum, Salpingoeca marina, Scenedesmus almeriensis, Nannochloropsis oculata*	Anti-oxidant, anti-viral, against obesity, diabetes, cancer, osteoporotic, inflammatory, anti-proliferative, rheumatoid arthritis, and diabetes controlling nutraceutical	McClure et al. (2018), Sathasivam and Ki (2018), Liu et al. (2017)
Lutein ($C_{40}H_{56}O_2$)	568.87	*Chlorella fusca, C. sorokiniana, Ankistrodesmus braunii, Chlamydomonas acidophila, Chlorococcum* sp., *Scenedesmus obliquus Muriellopsis, Tetraselmis suecica*	Antioxidative properties and anti-cancerous activity, protects macular disintegration, diabetic retinopathy, and age related retinal degeneration, regulation, cardiovascular diseases, cancers, cognitive function	Blanco et al. (2007), Rasmussen and Johnson (2013), Liu et al. (2017), Sansone et al. (2017), Chan et al. (2013)
Lycopene ($C_{40}H_{56}$)	536.873	*Chlorella marina*	Anti-oxidant, anti-malignant, antiatherogenic	Bhalamurugan et al. (2018)
Phycocyanin ($C_{165}H_{185}N_{20}O_{30}$)	30 kDa	*Spirulina* spp. *Arthrospira platensis*	Required as fluorescent reagents hepatoprotection activity, neuroprotective, antioxidant, anti-inflammatory	de Morais et al. (2018), Sonani et al. (2016), Paswan et al. (2016)
Phycoerythrin	265 kDa	*Porphyridium* spp., *Polysiphonia morrowii*	Phycoerythrin-labelled streptavidin applied to find biotin labelled DNA and protein analyses	Sekar and Chandramohan (2008), Chakdar and Pabbi (2017)

(continued)

Table 16.3 (continued)

Pigments	Molecular Weight (g/mol)	Microalgae (strains)	Applications	References
Violaxanthin ($C_{40}H_{56}O_4$)	600.87	*Chlorella ellipsoidea*	Anti-inflammatory and anti-proliferative activities	Sathasivam and Ki (2018), Koller et al. (2014), Kiran and Venkata Mohan (2021)
β-carotene ($C_{40}H_{56}$)	536.87	*Botryococcus braunii, Chlamydomonas nivalis, Chlamydocapsa* sp., *Chlorococcum* sp., *Chlorella sorokiniana, Chondria striolata, Dunaliella tertiolecta, Dunaliella salina, Scenedesmus almeriensis, Dunaliella bardawil*	Anti-aging, predecessor of vitamin-A, cancer, immunity regulator, coronary illness prevention, retinal and sensory debility enhancement and less-density lipoprotein oxidation prevention, anti-oxidative, asthma, and cosmetics	Chakdar and Pabbi (2017), Andrade et al. (2018), Barkia et al. (2019), Gong and Bassi (2016), Xu et al. (2018), Kiran and Venkata Mohan (2021)

compounds are shown in Fig. 16.2. Driven by the high value of purified astaxanthin, β-carotene, and phycocyanin (Brennan and Owende 2010), the microalgae industry is experiencing rapid growth, attracting both established companies and new ventures (Caporgno and Mathys 2018). This growth encompasses both the production of end products and biomass for further processing, demonstrating the versatility and potential of this sector.

16.4.2 Microalgal Proteins

Proteins, usually identified as the "building blocks of the human body," are essential macronutrients that contribute to a distinct overall growth. Microalgae rely on proteins for two critical functions: building and maintaining their cell structure and metabolism. Additionally, proteins act as essential enzymes in many photosynthesis reactions, and they are a vital component of the membranes and light-harvesting complexes that capture sunlight for energy (Williams and Laurens 2010). Many microalgal species are protein rich in terms of quantity and quality that may match traditional protein sources. On a dry weight basis, some species of microalgae have been initiated to have extremely high protein contents, ranging from almost 42–70% cyanobacteria (Plaza et al. 2009; Milovanovic et al. 2012) and capable about 58% in *Chlorella vulgaris* (Becker 2007). Microalgae offer a promising solution to the

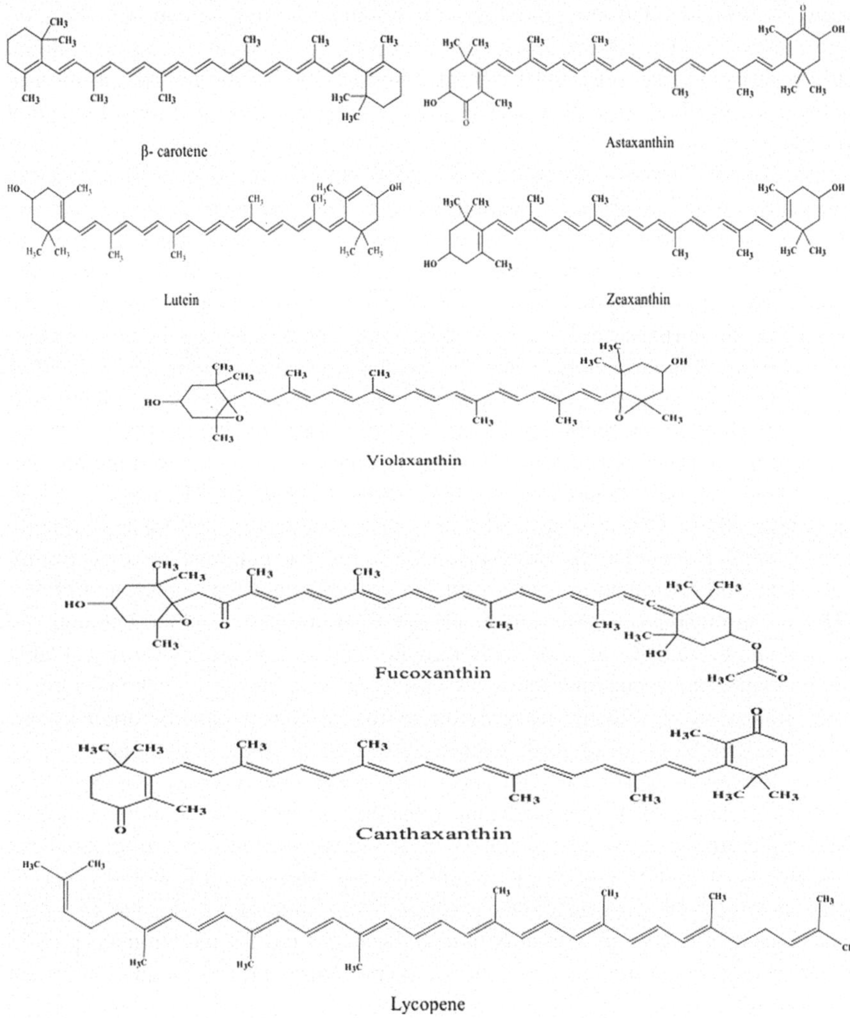

Fig. 16.2 Chemical structures of pigments attained from microalgae

world's protein needs. They boast several advantages over traditional sources like meat (pork, beef, and poultry). For example, producing a kilogram of protein from microalgae requires significantly less land – only 2.5 m^2 compared to land-intensive animal agriculture (Van Krimpen et al. 2013; De Vries and de Boer 2010). Even less land requirement than proteins established from plants used for human food and animal feed, for instance, soybean, pea, microalgae, required less land (Smetana et al. 2017). Other benefits include using non-arable soil for production and growth, low usage of freshwater, the ability to grow in marine water, and the capability to replace non-sustainable imports of soy (FAO 2010). For high protein quality, both *Chlorella* sp. and *Arthrospira* sp. produce have well-composed amino acid profiles based on WHO/UNU/FAO guidelines for human essential amino acid (EAAs) requirements (Becker 2007; Chronakis and Madsen 2011). Both species have amino acid components like eggs and soybeans (Becker 2007). Microalgae are high in quality because they consist of all the necessary amino acids that humans cannot synthesize.

Furthermore, amino acids profiles are stable and comparable to those seen in high-value protein sources, for example, egg albumin, lactoglobulin, and soy (Williams and Laurens 2010). The requirement for plant-based nutrients, particularly protein resources, has increased over time due to rising health interests and a global change in the present time for nutraceutical solutions that are accessible and provide a high nutritional value. *Chlorella* sp. and *Spirulina* sp. are the utmost globally required in the microalgae trade due to their excessive protein, about 50–70% protein by dry-weight, vitamins, minerals, lipids, polysaccharides, pigments, and other trace elements (Capelli and Cysewski 2010). In *Spirulina* sp. per unit mass, chlorophyll, vitamins, phycocyanin, vitamin E, omega-six fatty acids, minerals, and up to ten times more beta-carotene than carrots are present (Hemantkumar and Rahimbhai 2019). *Chlorella* sp. and *Spirulina* sp. are recognized for high-quality protein, which is linked to their well-balanced amino acid composition, digestibility coefficient, and natural amino acids value engrossed from food, according to FAO/WHO recommendations (Fujitani et al. 2001; Borowitzka 1995). Although no recombinant proteins from genetically modified microalgae are currently available commercially, transgenic microalgae are emerging as a promising method for large-scale production of valuable proteins for industrial and therapeutic applications. Notably, research has successfully expressed over 20 therapeutically important proteins in microalgae, particularly the species *Chlamydomonas reinhardtii* (Specht et al. 2010). Because of its many advantageous characteristics, such as the constant and comparatively easy conversion of chloroplast, nuclear, and mitochondrial organelles, fast initiation of positive transformants, substitute of photo-trophic as well as heterotrophic growth, and the accessibility of numerous molecular kit, green algae *Chlamydomonas sp.* which is unicellular is desired for recombinant protein production. Essential functions of some of the proteins expressed in microalgae are shown in Table 16.4.

16.4.3 Poly-Unsaturated Fatty Acids

PUFAs, hydrocarbon chains with two or more double bonds, terminate with a methyl group on one end and a carboxyl group on the other. They react with alcohol groups to form various lipid-based compounds, such as phospholipids and triglycerides. Based on the number and location of their double bonds, PUFAs fall into two main categories: omega-3 and omega-6, also known as n-3 and n-6 PUFAs, respectively. Linoleic acid (LA, 18:2), an essential fatty acid precursor, is used to produce n-6 PUFAs, while α-linolenic acid (ALA, 18:3), another essential fatty acid, is utilized for n-3 PUFA synthesis. Due to a predisposed hereditary absence of the Δ-15 desaturase enzyme (Sprecher et al. 1999), LA and ALA are denoted as "essential" unable to synthesize de novo in animals. As a result, LA and ALA are essentially

Table 16.4 Protein production and expression in microalgae

Microalgae	Expressed protein	Functions	Applications	References
Chlamydomonas reinhardtii	83K7C	IgG1 human monoclonal antibody, anti-anthrax defensive antigen	Therapeutics	Tran et al. (2009)
Chlamydomonas reinhardtii	M-SAA	Bovine mammary-linked serum amyloid; promote mucin making cell lines in human gut epithelial	Therapeutics, oral delivery	Manuell et al. (2007)
Chlamydomonas reinhardtii	VP28	White spot syndrome virus protein	Vaccine	Surzycki et al. (2009)
Chlamydomonas reinhardtii	Human glutamic acid decarboxylase 65 (hGAD65)	Autoantigen in type 1 diabetes	Diagnostics and therapeutics	Wang et al. (2008)
Chlamydomonas reinhardtii	VEGF	Healing for pulmonary emphysema	Therapeutics	Rasala et al. (2010)
Chlamydomonas reinhardtii	Green fluorescent protein (GFP)	Reporter gene		Franklin et al. (2002)
Chlamydomonas reinhardtii	HSV8-lsc	Mammalian protein articulated	Pharmaceutical	Mayfield et al. (2003)
Chlamydomonas reinhardtii	CTB-VP1	Cholera toxin B subunit enclosed to foot and mouth disease	Vaccine	Sun et al. (2003)
Chlamydomonas reinhardtii	hMT-2	Human metallothionine-2	Pharmaceutical, UV-protection	Zhang et al. (2006)
Chlamydomonas reinhardtii	hTRAIL	Human tumor necrosis factor-related apoptosis	Pharmaceutical	Yang et al. (2006)
Chlamydomonas reinhardtii	CSFV-E2	Swine fever viral protein	Vaccine	He et al. (2007)
Chlamydomonas reinhardtii	ARS2-crEpo-his6	Human erythropoietin bonded to ARS2 export sequence w/6xhis tag	Pharmaceutical, protein trade	Eichler-Stahlberg et al. (2009)
Chlamydomonas reinhardtii	IgG1	Murine and human antibodies	Therapeutics	Tran et al. (2009)

(continued)

Table 16.4 (continued)

Microalgae	Expressed protein	Functions	Applications	References
Chlamydomonas reinhardtii	CTB-D2	D2 fibronectin-binding field of *Staphylococcus aureus* bonded along with the cholera toxin B subunit	Vaccine (oral)	Dreesen et al. (2010)
Chlamydomonas reinhardtii	10NF3	Human fibronectin, antibody imitators	Therapeutics	Rasala et al. (2010)
Chlamydomonas reinhardtii	M-SAA-Interferon b1	Numerous sclerosis treatment	Therapeutics	Rasala et al. (2010)
Chlamydomonas reinhardtii	HMGB1	High movement group protein B1	Therapeutics	Rasala et al. (2010)
Phaeodactylum tricornutum	CL4mAb	Monoclonal human IgG antibodies against hepatitis B virus		Hempel et al. (2011)
Phaeodactylum tricornutum, Dunaliella salina	HbsAg	Vaccine against hepatitis B		Hempel et al. (2011), Geng et al. (2003)
Synechococcus Synechocystis Spirulina platensis	Phycobiliproteins	Marker pigments; food colorants and cosmetics		Viskari and Colyer (2003), Zolla and Bianchetti (2001), Simó et al. (2005)

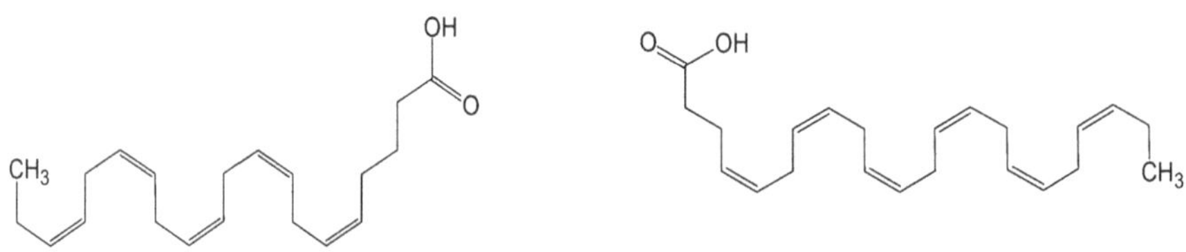

EPA ($C_{20}H_{30}O_2$) DHA ($C_{22}H_{32}O_2$)

Fig. 16.3 Chemical structure of EPA and DHA

acquired from nutrition. Plant oils, for instance, canola, soybean, flaxseed, and sunflower oils are high in these two fatty acids. However, they lack the enzyme ability to convert LA and ALA into more extended chain compounds; therefore, they do not many. LA and ALA endure a sequence of elongation and desaturation processes in mammalian tissues, resulting in long-chain PUFA. LA and ALA undergo a sequence of extension and desaturation processes in mammalian tissues, resulting in longer chain PUFA. The synthesis of ALA and LA PUFAs occurs through two distinct, yet interconnected pathways located in proximity. Each pathway utilizes specific enzymes to produce its signature product: the n-6 pathway yields Arachidonic acid (AA, 20:4), while the n-3 pathway generates eicosapentaenoic acid (EPA, 20:5) and docosahexaenoic acid (DHA, 22:6) (Robertson et al. 2013). The precise chemical assembly of EPA and DHA is illustrated in Fig. 16.3.

Long-chain PUFA can be attained from non-marine plant sources by converting LA and ALA to extended PUFA chain. Although EPA and DHA may be produced from ALA, fish oils remain the most commercially accessible supply of EPA and DHA because humans are inefficient in biological conversions in their bodies (Peltomaa et al. 2018). Dietary PUFA intake becomes even more critical in old age and ill conditions. The essential fatty acid DHA, crucial for brain and retina function and development (Auestad et al. 2003), must be primarily acquired from food due to limited human biochemical conversion (Gao et al. 2013; Le et al. 2009). Furthermore, EPA and DHA have been linked with lower risks of various chronic diseases, including hypertension, stroke, heart disease, Alzheimer's, dementia, and depression (Das 2008). Long-chain EPA and DHA are essential dietary elements constructed by plants, but their maximum production from microalgae is an opportunity (John et al. 2011). These fatty acids are also commercially generated from *Crypthecodinium cohnii*, a heterotrophic microalga, photoautotrophic microalgae *Nannochloropsis* sp., and *Phaeodactylum tricornutum*. As a result, a new "vegetarian source" for the fatty acids devised in Table 16.5.

16.4.4 Polypeptides

Naturally occurring peptides have fewer adverse influences when opposed to manufactured supplements, which may cause inhibition with immune system function and defensive processes, resulting in pro-oxidant outcomes (Poljsak and Milisav 2012). The vast range of uses for bioactive peptides as prospective therapies for various clinical disorders, including hypertension, inflammation, diabetes, and oxidative stress, has sparked global attention (Lemes et al. 2016). Plant and animal proteins are the most common cause of bioactive peptides nowadays (Ngo et al. 2012; Maestri et al. 2016), with microalgae being a rare exception. Genera, for instance, *Chlorella, Tetraselmis*, *Nitzschia* and *Navicula,* have been studied but the detail studies are limited. These genera generate a variety of peptides which are bioactives, and having antioxidative capabilities (Ejike et al. 2017; Alzahrani et al. 2018). Antioxidant peptides are often prepared using enzymes to break down proteins. These peptides can be studied in their whole form (crude hydrolysates) or after purification. Research has

Table 16.5 PUFA content in autotrophic, mixotrophic, and heterotrophic microalgal species

Taxonomic group	Species	Culture mode	PUFA (% total fatty acids)	References
Diatoms (Bacillariophyceae)	*Phaeodactylum tricornutum*	Autotrophic	EPA (9–57%)	Yongmanitchai and Ward (1991), Fernández et al. (2000)
	Nitzschia laevis	Heterotrophic	EPA (~10% to 11%)	Chen et al. (2007), Tan and Johns (1996)
	Odontella aurita	Autotrophic	EPA (~25% to 27%)	Guihéneuf et al. (2010)
Prasinophyceae (Green algae)	*Ostreococcus tauri*	Autotrophic	SDA (~17% to 18%) EPA (~12 to 13%)	Wagner et al. (2010)
Trebouxiophyceae (Green algae)	*Parietochloris incisa*	Autotrophic	AA (~40% 60%)	Khozin-Goldberg et al. (2002)
	P. incisa (Δ5-des mutant)	Autotrophic	DGLA (~16% to 32%)	Iskandarov et al. (2011)
Dinophyceae	*Crypthecodinium cohnii*	Heterotrophic	DHA (~36% to 44%)	de Swaaf et al. (1999)
Prymnesiophyceae	*Isochrysis galbana*	Autotrophic	SDA (~7% to 15%) EPA (~12% to 27%) DHA (~5% to 14%)	Fidalgo et al. (1998), Qi et al. (2002)
Pavlovophyceae	*Pavlova lutheri*	Autotrophic mixotrophic	SDA (~5% to 11%) EPA (~22% to 29%) DHA (~13% to 19%)	Guihéneuf et al. (2009)

(continued)

Table 16.5 (continued)

Taxonomic group	Species	Culture mode	PUFA (% total fatty acids)	References
Eustigmatophyceae	*Nannochloropsis* sp. *N. gaditana* *N. gaditana*	Autotrophic mixotrophic heterotrophic	EPA (~9% to 27%) EPA (~21% to 36%) EPA (~7.4% to 28.8%) EPA (~20.5% to 37.83%)	Sukenik (1991), Fang et al. (2004), Pal et al. (2011), Mitra and Mishra (2019), Beacham et al. (2014), Khozin-Goldberg and Boussiba (2011), Mitra et al. (2015)
	Monodus subterraneus	Autotrophic	EPA (~15% to 28%)	Cohen (1994), Khozin-Goldberg and Cohen (2006)
Porphyridiophyceae (Red algae)	*Porphyridium* sp.	Autotrophic	AA (~22% to 35%) EPA (~3 to 27%)	Nuutila et al. (1997), Khozin et al. (1997), Fábregas et al. (1998)
Labyrinthulea	*Thraustochytrium* sp.	Heterotrophic	DHA (~35% to 40%)	Singh et al. (1996), Min et al. (2012)
	Schizochytrium sp.	Heterotrophic	DHA (~33% to 37%)	Yaguchi et al. (1997)

shown that crude hydrolysates from certain green algae, like *Chlorella pyrenoidosa* (Shih and Cherng 2012), *Chlorella ellipsiodea*, and *Tetraselmis suecica* (Lee et al. 2009), demonstrate powerful antioxidant effects in lab tests and on skin cells. Even more impressive antioxidant properties were observed in purified peptides derived from these algae using pepsin hydrolysis. For instance, the peptides Val-Glu-Cys-Tyr-Gly-Pro-Asn-Arg-Pro-Glu-Phe and Leu-Asn-Gly-Asp-Val-Trp, isolated from *Chlorella vulgaris* and *Chlorella ellipsiodea* respectively, exhibited strong abilities to scavenge superoxide and DPPH radicals (Sheih et al. 2009; Ko et al. 2012).

Similarly, two bioactive peptides extracted from *Navicula incerta* papain hydrolysate, Pro-Gly-Trp-Asn-Gln-Trp-Phe-Leu and Val-Glu-Val-Leu-Pro-Pro-Ala-Glu-Leu, displayed significant antioxidant activities in HepG2/CYP2E1 cells (Kang et al. 2012). Several features of peptides' chemical structure were found to have a positive influence on their antioxidant activity. These features include small molecular weight, amino acid configuration and sequence, hydrophobicity, the presence of an indole/pyrrolidine/imidazole ring, and steric assembly at the C- and N-regions (Zou et al. 2016; Tian et al. 2015). ACE-constraining peptides have been produced from *Chlorella vulgaris*, *Chlorella ellipsoidea*, *Nannochloropsis oculate*, and

Arthrospira platensis. Animal research has widely studied their capability to lower blood pressure (Barkia et al. 2019).

16.5 Role of Bioactive Compounds in Human Health

Microalgae boast a diverse repertoire of pigments with therapeutic potential, including astaxanthin from *Chlorella* spp. and *Hematococcus pluvialis*, β-carotene from *Dunaliella salina*, fucoxanthin from *Muriellopsis* spp., zeaxanthin obtained from *Dunaliella salina*, and phycoerythrin from *Anabaena variabilis* (Chakdar and Pabbi 2012; Liu et al. 2018; Shukla and Kumar 2018). These pigments, like β-carotene, lutein, and astaxanthin, offer promising avenues for treating various health issues. Phycobiliproteins (phycoerythrin and phycocyanin) are utilized in immunological testing and research as biochemical tags and markers (Dasgupta 2015). Carotenoids are the most marketable bio-pigments because of their numerous health and industrial uses. Antioxidant properties are known to exist in these lipophilic substances, which serve to protect the cell from oxidative damage. Bio-pigments also influence many biochemical processes and cell signaling systems (Christaki et al. 2013; Matos et al. 2017). Because of their future application in the food industry, lutein, β-carotene, and astaxanthin are significantly valuable, as shown in Table 16.2. Natural β-carotene is becoming more popular than synthetic β-carotene because natural pigments have numerous qualities that synthetic pigments lack. According to Jayappriyan et al. (2013), naturally occurring β-carotene, instead of manufactured β-carotene, improves growth inhibition of cancerous prostate cells. As a result, people are paying greater attention to the natural resource of carotenoids other than plants these days, and microalgae are one intriguing pigment synthesis source (Mata et al. 2010; Mehta et al. 2018). The *Spirulina* genus has a significant protein, particularly phycocyanin, a blue pigment with numerous health advantages. Microalgae contend with traditional protein resources such as soybeans, eggs, and other plant and animal-based foods due to their high protein concentration. In comparison to plant products, *Dunaliella* sp. may generate 50–80% protein (Ejike et al. 2017), while *Spirulina* sp. has around 50–70% protein (Plaza et al. 2009). Studies have shown that the peptide VECYGPNRPQF, isolated from the green algae *Chlorella vulgaris*, possesses antioxidant properties. This peptide is also resistant to digestion in the gastrointestinal tract and non-toxic to human lung fibroblast WI-38 cells (Sheih et al. 2009). Similarly, researchers have identified antioxidant activity in the peptide LNGDVW derived from the green alga *C. ellipsoidea* (Ko et al. 2012). Additionally, two peptides, NIPP-1 (PGWNQWFL) and NIPP-2 (VEVLPPAEL), isolated from the benthic diatom *Navicula incerta*, have been found to exhibit antioxidant properties (Kang et al. 2012). Microalgae have yielded promising anticancer and antihypertensive peptides. Sheih et al. (2010) identified the anticancer peptide VECYGPNRPQF from *Chlorella vulgaris*, while the polypeptide CPAP from *Chlorella pyrenoidosa* displayed antiproliferative activity against liver cancer cells (Sheih et al. 2009). Additionally, peptides from *Nannochloropsis oculata*, like VECYGPNRPQF and GMNNLTP/LEQ, exhibited antihypertensive properties (Samarakoon et al. 2013). *Nannochloropsis oculata's* MPDW peptide emerged as a

Table 16.6 Active peptides recognized from microalgae and suggested action mechanism

Microalgae	Compound	Mechanism	References
Arthrospira platensis (Spirulina)	Enzymatic hydrolysates	Promoted hydration in skin cells and Filaggrin expression; responded osmotic stress damage by modulating Smith gene expression	De Lucia et al. (2018)
Chlorella ellipsoidea	LNGDVW	Scavenging of free radicals	Ko et al. (2012)
Chlorella pyrenoidosa	Polypeptide CPAP	HepG2 cells	Wang and Zhang (2013)
Chlorella sp.	VECYGPNRPQF	Gene expression of vascular adhesion molecules (E-selectin, ICAM, VCAM, MCP-1 and ET-1) inhibition	Shih et al. (2013)
Chlorella sp.	Chlorella produced peptide	Inhibition of MMP-1, CYR61	Chen et al. (2011)
Chlorella vulgaris	VECYGPNRPQF	Quenching of superoxide radical	Sheih et al. (2009)
Chlorella vulgaris	VECYGPNRPQF	AGS cells	Sheih et al. (2010)
Chlorella vulgaris	VECYGPNRPQF	ACE inhibitor	Sheih et al. (2009)
Nannochloropsis oculata	GMNNLTP; LEQ	ACE inhibitor	Samarakoon et al. (2013)
Nannochloropsis oculata	MPDW	Increase of ALP, osteocalcin, collagen type I, phosphorylation of MAPK pathways	Nguyen et al. (2013)
Navicula incerta	PGWNQWFL, VEVLPPAEL	Cytotoxicity in HepG2/CYP2E1 cells	Kang et al. (2012)

powerful anti-osteoporosis agent, stimulating osteoblast proliferation with remarkable efficacy (Table 16.6).

Polyunsaturated fatty acids (PUFAs) have been well correlated to the preclusion of chronic illness, for instance, cancer and coronary heart disease, and the improvement of inflammatory disorders like arthritis, the reduction of plasma triacylglycerol levels, and the reduction of blood pressure (Hibbeln et al. 2006). As a result, PUFAs and their derivatives and equivalents are key nutraceuticals with growing nutritional interest (McClements et al. 2009; Barba et al. 2012). PUFAs derived from microalgae are likewise regarded as high-value commercial materials (Mata et al. 2010). Some microalgae species (e.g., *Cryptothecodininum cohnii* and *Schizochytrium limacinum*) grown heterotrophically in industries in fermenters to produce docosahexaenoic acid (DHA) (Norton and Sun 2008; Plaza et al. 2012). Omega-3 fatty acids, particularly long-chain PUFAs like EPA and DHA, are increasingly used as a valuable dietary component in baby food prescriptions. This booming $9 billion industry is driven by the potential of these fatty acids to treat inflammatory conditions like Alzheimer's disease, lupus, and arthritis (Kong et al. 2014; Yates et al. 2014). Studies in mice and human cells suggest that PUFAs, including those extracted from the brown seaweed *Ishige okamurae*, can effectively reduce inflammation by lowering histamine release and modifying inflammatory cytokine production (Vo et al. 2011). This makes PUFA-rich lipid extracts from micro- and macroalgae promising candidates for functional foods with anti-inflammatory properties.

16.6 Microalgae: Indian Perspectives

Globally, India is the largest user of antibiotics and a significant source of drug-resistant bacteria (Dixit et al. 2019). With an ever-increasing population, producing healthy and sufficient food quantities is an ongoing struggle. For example, the Brahmaputra valley in Assam, North-East India, has documented and taxonomically classified six species of *Spirulina* (Baruah et al. 2014). *Spirulina* can combat India's malnutrition, according to experts. On the other hand, food businesses have several hurdles in developing and marketing value-added microalgae products, such as viz. (1) Efficient biomass and downstream recovery, (2) Developments in photobioreactors, (3) System and synthetic biology methodologies, (4) Efficient biomass and downstream recovery, (5) Checking for biodiversity, (6) Environmental sustainability, (7) Food engineering, (8) Consumer acceptability, (9) Safety and regulatory problems, (10) Marketing nutraceuticals and bioactive compounds. Recognizing the potential of the Indian food industry, the Indian government has launched several initiatives to overcome challenges and introduce innovative products and processes. A major step in this direction was the signing of a Memorandum of Understanding between the Council of Scientific and Industrial Research (CSIR), CSIR-CFTRI (Centre Food Technology Research Institute), Mysore, Karnataka (India) and the Agricultural and Processed Food Products Export Development Authority (APEDA) to establish a liaison office in Assam. This collaboration aims to unlock the potential of value-added food products, including microalgae like *Spirulina*. Even though India has abundant algal resources, the commercialization of value-added products is still needs lots of improvement and market reach. The following processes should be optimized for a thriving economy based on value-added products from algal resources such as identification and preservation, cost-effective and high-quality large-scale cultivation, efficient harvesting and commercialization. This, however, necessitates awareness, training, and the creation of trained infrastructure (Medhi and Kalita 2020).

16.7 Conclusion

As significant bio-factories, microalgae can meet many of the worldwide demands of nutraceuticals and other related industries (e.g., biofuel, electricity, and animal feed). The World Health Organization (Udayan et al. 2017) recognized the usage of *Arthrospira (Spirulina)* as a superfood of the globe based on the evidence acquired from multiple study findings. Microalgae may provide efficient and sustainable alternatives for vegetarian and non-vegetarian commodities in the perspective of worldwide population expansion and the accessibility of terrestrial food stocks. However, only a few strains, mostly the cyanobacterium *Arthrospira* sp. and a few microalgae (green), for instance, *Chlorella* sp., *Haematococcus* sp., and *Dunaliella* sp., have been used for industrial making of a small number of goods so far. Algal products' transition from a local to universal market usage as food products would need substantial research and development. This would entail the generation of microalgae with increased specific metabolites production, and the showing novel species or the enhancement of current strains obtained by genetic engineering. Inexpensive manufacturing involving microalgae, on par with synthetic or biological alternatives from other sources, hinges on innovation in economical cultivation, harvesting, and biomass processing.

Functional food products and nutraceuticals with specific health advantages are becoming increasingly popular. Nutraceuticals and functional food components contain complete microalgal biomass and derived co-products of high value now. Purified compounds are also desirable since complete microalgal biomass contains numerous unknown biochemicals, possible toxins, solvent traces, and heavy metals that might be dangerous to consume and cause significant health problems. Furthermore, through literature, customers must be informed about the substance and acceptance of microalgal products. The total production expenses include cultivation system cost and maintenance, restricted or lower culture/biomass productivity, and bio-refining procedures, raising constraints to the production scale-up of valuable algal products. Despite the difficulties of extensive growth, microalgal components are in high demand in meals, vitamins, and future medications. Further exploration into the biochemical makeup of various microalgae strains is necessary to unlock their full potential. This knowledge will be instrumental in harnessing their benefits for applications like biofuel production or environmental remediation while mitigating any potential risks associated with their cultivation or use. The limited research conducted on a specific subset of microalgae strains highlights the substantial opportunity for identifying novel bioactive metabolites, including those with antimicrobial, antifungal, and antitumorigenic properties, within the vast array of unexplored or underexplored strains. Integration of multidisciplinary research tools, encompassing bioinformatics, molecular biology, systems biology, and artificial intelligence, can synergistically enhance our understanding of microalgal production at a systemic level, ultimately leading to the development of more efficient strains for industrial utilization.

References

Alzahrani MAJ, Perera CO, Hemar Y (2018) Production of bioactive proteins and peptides from the diatom Nitzschia laevis and comparison of their in vitro antioxidant activities with those from Spirulina platensis and Chlorella vulgaris. Int J Food Sci Technol 53(3):676–682

Ambati RR, Phang SM, Ravi S, Aswathanarayana RG (2014) Astaxanthin: sources, extraction, stability, biological activities and its commercial applications—a review. Mar Drugs 12(1):128–152

Ampofo J, Abbey L (2022) Microalgae: bioactive composition, health benefits, safety and prospects as potential high-value ingredients for the functional food industry. Foods 11(12):1744

Andrade LM, Andrade CJ, Dias M, Nascimento C, Mendes MA (2018) Chlorella and spirulina microalgae as sources of functional foods, nutraceuticals, and food supplements. MOJ Food Process Technol 6(1):45–58

Auestad N, Scott DT, Janowsky JS, Jacobsen C, Carroll RE, Montalto MB et al (2003) Visual, cognitive, and language assessments at 39 months: a follow-up study of children fed formulas containing long-chain polyunsaturated fatty acids to 1 year of age. Pediatrics 112(3):e177–e183

Bajpai VK, Shukla S, Kang SM, Hwang SK, Song X, Huh YS, Han YK (2018) Developments of cyanobacteria for nano-marine drugs: relevance of nanoformulations in cancer therapies. Mar Drugs 16(6):179

Balasubramaniam V, Gunasegavan RDN, Mustar S, Lee JC, Mohd Noh MF (2021) Isolation of industrial important bioactive compounds from microalgae. Molecules 26(4):943

Barba FJ, Esteve MJ, Frigola A (2012) Impact of high-pressure processing on vitamin E (α-, γ-, and δ-tocopherol), vitamin D (cholecalciferol and ergocalciferol), and fatty acid profiles in liquid foods. J Agric Food Chem 60(14):3763–3768

Barkia I, Saari N, Manning SR (2019) Microalgae for high-value products towards human health and nutrition. Mar Drugs 17(5):304

Baruah PP, Baruah R, Das P (2014) A preliminary study on diversity and distribution of Spirulina, Arthrospira and Glaucospira (Cyanobacteria) in the Brahmaputra Valley of Assam (India). Feddes Repert 125(3–4):85–92

Beacham TA, Bradley C, White DA, Bond P, Ali ST (2014) Lipid productivity and cell wall ultrastructure of six strains of nannochloropsis: implications for biofuel production and downstream processing. Algal Res 6:64–69

Becker W (2004) Chapter 18: Microalgae in human and animal nutrition. In: Handbook of microalgal culture: biotechnology and applied phycology, vol 312. Wiley Online Library, Hoboken

Becker EW (2007) Micro-algae as a source of protein. Biotechnol Adv 25(2):207–210

Beetul K, Gopeechund A, Kaullysing D, Mattan-Moorgawa S, Puchooa D, Bhagooli R (2016) Challenges and opportunities in the present era of marine algal applications. In: Thajuddin N, Dhanasekaran D (eds) Algae-organisms for imminent biotechnology. IntechOpen, London, pp 237–276

Begum H, Yusoff FM, Banerjee S, Khatoon H, Shariff M (2016) Availability and utilization of pigments from microalgae. Crit Rev Food Sci Nutr 56(13):2209–2222

Bhalamurugan GL, Valerie O, Mark L (2018) Valuable bioproducts obtained from microalgal biomass and their commercial applications: a review. Environ Eng Res 23(3):229–241

Bhattacharjee M (2016) Pharmaceutically valuable bioactive compounds of algae. Asian J Pharm Clin Res 9:43–47

Bioeconomy BW (2020) Microalgae can produce more than just fuel. https://www.biooekonomie-bw.de/en/articles/news/microalgae-can-produce-more-than-just-fuel. Accessed 10 Sep 2020

Blanco AM, Moreno J, Del Campo JA, Rivas J, Guerrero MG (2007) Outdoor cultivation of lutein-rich cells of Muriellopsis sp. in open ponds. Appl Microbiol Biotechnol 73(6):1259–1266

Borowitzka MA (1995) Microalgae as sources of pharmaceuticals and other biologically active compounds. J Appl Phycol 7(1):3–15

Borowitzka MA (2013) High-value products from microalgae—their development and commercialisation. Journal of applied phycology 25:743–756

Brennan L, Owende P (2010) Biofuels from microalgae—a review of technologies for production, processing, and extractions of biofuels and co-products. Renew Sust Energ Rev 14(2):557–577

Brillatz T, Lauritano C, Jacmin M, Khamma S, Marcourt L, Righi D et al (2018) Zebrafish-based identification of the antiseizure nucleoside inosine from the marine diatom Skeletonema marinoi. PLoS One 13(4):e0196195

Cai X, Chen Y, Xie X, Yao D, Ding C, Chen M (2019) Astaxanthin prevents against lipopolysaccharide-induced acute lung injury and sepsis via inhibiting activation of MAPK/NF-κB. Am J Transl Res 11(3):1884

Capelli B, Cysewski GR (2010) Potential health benefits of spirulina microalgae. Nutrafoods 9(2):19–26
Capelli B, Talbott S, Ding L (2019) Astaxanthin sources: suitability for human health and nutrition. Funct Foods Health Dis 9(6):430–445
Caporgno MP, Mathys A (2018) Trends in microalgae incorporation into innovative food products with potential health benefits. Front Nutr 5:58
Chakdar H, Pabbi S (2012) Extraction and purification of phycoerythrin from Anabaena variabilis (CCC421). Phykos 42(1):25–31
Chakdar H, Pabbi S (2017) Algal pigments for human health and cosmeceuticals. In: Algal green chemistry. Elsevier, Amsterdam, pp 171–188
Chan MC, Ho SH, Lee DJ, Chen CY, Huang CC, Chang JS (2013) Characterization, extraction and purification of lutein produced by an indigenous microalga Scenedesmus obliquus CNW-N. Biochem Eng J 78:24–31
Chang CH, Chen CY, Chiou JY, Peng RY, Peng CH (2010) Astaxanthine secured apoptotic death of PC12 cells induced by β-amyloid peptide 25–35: its molecular action targets. J Med Food 13(3):548–556
Chen GQ, Jiang Y, Chen F (2007) Fatty acid and lipid class composition of the eicosapentaenoic acid-producing microalga, Nitzschia laevis. Food Chem 104(4):1580–1585
Chen CL, Liou SF, Chen SJ, Shih MF (2011) Protective effects of Chlorella-derived peptide on UVB-induced production of MMP-1 and degradation of procollagen genes in human skin fibroblasts. Regul Toxicol Pharmacol 60(1):112–119
Chisti Y (2007) Biodiesel from microalgae. Biotechnol Adv 25(3):294–306
Chisti Y (2013) Constraints to commercialization of algal fuels. J Biotechnol 167(3):201–214
Christaki E, Bonos E, Giannenas I, Florou-Paneri P (2013) Functional properties of carotenoids originating from algae. J Sci Food Agric 93(1):5–11
Chronakis IS, Madsen M (2011) Algal proteins. In: Handbook of food proteins. Woodhead Publishing, Sawston, pp 353–394
Coesel SN, Baumgartner AC, Teles LM, Ramos AA, Henriques NM, Cancela L, Varela JCS (2008) Nutrient limitation is the main regulatory factor for carotenoid accumulation and for Psy and Pds steady state transcript levels in Dunaliella salina (Chlorophyta) exposed to high light and salt stress. Mar Biotechnol 10(5):602–611
Cohen Z (1994) Production potential of eicosapentaenoic acid by Monodus subterraneus. J Am Oil Chem Soc 71(9):941–945
Das UN (2008) Folic acid and polyunsaturated fatty acids improve cognitive function and prevent depression, dementia, and Alzheimer's disease—but how and why? Prostaglandins Leukot Essent Fat Acids 78(1):11–19
Dasgupta CN (2015) Algae as a source of phycocyanin and other industrially important pigments. In: Algal biorefinery: an integrated approach. Springer, Cham, pp 253–276
de Jesus Raposo MF, De Morais AMB, De Morais RMSC (2015) Marine polysaccharides from algae with potential biomedical applications. Mar Drugs 13(5):2967–3028
De Lucia A, Zappelli C, Angelillo M, Langelotti AL, Fogliano V, Cucchiara M (2018) A novel biotechnological active ingredient, derived from the microalga Spirulina, increases hydration and reduces osmotic stress in skin cells. H&PC Today 13:60
de Morais MG, da Fontoura Prates D, Moreira JB, Duarte JH, Costa JAV (2018) Phycocyanin from microalgae: properties, extraction and purification, with some recent applications. Ind Biotechnol 14(1):30–37
de Swaaf ME, de Rijk TC, Eggink G, Sijtsma L (1999) Optimisation of docosahexaenoic acid production in batch cultivations by Crypthecodinium cohnii. In: Progress in industrial microbiology, vol 35. Elsevier, Amsterdam, pp 185–192
De Vries M, de Boer IJ (2010) Comparing environmental impacts for livestock products: a review of life cycle assessments. Livest Sci 128(1–3):1–11
Del Mondo A, Smerilli A, Sané E, Sansone C, Brunet C (2020) Challenging microalgal vitamins for human health. Microb Cell Factories 19(1):1–23

Denery JR, Dragull K, Tang CS, Li QX (2004) Pressurized fluid extraction of carotenoids from Haematococcus pluvialis and Dunaliella salina and kavalactones from Piper methysticum. Anal Chim Acta 501(2):175–181

Dixit A, Kumar N, Kumar S, Trigun V (2019) Antimicrobial resistance: progress in the decade since emergence of New Delhi metallo-β-lactamase in India. Indian J Community Med 44(1):4

Dreesen IA, Charpin-El Hamri G, Fussenegger M (2010) Heat-stable oral alga-based vaccine protects mice from Staphylococcus aureus infection. J Biotechnol 145(3):273–280

Dufossé L, Galaup P, Yaron A, Arad SM, Blanc P, Murthy KNC, Ravishankar GA (2005) Microorganisms and microalgae as sources of pigments for food use: a scientific oddity or an industrial reality? Trends Food Sci Technol 16(9):389–406

Eichler-Stahlberg A, Weisheit W, Ruecker O, Heitzer M (2009) Strategies to facilitate transgene expression in Chlamydomonas reinhardtii. Planta 229(4):873–883

Ejike CE, Collins SA, Balasuriya N, Swanson AK, Mason B, Udenigwe CC (2017) Prospects of microalgae proteins in producing peptide-based functional foods for promoting cardiovascular health. Trends Food Sci Technol 59:30–36

Elagoz AM, Ambrosino L, Lauritano C (2020) De novo transcriptome of the diatom Cylindrotheca closterium identifies genes involved in the metabolism of anti-inflammatory compounds. Sci Rep 10(1):1–9

Enzing C, Ploeg M, Barbosa M, Sijtsma L (2014) Microalgae-based products for the food and feed sector: an outlook for Europe. JRC scientific and policy reports, pp 19–37

Fabregas J, Herrero C (1990) Vitamin content of four marine microalgae. Potential use as source of vitamins in nutrition. J Ind Microbiol Biotechnol 5(4):259–263

Fábregas J, García D, Morales E, Domínguez A, Otero A (1998) Renewal rate of semicontinuous cultures of the microalga Porphyridium cruentum modifies phycoerythrin, exopolysaccharide and fatty acid productivity. J Ferment Bioeng 86(5):477–481

Fang X, Wei C, Zhao-Ling C, Fan O (2004) Effects of organic carbon sources on cell growth and eicosapentaenoic acid content of Nannochloropsis sp. J Appl Phycol 16(6):499–503

FAO Aquatic Biofuels Working Group (2010) Algae-based biofuels: application and co-product. National Natural Resources Management System, p 44

Farag MR, Alagawany M, El-Hack MEA, Dhama K (2016) Nutritional and healthical aspects of Spirulina (Arthrospira) for poultry, animals and human. Int J Pharmacol 12(1):36–51

Fernández FA, Pérez JS, Sevilla JF, Camacho FG, Grima EM (2000) Modeling of eicosapentaenoic acid (EPA) production from Phaeodactylum tricornutum cultures in tubular photobioreactors. Effects of dilution rate, tube diameter, and solar irradiance. Biotechnol Bioeng 68(2):173–183

Fidalgo JP, Cid A, Torres E, Sukenik A, Herrero C (1998) Effects of nitrogen source and growth phase on proximate biochemical composition, lipid classes and fatty acid profile of the marine microalga Isochrysis galbana. Aquaculture 166(1–2):105–116

Franklin S, Ngo B, Efuet E, Mayfield SP (2002) Development of a GFP reporter gene for Chlamydomonas reinhardtii chloroplast. Plant J 30(6):733–744

Fujitani N, Sakaki S, Yamaguchi Y, Takenaka H (2001) Inhibitory effects of microalgae on the activation of hyaluronidase. J Appl Phycol 13(6):489–492

Galasso C, Gentile A, Orefice I, Ianora A, Bruno A, Noonan DM et al (2019) Microalgal derivatives as potential nutraceutical and food supplements for human health: a focus on cancer prevention and interception. Nutrients 11(6):1226

Gammone MA, Riccioni G, D'Orazio N (2015) Marine carotenoids against oxidative stress: effects on human health. Mar Drugs 13(10):6226–6246

Gani P, Sunar NM, Matias-Peralta H, Mohamed RMSR, Latiff AAA, Parjo UK (2017) Extraction of hydrocarbons from freshwater green microalgae (Botryococcus sp.) biomass after phycoremediation of domestic wastewater. Int J Phytoremediation 19(7):679–685

Gao F, Taha AY, Ma K, Chang L, Kiesewetter D, Rapoport SI (2013) RETRACTED ARTICLE: aging decreases rate of docosahexaenoic acid synthesis-secretion from circulating unesterified α-linolenic acid by rat liver. Age 35(3):597–608

Gayatri K, Soundhari C, Pavithra BP (2019) Biofilm inhibitory effect of Chlorella extracts on pseudomonas aeruginosa. Int J Pharm Sci Res 10:1966–1971

GCMI (2018) Global chlorella market information: by origin (organic, conventional), by form (powder, liquid, tablet capsules and others), by application (food & beverages, pharmaceuticals, animal feed, others) and region forecast till 2023. https://www.marketresearchfuture.com/report_info.pdf?report_id=4413. Accessed 26 Nov 2018

Geng D, Wang Y, Wang P, Li W, Sun Y (2003) Stable expression of hepatitis B surface antigen gene in Dunaliella salina (Chlorophyta). J Appl Phycol 15(6):451–456

Giordano D, Costantini M, Coppola D, Lauritano C, Pons LN, Ruocco N et al (2018) Biotechnological applications of bioactive peptides from marine sources. Adv Microb Physiol 73:171–220

Goiris K, Muylaert K, Fraeye I, Foubert I, De Brabanter J, De Cooman L (2012) Antioxidant potential of microalgae in relation to their phenolic and carotenoid content. J Appl Phycol 24(6):1477–1486

Gong M, Bassi A (2016) Carotenoids from microalgae: a review of recent developments. Biotechnol Adv 34(8):1396–1412

Greene CH, Huntley ME, Archibald I, Gerber LN, Sills DL, Granados J et al (2016) Marine microalgae: climate, energy, and food security from the sea. Oceanography 29(4):10–15

Guihéneuf F, Mimouni V, Ulmann L, Tremblin G (2009) Combined effects of irradiance level and carbon source on fatty acid and lipid class composition in the microalga Pavlova lutheri commonly used in mariculture. J Exp Mar Biol Ecol 369(2):136–143

Guihéneuf F, Fouqueray M, Mimouni V, Ulmann L, Jacquette B, Tremblin G (2010) Effect of UV stress on the fatty acid and lipid class composition in two marine microalgae Pavlova lutheri (Pavlovophyceae) and Odontella aurita (Bacillariophyceae). J Appl Phycol 22(5):629–638

Habib M, Ahsan B (2008) Review on culture, production and use of Spirulina as food for humans and feeds for domestic animals and fish. Food and Agriculture Organization of the United Nations, Rome

Haimeur A, Ulmann L, Mimouni V, Guéno F, Pineau-Vincent F, Meskini N, Tremblin G (2012) The role of Odontella aurita, a marine diatom rich in EPA, as a dietary supplement in dyslipidemia, platelet function and oxidative stress in high-fat fed rats. Lipids Health Dis 11(1):1–13

Haque F, Dutta A, Thimmanagari M, Chiang YW (2016) Intensified green production of astaxanthin from Haematococcus pluvialis. Food Bioprod Process 99:1–11

He DM, Qian KX, Shen GF, Zhang ZF, Yi-Nü LI, Su ZL, Shao HB (2007) Recombination and expression of classical swine fever virus (CSFV) structural protein E2 gene in Chlamydomonas reinhardtii chroloplasts. Colloids Surf B: Biointerfaces 55(1):26–30

Hemantkumar JN, Rahimbhai MI (2019) Microalgae and its use in nutraceuticals and food supplements. In: Microalgae—from physiology to application, vol 10. IntechOpen, London

Hempel F, Lau J, Klingl A, Maier UG (2011) Algae as protein factories: expression of a human antibody and the respective antigen in the diatom Phaeodactylum tricornutum. PLoS One 6(12):e28424

Hibbeln JR, Nieminen LR, Blasbalg TL, Riggs JA, Lands WE (2006) Healthy intakes of n–3 and n–6 fatty acids: estimations considering worldwide diversity. Am J Clin Nutr 83(6):1483S–1493S

Ibañez E, Cifuentes A (2013) Benefits of using algae as natural sources of functional ingredients. J Sci Food Agric 93(4):703–709

Iskandarov U, Khozin-Goldberg I, Cohen Z (2011) Selection of a DGLA-producing mutant of the microalga Parietochloris incisa: I. Identification of mutation site and expression of VLC-PUFA biosynthesis genes. Appl Microbiol Biotechnol 90(1):249–256

Islam MN, Alsenani F, Schenk PM (2017) Microalgae as a sustainable source of nutraceuticals. In: Microbial functional foods and nutraceuticals, vol 1. Wiley, New York

Jain R, Raghukumar S, Tharanathan R, Bhosle NB (2005) Extracellular polysaccharide production by thraustochytrid protists. Mar Biotechnol 7(3):184–192

Jayappriyan KR, Rajkumar R, Venkatakrishnan V, Nagaraj S, Rengasamy R (2013) In vitro anticancer activity of natural β-carotene from Dunaliella salina EU5891199 in PC-3 cells. Biomed Prev Nutr 3(2):99–105

Jha D, Jain V, Sharma B, Kant A, Garlapati VK (2017) Microalgae-based pharmaceuticals and nutraceuticals: an emerging field with immense market potential. ChemBioEng Reviews 4(4):257–272

John RP, Anisha GS, Nampoothiri KM, Pandey A (2011) Micro and macroalgal biomass: a renewable source for bioethanol. Bioresour Technol 102(1):186–193

Jung F, Krüger-Genge A, Waldeck P, Küpper JH (2019) Spirulina platensis, a super food? J Cell Biotechnol 5(1):43–54

Kagan ML, West AL, Zante C, Calder PC (2013) Acute appearance of fatty acids in human plasma—a comparative study between polar-lipid rich oil from the microalgae Nannochloropsis oculata and krill oil in healthy young males. Lipids Health Dis 12(1):1–10

Kagan ML, Levy A, Leikin-Frenkel A (2015) Comparative study of tissue deposition of omega-3 fatty acids from polar-lipid rich oil of the microalgae Nannochloropsis oculata with krill oil in rats. Food Funct 6(1):185–191

Kang KH, Qian ZJ, Ryu B, Karadeniz F, Kim D, Kim SK (2012) Antioxidant peptides from protein hydrolysate of microalgae Navicula incerta and their protective effects in HepG2/CYP2E1 cells induced by ethanol. Phytother Res 26(10):1555–1563

Karnaouri A, Chalima A, Kalogiannis KG, Varamogianni-Mamatsi D, Lappas A, Topakas E (2020) Utilization of lignocellulosic biomass towards the production of omega-3 fatty acids by the heterotrophic marine microalga Crypthecodinium cohnii. Bioresour Technol 303:122899

Khan Z, Bhadouria P, Bisen PS (2005) Nutritional and therapeutic potential of Spirulina. Curr Pharm Biotechnol 6(5):373–379

Khanra S, Mondal M, Halder G, Tiwari ON, Gayen K, Bhowmick TK (2018) Downstream processing of microalgae for pigments, protein and carbohydrate in industrial application: a review. Food Bioprod Process 110:60–84

Khozin I, Adlerstein D, Bigongo C, Heimer YM, Cohen Z (1997) Elucidation of the biosynthesis of eicosapentaenoic acid in the microalga Porphyridium cruentum (II. Studies with radiolabeled precursors). Plant Physiol 114(1):223–230

Khozin-Goldberg I, Boussiba S (2011) Concerns over the reporting of inconsistent data on fatty acid composition for microalgae of the genus Nannochloropsis (Eustigmatophyceae). J Appl Phycol 23(5):933–934

Khozin-Goldberg I, Cohen Z (2006) The effect of phosphate starvation on the lipid and fatty acid composition of the fresh water eustigmatophyte Monodus subterraneus. Phytochemistry 67(7):696–701

Khozin-Goldberg I, Bigogno C, Shrestha P, Cohen Z (2002) Nitrogen starvation induces the accumulation of arachidonic acid in the freshwater green alga Parietochloris incisa (Trebuxiophyceae) 1. J Phycol 38(5):991–994

Kiran BR, Venkata Mohan S (2021) Microalgal cell biofactory—therapeutic, nutraceutical and functional food applications. Plants 10(5):836

Ko SC, Kim D, Jeon YJ (2012) Protective effect of a novel antioxidative peptide purified from a marine Chlorella ellipsoidea protein against free radical-induced oxidative stress. Food Chem Toxicol 50(7):2294–2302

Kobylewski S, Jacobson MF (2010) Food dyes: a rainbow of risks. Center for Science in the Public Interest, Washington, DC

Koller M, Muhr A, Braunegg G (2014) Microalgae as versatile cellular factories for valued products. Algal Res 6:52–63

Kong W, Liu N, Zhang J, Yang Q, Hua S, Song H, Xia C (2014) Optimization of ultrasound-assisted extraction parameters of chlorophyll from Chlorella vulgaris residue after lipid separation using response surface methodology. J Food Sci Technol 51(9):2006–2013

Koyande AK, Chew KW, Rambabu K, Tao Y, Chu DT, Show PL (2019) Microalgae: a potential alternative to health supplementation for humans. Food Sci Human Wellness 8(1):16–24

Kratzer R, Murkovic M (2021) Food ingredients and nutraceuticals from microalgae: main product classes and biotechnological production. Foods 10(7):1626

Kumudha A, Selvakumar S, Dilshad P, Vaidyanathan G, Thakur MS, Sarada R (2015) Methylcobalamin—a form of vitamin B12 identified and characterised in Chlorella vulgaris. Food Chem 170:316–320
Lauritano C, Ianora A (2016) Marine organisms with anti-diabetes properties. Mar Drugs 14(12):220
Lauritano C, Andersen JH, Hansen E, Albrigtsen M, Escalera L, Esposito F et al (2016) Bioactivity screening of microalgae for antioxidant, anti-inflammatory, anticancer, anti-diabetes, and antibacterial activities. Front Mar Sci 3:68
Lauritano C, Martín J, de la Cruz M, Reyes F, Romano G, Ianora A (2018) First identification of marine diatoms with anti-tuberculosis activity. Sci Rep 8(1):1–10
Le HD, Meisel JA, de Meijer VE, Gura KM, Puder M (2009) The essentiality of arachidonic acid and docosahexaenoic acid. Prostaglandins Leukot Essent Fatty Acids 81(2–3):165–170
Lee SH, Chang DU, Lee BJ, Jeon YJ (2009) Antioxidant activity of solubilized Tetraselmis suecica and Chlorella ellipsoidea by enzymatic digests. Prev Nutr Food Sci 14(1):21–28
Lemes AC, Sala L, Ores JDC, Braga ARC, Egea MB, Fernandes KF (2016) A review of the latest advances in encrypted bioactive peptides from protein-rich waste. Int J Mol Sci 17(6):950
LewisOscar F, Nithya C, Bakkiyaraj D, Arunkumar M, Alharbi NS, Thajuddin N (2017) Biofilm inhibitory effect of Spirulina platensis extracts on bacteria of clinical significance. Proc Natl Acad Sci, India Sect B Biol Sci 87(2):537–544
Li S, Ji L, Shi Q, Wu H, Fan J (2019) Advances in the production of bioactive substances from marine unicellular microalgae Porphyridium spp. Bioresour Technol 292:122048
Liu J, Sun Z, Gerken H, Liu Z, Jiang Y, Chen F (2014) Chlorella zofingiensis as an alternative microalgal producer of astaxanthin: biology and industrial potential. Mar Drugs 12(6):3487–3515
Liu T, Liu WH, Zhao JS, Meng FZ, Wang H (2017) Lutein protects against β-amyloid peptide-induced oxidative stress in cerebrovascular endothelial cells through modulation of Nrf-2 and NF-κb. Cell Biol Toxicol 33(1):57–67
Liu ZW, Zeng XA, Cheng JH, Liu DB, Aadil RM (2018) The efficiency and comparison of novel techniques for cell wall disruption in astaxanthin extraction from Haematococcus pluvialis. Int J Food Sci Technol 53(9):2212–2219
López Y, Soto SM (2020) The usefulness of microalgae compounds for preventing biofilm infections. Antibiotics 9(1):9
Low KL, Idris A, Yusof NM (2020) Novel protocol optimized for microalgae lutein used as food additives. Food chemistry 307:125631
Maestri E, Marmiroli M, Marmiroli N (2016) Bioactive peptides in plant-derived foodstuffs. J Proteome 147:140–155
Manuell AL, Beligni MV, Elder JH, Siefker DT, Tran M, Weber A et al (2007) Robust expression of a bioactive mammalian protein in Chlamydomonas chloroplast. Plant Biotechnol J 5(3):402–412
Marcati A, Ursu AV, Laroche C, Soanen N, Marchal L, Jubeau S et al (2014) Extraction and fractionation of polysaccharides and B-phycoerythrin from the microalga Porphyridium cruentum by membrane technology. Algal Res 5:258–263
Martínez Andrade KA, Lauritano C, Romano G, Ianora A (2018) Marine microalgae with anticancer properties. Mar Drugs 16(5):165
Mata TM, Martins AA, Caetano NS (2010) Microalgae for biodiesel production and other applications: a review. Renew Sust Energ Rev 14(1):217–232
Matos J, Cardoso C, Bandarra NM, Afonso C (2017) Microalgae as healthy ingredients for functional food: a review. Food Funct 8(8):2672–2685
Mayfield SP, Franklin SE, Lerner RA (2003) Expression and assembly of a fully active antibody in algae. Proc Natl Acad Sci 100(2):438–442
McClements DJ, Decker EA, Park Y, Weiss J (2009) Structural design principles for delivery of bioactive components in nutraceuticals and functional foods. Crit Rev Food Sci Nutr 49(6):577–606

McClure DD, Luiz A, Gerber B, Barton GW, Kavanagh JM (2018) An investigation into the effect of culture conditions on fucoxanthin production using the marine microalgae Phaeodactylum tricornutum. Algal Res 29:41–48

Medhi J, Kalita MC (2020) An emerging aquatic green gold for food and medicine: a review of algae from north east India. Int J Pharm Pharm Sci:7–15

Mehta P, Singh D, Saxena R, Rani R, Gupta RP, Puri SK, Mathur AS (2018) High-value coproducts from algae—an innovational way to deal with advance algal industry. In: Waste to wealth. Springer, Singapore, pp 343–363

Mendes A, Reis A, Vasconcelos R, Guerra P, Lopes da Silva T (2009) Crypthecodinium cohnii with emphasis on DHA production: a review. J Appl Phycol 21(2):199–214

Merlo S, Gabarrell Durany X, Pedroso Tonon A, Rossi S (2021) Marine microalgae contribution to sustainable development. Water 13(10):1373

Milovanovic I, Misan A, Saric B, Kos J, Mandic A, Simeunovic J, Kovac D (2012, May) Evaluation of protein and lipid content and determination of fatty acid profile in selected species of cyanobacteria. In Proceedings of the 6th Central European Congress on Food, CEFood2012, Novi Sad, Serbia (pp. 23–26)

Min KH, Lee HH, Anbu P, Chaulagain BP, Hur BK (2012) The effects of culture condition on the growth property and docosahexaenoic acid production from Thraustochytrium aureum ATCC 34304. Korean J Chem Eng 29(9):1211–1215

Minhas AK, Hodgson P, Barrow CJ, Adholeya A (2016) A review on the assessment of stress conditions for simultaneous production of microalgal lipids and carotenoids. Front Microbiol 7:546

Mishra VK, Bacheti RK, Husen A (2011) Medicinal uses of chlorophyll: a critical overview. In: Chlorophyll: structure, function and medicinal uses. Nova Science, Hauppauge, pp 177–196

Mitra M, Mishra S (2019) A comparative analysis of different extraction solvent systems on the extractability of eicosapentaenoic acid from the marine eustigmatophyte Nannochloropsis oceanica. Algal Res 38:101387

Mitra M, Patidar SK, George B, Shah F, Mishra S (2015) A euryhaline Nannochloropsis gaditana with potential for nutraceutical (EPA) and biodiesel production. Algal Res 8:161–167

Mobin S, Alam F (2017) Some promising microalgal species for commercial applications: a review. Energy Procedia 110:510–517

Molino A, Iovine A, Casella P, Mehariya S, Chianese S, Cerbone A et al (2018a) Microalgae characterization for consolidated and new application in human food, animal feed and nutraceuticals. Int J Environ Res Public Health 15(11):2436

Molino A, Rimauro J, Casella P, Cerbone A, Larocca V, Chianese S et al (2018b) Extraction of astaxanthin from microalga Haematococcus pluvialis in red phase by using generally recognized as safe solvents and accelerated extraction. J Biotechnol 283:51–61

Mourelle ML, Gómez CP, Legido JL (2017) The potential use of marine microalgae and cyanobacteria in cosmetics and thalassotherapy. Cosmetics 4(4):46

Musa M, Ayoko GA, Ward A, Rösch C, Brown RJ, Rainey TJ (2019) Factors affecting microalgae production for biofuels and the potentials of chemometric methods in assessing and optimizing productivity. Cells 8(8):851

Nakano S, Takekoshi H, Nakano M (2010) Chlorella pyrenoidosa supplementation reduces the risk of anemia, proteinuria and edema in pregnant women. Plant Foods Hum Nutr 65(1):25–30

Nakazawa Y, Sashima T, Hosokawa M, Miyashita K (2009) Comparative evaluation of growth inhibitory effect of stereoisomers of fucoxanthin in human cancer cell lines. J Funct Foods 1(1):88–97

Neumann U, Derwenskus F, Gille A, Louis S, Schmid-Staiger U, Briviba K, Bischoff SC (2018) Bioavailability and safety of nutrients from the microalgae Chlorella vulgaris, Nannochloropsis oceanica and Phaeodactylum tricornutum in C57BL/6 mice. Nutrients 10(8):965

Ngo DH, Vo TS, Ngo DN, Wijesekara I, Kim SK (2012) Biological activities and potential health benefits of bioactive peptides derived from marine organisms. Int J Biol Macromol 51(4):378–383

Nguyen MHT, Qian ZJ, Nguyen VT, Choi IW, Heo SJ, Oh CH et al (2013) Tetrameric peptide purified from hydrolysates of biodiesel byproducts of Nannochloropsis oculata induces osteo-

blastic differentiation through MAPK and Smad pathway on MG-63 and D1 cells. Process Biochem 48(9):1387–1394
Norton T, Sun DW (2008) Recent advances in the use of high pressure as an effective processing technique in the food industry. Food Bioprocess Technol 1(1):2–34
Norton TA, Melkonian M, Andersen RA (1996) Algal biodiversity. Phycologia 35(4):308–326
Nuutila AM, Aura AM, Kiesvaara M, Kauppinen V (1997) The effect of salinity, nitrate concentration, pH and temperature on eicosapentaenoic acid (EPA) production by the red unicellular alga Porphyridium purpureum. J Biotechnol 55(1):55–63
Odjadjare EC, Mutanda T, Olaniran AO (2017) Potential biotechnological application of microalgae: a critical review. Crit Rev Biotechnol 37(1):37–52
Olaizola M (2000) Commercial production of astaxanthin from Haematococcus pluvialis using 25,000-liter outdoor photobioreactors. J Appl Phycol 12(3):499–506
Pal D, Khozin-Goldberg I, Cohen Z, Boussiba S (2011) The effect of light, salinity, and nitrogen availability on lipid production by Nannochloropsis sp. Appl Microbiol Biotechnol 90(4):1429–1441
Panis G, Carreon JR (2016) Commercial astaxanthin production derived by green alga Haematococcus pluvialis: a microalgae process model and a techno-economic assessment all through production line. Algal Res 18:175–190
Park JS, Chyun JH, Kim YK, Line LL, Chew BP (2010) Astaxanthin decreased oxidative stress and inflammation and enhanced immune response in humans. Nutr Metab 7(1):1–10
Paswan MB, Chudasama MM, Mitra M, Bhayani K, George B, Chatterjee S, Mishra S (2016) Fluorescence quenching property of C-phycocyanin from Spirulina platensis and its binding efficacy with viable cell components. J Fluoresc 26(2):577–583
Pathak J, Maurya PK, Singh SP, Häder DP, Sinha RP (2018) Cyanobacterial farming for environment friendly sustainable agriculture practices: innovations and perspectives. Front Environ Sci 6:7
Patidar SK, Mishra S (2017) Carbon sequestration by microalgae: a green approach for climate change mitigation. In: Encyclopedia of sustainable technologies, vol 3. Elsevier, Amsterdam, pp 477–483
Patidar SK, Mitra M, George B, Soundarya R, Mishra S (2014) Potential of Monoraphidium minutum for carbon sequestration and lipid production in response to varying growth mode. Bioresour Technol 172:32–40
Peltomaa E, Johnson MD, Taipale SJ (2018) Marine cryptophytes are great sources of EPA and DHA. Mar Drugs 16(1):3
Peng J, Yuan JP, Wu CF, Wang JH (2011) Fucoxanthin, a marine carotenoid present in brown seaweeds and diatoms: metabolism and bioactivities relevant to human health. Mar Drugs 9(10):1806–1828
Pereira AG, Jimenez-Lopez C, Fraga M, Lourenço-Lopes C, García-Oliveira P, Lorenzo JM et al (2020) Extraction, properties, and applications of bioactive compounds obtained from microalgae. Curr Pharm Des 26(16):1929–1950
Plaza M, Herrero M, Cifuentes A, Ibanez E (2009) Innovative natural functional ingredients from microalgae. J Agric Food Chem 57(16):7159–7170
Plaza M, Santoyo S, Jaime L, Avalo B, Cifuentes A, Reglero G et al (2012) Comprehensive characterization of the functional activities of pressurized liquid and ultrasound-assisted extracts from Chlorella vulgaris. LWT-Food Sci Technol 46(1):245–253
Poljsak B, Milisav I (2012) The neglected significance of "antioxidative stress". Oxid Med Cell Longevity 2012:480895
Pouvreau JB, Taran F, Rosa P, Pin S, Fleurence J, Pondaven P (2008) Antioxidant and free radical scavenging properties of marennine, a blue-green polyphenolic pigment from the diatom Haslea ostrearia (Gaillon/Bory) Simonsen responsible for the natural greening of cultured oysters. J Agric Food Chem 56(15):6278–6286
Pulz O, Gross W (2004) Valuable products from biotechnology of microalgae. Appl Microbiol Biotechnol 65(6):635–648

Puyfoulhoux G, Rouanet JM, Besançon P, Baroux B, Baccou JC, Caporiccio B (2001) Iron availability from iron-fortified spirulina by an in vitro digestion/Caco-2 cell culture model. J Agric Food Chem 49(3):1625–1629

Qi B, Beaudoin F, Fraser T, Stobart AK, Napier JA, Lazarus CM (2002) Identification of a cDNA encoding a novel C18-Δ9 polyunsaturated fatty acid-specific elongating activity from the docosahexaenoic acid (DHA)-producing microalga, Isochrysis galbana. FEBS Lett 510(3):159–165

Raposo MFDJ, De Morais RMSC, Bernardo de Morais AMM (2013) Bioactivity and applications of sulphated polysaccharides from marine microalgae. Mar Drugs 11(1):233–252

Raposo MFDJ, De Morais AMMB, De Morais RMSC (2015) Carotenoids from marine microalgae: a valuable natural source for the prevention of chronic diseases. Mar Drugs 13(8):5128–5155

Rasala BA, Muto M, Lee PA, Jager M, Cardoso RM, Behnke CA et al (2010) Production of therapeutic proteins in algae, analysis of expression of seven human proteins in the chloroplast of Chlamydomonas reinhardtii. Plant Biotechnol J 8(6):719–733

Rasmussen HM, Johnson EJ (2013) Nutrients for the aging eye. Clin Interv Aging 8:741

Reynolds CS (2006) The ecology of phytoplankton. Cambridge University Press, Cambridge

Riccio G, Lauritano C (2020) Microalgae with immunomodulatory activities. Mar Drugs 18(1):2

Robertson R, Guihéneuf F, Schmid M, Stengel DB, Fitzgerald G, Ross P, Stanton C (2013) Algae-derived polyunsaturated fatty acids: implications for human health. In: Polyunsaturated fatty acids: sources, antioxidant properties and health benefits, pp 45–99

Saha SK, Murray P (2018) Exploitation of microalgae species for nutraceutical purposes: cultivation aspects. Fermentation 4(2):46

Saha SK, McHugh E, Murray P, Walsh DJ (2015) Microalgae as a source of nutraceuticals. In: Botana LM, Alfonso A (eds) Phycotoxins: chemistry and biochemistry, 2nd edn. Wiley, New York, pp 255–292

Saide A, Martínez KA, Ianora A, Lauritano C (2021) Unlocking the health potential of microalgae as sustainable sources of bioactive compounds. Int J Mol Sci 22(9):4383

Saifullah AZA, Karim MA, Ahmad-Yazid A (2014) Microalgae: an alternative source of renewable energy. Am J Eng Res 3(3):330–338

Saini DK, Chakdar H, Pabbi S, Shukla P (2020) Enhancing production of microalgal biopigments through metabolic and genetic engineering. Crit Rev Food Sci Nutr 60(3):391–405

Samarakoon KW, O-Nam K, Ko JY, Lee JH, Kang MC, Kim D et al (2013) Purification and identification of novel angiotensin-I converting enzyme (ACE) inhibitory peptides from cultured marine microalgae (Nannochloropsis oculata) protein hydrolysate. J Appl Phycol 25(5):1595–1606

Sansone C, Galasso C, Orefice I, Nuzzo G, Luongo E, Cutignano A et al (2017) The green microalga Tetraselmis suecica reduces oxidative stress and induces repairing mechanisms in human cells. Sci Rep 7(1):1–12

Sathasivam R, Ki JS (2018) A review of the biological activities of microalgal carotenoids and their potential use in healthcare and cosmetic industries. Mar Drugs 16(1):26

Sathasivam R, Radhakrishnan R, Hashem A, Abd Allah EF (2019) Microalgae metabolites: a rich source for food and medicine. Saudi J Biol Sci 26(4):709–722

Sekar S, Chandramohan M (2008) Phycobiliproteins as a commodity: trends in applied research, patents and commercialization. J Appl Phycol 20(2):113–136

Shah M, Mahfuzur R, Liang Y, Cheng JJ, Daroch M (2016) Astaxanthin-producing green microalga Haematococcus pluvialis: from single cell to high value commercial products. Front Plant Sci 7:531

Sheih IC, Wu TK, Fang TJ (2009) Antioxidant properties of a new antioxidative peptide from algae protein waste hydrolysate in different oxidation systems. Bioresour Technol 100(13):3419–3425

Sheih IC, Fang TJ, Wu TK, Lin PH (2010) Anticancer and antioxidant activities of the peptide fraction from algae protein waste. J Agric Food Chem 58(2):1202–1207

Shih MF, Cherng JY (2012) Protective effects of Chlorella-derived peptide against UVC-induced cytotoxicity through inhibition of caspase-3 activity and reduction of the expression of phosphorylated FADD and cleaved PARP-1 in skin fibroblasts. Molecules 17(8):9116–9128

Shih MF, Chen LC, Cherng JY (2013) Chlorella 11-peptide inhibits the production of macrophage-induced adhesion molecules and reduces endothelin-1 expression and endothelial permeability. Mar Drugs 11(10):3861–3874
Shukla M, Kumar S (2018) Algal biorefineries for biofuels and other value-added products. In: Biorefining of biomass to biofuels. Springer, Cham, pp 305–341
Simó C, Herrero M, Neusüß C, Pelzing M, Kenndler E, Barbas C et al (2005) Characterization of proteins from Spirulina platensis microalga using capillary electrophoresis-ion trap-mass spectrometry and capillary electrophoresis-time of flight-mass spectrometry. Electrophoresis 26(13):2674–2683
Singh A, Wilson S, Ward OP (1996) Docosahexaenoic acid (DHA) production by Thraustochytrium sp. ATCC 20892. World J Microbiol Biotechnol 12(1):76–81
Singh B, Guldhe A, Singh P, Singh A, Rawat I, Bux F (2015) Sustainable production of biofuels from microalgae using a biorefinary approach. In: Applied environmental biotechnology: present scenario and future trends. Springer, New Delhi, pp 115–128
Smetana S, Sandmann M, Rohn S, Pleissner D, Heinz V (2017) Autotrophic and heterotrophic microalgae and cyanobacteria cultivation for food and feed: life cycle assessment. Bioresour Technol 245:162–170
Sonani RR, Rastogi RP, Patel R, Madamwar D (2016) Recent advances in production, purification and applications of phycobiliproteins. World J Biol Chem 7(1):100
Specht E, Miyake-Stoner S, Mayfield S (2010) Micro-algae come of age as a platform for recombinant protein production. Biotechnol Lett 32(10):1373–1383
Spolaore P, Joannis-Cassan C, Duran E, Isambert A (2006) Commercial applications of microalgae. J Biosci Bioeng 101(2):87–96
Sprecher H, Chen Q, Yin FQ (1999) Regulation of the biosynthesis of 22: 5n-6 and 22: 6n-3: a complex intracellular process. Lipids 34:S153–S156
Sukenik A (1991) Ecophysiological considerations in the optimization of eicosapentaenoic acid production by Nannochloropsis sp.(Eustigmatophyceae). Bioresour Technol 35(3):263–269
Sun M, Qian K, Su N, Chang H, Liu J, Shen G (2003) Foot-and-mouth disease virus VP1 protein fused with cholera toxin B subunit expressed in Chlamydomonas reinhardtii chloroplast. Biotechnol Lett 25(13):1087–1092
Sun Z, Li T, Zhou ZG, Jiang Y (2015) Microalgae as a source of lutein: chemistry, biosynthesis, and carotenogenesis. In: Microalgae biotechnology. Springer, Cham, pp 37–58
Surzycki R, Greenham K, Kitayama K, Dibal F, Wagner R, Rochaix JD et al (2009) Factors effecting expression of vaccines in microalgae. Biologicals 37(3):133–138
Talukdar J, Dasgupta S, Nagle V, Bhadra B (2020) COVID-19: potential of microalgae derived natural astaxanthin as adjunctive supplement in alleviating cytokine storm. SSRN 3579738
Tan CK, Johns MR (1996) Screening of diatoms for heterotrophic eicosapentaenoic acid production. J Appl Phycol 8(1):59–64
Tang G, Suter PM (2011) Vitamin A, nutrition, and health values of algae: spirulina, chlorella, and dunaliella. J Pharm Nutr Sci 1(2):111–118
Tian M, Fang B, Jiang L, Guo H, Cui J, Ren F (2015) Structure-activity relationship of a series of antioxidant tripeptides derived from β-Lactoglobulin using QSAR modeling. Dairy Sci Technol 95(4):451–463
Tran M, Zhou B, Pettersson PL, Gonzalez MJ, Mayfield SP (2009) Synthesis and assembly of a full-length human monoclonal antibody in algal chloroplasts. Biotechnol Bioeng 104(4):663–673
Udayan A, Arumugam M, Pandey A (2017) Nutraceuticals from algae and cyanobacteria. In: Algal green chemistry. Elsevier, Amsterdam, pp 65–89
Van Krimpen MM, Bikker P, Van der Meer IM, Van der Peet-Schwering CMC, Vereijken JM (2013) Cultivation, processing and nutritional aspects for pigs and poultry of European protein sources as alternatives for imported soybean products (No. 662). Wageningen UR Livestock Research, Wageningen
Varshney P, Mikulic P, Vonshak A, Beardall J, Wangikar PP (2015) Extremophilic micro-algae and their potential contribution in biotechnology. Bioresour Technol 184:363–372

Viskari PJ, Colyer CL (2003) Rapid extraction of phycobiliproteins from cultured cyanobacteria samples. Anal Biochem 319(2):263–271
Vo TS, Kim J, Wijesekara I, Kong CS, Kim SK (2011) Potent effect of brown algae (Ishige okamurae) on suppression of allergic inflammation in human basophilic KU812F cells. Food Sci Biotechnol 20(5):1227–1234
Vonshak A, Guy R (1992) Photoadaptation, photoinhibition and productivity in the blue-green alga, Spirulina platensis grown outdoors. Plant Cell Environ 15(5):613–616
Wagner M, Hoppe K, Czabany T, Heilmann M, Daum G, Feussner I, Fulda M (2010) Identification and characterization of an acyl-CoA: diacylglycerol acyltransferase 2 (DGAT2) gene from the microalga O. tauri. Plant Physiol Biochem 48(6):407–416
Wang X, Zhang X (2013) Separation, antitumor activities, and encapsulation of polypeptide from Chlorella pyrenoidosa. Biotechnol Prog 29(3):681–687
Wang X, Brandsma M, Tremblay R, Maxwell D, Jevnikar AM, Huner N, Ma S (2008) A novel expression platform for the production of diabetes-associated autoantigen human glutamic acid decarboxylase (hGAD65). BMC Biotechnol 8(1):1–13
Watanabe F, Takenaka S, Kittaka-Katsura H, Ebara S, Miyamoto E (2002) Characterization and bioavailability of vitamin B12-compounds from edible algae. J Nutr Sci Vitaminol 48(5):325–331
Williams PJLB, Laurens LM (2010) Microalgae as biodiesel & biomass feedstocks: review & analysis of the biochemistry, energetics & economics. Energy Environ Sci 3(5):554–590
Wu Q, Zhang XS, Wang HD, Zhang X, Yu Q, Li W et al (2014) Astaxanthin activates nuclear factor erythroid-related factor 2 and the antioxidant responsive element (Nrf2-ARE) pathway in the brain after subarachnoid hemorrhage in rats and attenuates early brain injury. Mar Drugs 12(12):6125–6141
Xin Y, Lu Y, Lee YY, Wei L, Jia J, Wang Q et al (2017) Producing designer oils in industrial microalgae by rational modulation of co-evolving type-2 diacylglycerol acyltransferases. Mol Plant 10(12):1523–1539
Xu Y, Ibrahim IM, Wosu CI, Ben-Amotz A, Harvey PJ (2018) Potential of new isolates of Dunaliella salina for natural β-carotene production. Biology 7(1):14
Yaguchi T, Tanaka S, Yokochi T, Nakahara T, Higashihara T (1997) Production of high yields of docosahexaenoic acid by Schizochytrium sp. strain SR21. J Am Oil Chem Soc 74(11):1431–1434
Yang Z, Chen F, Li D, Zhang Z, Liu Y, Zheng D et al (2006) Expression of human soluble TRAIL in Chlamydomonas reinhardtii chloroplast. Chin Sci Bull 51(14):1703–1709
Yates CM, Calder PC, Rainger GE (2014) Pharmacology and therapeutics of omega-3 polyunsaturated fatty acids in chronic inflammatory disease. Pharmacol Ther 141(3):272–282
Yongmanitchai W, Ward OP (1991) Growth of and omega-3 fatty acid production by Phaeodactylum tricornutum under different culture conditions. Appl Environ Microbiol 57(2):419–425
Zhang YK, Shen GF, Ru BG (2006) Survival of human metallothionein-2 transplastomic Chlamydomonas reinhardtii to ultraviolet B exposure. Acta Biochim Biophys Sin 38(3):187–193
Zolla L, Bianchetti M (2001) High-performance liquid chromatography coupled on-line with electrospray ionization mass spectrometry for the simultaneous separation and identification of the Synechocystis PCC 6803 phycobilisome proteins. J Chromatogr A 912(2):269–279
Zou TB, He TP, Li HB, Tang HW, Xia EQ (2016) The structure-activity relationship of the antioxidant peptides from natural proteins. Molecules 21(1):72

Unveiling Nature's Treasures: Exploring Bioactive Compounds from Algae for Extraction, Refinement, and Diverse Applications

17

Andhere Animish and M. A. Jayasri

Contents

A. Animish · M. A. Jayasri (✉)
Marine Biotechnology and Bioproducts Laboratory, School of Biosciences and Technology, Vellore Institute of Technology, Vellore, Tamil Nadu, India
e-mail: jayasri.ma@vit.ac.in

S. K. Arya et al. (eds.), *Value Added Products From Bioalgae Based Biorefineries: Opportunities and Challenges*,
https://doi.org/10.1007/978-981-97-1662-3_17

Abstract

Exploring bioactive compounds from algae has emerged as a captivating research focus in recent years. Studies have demonstrated that algae harbor a reservoir of intrinsic natural substances encompassing pigments, polyphenols, terpenes, polysaccharides, vitamins, and other bioactive entities, each bestowing distinctive physiological properties and multiple applications. Pigments, recognized for their antioxidative potency, confer cellular protection against oxidative stress and inflammation, showcasing their potential as therapeutic agents. Compounds like polyphenols and flavonoids have gained scientific attention due to their potential contributions to treating cardiovascular health, cognitive well-being, and other major diseases. Advancements in extraction methodologies, such as supercritical fluid extraction and enzyme-assisted extraction, have been harnessed to unlock these bioactive compounds from algae. Subsequent refinement endeavors, encompassing chromatographic techniques and filtration, ensure the attainment of high-purity compounds. These bioactive compounds have myriad applications in nutraceuticals, pharmaceuticals, food and cosmetics, biofuels, animal feed, agrochemicals, and environmental remediation. These compounds exhibit the potential to promote overall well-being, propelling advancements in medical research with other diverse applications. Despite the promising effects, challenges persist, necessitating sustainable extraction practices and new methods for refining these bioactives and evaluating their potential for diverse applications. In addition to evaluating our existing knowledge, we distinctly emphasize the latest advancements in extraction, purification, and refinement methodologies. Exploration of bioactive compounds using eco-friendly and/or cost-effective extraction techniques and innovative purification strategies for versatile applications holds the key to a harmonious future where algal bioactive compounds harmonize health, industry, and the environment.

Keywords

Algal bioactives · Extraction techniques · Purification strategies · Algal applications · Industrial utilization · Commercial bioproducts

17.1 Introduction: Algae's Hidden Gems

17.1.1 The Enigmatic World of Algal Bioactive Compounds

Algae, often overshadowed by their terrestrial counterparts, have emerged as a source of innovation with profound implications across various sectors. The study of algal bioactives has gained remarkable momentum in recent years, revealing a treasure trove of compounds with pharmaceutical, nutraceutical, agricultural, and environmental significance. Algal biomasses offer a rich reservoir of varied bioactive elements like polysaccharides, lipids, vitamins, polyphenols, terpenes, pigments, and additional bioactive substances that emerge during their growth and development. These components have been reported to exhibit antibacterial, anti-inflammatory, antifungal, antiviral, anticancer, antioxidant, chemoprotective, and neuroprotective properties (Eze et al. 2023). Algae have different morphological varieties and inhabit various aquatic environments, thriving in habitats containing harmful metals, sewage, chemicals, salts, and pesticides, with varied temperatures, ultraviolet-B radiation, fluctuating light, and pH conditions (Rai and Gaur 2001; Xue et al. 2005). Besides, they serve as primary producers of the ecosystems of various water sources and possess elevated tolerance capacity to endure a wide array of stressors (Rai and Gaur 2001; Saha and Fink 2022). In response to these challenging circumstances, they synthesize a multitude of biologically active compounds to ensure their survival (Akula and Ravishankar 2011; Kolackova et al. 2023).

Essential medicinal pigments like phycobilins, chlorophylls, and carotenoids (fucoxanthins, astaxanthin, zeaxanthin, canthaxanthin, and lutein) possess a wide range of uses, including antioxidant, anti-inflammatory, immunoprophylactic, anti-obesity, and antitumor properties, among various others (Chen et al. 2023). Algae stand out as prominent and abundant reservoirs of polysaccharides with diverse structures, like carrageenans, xylogalactans, fucoidans, galactans, porphyran, floridean-starch, agar, laminarins, and alginates (Hentati et al. 2020). Moreover, algae are also renowned for their lipids with significant nutritional worth exhibiting multiple applications due to the presence of essential fatty acids (n-6 and n-3 polyunsaturated fatty acids, glycolipids, phospholipids, derivatives of hexaenoic and pentaenoic acids, monounsaturated and unsaturated fatty acids) (Domingues and Calado 2022; Zhou et al. 2022a). Besides, it is also reported that algal lipids hold the potential in the realm of biofuel generation (Ramesh Kumar et al. 2019).

Utilized across diverse industries, lectins sourced from macroalgae and phycobiliproteins obtained from microalgae stand out as two classes of bioactive algal proteins (Bleakley and Hayes 2017). Lectins exhibit functions in blood group classification, identification of cancer indicators, antiviral treatments, and identification of cancer biomarkers, and are also reported to serve as targets for efficient drug delivery (Weis and Drickamer 1996; Ziółkowska and Wlodawer 2006; Naeem et al. 2007). Phycobiliproteins, encompassing phycocyanin, phycoerythrin, allophycocyanin, and phycoerythrocyanin, serve roles in fluorescent tagging, flow cytometry, fluorescent microscopy, and immunochemical processes, and are commercially also used as a blue pigment in food, dairy goods, and cosmetics (Kronick and Grossman

1983; Glazer 1994; Spolaore et al. 2006; Sekar and Chandramohan 2008). In terms of trace elements, including minerals and vitamins, algae offer a valuable repository. They provide essential macronutrients and microelements such as zinc, iron, potassium, magnesium, selenium, calcium, and sodium, which play a vital role in metabolism (Demarco et al. 2022). Additionally, various studies have also revealed the presence of Vitamin A, Vitamin B1, Vitamin B12, Vitamin C, Vitamin D, and Vitamin E (Biris-Dorhoi et al. 2020; Del Mondo et al. 2020). Prominent compounds such as phlorotannins, phloroglucinol, phycocolloids, eckol, dieckol, 7-phloroeckol, fucosterols, pheophorbide-A, phlorofucofuroeckol-A, alginic acid, as well as sargaquinoic and sargahydroquinoic acids have been documented for their robust potential as effective antidiabetic agents (Jones et al. 1988; Nwosu et al. 2011; Kellogg et al. 2014; Lee et al. 2015; Remya and Radhika Rajasree 2016; Gunathilaka et al. 2020; Animish and Jayasri 2023).

This chapter lists the intricate techniques involved in extracting and refining valuable compounds from diverse algal sources. It also describes the methodologies employed to obtain purified and potent bioactive compounds. Also, the chapter aims to shed light on the transformative applications that arise from these extracted and refined compounds, ranging from pharmaceutical advancements to innovative solutions in agriculture, materials science, and beyond. Through a systemic and insightful examination, the chapter emphasizes the practical and theoretical dimensions that underscore the importance of extraction, refinement, and application in contemporary scientific endeavors. Both macro- and microalgae have proven to be a prosperous source of secondary metabolites and bioactive compounds, each holding significant promise in fields ranging from medicine to agriculture, pharmaceutical, cosmetics, and industrial applications. The diversity and complexity of these compounds demonstrate that algae possess an intricate biochemical reserve that continues to inspire innovative research and the development of novel solutions for various challenges, promising a future rich with discoveries and transformative applications, occupying a central role in advanced research and development.

17.2 Algal Biomolecules: Diverse Bioactive Compounds

17.2.1 Nature's Palette: Pigments, Antioxidants, and Beyond

The captivating domain of algal pigments and antioxidants has garnered significant attention due to their multiple applications and potential benefits. Algae, often overlooked, are abundant sources of these bioactive compounds, offering a diverse spectrum of colors and properties that extend beyond their aesthetic appeal. These pigments, such as chlorophylls, carotenoids, and phycobilins, not only drive photosynthesis but also exhibit remarkable antioxidant capabilities (Negreanu-Pirjol et al. 2020; Yalçın et al. 2021b; Sanger et al. 2022). Vital therapeutic carotenoids such as fucoxanthins, astaxanthin, zeaxanthin, phycoerythrin, canthaxanthin, and lutein offer a diverse spectrum of applications, encompassing antioxidative, anti-inflammatory, immune-protective, antiobesity, and anticancer attributes, among numerous others (Chen et al. 2023). The incorporation of double bonds into their molecular structure enables these pigments to offer extensive health advantages, shielding other molecules from the oxidative stress triggered by free radicals (Patel et al. 2022). The antioxidant properties of carotenoids play a pivotal role in shaping their utilization as ingredients in cosmetics and as safe colorants (Singh and Purwar 2022). These antioxidants play a pivotal role in mitigating oxidative stress and its associated health implications, thus paving the way for applications in pharmaceuticals, nutraceuticals, and functional foods. Moreover, the exploration of algal pigments and antioxidants extends to cosmetics, where their natural and vibrant hues find use in various beauty and personal care products. These convergences of vibrant colors, potent antioxidants, and diverse applications showcase the versatility and untapped potential of algal compounds, underscoring their importance in shaping various industries while also contributing to understanding natural chemical diversity.

17.2.2 Polyphenols and Flavonoids: Guardians of Health

Due to their diverse biological activities, algal polyphenols and flavonoids have drawn favorable attention for their promising therapeutic potential. These compounds, derived from various algal species, exhibit antioxidant, antiviral, anti-inflammatory, antimicrobial, and anticancer properties, among others (Besednova et al. 2021; Ferdous and Balia Yusof 2021; Pereira and Cotas 2023). Extraction and refinement of these bioactive compounds involve techniques such as solvent extraction, ultrasound, solid-phase extraction, microwaves, and chromatography, aiming to isolate and purify them effectively (Chaves et al. 2020; Kim et al. 2022). Once refined, algal polyphenols and flavonoids find applications in various industries. In the pharmaceutical sector, they are explored for their potential in drug development and disease prevention; at the same time, in the nutraceutical and functional food industries, they add value as natural additives that promote health and wellness (Kumari et al. 2023). The cosmetic industry benefits from their antiaging and skin-enhancing, whitening properties, which are increasingly sought after by consumers (Thomas and Kim 2013; Thiyagarasaiyar et al. 2020). Thus, algal polyphenols and flavonoids stand as valuable resources in the pursuit of novel therapeutic agents with a broad spectrum of applications across different sectors.

17.2.3 Terpenes and Sterols: Aromatic Wonders

Algal terpenes and sterols encompass a broad and extensively studied category of natural compounds characterized by distinct structures with different applications. Within the microalgal diversity, brown algae stand out as a prime reservoir of terpenoids that hold significant biological and ecological importance (Gaysinski et al. 2015). Terpenes are categorized into mono-, di-, tri-, tetra-, and sesquiterpenes based on the count of isoprene units they contain (Gaysinski et al. 2015). These compounds exhibit a plethora of potential benefits, including antioxidants, antiplasmodial, antiviral, anticancer, antidiabetic, anti-inflammatory, and other properties (Lee et al. 2013; Yang et al. 2020; Tamanna Ferdous and Norhana Balia Yusof 2021). Terpenes are also reported to play vital roles in the metabolism of microalgae, such as participating in respiration via plastoquinone and ubiquinone, forming membrane components, and acting as antioxidants (Borisova-Mubarakshina et al. 2019). Terpenes from microalgae have also been explored for a wide range of applications, from functional foods to biofuel production (Costa et al. 2022). Algae also serve as abundant sources of phytosterols, with dietary significance that effectively lowers blood-cholesterol levels and exhibits potential in curbing colon cancer, anti-inflammatory, antibacterial, antifungal, antiulcerative, and antitumor effects (Lopes et al. 2013). Thus, algae emerge as intriguing repositories of dietary phytosterols and terpenes with immense potential and novel applications in health, cosmetics, agriculture, and beyond.

17.2.4 Polysaccharides and Proteins: Structural Wonders

Algal polysaccharides and proteins exhibit diverse properties and applications that underscore their significance across various sectors. Derived from a range of algal species, these compounds offer emulsifying, gelling, and antioxidant qualities, rendering them valuable components within the food and pharmaceutical industries (Bixler and Porse 2011; Soto-Sierra et al. 2021). Notable algal polysaccharides like carrageenans, agar, and alginate are pivotal in food products as stabilizers, thickeners, gelling agents, and tooth implants (Bixler and Porse 2011; Rajendran 2020), while algal proteins such as spirulina protein and seaweed-derived protein hydrolysates hold promise for their nutritional and bioactive attributes (Karkos et al. 2011; Nuhu 2013; Soto-Sierra et al. 2021). Moreover, the biodegradable and bioactive nature of these compounds contributes to the development of economically friendly packaging materials (Perera et al. 2021). Literature suggests that algal polysaccharides offer multiple applications, serving as bio-stimulants for enhanced plant growth, sustainable fertilizers, bio-elicitors triggering plant defense, wound dressings with antimicrobial properties, biomaterials for tissue engineering scaffolds, and controlled drug delivery systems (Bouissil et al. 2019). As the scope of applications for algal polysaccharides and proteins continues to broaden, their versatile characteristics stand poised to reshape industries and drive sustainable solutions.

17.2.5 Vitamins, Minerals, and Beyond: The Nutrient Ensemble

Within the Kingdom Plantae, marine algae exhibit the capacity to generate a wide array of vitamins, and, with microalgae having rapid growth rates, they hold the potential as valuable sources of vitamins (Del Mondo et al. 2020). An example of this is the widely recognized *Spirulina platensis*, renowned for its vitamin richness, thus earning its status as a "superfood" (Jung et al. 2019). Microalgae exhibit the capability to contain specific vitamins that are absent in higher plants, including Vitamin D, Vitamin K, and Vitamin B12 (Smith et al. 2007; Edelmann et al. 2019). Studies have reported that Vitamin D is notably high in edible marine fishes, which acquire it from algal-based diets, as they cannot synthesize it themselves (Lock et al. 2010; Lehmann et al. 2016). Microalgae serve as valuable reservoirs of vitamins, including Vitamin A, Vitamin B family (biotin, pantothenic acid, niacin), Vitamin C, and Vitamin E, along with essential minerals like magnesium, iron, potassium, calcium, and iodine (Becker 2013; Koyande et al. 2019). Findings through acid digestion and atomic absorption spectroscopy of seaweeds revealed the presence of zinc, calcium, copper, iron, magnesium, potassium, and manganese, indicating their potential as a rich mineral source for enhanced human nutrition in the future (Xavier and Jose 2020). Algal vitamins and minerals are set to revolutionize nutrition, health, and technology, opening doors to a future of limitless possibilities.

Figure 17.1 demonstrates a diverse spectrum of bioactive compounds from algae including important pigments, polyphenols, flavonoids, terpenes, terpenoids, sterols, lipids, polysaccharides, proteins, vitamins, and minerals.

17.3 Extraction Techniques: Unveiling Algal Treasures

17.3.1 Solvent Extraction

Solvent extraction is a widely employed technique for the isolation of bioactive compounds. This method entails the use of organic solvents to extract various valuable compounds, such as pigments, lipids, phenols, and other valuable bioactives, from algal biomass. The selection of solvent is crucial in determining the efficiency of extraction, as it affects the selectivity and yield of the target bioactives (Sasidharan et al. 2011). Common solvents include chloroform, methanol, ethanol, acetone, hexane, and water, each offering specific advantages based on the compounds of interest (Truong et al. 2019). For instance, chloroform effectively extracts lipophilic compounds like lipids, while methanol is often used to extract pigments like chlorophyll and carotenoids (Simon and Helliwell 1998; Da Silva Nonato et al. 2020). It is important to note that optimization of extraction parameters, such as solvent-to-biomass ratio, extraction time, and temperature, can significantly influence the yield and purity of extracted bioactives. The technique's effectiveness and versatility have led to its widespread use in both laboratory-scale research and industrial applications. Considering that the compounds of interest can range from non-polar to polar and can be sensitive to heat, it becomes crucial to carefully select appropriate extraction methods (Sasidharan et al. 2011). Commonly employed techniques include Soxhlet extraction, sonication, cold maceration, infusion, decoction, and reflux heating (The United States Pharmacopeia 2004; Chinese Pharmacopoeia Commission 2015; The Ayurvedic Pharmacopoeia of India 2016). While these methods have a history of effective use, it is worth noting that traditional solvents used in these processes are environmentally and potentially harmful to human

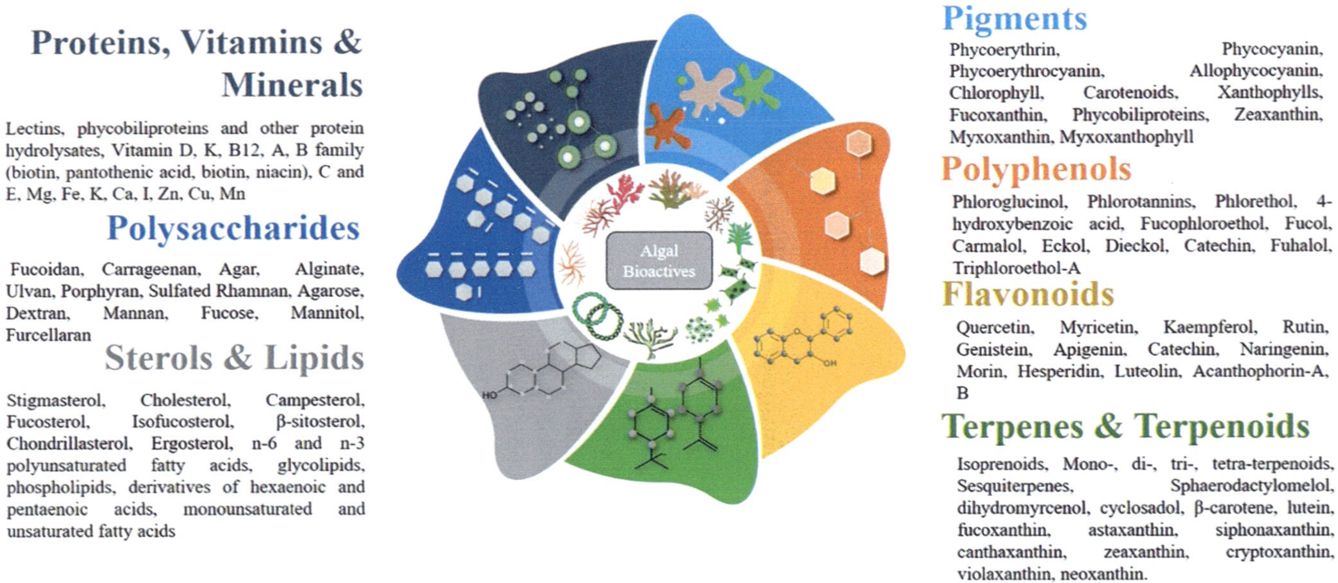

Fig. 17.1 Diverse Algal Bioactives

health due to their toxicity. Substitute solvents encompass a category of environmentally friendly options, including bio-derived solvents like ionic liquids, terpenes, liquid polymers such as polyethylene glycol, and deep eutectic solvents (Corrêa et al. 2020). Therefore, there is a growing need to align extraction practices with sustainability and safety considerations.

17.3.2 Supercritical Fluid Extraction

Supercritical fluid extraction has emerged as an innovative technique for the extraction of bioactive compounds. This method employs supercritical fluids, such as carbon dioxide (CO_2), which exist in a state between a gas and a liquid under specific temperature and pressure conditions (Tirado et al. 2023). The unique properties of supercritical fluids, including their high diffusivity and low viscosity, make them excellent solvents for extracting a wide range of compounds. In the context of algal bioactives, supercritical fluid extraction offers advantages such as reduced solvent usage, minimized thermal degradation, and the ability to extract temperature-sensitive compounds. Notably, supercritical CO_2 is considered safe, non-toxic, and eco-friendly (Montalbán and Víllora 2022). Researchers have successfully applied this technique to extract lipids, pigments, bio-oils, and other valuable components from various algal species (Mendes et al. 1995; Goto et al. 2015; Patil et al. 2018; Georgiopoulou et al. 2022). This method's versatility and potential for producing high-quality extracts have contributed to its growing relevance in various applications. When contrasting supercritical fluid extraction with traditional ethanol extraction, researches have reported distinct advantages in terms of extraction speed, efficacy, carotenoid content, and sensory attributes like odor and color despite yielding lower quantities (Kader and Rahman 2021; Georgiopoulou et al. 2022). This comparison underscores supercritical fluid extraction's superiority in extracting valuable bioactives from algal sources, showcasing its potential for producing high-quality extracts for various applications.

17.3.3 Microwave-Assisted Extraction

Microwave-assisted extraction has emerged as a promising method for the extraction of phytochemicals. This technique employs microwave radiation to enhance the extraction process by promoting the rapid heating and release of target compounds from algae. Microwaves trigger the rotation of dipoles within the organic molecules, accompanied by heating, which disrupts hydrogen bonding, which prompts the movements of ions, generating a heating impact owing to high ion kinetic energies and the friction arising from their ceaseless motion and altered trajectories (Hudaib et al. 2003; Rao et al. 2014). The breakdown of hydrogen bonding further augments the solvent's ability to permeate the plant material (Akhtar et al. 2019). Besides, microwave energy facilitates the disruption of algal cell walls, allowing for the rapid release of metabolites into the solvent (Kapoore et al. 2018). Optimizing the process's effectiveness involves taking into account various factors, including the nature of the material, choice of the solvent, the ratio of solvent to plant material, the microwave's power, temperature, the extraction duration, instrument type (closed or open type), and the characteristics of both the algal material and the desired molecule. Notable classes of algal metabolites extracted using microwave-assisted extraction include pigments (carotenoids, chlorophylls) from microalgae, lipids, antioxidants, and polysaccharides (Pasquet et al. 2011; Kapoore et al. 2018; Sarkar et al. 2020). Additionally, lipids from various algal species, including *Chlorella* and *Nannochloropsis*, are also extracted using this technique, demonstrating its potential for lipid recovery (Zghaibi et al. 2019; Krishnan et al. 2020; Rezaei Motlagh et al. 2020). The versatility of microwave-assisted extraction in targeting different algal metabolites underscores its significance in optimizing extraction processes for algal bioactives.

17.3.4 Enzyme-Assisted Extraction

Enzyme-assisted extraction stands out as a potent technology for yielding bioactives, as it effectively weakens/disrupts the rigid heterogenous cell wall structures, enabling the release of desired bioactive compounds into the extraction solvent (Teixeira-Guedes et al. 2023). The cell walls of seaweed consist of intricate and chemically diverse biomolecules, including sulfated polysaccharides, branched polysaccharides, bound proteins, various ions, and other complex components that pose challenges to extracting bioactive compounds (Domozych et al. 2012; Shao and Duan 2022). While organic solvents like diethyl ether, benzene, and acetonitrile yield higher quantities of desired compounds, these methods are not suitable for food applications and can be environmentally harmful (Wijesinghe and Jeon 2012; Rodrigues et al. 2015). Utilizing food-grade enzymes like carbohydrases and proteases to break down seaweed cell walls has gained much attention (Kadam et al. 2013; Ummat et al. 2021). Notable classes of algal metabolites extracted using enzyme-assisted extraction include antioxidant and antiviral extracts, bioactive peptides, phlorotannins, phenols, and others (Sánchez-Camargo et al. 2016). In the area of macroalgae-based biorefinery, a prevailing approach involves the extraction of novel, specialized enzymes from marine life forms. These enzymes serve to break down complex polysaccharides such as ulvan, laminarin, glucuronan, porphyran, carrageenan, and agar (Holdt and Kraan 2011). In order to achieve a higher yield of the desired bioactives, this technique can be used in combination with other methods such as solvent extraction, microwave-assisted extraction, ultrasonication, among others (Wen et al. 2020; Li et al. 2021). This enzyme-based approach offers eco-friendly and non-toxic advantages by minimizing the need of solvents in the extraction process.

17.3.5 Subcritical Water Extraction

Subcritical water extraction has emerged as a promising new method for obtaining bioactive compounds from algae. This technique involves utilizing water at temperatures and pressures below its critical point, enhancing its solvent properties, and facilitating the extraction of valuable components (Radovanović et al. 2022). It offers distinct advantages such as reducing extraction times, minimal solvent usage, and preserving thermally sensitive compounds (Radovanović et al. 2022). An intriguing and valuable attribute of subcritical water lies in its ability to significantly reduce its polarity as temperature rises. As a result, subcritical water can exhibit resemblances to solvents like ethanol/methanol, and this quality positions subcritical water as an environmentally friendly extraction solvent employed across diverse organic compounds (Cheng et al. 2021). In the context of algal bioactives, this method has been shown to effectively recover various compounds, including pigments, phenolic compounds, lipids, biofuels, and polysaccharides (Herrero et al. 2006; Thiruvenkadam et al. 2015; Gallego et al. 2019; Sarkar et al. 2022). Additionally, subcritical water extraction has demonstrated potential in the extraction of lipids from algae for biodiesel production, with lipid yields comparable to conventional solvent extraction methods (Huang et al. 2019). These findings underscore the utility of subcritical water extraction as an eco-friendly and efficient technique for obtaining bioactive compounds from algae.

17.3.6 Pressurized-Liquid Extraction

Pressurized-liquid extraction, also known as accelerated solvent extraction, has emerged as a potent technique for extracting bioactive compounds. This method involves subjecting the sample to elevated pressures and temperatures, leading to the solubilization of target compounds within a solvent. It offers distinct advantages such as shorter extraction durations, lower solvent consumption, and elevated extraction efficacy. Optimizing the process's effectiveness involves taking into account various factors, including the nature of the material, choice of the solvent, number of cycles, temperature, and the characteristics of both the algal material and the desired molecule. Recent research suggests that the fusion of pressurized-liquid extraction and supercritical carbon dioxide fluid extraction is the most favorable choice, encompassing improved yields and the economic viability of potential expansion (Perez-Vazquez et al. 2023).

In contrast to supercritical fluid extraction, pressurized-liquid extraction functions without reaching the critical point of the liquid solvent and permits the utilization of a wider solvent array (Nieto et al. 2010; Kadam et al. 2015). Reports suggest that pressurized-liquid extraction coupled with ethanol displayed enhanced lipid extraction efficiency compared to both the traditional extraction methods, such as Folch techniques and Soxhlet extraction technique, and pressurized-liquid extraction employing hexane (Perez-Vazquez et al. 2023). Pressurized-liquid extraction has been effectively used for the extraction of oleic acid, eicosapentaenoic acid, arachidonic acid, lipids, polysaccharides, proteins, carotenoids, and other bioactive compounds (Otero et al. 2018; He et al. 2019; Gallego et al. 2021). These observations underscore the significance of pressurized-liquid extraction as a robust and efficacious avenue to amass bioactive compounds from algae.

17.3.7 Ultrasound-Assisted Extraction

Ultrasound-assisted extraction has emerged as a promising method for extracting bioactive compounds from algae using ultrasound, owing to multiple advantages over conventional techniques. This technique employs ultrasonic waves to generate cavitation and micro-turbulence within the solvent, leading to enhanced mass transfer and improved extraction efficiency. The ultrasounds have demonstrated successful utility in extracting polyphenols, carotenoids, proteins, fucoxanthin, lipids, carrageenans, and others (Carreira-Casais et al. 2021). An inherent benefit of this method is its non-thermal nature, and, unlike microwave-assisted extraction, this approach also enables the separation of heat-sensitive substances. It offers several advantages, including reduced extraction time and preservation of thermally sensitive compounds. The methodology operates based on the principle of degassing induced by the simultaneous compression and rarefaction of ultrasonic waves (Chemat et al. 2017). This method utilizes ultrasonic waves with frequencies ranging from 20 kHz to 10 MHz (power ultrasound—that finds application in processes like extraction) and 100 kHz to 10 MHz (diagnostic or signal ultrasound—that finds application in quality control and medical diagnostics) (Tiwari 2015). The utilization of ultrasound probes and ultrasonic baths stands as a prevalent practice in ultrasound-assisted extraction (Panda and Manickam 2019; Carreira-Casais et al. 2021). Despite possessing individual merits and drawbacks, the trend is leaning toward employing probe systems due to their capacity to optimize the extraction variables (Mehta et al. 2022). Moreover, this approach proves fitting for advancing a green, sustainable, and efficient approach to the extraction of bioactive compounds.

17.3.8 Comparison and Selection of Extraction Methods

When it comes to extracting bioactive compounds from algae, selecting an appropriate extraction method plays a crucial role in determining the process's yield, quality, and efficiency. Several extraction techniques have been employed and compared to determine the most suitable approach for algal bioactives. Among these methods, solvent-based extractions such as maceration, solid–liquid extraction, and Soxhlet extraction are commonly used due to their simplicity, availability, and versatility (Rasul 2018). These methods involve the use of various organic solvents to extract bioactive compounds from algal biomass. On the other hand, more advanced techniques, such as supercritical fluid extraction, microwave-assisted extraction, enzyme-assisted extraction, ultrasound-assisted extraction, pressurized-liquid extraction, and subcritical water extraction, offer enhanced efficiency, minimum solvent residue, high selectivity, and reduced environmental impact (Zhang et al. 2018). These methods are combined with each other to yield better extraction; however, some of these require specialized equipment. Among these methods, supercritical fluid extraction has emerged as a promising technique for extraction, as it offers higher extraction efficiency, better preservation of thermally sensitive compounds, and shorter extraction time compared to traditional solvent-based methods.

Additionally, it has the advantage of using carbon dioxide (CO_2) as a solvent, which is non-toxic, non-flammable, and environmentally friendly. Based on extensive literature study, the comparison of various extraction methods for algal bioactives suggests that supercritical fluid extraction holds the most potential (Smith 1999; Girotra et al. 2013; Khaw et al. 2017; Uwineza and Waśkiewicz 2020). Its ability to provide high extraction efficiency and environmentally friendly operation makes it attractive for extracting valuable compounds from algae. Extraction of bioactive compounds from algal sources involves selecting a proper extraction technique that has stability between the quality of the product, cost of production, efficiency process, and an eco-friendly approach.

Figure 17.2 displays the most employed conventional extraction techniques, along with the advanced extraction techniques and combined techniques for extracting bioactive compounds from algae. Recently, the utilization of combined techniques has proven highly effective in achieving maximum yields.

17.4 Purification Strategies for Bioactive Compounds

17.4.1 Filtration and Centrifugation

In the purification of bioactive compounds derived from algae, filtration and centrifugation are two prevalent techniques employed to achieve high-quality extracts. Filtration technique involves the separation of solid and liquid components by passing the mixture through a porous medium, such as filter paper or membrane (Saxena et al. 2009; Barsan et al. 2020). This method is effective in removing larger particulate matter, resulting in a liquid extract that contains the desired bioactive compounds. Filtration involves separation of algal biomass from the liquid growth medium by utilizing a permeable membrane with diverse particle size categories, using three distinct methods to execute this: *conventional filtration*, for larger microalgae greater than 70 μm, such as *Spirulina* and *Coelastrum*; conversely, *ultrafiltration* and *microfiltration* are harnessed to capture smaller microalgae (Show et al. 2013; Gultom and Hu 2013). On the other hand, centrifugation utilizes the principles of sedimentation to separate components based on their density, which involves the retrieval of algal biomass from the growth medium by applying gravitational force using a centrifuge (Najjar and Abu-Shamleh 2020). By subjecting the mixture to high centrifugal forces, heavier particles or solids settle at bottom of the centrifuge tube while the supernatant, containing the bioactive compounds, remains above. This method is characterized by its swiftness, simplicity, and effectiveness. Centrifugation can efficiently remove larger debris and cells from the extract,

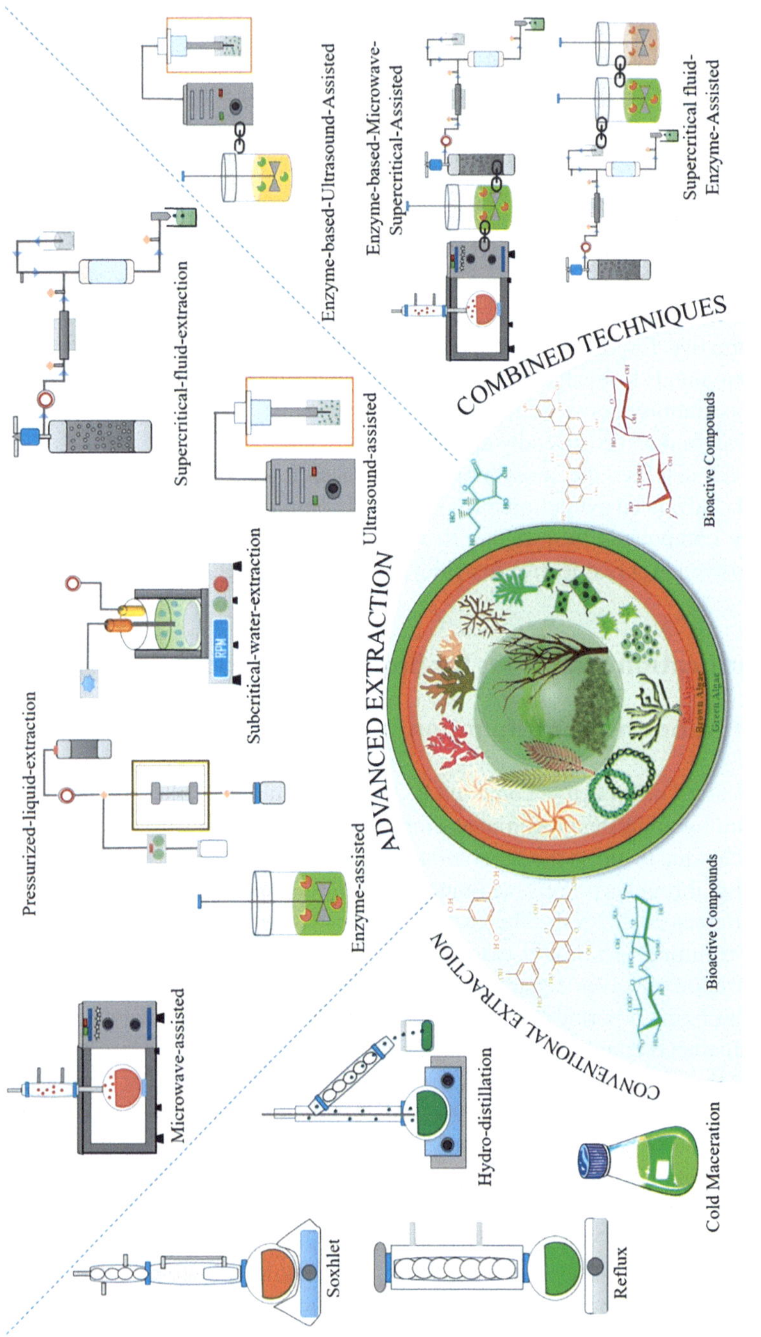

Fig. 17.2 Most employed extraction methods

producing more refined products. However, its potential drawback lies in the potential increase in costs due to the substantial energy consumption and maintenance demands (Molina Grima et al. 2003). Both filtration and centrifugation are valuable purification techniques, with their applicability depending on the nature of the algae biomass and the desired bioactive compounds. Filtration is particularly useful for initial clarifying steps, while centrifugation can effectively remove heavier solids and cells. Often, a combination of both methods is utilized in a purification protocol to achieve the desired level of purity (Balasubramaniam et al. 2021). Careful selection and optimization of these techniques contribute to obtaining bioactive compounds with enhanced quality and potency.

17.4.2 Chromatographic Techniques

Chromatographic techniques have emerged as powerful tools for the purification of algal bioactives, facilitating the isolation of compounds with high purity and specificity. These methods rely on the differential interactions between the components of a mixture and a stationary phase, enabling the separation of compounds based on their chemical properties (Batool and Menaa 2020). Several chromatographic methods can be utilized to purify algal bioactives, with the selection of the appropriate technique depending upon factors such as the inherent nature of the sample, its solubility, volatility, required accuracy, and specific objectives. Notably, these chromatographic methodologies extend beyond mere purification and also enable the precise quantification of important bioactives derived from algae. High-performance liquid chromatography is widely employed due to its efficiency, accuracy, and ability to handle a variety of compounds present in algal extracts (Karimi and Jaafar 2011; Yalçın et al. 2021a; Lourenço-Lopes et al. 2022). Additionally, gas chromatography proves valuable for volatile compound analysis, while thin-layer chromatography offers simplicity and quick qualitative insights into the composition of algal extracts (Jaime et al. 2005; Rajauria and Abu-Ghannam 2013; Yamamoto et al. 2014; Garicano Vilar et al. 2020; Herman et al. 2023). Furthermore, preparative liquid chromatography aids in isolating substantial quantities of bioactive compounds such as peptides, polyphenol fractions, terpenoids, and important pigments for further characterization and testing (Sun et al. 2019; Hu et al. 2019; Nie et al. 2021).

The advanced chromatographic approach helps achieve both purification and identification of bioactives using methods like liquid chromatography coupled with mass spectroscopy or gas chromatography coupled with mass spectroscopy, among others (Shobier et al. 2016; Nayaka et al. 2020; Teoh et al. 2023). In addition, utilization of advanced counter-current chromatographic techniques such as high-speed counter-current chromatography has been employed for the isolation of anti-inflammatory compounds and zeaxanthin from algae (Chen et al. 2005; Kim et al. 2021). Elusion-extrusion counter-current is another advanced counter-current chromatographic technique that has been used for the separation of fucoxanthin, chlorophyll, and pheophytin from edible algae belonging to Phaeophyta (Chen et al.

2022). Another innovative chromatographic technique gaining prominence is expanded bed adsorption, which involves adsorbing the bioactives onto solid supports within a column, allowing for efficient separation and purification by exploiting differences in adsorption behavior and particle size. It is used to purify phycocyanin and phycoerythrin (Wang et al. 2002; Bermejo et al. 2006). Supercritical fluid chromatography has emerged as a valuable technique for the purification of algal bioactives by utilizing supercritical fluids as the mobile phase and providing advantages such as enhanced separation, lower solvent consumption, and reduced analysis time, compared to traditional chromatographic methods (Montañés and Tallon 2018; Teboul et al. 2023). Additionally, the combination of supercritical fluid chromatography with tandem mass spectrometry further enhances the capabilities of bioactive purification by precise identification and characterization of compounds by offering insights into their molecular composition, leading to a deeper understanding of their potential applications (Ubhayasekera et al. 2018; van de Velde et al. 2020).

Column chromatography involves the separation of compounds carried by the solvent based on the interaction with the stationary phase in the column and based on polarity. Extensive literature suggests that silica column is used effectively in the purification of monounsaturated and polyunsaturated fatty acids, tridocosahexaenoyl glycerol, bio-oil, gamma-linolenic acid, arachidonic acid, fucoxanthin, and eicosapentaenoic acid, from different algal sources (Guil-Gucrrcro et al. 2000; Venegas-Venegas et al. 2014; Ronda and Lele 2007; González-Fernández et al. 2017). Hydroxyapatite chromatography is a robust and selective technique to separate proteins and nucleic acid based on their affinity to hydroxyapatite resin, and it is controlled by a number of variables, including size, charge, and hydrophobic interactions, making it an efficient method for separation and purification of algal phycobiliproteins, enzymes, and phosphoproteins (Reuter et al. 1990; Lojero et al. 1997; Rossano et al. 2003; Cai et al. 2011). Ion exchange chromatography is employed for charged bioactive molecules for separation based on their affinity for ion exchange resins, exploiting differences in charge and polarity by modulating the pH and composition of elution buffer in order to elute the target bioactive such as peptides, and complex pigment-proteins such as phycocyanin and phycoerythrin, from algal extracts (Tang et al. 2016; Chen et al. 2019). Gel permeating chromatography separates compounds on the principle of size exclusion, allowing for the isolation of different-sized molecules from complex algal matrices, and is particularly well suited for the purification of polysaccharides for therapeutic applications (Zhang et al. 2019). Affinity chromatography finds wide application in extracting various proteins, enzymes, and receptors while maintaining their biological activity based on highly specific interactions between target molecules and immobilized ligands (Urh et al. 2009).

The synergistic integration of diverse chromatographic techniques has emerged as a powerful strategy for the purification of algal bioactives, and this amalgamation not only expedites the purification process but also improves the purity and yield of bioactives. Researchers can harness a broader spectrum of separation mechanisms through this synergistic chromatographic approach, resulting in more efficient and effective purification strategies. Numerous reports underscore the pivotal role of chromatographic techniques in not only establishing the analytical and technological viability but also improving the economic productivity by upscaling the production of polyphenols, important pigments, lipids, and other valuable fractions for multifarious applications.

17.4.3 Membrane Separation Processes

Membrane separation processes are one of the crucial techniques employed to purify algal bioactives, offering efficient and selective separation based on molecular size, charge, and solubility. These processes encompass microfiltration, nanofiltration, and ultrafiltration, governed by sieving mechanisms, and reverse osmosis, governed by pressure-driven solute diffusion, each tailored to distinct molecular ranges (Gerardo et al. 2014). For instance, microfiltration is adept at separating larger particles, while ultrafiltration targets macromolecules such as polysaccharides and proteins, and nanofiltration and reverse osmosis excel in desalting and concentration of smaller solutes (Castro-Muñoz and García-Depraect 2021). Another innovative and controlled approach for the separation of bioactives is electromembrane filtration, which utilizes an electric field to facilitate the transport of charged molecules through selective membranes, enabling the filtration of proteins, nucleic acids, and essential pigments (Bargeman et al. 2002a, b). Owing to the versatility and sustainability of these membrane separation techniques, they can be considered valuable processes for the purification of algal bioactives, contributing to applications in industries spanning pharmaceuticals, functional foods, and cosmetics.

17.4.4 Precipitation and Crystallization

Utilizing precipitation as a purification technique holds substantial promise in the isolation of algal metabolites, by altering the solubility of target compounds through controlled temperature, pH, or the addition of precipitating agents like ammonium sulfate, which induces the selective formation of solid particles, which are subsequently separated, resulting in enrichment of desired bioactive compounds. The simplicity and versatility of adding a precipitating agent into the crude aqueous extract containing protein triggers a dissociation process wherein the salt separates into ions. As a result, the surrounding water molecules, which previously had a strong solvating affinity toward proteins, preferentially solvate these dissociated ions, leading to subsequent precipitation and aggregation—a phenomenon referred to as "salting out," as ammonium sulfate often emerges as a preferred salt due to its advantageous combination of high water solubility and cost-effectiveness (Duong-Ly and Gabelli 2014). On the other hand, crystallization involves a more gradual process where solutes are brought out of the solution to form well-defined crystals by controlling temperature, solvent composition, and nucleation conditions (Chia et al.

2019). However, one notable disadvantage of the crystallization process is the potential loss of yield, as not all the target compounds may readily crystallize and/or form impurities or mixed crystals, leading to challenges in achieving the desired purity level (Ulrich and Frohberg 2013). Through careful optimization and evaluation of its suitability for multiple applications, one can select these techniques to purify algal bioactives.

17.4.5 Aqueous Two-Phase System and Three-Phase Partitioning

These are innovative techniques for the purification of algal bioactives. The aqueous two-phase system involves the creation of a biphasic environment within an aqueous solution, where two immiscible polymer-rich phases are formed and, by optimizing factors such as polymer type, concentration, and pH, selective partitioning of target bioactive compounds can be achieved (Iqbal et al. 2016; Zafarani-Moattar et al. 2022). Similarly, three-phase partitioning leverages the formation of a third organic phase in addition to the two aqueous phases, offering enhanced separation capabilities. Both these techniques exploit the differing solubilities and the interactions of bioactives with the phases, enabling their efficient isolation (Dennison and Lovrien 1997; Gagaoua and Hafid 2016). These techniques have shown promise in extracting proteins, enzymes, pigments, and other biomolecules (Chew et al. 2019; Suarez Ruiz et al. 2020), offering efficient tools for bioactives' purification with diverse applications in research and industry.

17.4.6 Quality Control and Analysis of Purified Compounds

Quality control techniques and the subsequent analysis of purified algal compounds play a critical role in ensuring the efficacy and safety of extracted bioactive substances. These techniques encompass a range of methods aimed at assessing the identity, purity, potency, and consistency of the isolated compounds. As discussed in the above sections, analytical techniques, including high-performance liquid chromatography, gas chromatography, and others, are commonly employed to characterize the chemical composition and structural integrity of purified algal compounds. Nuclear magnetic resonance spectroscopy is also employed for analyzing the algal bioactives by subjecting purified compounds to radiofrequency radiations in a magnetic field to elucidate the chemical structure, purity, dynamic properties, interactions, and confirmations of algal biomolecules. Quantitative assays also provide valuable insights into the concentration and bioactivity of the target components.

Additionally, spectrophotometric assays offer quantitative insights into compound concentrations and activities, contributing to potency assessment. Furthermore, microbial and heavy metal testing help ascertain the safety of the extracted compounds for various applications. Quality control measures are essential to validate the purification process's success and ensure that the obtained algal bioactives meet the desired specifications, facilitating their utilization in industries such as pharmaceuticals, cosmetics, and functional foods.

17.5 Bioactive Compounds in Diverse Spheres (Applications)

17.5.1 Nutraceutical and Functional Foods: Fueling Wellness

Algae are rich sources of valuable compounds, including proteins, vitamins, omega-3 fatty acids, polysaccharides, lipids, and pigments, which have been harnessed to enhance food's nutritional content and functional properties (Matos et al. 2022). Incorporating algal ingredients into nutraceutical formulations has led to the development of supplements with potential benefits for cardiovascular health, immune support, and cognitive function (Kumari et al. 2023). In functional foods, algal-derived compounds contribute to texture enhancement, coloration, and natural preservation (Ścieszka and Klewicka 2019). Moreover, the bioactive content of algae offers antioxidant, anti-inflammatory, and antimicrobial properties, which further elevate their appeal in promoting overall well-being (Zhou et al. 2022b; Ahmad and Ashraf 2023). As consumer demand for health-enhancing products grows, algal bioactives continue to pave the way for innovative and sustainable solutions in the nutraceutical and functional food industries.

17.5.2 Pharmaceuticals and Medicinal Products: Healing Harmonie

The applications of algal bioactives in the pharmaceutical and medicinal sectors have garnered significant attention for their potential therapeutic benefits. The polyphenols, flavonoids, polysaccharides, and other secondary metabolites exhibit a range of pharmacological activities (Dai et al. 2022). These bioactives hold promise in drug development, serving as sources of novel molecules with various therapeutic properties (Khavari et al. 2021). Algal compounds also demonstrated potential in wound healing, skincare, and treatment of various diseases (Thiyagarasaiyar et al. 2020; Ajayi et al. 2023). With advancements in research and technology, algal bioactives are being employed for targeted drug delivery systems and the formulation of pharmaceutical products (Zhang et al. 2022). Integrating algal-derived compounds into pharmaceutical and medicinal products offers exciting prospects for effective therapeutic interventions.

17.5.3 Cosmetics and Personal Care Products

The applications of algal bioactives in cosmetics and personal care products have surged due to their diverse benefits and natural origins. Algal bioactives, including polysaccharides, peptides, pigments, polyphenols, and antioxidants, offer a wide array of skincare and cosmetic advantages (Thiyagarasaiyar et al. 2020). These bioactives possess moisturizing, antiaging, and antioxidant properties, making them ideal for the formulations targeting hydration, wrinkle reduction, and protection against environmental stressors (Wang et al. 2015; López-Hortas et al. 2021). Algal extracts and oils find use in products like creams, serums, and sunscreens, contributing to improved skin textures and radiance (Mourelle et al. 2017). The natural colors derived from algal pigments are employed in cosmetic products, providing safer alternatives to synthetic dyes (Ariede et al. 2017). Algal bioactives thus offer a compelling avenue of clean, sustainable, and effective personal care solutions, aligning with consumer preferences for natural and eco-friendly products.

17.5.4 Agrochemicals and Crop Enhancers

The utilization of algal bioactives in agrochemicals and crop enhancers holds immense potential to revolutionize modern agricultural practices. Using the diverse array of algal bioactive compounds offers a new direction for sustainable and effective approaches to crop cultivation (Kokabi and Yousefzadi 2022). By tapping into the unique properties of these compounds, agricultural industries can unlock innovative solutions that enhance plant growth, suppress pests and weeds, produce biofertilizers, and optimize nutrient utilization (Chatterjee et al. 2017; Asimakis et al. 2022). Algal bioactives stand at the forefront of transformative advancements that can reshape the future of global food production.

17.5.5 Biofuel and Biodiesel Production

Algae are rich sources of carbohydrates and lipids that can be efficiently converted into biofuels through processes like transesterification (Makareviciene and Skorupskaite 2019; Ghedini et al. 2021). These bioactives offer a renewable alternative to fossil fuels, reducing greenhouse gas emissions and mitigating the environmental impact of energy production, and these bioactives not only enable the generation of biodiesel but also contribute to the production of bioethanol and biogas (Adeniyi et al. 2018). Algae also have the potential to be converted into jet fuels, gasoline, biodiesel, bioethanol, and biogas varieties (methane, hydrogen, and syngas), which can be subsequently utilized for the purpose of producing energy (Ghedini et al. 2021). The rapid growth rates and high oil content of certain algal strains make them particularly suitable for large-scale cultivation and energy generation (Adeniyi et al. 2018; Alam 2022). As the world seeks transition toward cleaner and more sustainable energy sources, algal bioactives are a promising resource for addressing energy demands while reducing carbon footprint.

17.5.6 Environmental Applications (Bioremediation, Biomaterials)

Algal bioactives have also gained substantial attention for their far-reaching applications in environmental remediation and the creation of biomaterials. With their innate ability to accumulate heavy metals and toxins, algae are being explored as biofilters for water treatment and bioremediation purposes. These bioactives can effectively sequester pollutants, contributing to the restoration of contaminated ecosystems. Moreover, algal compounds such as polysaccharides and proteins possess qualities that qualify them as potential candidates for developing biodegradable plastics, sustainable packaging, and wound dressings. By repurposing algal biomass and its constituents, we can address critical environmental challenges and minimize the ecological footprint of various industries. The integration of algal bioactives in both bioremediation strategies and the creation of eco-friendly biomaterials exemplifies the multifaceted potential of algae in fostering a more sustainable and harmonious relationship between human activities and the environment.

17.5.7 Animal Feed

The applications of algae as animal feed have gained increasing recognition for their potential to revolutionize the agricultural and aquaculture industries. Microalgae and seaweeds offer a rich source of nutrients, including proteins, vitamins, and lipids, making them highly valuable components in animal diets. Algae can serve as sustainable alternatives to conventional feed ingredients, reducing the ecological footprint of livestock and aquaculture production (Healy et al. 2023). Additionally, algae-based feeds enhance the nutritional profile of animal products, improving the quality of meat, milk, and fish. Furthermore, algae cultivation for animal feed can contribute to nutrient recycling and wastewater treatment, promoting eco-friendly farming practices. As global demand for high-quality animal products rises, integrating algae into animal diets represents a promising avenue for enhancing both the sustainability and productivity of animal agriculture.

17.5.8 Transforming Challenges into Opportunities: Harnessing Algal Blooms for Multifaceted Applications

Leveraging algal blooms for various applications has emerged as a potential strategy to transform an environmental challenge into valuable opportunities. Rather than viewing algal blooms solely as problems, researchers and industries are exploring ways to harness their biomass for beneficial purposes. Algal blooms can serve as feedstocks for biofuel production, as their rapid growth can yield substantial amounts of lipids suitable for biodiesel. Additionally, algal blooms can be a resource for nutraceuticals, pharmaceuticals, and bioactive compounds used in cosmetics and functional foods (Kim et al. 2015). Moreover, algal blooms can play a role in

Fig. 17.3 Application of algal bioactive compounds across diverse sectors

bioremediation by sequestering excess nutrients from water bodies, helping to restore ecosystem health. By adopting a holistic approach that capitalizes on the potential of algal blooms, we can not only mitigate environmental issues but also create sustainable solutions that address energy, health, and environmental challenges.

Figure 17.3 illustrates the versatile applications of algae in various sectors.

Table 17.1 provides a detailed overview of various algal species, highlighting the presence of diverse bioactive compounds with multiple applications. It serves as a valuable reference for quickly identifying the list of explored algal species for various bioactive compounds.

17.6 Conclusion

In conclusion, exploring bioactive compounds from algae carries substantial scientific significance and profound applications. The synergy between advancements in extraction and refinement techniques and our expanding understanding of algae's intricate biochemical composition provides a solid foundation for pioneering breakthroughs. As we venture into unexplored realms, tapping into the rich biodiversity of algal species, we are presented with novel prospects to harness these compounds for applications spanning energy, cosmetics, agriculture, and pharmaceuticals. The progressive interactions between the biological, chemical, and engineering elements underscore the multidisciplinary nature of this endeavor, emphasizing the need for continued collaboration among researchers, technologists, and industries. This pursuit not only seeks to enhance the efficiency of the extraction process but also aligns with sustainable practices. Although extensive research has been

Table 17.1 List of algal species with their isolated bioactive compounds utilized for multiple applications

Algal species	Bioactive compounds	Application	References
Green algae: *Cladophora vagabunda*, *Ulva lactuca*, *Flabellia petiolate*, *Dunaliella salina*, *Spirulina platensis*, *Spirulina maxima*, *Haematococcus lacustris*, *Scenedesmus* sp., *Isochrysis* sp., *Schizochytrium* sp., and *Nannochloropsis salina* **Brown algae:** *Sargassum vulgare*, *Sargassum fulvellum*, *Cystosiera barbata*, *Sargassum* sp., *Kappaphycus alvarezii*, and *Laminaria digitata* ***Red algae:*** *Porphyridium cruentum*, *Porphyridium purpureum*, *Mastocarpus stellatus*, *Chondrus crispus*, *Porphyra* sp., *Pyropia yezoensis*, *Palmaria palmata*, *Gracilaria verrucosa*, *Gracilaria gracilis*, *Gelidium pusillum*, and *Grateloupia turuturu*	**Pigments:** Phycoerythrin, phycocyanin, phycoerythrocyanin, allophycocyanin, chlorophyll, carotenoids, xanthophylls, fucoxanthin, phycobiliproteins, zeaxanthin, myxoxanthin, and myxoxanthophyll	Food colorant, fluorescent tagging, flow cytometry, fluorescent microscopy, immunochemical processes, and commercially also used as pigment in food, dairy goods, and cosmetics; antioxidative, anti-inflammatory, immune-protective, antiobesity, and anticancer attributes	Negreanu-Pirjol et al. (2020), Yalçın et al. (2021b), Singh and Purwar (2022), Sanger et al. (2022), Chen et al. (2023)

(continued)

Table 17.1 (continued)

Algal species	Bioactive compounds	Application	References
Blue-green algae: *Nostoc ellipsosporum*, *Hapalosiphon fontinalis*, *Microchaete tenera*, *Limnothrix obliqueacuminata*, *Mastigocladus laminosus*, and *Lyngbya* sp. **Green algae:** *Dunaliella tertiolecta*, *Nannochloropsis* sp., *Dunaliella tertiolecta*, *Caulerpa serrulata*, *Spirulina platensis*, *Euglena cantabrica*, and *Haematococcus pluvialis* **Brown algae:** *Hizikia fusiformis* and *Undaria pinnatifida* **Red algae:** *Porphyra tenera*, *Gracilaria textorii*, and *Acanthophora spicifera*	**Flavonoids:** Quercetin, myricetin, kaempferol, rutin, genistein, apigenin, catechin, naringenin, morin, hesperidin, luteolin, and acanthophorin-A, B	Antiallergic, antioxidant, antiviral, antidiabetic, anti-inflammatory, antibacterial, antiosteoporosis, cardioprotective, nephroprotective, antimicrobial, neuroprotective, antimicrobial, and anticancer properties; hepatoprotective and ultraviolet (UV) protection activity	Zeng et al. (2010), Thomas and Kim (2013), Murray et al. (2018), Thiyagarasaiyar et al. (2020), Besednova et al. (2020, 2021), Chaves et al. (2020), Ferdous and Balia Yusof (2021), Kim et al. (2022), Kumari et al. (2023), Pereira and Cotas (2023)

Green algae: *Cladophora socialis* **Brown algae:** *Ecklonia arborea*, *Ecklonia kurome*, *Ecklonia stolonifera*, *Eisenia bicyclis*, *Cystoseira nodicaulis*, *Saccharina japonica*, *Fucus evanescens*, *Fucus vesiculosus*, *Fucus serratus*, *Fucus spiralis*, *Ishige sinicola*, *Bifurcaria bifurcata*, *Thalassiophyllum clathrus*, *Stephanocystis crassipes*, *Sargassum thunbergii*, *Laminaria digitata*, *Padina tetrastromatica*, *Padina gymnospora*, *Ascophyllum nodosum*, *Himanthalia elongata*, and *Pelvetia canaliculata* **Red algae:** *Solieria filiformis*, *Ahnfeltia chinensis*, *Symphyocladia latiuscula*, and *Kappaphycus alvarezii*	**Polyphenols:** Phlorofucofuroeckol A, phloroglucinol, phlorotannins, phlorethol, 4-hydroxybenzoic acid, fucophlorethol, fucol, carmalol, eckol, dieckol, catechin, fuhalol, and triphlorethol-A	Antioxidant, antiviral, anti-inflammatory, neuroprotective, antimicrobial, and anticancer properties; hepatoprotective, antiaging, skin-enhancing, and whitening properties	Thomas and Kim (2013), Murray et al. (2018), Thiyagarasaiyar et al. (2020), Besednova et al. (2020, 2021), Chaves et al. (2020), Ferdous and Balia Yusof (2021), Kim et al. (2022), Kumari et al. (2023), Pereira and Cotas (2023)

(continued)

Table 17.1 (continued)

Algal species	Bioactive compounds	Application	References
Green algae: *Chlamydomonas reinhardtii*, *Synechocystis* sp., *Chlorella vulgaris*, *Scenedesmus obliquus*, *Botryococcus braunii*, *Dunaliella salina*, *Nannochloropsis oculata*, *Isochrysis galbana*, *Codium fragile*, and *Ulva fasciata* **Brown algae:** *Cystoseira amentacea* var. *stricta*, *Sargassum wightii*, *Padina boergesenii*, *Kjellmaniella crassifolia*, *Stypopodium flabelliforme*, and *Sargassum muticum* **Red algae:** *Plocamium cartilagineum*, *Sargassum ringgoldianum*, *Gracilaria salicornia*, *Bifurcaria bifurcata*, *Dictyota dichotoma*, *Sphaerococcus coronopifolius*, *Eucheuma cottonii*, *Laurencia obtusa*, *Laurencia mariannensis*, *Laurencia viridis*, *Porphyridium aerugineum*, *Laurencia composita*, *Kappaphycus alvarezii*, and *Hypnea musciformis*	**Terpenes and terpenoids:** Isoprenoids, mono-, di-, tri-, and tetra-terpenoids, sesquiterpenes, sphaerodactylomelol, dihydromyrcenol, cyclosadol, β-carotene, lutein, fucoxanthin, astaxanthin, siphonaxanthin, canthaxanthin, zeaxanthin, cryptoxanthin, violaxanthin, cyclohexane, neoxanthin, pirene, furoplocamioid C, prefuroplocamioid, lactone, eleganolone, eleganonal, 3-epicyclomusalenol, and dihydrocimicifugenol	Antioxidant, antiplasmodial, antiviral, anticancer, antidiabetic, anti-inflammatory properties; DNA repair mechanism; neuroprotective effect; inhibition of angiogenesis and metastasis in cancer; and chemoprotective properties	Lee et al. (2013), Gaysinski et al. (2015), Yang et al. (2020), Tamanna Ferdous and Norhana Balia Yusof (2021), Costa et al. (2022)

Green algae: *Codium intricatum*, *Bryopsis*, *Halimeda*, *Caulerpa racemosa*, *Derbesia*, *Cladophora*, *Chaetomorpha*, *Gayralia oxysperma*, *Dichotomosiphon*, *Acetabularia acetabulum*, *Ulva lactuca*, *Ulva compressa*, *Ulva prolifera*, *Ulva intestinalis*, *Ulva linza*, and *Spongomorpha indica* **Brown algae:** *Laminaria japonica*, *Laminaria hyperborean*, *Ascophyllum nodosum*, *Macrocystis pyrifera*, *Undaria pinnatifida*, *Cladosiphon okamuranus*, *Adenocystis utricularis*, *Stoechospermum marginatum*, *Fucus vesiculosus*, and *Cystoseira indica* **Red algae:** *Gelidium amansii*, *Gelidium crinale*, *Chondrus ocellatus*, *Agardhiella tenera*, *Gigartina skottsbergii*, *Schizymenia binderi*, *Callophyllis variegate*, *Cryptonemia crenulata*, and *Schizymenia pacifica*	**Polysaccharides:** Fucoidan, carrageenan, agar, alginate, ulvan, porphyran, sulfated rhamnan, agarose, dextran, mannan, fucose, mannitol, and furcellaran	Pharmaceuticals, prebiotics, functional food, animal feed, nutraceuticals, health care, biomedicines, biofuels, bio-stimulants, tooth implants, thickeners, stabilizers and emulsifiers, bio-stimulants for enhanced plant growth, sustainable fertilizers, bio-elicitors triggering plant defense, wound dressings with antimicrobial properties, biomaterials for tissue engineering scaffolds, and controlled drug delivery systems	Karkos et al. (2011), Bixler and Porse (2011), Nuhu (2013), Ahmadi et al. (2015), Pankiewicz et al. (2016), Bouissil et al. (2019), Ciancia et al. (2020), Soto-Sierra et al. (2021), Perera et al. (2021)

(continued)

Table 17.1 (continued)

Algal species	Bioactive compounds	Application	References
Blue-green algae: *Dunaliella tertiolecta*, *Dunaliella salina*, *Nostoc*, *Spirulina platensis*, *Chlorella* sp., *Aphanizomenon flos-aquae*, and *Nannochloropsis* **Green algae:** *Cladophora rupestris*, *Ulva intestinalis*, and *Ulva lactuca* **Brown algae:** *Undaria pinnatifida*, *Himanthalia elongata*, *Laminaria ochroleuca*, *Laminaria digitata*, *Palmaria* sp., *Porphyra* sp., *Ascophyllum nodosum*, *Fucus serratus*, *Fucus vesiculosus*, and *Alaria esculenta* **Red algae:** *Ceramium diaphanum*, *Kappaphycus spinosum*, *Kappaphycus striatum*, *Kappaphycus alvarezii*, *Corallina mediterranea*, *Gelidium* sp., *Gracilaria* sp., *Hypnea* sp., *Laurencia filiformis*, *Osmundea pinnatifida*, *Porphyra* sp., *Pyropia* sp., and *Eucheuma cottonii*	**Proteins, vitamins, and minerals:** Lectins, phycobiliproteins, and other protein hydrolysates; vitamins D, K, B12, A, B family (biotin, pantothenic acid, biotin, niacin), C, and E; Mg, Fe, K, Ca, I, Zn, Cu, and Mn	Food ingredients, aquaculture, animal feed, nutraceuticals, food supplements, cosmetic applications, and pharmaceuticals	Koyande et al. (2019), Afonso et al. (2019), Gamero-Vega et al. (2020), Healy et al. (2023), Ilyas et al. (2023)

Blue-green algae: *Nostoc commune*, *Phormidium luridum*, *Phormidium autumnale*, and *Scenedesmus* **Green algae:** *Chlorella vulgaris*, *Ulva lactuca*, *Eudorina unicocca*, *Isochrysis galbana*, *Botryococcus braunii*, and *Volvox aureus* **Brown algae:** *Cystoseira adriatica*, *Cystoseira tamariscifolia*, *Padina pavonica*, and *Sargassum vulgare* **Red algae:** *Sphaerococcus coronopifolius*, *Plocamium cartilagineum*, and *Osmundea pinnatifida*	**Sterols:** Stigmasterol, cholesterol, campesterol, fucosterol, isofucosterol, β-sitosterol, chondrillasterol, ergosterol, n-6 and n-3 polyunsaturated fatty acids, glycolipids, phospholipids, derivatives of hexaenoic and pentaenoic acids, and monounsaturated and unsaturated fatty acids	Effectively lowers blood-cholesterol levels and exhibits potential in curbing colon cancer; anti-inflammatory, antibacterial, antifungal, antiulcerative, and antitumor effects; cholesterol-lowering effect; vitamin absorption, anticancer activity, supplements, and bioremediation	Lopes et al. (2013), Bittencourt Fagundes and Wagner (2021)

dedicated to algae, translating these efforts into commercially viable products remains challenging. This highlights the need to bridge the gap between research and industrial implementation, fostering the scale-up of algal bioproducts for diverse uses. As we continue to explore the algae's biochemical repository, armed with unique secondary metabolites and significant bioactive compounds, the aim of unlocking the full potential of algal bioactives for diverse applications signifies both human resourcefulness and nature's limitless offerings.

17.7 Steering Through Current Trends: Addressing Challenges and Future Directions

In the context of algal bioactives, several key challenges and promising future directions hold attention. Firstly, sustainability and scalability remain critical issues, as the efficient cultivation and harvesting of algae on a larger scale warrant innovative solutions. Secondly, the economic viability and evolving market trends underline the necessity of aligning economic realities with the potential of algal bioactives. Thirdly, significant advances in extraction and purification technologies are expected, paving the way for higher yields and better product quality. Meanwhile, stringent regulations and safety considerations sound the alarm for responsible utilization of algal sources without overexploiting them. Collaboration and interdisciplinary research are vital enablers, fostering diverse expertise and collective progress. As we navigate these challenges, the outlook for bioactive compounds derived from algae remains bright, holding transformative potential across various industries and applications.

Acknowledgments The authors would like to express heartfelt gratitude to the Vellore Institute of Technology for their invaluable support and resources during this study.

References

Adeniyi OM, Azimov U, Burluka A (2018) Algae biofuel: current status and future applications. Renew Sust Energ Rev 90:316–335. https://doi.org/10.1016/j.rser.2018.03.067

Afonso NC, Catarino MD, Silva AMS, Cardoso SM (2019) Brown macroalgae as valuable food ingredients. Antioxidants 8:365. https://doi.org/10.3390/antiox8090365

Ahmad A, Ashraf SS (2023) Sustainable food and feed sources from microalgae: food security and the circular bioeconomy. Algal Res 74:103185. https://doi.org/10.1016/j.algal.2023.103185

Ahmadi A, Zorofchian Moghadamtousi S, Abubakar S, Zandi K (2015) Antiviral potential of algae polysaccharides isolated from marine sources: a review. Biomed Res Int 2015:1–10. https://doi.org/10.1155/2015/825203

Ajayi EIO, Oladele JO, Nkumah AO (2023) Application of algae in wound healing. In: Next-generation algae. Wiley, New York, pp 251–284

Akhtar I, Javad S, Yousaf Z et al (2019) Microwave assisted extraction of phytochemicals an efficient and modern approach for botanicals and pharmaceuticals. Pak J Pharm Sci 32:223–230

Akula R, Ravishankar GA (2011) Influence of abiotic stress signals on secondary metabolites in plants. Plant Signal Behav 6:1720–1731. https://doi.org/10.4161/psb.6.11.17613

Alam S (2022) Algae: an emerging feedstock for biofuels production. In: Algal biotechnology. Elsevier, Amsterdam, pp 165–185

Animish A, Jayasri MA (2023) A retrospective review of marine algae and the strategies employed for prospective diabetes management. Algal Res 74:103209. https://doi.org/10.1016/j.algal.2023.103209

Ariede MB, Candido TM, Jacome ALM et al (2017) Cosmetic attributes of algae—a review. Algal Res 25:483–487. https://doi.org/10.1016/j.algal.2017.05.019

Asimakis E, Shehata AA, Eisenreich W et al (2022) Algae and their metabolites as potential biopesticides. Microorganisms 10:307. https://doi.org/10.3390/microorganisms10020307

Balasubramaniam V, Gunasegavan RD-N, Mustar S et al (2021) Isolation of industrial important bioactive compounds from microalgae. Molecules 26:943. https://doi.org/10.3390/molecules26040943

Bargeman G, Houwing J, Recio I et al (2002a) Electro-membrane filtration for the selective isolation of bioactive peptides from an alpha(s2)-casein hydrolysate. Biotechnol Bioeng 80:599–609. https://doi.org/10.1002/bit.10419

Bargeman G, Koops G-H, Houwing J et al (2002b) The development of electro-membrane filtration for the isolation of bioactive peptides: the effect of membrane selection and operating parameters on the transport rate. Desalination 149:369–374. https://doi.org/10.1016/S0011-9164(02)00824-X

Barsan N, Zaharia A, Chitimus D, et al (2020) Filtration theory and techniques. A short review on the filtration process. In: 2020 7th international conference on energy efficiency and agricultural engineering (EE&AE). IEEE, pp 1–6

Batool A, Menaa F (2020) Concentration and purification of seaweed components by chromatography methods. In: Sustainable seaweed technologies. Elsevier, Amsterdam, pp 315–370

Becker EW (2013) Microalgae for human and animal nutrition. In: Handbook of microalgal culture. Wiley, Oxford, pp 461–503

Bermejo R, Felipe MA, Talavera EM, Alvarez-Pez JM (2006) Expanded bed adsorption chromatography for recovery of phycocyanins from the microalga spirulina platensis. Chromatographia 63:59–66. https://doi.org/10.1365/s10337-005-0702-9

Besednova NN, Andryukov BG, Zaporozhets TS et al (2020) Algae polyphenolic compounds and modern antibacterial strategies: current achievements and immediate prospects. Biomedicines 8:342. https://doi.org/10.3390/biomedicines8090342

Besednova NN, Andryukov BG, Zaporozhets TS et al (2021) Antiviral effects of polyphenols from marine algae. Biomedicines 9:200. https://doi.org/10.3390/biomedicines9020200

Biris-Dorhoi E-S, Michiu D, Pop CR et al (2020) Macroalgae—a sustainable source of chemical compounds with biological activities. Nutrients 12:3085. https://doi.org/10.3390/nu12103085

Bittencourt Fagundes M, Wagner R (2021) Sterols biosynthesis in algae. In: Biosynthesis. IntechOpen, London

Bixler HJ, Porse H (2011) A decade of change in the seaweed hydrocolloids industry. J Appl Phycol 23:321–335. https://doi.org/10.1007/s10811-010-9529-3

Bleakley S, Hayes M (2017) Algal proteins: extraction, application, and challenges concerning production. Food Secur 6:33. https://doi.org/10.3390/foods6050033

Borisova-Mubarakshina MM, Vetoshkina DV, Ivanov BN (2019) Antioxidant and signaling functions of the plastoquinone pool in higher plants. Physiol Plant 166:181–198. https://doi.org/10.1111/ppl.12936

Bouissil S, Pierre G, El Alaoui-Talibi Z et al (2019) Applications of algal polysaccharides and derivatives in therapeutic and agricultural fields. Curr Pharm Des 25:1187–1199. https://doi.org/10.2174/1381612825666190425162729

Cai C, Wu L, Li C et al (2011) Purification, crystallization and preliminary X-ray analysis of phycocyanin and phycoerythrin from Porphyra yezoensis Ueda. Acta Crystallogr Sect F Struct Biol Cryst Commun 67:579–583. https://doi.org/10.1107/S1744309111007925

Carreira-Casais A, Otero P, Garcia-Perez P et al (2021) Benefits and drawbacks of ultrasound-assisted extraction for the recovery of bioactive compounds from marine algae. Int J Environ Res Public Health 18:9153. https://doi.org/10.3390/ijerph18179153

Castro-Muñoz R, García-Depraect O (2021) Membrane-based harvesting processes for microalgae and their valuable-related molecules: a review. Membranes (Basel) 11:585. https://doi.org/10.3390/membranes11080585

Chatterjee A, Singh S, Agrawal C et al (2017) Role of algae as a biofertilizer. In: Algal green chemistry. Elsevier, Amsterdam, pp 189–200

Chaves JO, de Souza MC, da Silva LC et al (2020) Extraction of flavonoids from natural sources using modern techniques. Front Chem 8:507887. https://doi.org/10.3389/fchem.2020.507887

Chemat F, Rombaut N, Sicaire A-G et al (2017) Ultrasound assisted extraction of food and natural products. Mechanisms, techniques, combinations, protocols and applications. A review. Ultrason Sonochem 34:540–560. https://doi.org/10.1016/j.ultsonch.2016.06.035

Chen F, Li H-B, Wong RN-S et al (2005) Isolation and purification of the bioactive carotenoid zeaxanthin from the microalga microcystis aeruginosa by high-speed counter-current chromatography. J Chromatogr A 1064:183–186. https://doi.org/10.1016/j.chroma.2004.12.065

Chen K-H, Wang SS-S, Show P-L et al (2019) Rapid and efficient recovery of C-phycocyanin from highly turbid Spirulina platensis algae using stirred fluidized bed ion exchange chromatography. Sep Purif Technol 209:636–645. https://doi.org/10.1016/j.seppur.2018.08.057

Chen D, Jin Y, Hu D et al (2022) One-step preparative separation of fucoxanthin from three edible brown algae by elution-extrusion countercurrent chromatography. Mar Drugs 20:257. https://doi.org/10.3390/md20040257

Chen Z, Wu W, Wen Y et al (2023) Recent advances of natural pigments from algae. Food Prod Process Nutr 5:39. https://doi.org/10.1186/s43014-023-00155-y

Cheng Y, Xue F, Yu S et al (2021) Subcritical water extraction of natural products. Molecules 26:4004. https://doi.org/10.3390/molecules26134004

Chew KW, Chia SR, Lee SY et al (2019) Enhanced microalgal protein extraction and purification using sustainable microwave-assisted multiphase partitioning technique. Chem Eng J 367:1–8. https://doi.org/10.1016/j.cej.2019.02.131

Chia SR, Mak KY, Khaw YJ et al (2019) An efficient and rapid method to extract and purify protein—liquid triphasic flotation system. Bioresour Technol 294:122158. https://doi.org/10.1016/j.biortech.2019.122158

Chinese Pharmacopoeia Commission (2015) Pharmacopoeia of the People's Republic of China. Stationery Office, London

Ciancia M, Fernández PV, Leliaert F (2020) Diversity of sulfated polysaccharides from cell walls of Coenocytic green algae and their structural relationships in view of green algal evolution. Front Plant Sci 11:554585. https://doi.org/10.3389/fpls.2020.554585

Corrêa PS, Morais Júnior WG, Martins AA et al (2020) Microalgae biomolecules: extraction, separation and purification methods. Processes 9:10. https://doi.org/10.3390/pr9010010

Costa JAV, Cassuriaga APA, Moraes L, Morais MG (2022) Biosynthesis and potential applications of terpenes produced from microalgae. Bioresour Technol Rep 19:101166. https://doi.org/10.1016/j.biteb.2022.101166

Da Silva Nonato N, Nottar Escobar EL, Kochepka DM et al (2020) Extraction of Muriella decolor lipids using conventional and pressurized solvents and characterization of their fatty acid profile for biodiesel applications. J Supercrit Fluids 158:104750. https://doi.org/10.1016/j.supflu.2020.104750

Dai N, Wang Q, Xu B, Chen H (2022) Remarkable natural biological resource of algae for medical applications. Front Mar Sci 9:1060. https://doi.org/10.3389/fmars.2022.912924

Del Mondo A, Smerilli A, Sané E et al (2020) Challenging microalgal vitamins for human health. Microb Cell Factories 19:201. https://doi.org/10.1186/s12934-020-01459-1

Demarco M, Oliveira de Moraes J, Matos ÂP et al (2022) Digestibility, bioaccessibility and bioactivity of compounds from algae. Trends Food Sci Technol 121:114–128. https://doi.org/10.1016/j.tifs.2022.02.004

Dennison C, Lovrien R (1997) Three phase partitioning: concentration and purification of proteins. Protein Expr Purif 11:149–161. https://doi.org/10.1006/prep.1997.0779

Domingues MR, Calado R (2022) Lipids of marine algae—biomolecules with high nutritional value and important bioactive properties. Biomol Ther 12:134. https://doi.org/10.3390/biom12010134
Domozych DS, Ciancia M, Fangel JU et al (2012) The cell walls of green algae: a journey through evolution and diversity. Front Plant Sci 3:82. https://doi.org/10.3389/fpls.2012.00082
Duong-Ly KC, Gabelli SB (2014) Salting out of proteins using ammonium sulfate precipitation. Methods Enzymol 541:85–94
Edelmann M, Aalto S, Chamlagain B et al (2019) Riboflavin, niacin, folate and vitamin B12 in commercial microalgae powders. J Food Compos Anal 82:103226. https://doi.org/10.1016/j.jfca.2019.05.009
Eze CN, Onyejiaka CK, Ihim SA et al (2023) Bioactive compounds by microalgae and potentials for the management of some human disease conditions. AIMS Microbiol 9:55–74. https://doi.org/10.3934/microbiol.2023004
Ferdous UT, Balia Yusof ZN (2021) Insight into potential anticancer activity of algal flavonoids: current status and challenges. Molecules 26:6844. https://doi.org/10.3390/molecules26226844
Gagaoua M, Hafid K (2016) Three phase partitioning system, an emerging non-chromatographic tool for proteolytic enzymes recovery and purification. Biosens J 5:100134. https://doi.org/10.4172/2090-4967.1000134
Gallego R, Bueno M, Herrero M (2019) Sub- and supercritical fluid extraction of bioactive compounds from plants, food-by-products, seaweeds and microalgae—an update. TrAC Trends Anal Chem 116:198–213. https://doi.org/10.1016/j.trac.2019.04.030
Gallego R, Bueno M, Chourio AM et al (2021) Use of high and ultra-high pressure based-processes for the effective recovery of bioactive compounds from Nannochloropsis oceanica microalgae. J Supercrit Fluids 167:105039. https://doi.org/10.1016/j.supflu.2020.105039
Gamero-Vega G, Palacios-Palacios M, Quitral V (2020) Nutritional composition and bioactive compounds of red seaweed: a mini-review. J Food Nutr Res 8:431–440. https://doi.org/10.12691/jfnr-8-8-7
Garicano Vilar E, O'Sullivan MG, Kerry JP, Kilcawley KN (2020) Volatile compounds of six species of edible seaweed: a review. Algal Res 45:101740. https://doi.org/10.1016/j.algal.2019.101740
Gaysinski M, Ortalo-Magné A, Thomas OP, Culioli G (2015) Extraction, purification, and NMR analysis of terpenes from brown algae. Methods Mol Biol 1308:207–223
Georgiopoulou I, Tzima S, Louli V, Magoulas K (2022) Supercritical CO2 extraction of high-added value compounds from Chlorella vulgaris: experimental design, modelling and optimization. Molecules 27:5884. https://doi.org/10.3390/molecules27185884
Gerardo ML, Oatley-Radcliffe DL, Lovitt RW (2014) Integration of membrane technology in microalgae biorefineries. J Membr Sci 464:86–99. https://doi.org/10.1016/j.memsci.2014.04.010
Ghedini E, Taghavi S, Menegazzo F, Signoretto M (2021) A review on the efficient catalysts for algae transesterification to biodiesel. Sustain For 13:10479. https://doi.org/10.3390/su131810479
Girotra P, Singh SK, Nagpal K (2013) Supercritical fluid technology: a promising approach in pharmaceutical research. Pharm Dev Technol 18:22–38. https://doi.org/10.3109/10837450.2012.726998
Glazer AN (1994) Phycobiliproteins—a family of valuable, widely used fluorophores. J Appl Phycol 6:105–112. https://doi.org/10.1007/BF02186064
González-Fernández MJ, Ramos-Bueno RP, Rodríguez-García I, Guil-Guerrero JL (2017) Purification process for MUFA- and PUFA-based monoacylglycerols from edible oils. Biochimie 139:107–114. https://doi.org/10.1016/j.biochi.2017.06.002
Goto M, Kanda H, Wahyudiono MS (2015) Extraction of carotenoids and lipids from algae by supercritical CO2 and subcritical dimethyl ether. J Supercrit Fluids 96:245–251. https://doi.org/10.1016/j.supflu.2014.10.003
Guil-Gucrrcro JL, Belarbi EH, Rebolloso-Fuentes MM (2000) Eicosapentaenoic and arachidonic acids purification from the red microalga Porphyridium cruentum. Bioseparation 9:299–306. https://doi.org/10.1023/A:1011124624001/METRICS

Gultom S, Hu B (2013) Review of microalgae harvesting via co-pelletization with filamentous fungus. Energies 6:5921–5939. https://doi.org/10.3390/en6115921
Gunathilaka TL, Samarakoon K, Ranasinghe P, Peiris LDC (2020) Antidiabetic potential of marine brown algae—a mini review. J Diabetes Res 2020:1230218. https://doi.org/10.1155/2020/1230218
He Y, Huang Z, Zhong C et al (2019) Pressurized liquid extraction with ethanol as a green and efficient technology to lipid extraction of Isochrysis biomass. Bioresour Technol 293:122049. https://doi.org/10.1016/j.biortech.2019.122049
Healy LE, Zhu X, Pojić M et al (2023) Biomolecules from macroalgae—nutritional profile and bioactives for novel food product development. Biomol Ther 13:386. https://doi.org/10.3390/biom13020386
Hentati F, Tounsi L, Djomdi D et al (2020) Bioactive polysaccharides from seaweeds. Molecules 25:3152. https://doi.org/10.3390/molecules25143152
Herman LL, Padala SA, Ahmed I, Bashir K (2023) Angiotensin converting enzyme inhibitors (ACEI). StatPearls Publishing, Treasure Island
Herrero M, Cifuentes A, Ibanez E (2006) Sub- and supercritical fluid extraction of functional ingredients from different natural sources: plants, food-by-products, algae and microalgae: a review. Food Chem 98:136–148. https://doi.org/10.1016/j.foodchem.2005.05.058
Holdt SL, Kraan S (2011) Bioactive compounds in seaweed: functional food applications and legislation. J Appl Phycol 23:543–597. https://doi.org/10.1007/s10811-010-9632-5
Hu S, Fan X, Qi P, Zhang X (2019) Identification of anti-diabetes peptides from Spirulina platensis. J Funct Foods 56:333–341. https://doi.org/10.1016/j.jff.2019.03.024
Huang R, Cheng J, Qiu Y et al (2019) Solvent-free lipid extraction from microalgal biomass with subcritical water in a continuous flow reactor for acid-catalyzed biodiesel production. Fuel 253:90–94. https://doi.org/10.1016/j.fuel.2019.05.004
Hudaib M, Gotti R, Pomponio R, Cavrini V (2003) Recovery evaluation of lipophilic markers from Echinacea purpurea roots applying microwave-assisted solvent extraction versus conventional methods. J Sep Sci 26:97–104. https://doi.org/10.1002/jssc.200390021
Ilyas Z, Ali Redha A, Wu YS et al (2023) Nutritional and health benefits of the brown seaweed Himanthalia elongata. Plant Foods Hum Nutr 78:233–242. https://doi.org/10.1007/s11130-023-01056-8
Iqbal M, Tao Y, Xie S et al (2016) Aqueous two-phase system (ATPS): an overview and advances in its applications. Biol Proced Online 18:18. https://doi.org/10.1186/s12575-016-0048-8
Jaime L, Mendiola JA, Herrero M et al (2005) Separation and characterization of antioxidants from Spirulina platensis microalga combining pressurized liquid extraction, TLC, and HPLC-DAD. J Sep Sci 28:2111–2119. https://doi.org/10.1002/jssc.200500185
Jones AF, Winkles JW, Jennings PE et al (1988) Serum antioxidant activity in diabetes mellitus. Diabetes Res 7:89–92
Jung F, Krüger-Genge A, Waldeck P, Küpper J-H (2019) Spirulina platensis, a super food? J Cell Biotechnol 5:43–54. https://doi.org/10.3233/JCB-189012
Kadam SU, Tiwari BK, O'Donnell CP (2013) Application of novel extraction technologies for bioactives from marine algae. J Agric Food Chem 61:4667–4675. https://doi.org/10.1021/jf400819p
Kadam SU, Álvarez C, Tiwari BK, O'Donnell CP (2015) Extraction of biomolecules from seaweeds. In: Seaweed sustainability. Elsevier, Amsterdam, pp 243–269
Kader MS, Rahman MRT (2021) Supercritical fluid extraction (SFE), solid-phase micro extraction (SPME), and stir bar sorption extraction (SBSE) techniques. In: Techniques to measure food safety and quality. Springer International Publishing, Cham, pp 219–227
Kapoore R, Butler T, Pandhal J, Vaidyanathan S (2018) Microwave-assisted extraction for microalgae: from biofuels to biorefinery. Biology (Basel) 7:18. https://doi.org/10.3390/biology7010018
Karimi E, Jaafar HZE (2011) HPLC and GC-MS determination of bioactive compounds in microwave obtained extracts of three varieties of Labisia pumila Benth. Molecules 16:6791–6805. https://doi.org/10.3390/molecules16086791

Karkos PD, Leong SC, Karkos CD et al (2011) Spirulina in clinical practice: evidence-based human applications. Evid Based Complement Altern Med 2011:1–4. https://doi.org/10.1093/ecam/nen058

Kellogg J, Grace M, Lila M (2014) Phlorotannins from Alaskan seaweed inhibit carbolytic enzyme activity. Mar Drugs 12:5277–5294. https://doi.org/10.3390/md12105277

Khavari F, Saidijam M, Taheri M, Nouri F (2021) Microalgae: therapeutic potentials and applications. Mol Biol Rep 48:4757–4765. https://doi.org/10.1007/s11033-021-06422-w

Khaw K-Y, Parat M-O, Shaw PN, Falconer JR (2017) Solvent supercritical fluid technologies to extract bioactive compounds from natural sources: a review. Molecules 22:1186. https://doi.org/10.3390/molecules22071186

Kim JK, Kottuparambil S, Moh SH et al (2015) Potential applications of nuisance microalgae blooms. J Appl Phycol 27:1223–1234. https://doi.org/10.1007/s10811-014-0410-7

Kim H-S, Fernando IPS, Lee S-H et al (2021) Isolation and characterization of anti-inflammatory compounds from Sargassum horneri via high-performance centrifugal partition chromatography and high-performance liquid chromatography. Algal Res 54:102209. https://doi.org/10.1016/j.algal.2021.102209

Kim H-S, Je J-G, An H et al (2022) Isolation and characterization of efficient active compounds using high-performance centrifugal partition chromatography (CPC) from anti-inflammatory activity fraction of Ecklonia maxima in South Africa. Mar Drugs 20:471. https://doi.org/10.3390/md20080471

Kokabi M, Yousefzadi M (2022) Algal nanoparticles and their potential application in agriculture. In: Nano-enabled agrochemicals in agriculture. Elsevier, Amsterdam, pp 189–198

Kolackova M, Janova A, Dobesova M et al (2023) Role of secondary metabolites in distressed microalgae. Environ Res 224:115392. https://doi.org/10.1016/j.envres.2023.115392

Koyande AK, Chew KW, Rambabu K et al (2019) Microalgae: a potential alternative to health supplementation for humans. Food Sci Human Wellness 8:16–24. https://doi.org/10.1016/j.fshw.2019.03.001

Krishnan S, Ghani NA, Aminuddin NF et al (2020) Microwave-assisted lipid extraction from Chlorella vulgaris in water with 0.5%–2.5% of imidazolium based ionic liquid as additive. Renew Energy 149:244–252. https://doi.org/10.1016/j.renene.2019.12.063

Kronick MN, Grossman PD (1983) Immunoassay techniques with fluorescent phycobiliprotein conjugates. Clin Chem 29:1582–1586. https://doi.org/10.1093/clinchem/29.9.1582

Kumari A, Garima, Bharadvaja N (2023) A comprehensive review on algal nutraceuticals as prospective therapeutic agent for different diseases. 3 Biotech 13:44. https://doi.org/10.1007/s13205-022-03454-2

Lee J-C, Hou M-F, Huang H-W et al (2013) Marine algal natural products with anti-oxidative, anti-inflammatory, and anti-cancer properties. Cancer Cell Int 13:55. https://doi.org/10.1186/1475-2867-13-55

Lee SH, Kang SM, Sok CH et al (2015) Cellular activities and docking studies of eckol isolated from Ecklonia cava (Laminariales, Phaeophyceae) as potential tyrosinase inhibitor. Algae 30:163–170. https://doi.org/10.4490/algae.2015.30.2.163

Lehmann U, Riedel A, Hirche F et al (2016) Vitamin D3 supplementation: response and predictors of vitamin D3 metabolites—a randomized controlled trial. Clin Nutr 35:351–358. https://doi.org/10.1016/j.clnu.2015.04.021

Li Y, Zhao M, Gomez LP et al (2021) Investigation of enzyme-assisted methods combined with ultrasonication under a controlled alkali pretreatment for agar extraction from Gelidium sesquipedale. Food Hydrocoll 120:106905. https://doi.org/10.1016/j.foodhyd.2021.106905

Lock E-J, Waagbø R, Wendelaar Bonga S, Flik G (2010) The significance of vitamin D for fish: a review. Aquac Nutr 16:100–116. https://doi.org/10.1111/j.1365-2095.2009.00722.x

Lojero CG, Pérez-Gómez B, Krogmann DW, Peña-Diaz A (1997) The tricylindrical core of the phycobilisome of the cyanobacterium Arthrospira (Spirulina) maxima. Int J Biochem Cell Biol 29:959–970. https://doi.org/10.1016/S1357-2725(97)00042-3

Lopes G, Sousa C, Valentão P, Andrade PB (2013) Sterols in algae and health. In: Bioactive compounds from marine foods. Wiley, New York, pp 173–191

López-Hortas L, Flórez-Fernández N, Torres MD et al (2021) Applying seaweed compounds in cosmetics, cosmeceuticals and nutricosmetics. Mar Drugs 19:552. https://doi.org/10.3390/md19100552

Lourenço-Lopes C, Fraga-Corral M, Garcia-Perez P et al (2022) A HPLC-DAD method for identifying and estimating the content of fucoxanthin, β-carotene and chlorophyll a in brown algal extracts. Food Chem Adv 1:100095. https://doi.org/10.1016/j.focha.2022.100095

Makareviciene V, Skorupskaite V (2019) Transesterification of microalgae for biodiesel production. In: Second and third generation of feedstocks. Elsevier, Amsterdam, pp 469–510

Matos ÂP, Novelli E, Tribuzi G (2022) Use of algae as food ingredient: sensory acceptance and commercial products. Front Food Sci Technol 23:989801. https://doi.org/10.3389/frfst.2022.989801

Mehta N, Jeyapriya S, Kumar P et al (2022) Ultrasound-assisted extraction and the encapsulation of bioactive components for food applications. Food Secur 11:2973. https://doi.org/10.3390/foods11192973

Mendes RL, Coelho JP, Fernandes HL et al (1995) Applications of supercritical CO2 extraction to microalgae and plants. J Chem Technol Biotechnol 62:53–59. https://doi.org/10.1002/jctb.280620108

Molina Grima E, Belarbi E-H, Acién Fernández F et al (2003) Recovery of microalgal biomass and metabolites: process options and economics. Biotechnol Adv 20:491–515. https://doi.org/10.1016/S0734-9750(02)00050-2

Montalbán MG, Víllora G (2022) Supercritical fluids: properties and applications. In: Phase equilibria with supercritical carbon dioxide—application to the components of a biocatalytic process. IntechOpen, London

Montañés F, Tallon S (2018) Supercritical fluid chromatography as a technique to fractionate high-valued compounds from lipids. Separations 5:38. https://doi.org/10.3390/separations5030038

Mourelle M, Gómez C, Legido J (2017) The potential use of marine microalgae and cyanobacteria in cosmetics and thalassotherapy. Cosmetics 4:46. https://doi.org/10.3390/cosmetics4040046

Murray M, Dordevic AL, Ryan L, Bonham MP (2018) Phlorotannins and macroalgal polyphenols: potential as functional food ingredients and role in health promotion. In: Functional food and human health. Springer, Singapore, pp 27–58

Naeem A, Saleemuddin M, Hasan Khan R (2007) Glycoprotein targeting and other applications of lectins in biotechnology. Curr Protein Pept Sci 8:261–271. https://doi.org/10.2174/138920307780831811

Najjar YSH, Abu-Shamleh A (2020) Harvesting of microalgae by centrifugation for biodiesel production: a review. Algal Res 51:102046. https://doi.org/10.1016/j.algal.2020.102046

Nayaka SR, Anand JVS, Shareef SM, Usha NS (2020) Gas chromatography—mass spectroscopy [GC-MS] analysis and phytochemical screening for bioactive compounds in Caulerpa peltata (Greenalga). Biomed Pharmacol J 13:1921–1926. https://doi.org/10.13005/bpj/2069

Negreanu-Pirjol T, Sirbu R, Mirea M (2020) Antioxidant activity correlated with chlorophyll pigments and magnesium content of some green seaweeds. Eur J Med Nat Sci 3:107. https://doi.org/10.26417/393qel951

Nie J, Chen D, Ye J et al (2021) Preparative separation of three terpenoids from edible brown algae Sargassum fusiforme by high-speed countercurrent chromatography combined with preparative high-performance liquid chromatography. Algal Res 59:102449. https://doi.org/10.1016/j.algal.2021.102449

Nieto A, Borrull F, Pocurull E, Marcé RM (2010) Pressurized liquid extraction: a useful technique to extract pharmaceuticals and personal-care products from sewage sludge. TrAC Trends Anal Chem 29:752–764. https://doi.org/10.1016/j.trac.2010.03.014

Nuhu AA (2013) Spirulina (Arthrospira): an important source of nutritional and medicinal compounds. J Mar Biol 2013:1–8. https://doi.org/10.1155/2013/325636

Nwosu F, Morris J, Lund VA et al (2011) Anti-proliferative and potential anti-diabetic effects of phenolic-rich extracts from edible marine algae. Food Chem 126:1006–1012. https://doi.org/10.1016/j.foodchem.2010.11.111

Otero P, Quintana S, Reglero G et al (2018) Pressurized liquid extraction (PLE) as an innovative green technology for the effective enrichment of Galician algae extracts with high quality fatty acids and antimicrobial and antioxidant properties. Mar Drugs 16:156. https://doi.org/10.3390/md16050156

Panda D, Manickam S (2019) Cavitation technology—the future of greener extraction method: a review on the extraction of natural products and process intensification mechanism and perspectives. Appl Sci 9:766. https://doi.org/10.3390/app9040766

Pankiewicz R, Łęska B, Messyasz B et al (2016) First isolation of polysaccharidic ulvans from the cell walls of freshwater algae. Algal Res 19:348–354. https://doi.org/10.1016/j.algal.2016.02.025

Pasquet V, Chérouvrier J-R, Farhat F et al (2011) Study on the microalgal pigments extraction process: performance of microwave assisted extraction. Process Biochem 46:59–67. https://doi.org/10.1016/j.procbio.2010.07.009

Patel AK, Albarico FPJB, Perumal PK et al (2022) Algae as an emerging source of bioactive pigments. Bioresour Technol 351:126910. https://doi.org/10.1016/j.biortech.2022.126910

Patil PD, Dandamudi KPR, Wang J et al (2018) Extraction of bio-oils from algae with supercritical carbon dioxide and co-solvents. J Supercrit Fluids 135:60–68. https://doi.org/10.1016/j.supflu.2017.12.019

Pereira L, Cotas J (2023) Therapeutic potential of polyphenols and other micronutrients of marine origin. Mar Drugs 21:323. https://doi.org/10.3390/md21060323

Perera KY, Sharma S, Pradhan D et al (2021) Seaweed polysaccharide in food contact materials (active packaging, intelligent packaging, edible films, and coatings). Food Secur 10:2088. https://doi.org/10.3390/foods10092088

Perez-Vazquez A, Carpena M, Barciela P et al (2023) Pressurized liquid extraction for the recovery of bioactive compounds from seaweeds for food industry application: a review. Antioxidants 12:612. https://doi.org/10.3390/antiox12030612

Radovanović K, Gavarić N, Švarc-Gajić J et al (2022) Subcritical water extraction as an effective technique for the isolation of phenolic compounds of Achillea species. Processes 11:86. https://doi.org/10.3390/pr11010086

Rai LC, Gaur JP (eds) (2001) Algal adaptation to environmental stresses. Springer, Berlin

Rajauria G, Abu-Ghannam N (2013) Isolation and partial characterization of bioactive fucoxanthin from Himanthalia elongata brown seaweed: a TLC-based approach. Int J Anal Chem 2013:1–6. https://doi.org/10.1155/2013/802573

Rajendran I (2020) Marine algal polysaccharides and their applications. In: Encyclopedia of marine biotechnology. Wiley, New York, pp 1195–1208

Ramesh Kumar B, Deviram G, Mathimani T et al (2019) Microalgae as rich source of polyunsaturated fatty acids. Biocatal Agric Biotechnol 17:583–588. https://doi.org/10.1016/j.bcab.2019.01.017

Rao MA, Rizvi SSH, Datta AK (eds) (2014) Engineering properties of foods. CRC Press, Boca Raton

Rasul MG (2018) Conventional extraction methods use in medicinal plants, their advantages and disadvantages. Int J Basic Sci Appl Comput:10–14

Remya RR, Radhika Rajasree SR (2016) A study on bioactive compounds derived from brown seaweeds and their therapeutic applications towards various diseases. Res J Pharm Technol 9:369–372. https://doi.org/10.5958/0974-360X.2016.00066.4

Reuter W, Nickel C, Wehrmeyer W (1990) Isolation of allophycocyanin B from Rhodella violacea results in a model of the core from hemidiscoidal phycobilisomes of rhodophyceae. FEBS Lett 273:155–158. https://doi.org/10.1016/0014-5793(90)81073-W

Rezaei Motlagh S, Harun R, Biak DRA, Hussain SA (2020) Microwave assisted extraction of lipid from Nannochloropsis gaditana microalgae using [EMIM]Cl. IOP Conf Ser Mater Sci Eng 778:012164. https://doi.org/10.1088/1757-899X/778/1/012164

Rodrigues D, Sousa S, Silva A et al (2015) Impact of enzyme- and ultrasound-assisted extraction methods on biological properties of red, brown, and green seaweeds from the Central West

Coast of Portugal. J Agric Food Chem 63:3177–3188. https://doi.org/10.1021/JF504220E/ASSET/IMAGES/MEDIUM/JF-2014-04220E_0007.GIF

Ronda SR, Lele S (2007) Argentated silica gel chromatography for separation of γ-linolenic acid from microalgae. Biosci Biotechnol Res Asia 4:621–626

Rossano R, Ungaro N, D'Ambrosio A et al (2003) Extracting and purifying R-phycoerythrin from Mediterranean red algae Corallina elongata Ellis & Solander. J Biotechnol 101:289–293. https://doi.org/10.1016/S0168-1656(03)00002-6

Saha M, Fink P (2022) Algal volatiles—the overlooked chemical language of aquatic primary producers. Biol Rev 97:2162–2173. https://doi.org/10.1111/brv.12887

Sánchez-Camargo ADP, Montero L, Stiger-Pouvreau V et al (2016) Considerations on the use of enzyme-assisted extraction in combination with pressurized liquids to recover bioactive compounds from algae. Food Chem 192:67–74. https://doi.org/10.1016/j.foodchem.2015.06.098

Sanger G, Wonggo D, Montolalu LADY, Dotulong V (2022) Pigments constituents, phenolic content and antioxidant activity of brown seaweed Sargassum sp. In: IOP conference series: earth and environmental science, vol 1033. IOP Publishing, Bristol, p 012057

Sarkar S, Manna MS, Bhowmick TK, Gayen K (2020) Extraction of chlorophylls and carotenoids from dry and wet biomass of isolated Chlorella thermophila: optimization of process parameters and modelling by artificial neural network. Process Biochem 96:58–72. https://doi.org/10.1016/j.procbio.2020.05.025

Sarkar S, Sarkar S, Manna MS et al (2022) Extraction of carbohydrates and proteins from algal resources using supercritical and subcritical fluids for high-quality products. In: Innovative and emerging technologies in the bio-marine food sector. Elsevier, Amsterdam, pp 249–275

Sasidharan S, Chen Y, Saravanan D et al (2011) Extraction, isolation and characterization of bioactive compounds from plants' extracts. Afr J Tradit Complement Altern Med 8:1–10. https://doi.org/10.4314/ajtcam.v8i1.60483

Saxena A, Tripathi BP, Kumar M, Shahi VK (2009) Membrane-based techniques for the separation and purification of proteins: an overview. Adv Colloid Interf Sci 145:1–22. https://doi.org/10.1016/j.cis.2008.07.004

Ścieszka S, Klewicka E (2019) Algae in food: a general review. Crit Rev Food Sci Nutr 59:3538–3547. https://doi.org/10.1080/10408398.2018.1496319

Sekar S, Chandramohan M (2008) Phycobiliproteins as a commodity: trends in applied research, patents and commercialization. J Appl Phycol 20:113–136. https://doi.org/10.1007/s10811-007-9188-1

Shao Z, Duan D (2022) The cell wall polysaccharides biosynthesis in seaweeds: a molecular perspective. Front Plant Sci 13:902823. https://doi.org/10.3389/fpls.2022.902823

Shobier AH, Abdel Ghani SA, Barakat KM (2016) GC/MS spectroscopic approach and antifungal potential of bioactive extracts produced by marine macroalgae. Egypt J Aquat Res 42:289–299. https://doi.org/10.1016/j.ejar.2016.07.003

Show K-Y, Lee D-J, Chang J-S (2013) Algal biomass dehydration. Bioresour Technol 135:720–729. https://doi.org/10.1016/j.biortech.2012.08.021

Simon D, Helliwell S (1998) Extraction and quantification of chlorophyll a from freshwater green algae. Water Res 32:2220–2223. https://doi.org/10.1016/S0043-1354(97)00452-1

Singh S, Purwar V (2022) Role of algae in cosmetics

Smith RM (1999) Supercritical fluids in separation science—the dreams, the reality and the future. J Chromatogr A 856:83–115. https://doi.org/10.1016/S0021-9673(99)00617-2

Smith AG, Croft MT, Moulin M, Webb ME (2007) Plants need their vitamins too. Curr Opin Plant Biol 10:266–275. https://doi.org/10.1016/j.pbi.2007.04.009

Soto-Sierra L, Wilken LR, Mallawarachchi S, Nikolov ZL (2021) Process development of enzymatically-generated algal protein hydrolysates for specialty food applications. Algal Res 55:102248. https://doi.org/10.1016/j.algal.2021.102248

Spolaore P, Joannis-Cassan C, Duran E, Isambert A (2006) Commercial applications of microalgae. J Biosci Bioeng 101:87–96. https://doi.org/10.1263/jbb.101.87

Suarez Ruiz CA, Kwaijtaal J, Peinado OC et al (2020) Multistep fractionation of microalgal biomolecules using selective aqueous two-phase systems. ACS Sustain Chem Eng 8:2441–2452. https://doi.org/10.1021/acssuschemeng.9b06379

Sun P, Wong C-C, Li Y et al (2019) A novel strategy for isolation and purification of fucoxanthinol and fucoxanthin from the diatom Nitzschia laevis. Food Chem 277:566–572. https://doi.org/10.1016/j.foodchem.2018.10.133

Tamanna Ferdous U, Norhana Balia Yusof Z (2021) Algal terpenoids: a potential source of antioxidants for cancer therapy. In: Terpenes and terpenoids—recent advances. IntechOpen, London

Tang Z, Zhao JL, Ju B et al (2016) One-step chromatographic procedure for purification of B-phycoerythrin from Porphyridium cruentum. Protein Expr Purif 123:70–74. https://doi.org/10.1016/j.pep.2016.01.018

Teboul E, Tammekivi E, Batteau M et al (2023) Off-line two-dimensional separation involving supercritical fluid chromatography for the characterization of the wastewater from algae hydrothermal liquefaction. J Chromatogr A 1694:463907. https://doi.org/10.1016/j.chroma.2023.463907

Teixeira-Guedes C, Gomes-Dias JS, Cunha SA et al (2023) Enzymatic approach for the extraction of bioactive fractions from red, green and brown seaweeds. Food Bioprod Process 138:25–39. https://doi.org/10.1016/j.fbp.2022.12.005

Teoh WY, Yong YS, Razali FN et al (2023) LC-MS/MS and GC-MS analysis for the identification of bioactive metabolites responsible for the antioxidant and antibacterial activities of Lygodium microphyllum (Cav.) R. Br. Separations 10:215. https://doi.org/10.3390/separations10030215

The Ayurvedic Pharmacopoeia of India (2016) The Ayurvedic Pharmacopoeia of India part-I, volume-IX. Government of India Ministry of Ayush

The United States Pharmacopeia (2004) USP 27: the National Formulary: NF 22: by Authority of the United States Pharmacopeial Convention, Inc., Meeting at Washington, D.C., April 12-16, 2000. United States Pharmacopeial Convention, North Bethesda

Thiruvenkadam S, Izhar S, Yoshida H et al (2015) Process application of subcritical water extraction (SWE) for algal bio-products and biofuels production. Appl Energy 154:815–828. https://doi.org/10.1016/j.apenergy.2015.05.076

Thiyagarasaiyar K, Goh B-H, Jeon Y-J, Yow Y-Y (2020) Algae metabolites in cosmeceutical: an overview of current applications and challenges. Mar Drugs 18:323. https://doi.org/10.3390/md18060323

Thomas N, Kim S-K (2013) Beneficial effects of marine algal compounds in cosmeceuticals. Mar Drugs 11:146–164. https://doi.org/10.3390/md11010146

Tirado DF, Cabañas A, Calvo L (2023) Modelling and scaling-up of a supercritical fluid extraction of emulsions process. Processes 11:1063. https://doi.org/10.3390/pr11041063

Tiwari BK (2015) Ultrasound: a clean, green extraction technology. TrAC Trends Anal Chem 71:100–109. https://doi.org/10.1016/j.trac.2015.04.013

Truong D-H, Nguyen DH, Ta NTA et al (2019) Evaluation of the use of different solvents for phytochemical constituents, antioxidants, and in vitro anti-inflammatory activities of Severinia buxifolia. J Food Qual 2019:1–9. https://doi.org/10.1155/2019/8178294

Ubhayasekera SJKA, Acharya SR, Bergquist J (2018) A novel, fast and sensitive supercritical fluid chromatography-tandem mass spectrometry (SFC-MS/MS) method for analysis of arachidonic acid metabolites. Analyst 143:3661–3669. https://doi.org/10.1039/C8AN00788H

Ulrich J, Frohberg P (2013) Problems, potentials and future of industrial crystallization. Front Chem Sci Eng 7:1–8. https://doi.org/10.1007/s11705-013-1304-y

Ummat V, Sivagnanam SP, Rajauria G et al (2021) Advances in pre-treatment techniques and green extraction technologies for bioactives from seaweeds. Trends Food Sci Technol 110:90–106. https://doi.org/10.1016/j.tifs.2021.01.018

Urh M, Simpson D, Zhao K (2009) Chapter 26: Affinity chromatography. In: Methods in enzymology. Elsevier, Amsterdam, pp 417–438

Uwineza PA, Waśkiewicz A (2020) Recent advances in supercritical fluid extraction of natural bioactive compounds from natural plant materials. Molecules 25:3847. https://doi.org/10.3390/molecules25173847

van de Velde B, Guillarme D, Kohler I (2020) Supercritical fluid chromatography—mass spectrometry in metabolomics: past, present, and future perspectives. J Chromatogr B 1161:122444. https://doi.org/10.1016/j.jchromb.2020.122444
Venegas-Venegas E, Guil-Guerrero JL, Rincón-Cervera MÁ, Ramos-Bueno RP (2014) Tridocosahexaenoyl glycerol purification from docosahexaenoic acid single cell oil (DHASCO). Eur J Lipid Sci Technol 116:266–271. https://doi.org/10.1002/ejlt.201300350
Wang G, Sun H, Fan X, Tseng C (2002) Large-scale isolation and purification of R-phycoerythrin from red alga Palmaria palmata using the expanded bed adsorption method. J Integr Plant Biol 44:541–546
Wang H-MD, Chen C-C, Huynh P, Chang J-S (2015) Exploring the potential of using algae in cosmetics. Bioresour Technol 184:355–362. https://doi.org/10.1016/j.biortech.2014.12.001
Weis WI, Drickamer K (1996) Structural basis of lectin-carbohydrate recognition. Annu Rev Biochem 65:441–473. https://doi.org/10.1146/annurev.bi.65.070196.002301
Wen L, Zhang Z, Sun D-W et al (2020) Combination of emerging technologies for the extraction of bioactive compounds. Crit Rev Food Sci Nutr 60:1826–1841. https://doi.org/10.1080/10408398.2019.1602823
Wijesinghe WAJP, Jeon Y-J (2012) Enzyme-assistant extraction (EAE) of bioactive components: a useful approach for recovery of industrially important metabolites from seaweeds: a review. Fitoterapia 83:6–12. https://doi.org/10.1016/j.fitote.2011.10.016
Xavier J, Jose J (2020) Study of mineral and nutritional composition of some seaweeds found along the coast of Gulf of Mannar, India. Plant Sci Today 7:631–637. https://doi.org/10.14719/pst.2020.7.4.912
Xue L, Zhang Y, Zhang T et al (2005) Effects of enhanced ultraviolet-B radiation on algae and cyanobacteria. Crit Rev Microbiol 31:79–89. https://doi.org/10.1080/10408410590921727
Yalçın S, Karakaş Ö, Okudan EŞ et al (2021a) HPLC detection and antioxidant capacity determination of brown, red and green algal pigments in seaweed extracts. J Chromatogr Sci 59:325–337. https://doi.org/10.1093/chromsci/bmaa107
Yalçın S, Uzun M, Karakaş Ö et al (2021b) Determination of total antioxidant capacities of algal pigments in seaweed by the combination of high-performance liquid chromatography (HPLC) with a cupric reducing antioxidant capacity (CUPRAC) assay. Anal Lett 54:2239–2258. https://doi.org/10.1080/00032719.2020.1855439
Yamamoto M, Baldermann S, Yoshikawa K et al (2014) Determination of volatile compounds in four commercial samples of Japanese green algae using solid phase microextraction gas chromatography mass spectrometry. Sci World J 2014:1–8. https://doi.org/10.1155/2014/289780
Yang W, Chen X, Li Y et al (2020) Advances in pharmacological activities of terpenoids. Nat Prod Commun 15:1934578X2090355. https://doi.org/10.1177/1934578X20903555
Zafarani-Moattar MT, Shekaari H, Fattah SG, Mokhtarpour M (2022) Novel aqueous two-phase systems containing polymer-based deep eutectic solvent and citrate salts for high-performance extraction of dyes. J Mol Liq 360:119475. https://doi.org/10.1016/j.molliq.2022.119475
Zeng L-M, Wang C-J, Su J-Y et al (2010) Flavonoids from the red alga Acanthophora spicifera. Chinese J Chem 19:1097–1100. https://doi.org/10.1002/cjoc.20010191116
Zghaibi N, Omar R, Kamal SMM et al (2019) Microwave-assisted brine extraction for enhancement of the quantity and quality of lipid production from microalgae Nannochloropsis sp. Molecules 24:3581. https://doi.org/10.3390/molecules24193581
Zhang Q-W, Lin L-G, Ye W-C (2018) Techniques for extraction and isolation of natural products: a comprehensive review. Chin Med 13:20. https://doi.org/10.1186/s13020-018-0177-x
Zhang J, Liu L, Ren Y, Chen F (2019) Characterization of exopolysaccharides produced by microalgae with antitumor activity on human colon cancer cells. Int J Biol Macromol 128:761–767. https://doi.org/10.1016/j.ijbiomac.2019.02.009
Zhang S, Qamar SA, Junaid M et al (2022) Algal polysaccharides-based nanoparticles for targeted drug delivery applications. Starch - Stärke 74:7–8. https://doi.org/10.1002/star.202200014
Zhou J, Wang M, Saraiva JA et al (2022a) Extraction of lipids from microalgae using classical and innovative approaches. Food Chem 384:132236. https://doi.org/10.1016/j.foodchem.2022.132236

Zhou L, Li K, Duan X et al (2022b) Bioactive compounds in microalgae and their potential health benefits. Food Biosci 49:101932. https://doi.org/10.1016/j.fbio.2022.101932
Ziółkowska NE, Wlodawer A (2006) Structural studies of algal lectins with anti-HIV activity. Acta Biochim Pol 53:617–626

Machine Learning and Artificial Intelligence for Algal Cultivation, Harvesting Techniques, Wastewater Treatment, Nutrient Recovery, and Biofuel Production and Optimization

18

Iradat Hussain Mafat, Sridhar Palla, and Dadi Venkata Surya

Contents

Abstract

The applications of algae, such as in biofuel production, carbon capture and utilization, and assorted microalgae production due to their high nutrient content, wastewater treatment, and bioremediation, make algae cultivation extremely popular in industries. Various process parameters such as algal characteristics and operating conditions of the process affect the yield and productivity. The

I. H. Mafat · S. Palla (✉)
Department of Chemical Engineering, Indian Institute of Petroleum and Energy Visakhapatnam, Visakhapatnam, Andhra Pradesh, India
e-mail: sridharpalla.che@iipe.ac.in

D. V. Surya
Department of Chemical Engineering, Pandit Deendayal Energy University, Gandhinagar, Gujarat, India

S. K. Arya et al. (eds.), *Value Added Products From Bioalgae Based Biorefineries: Opportunities and Challenges*,
https://doi.org/10.1007/978-981-97-1662-3_18

operating conditions and feedstock characteristics are directly correlated to the product. For example, bio-oil yield produced is directly related to ultimate and proximate analysis; similarly wastewater treatment with algae is dependent upon the organic and inorganic content in the water. Therefore, it is essential to optimize these processes to enhance productivity and identify the efficient method for producing high-quality products with minimal wastage. This can be accomplished by using machine learning (ML), one of the most recently developed tools for modeling a process with multiple inputs to predict output accurately without conducting tedious experiments. ML is widely applied in predictive modeling for growth optimization, nutrient recovery, real-time decision support systems, quality control in algal biomass, energy efficiency optimization, and many more. The incorporation of ML is playing a critical role in the evolution of these algae farming applications. This chapter examines the different applications of artificial intelligence (AI) and ML-based algorithms for product enhancement, process optimization, and gaining important insights into algal biotechnology.

Keywords

Machine learning · Algal cultivation · Biofuel production · Wastewater treatment · Nutrient recovery

Abbreviations

AI	Artificial intelligence
ANN	Artificial neural network
DBSCAN	Density-Based Spatial Clustering of Applications with Noise
DT	Decision tree
HRAP	High-rate algal wastewater pond
KS	Kolmogorov-Smirnov
LIME	Local Interpretable Model-Agnostic Explanations
MAPE	Mean absolute percentage error
ML	Machine learning
MODIS	Moderate resolution imaging spectrometer
PLS	Partial Least Square
R^2	Coefficient of determination
RF	Random forest
RMSE	Root mean square error
SHAP	Shapely Additive Explanations
SVM	Support vector machine
TD	Temporal Difference
XGB	Extreme gradient boosting

18.1 Introduction

Algae are an extensive group of aquatic, photosynthetic organisms classified as either macroalgae or microalgae that produce most of the earth's oxygen that can feed aquatic life. Marine algae are a high-nutritional source of proteins, carbohydrates, vitamins, antioxidants, and polyunsaturated fatty acids. Algal biotechnology is a technology used to synthesize the product for specific application by tuning micro/macro algae or their derivatives (Borowitzka 2015). Algae has significant contribution in the remediation of different toxins since it grows collecting nutrients and carbon dioxide (CO_2) from their surroundings by nature (Leng et al. 2020a, b). Algal cultivation has the potential to reduce climate change, energy depletion, and environmental pollution (Savage 2012). Figure 18.1 depicts numerous methods for converting algal biomass to various energy sources, including anaerobic digestion, transesterification, fermentation, thermochemical conversion, and biohydrogen production.

Anaerobic digestion is an effective method for the degradation of microbial algae to convert into nontoxic product. Although anaerobic digestion is a cost-effective conversion technique of algae to energy, the scale-up of this process is a bottleneck (Veerabadhran et al. 2021). Transesterification is the catalytic reaction of oil/fat with alcohol to yield biodiesel, which is an economical and time-saving method (Sajjad et al. 2022). The fermentation is defined as the breakdown of organic compounds by microbes either in the presence or absence of oxygen, which results in the production of energy and can generate various commercial products (Ryu and Kim 2013). The higher yields and efficiency are strongly recommended for these processes to replace the existing technologies. To understand the fundamental phenomenon of each process, the effect of various parameters on product yield was investigated by modeling the system in the literature.

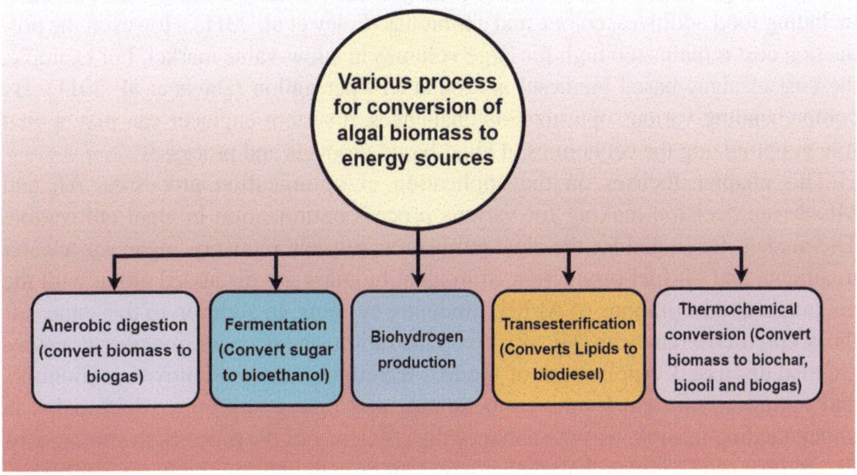

Fig. 18.1 Various energy generation processes from algal biomass

Robust and precise kinetic and machine learning (ML) models have been widely applied for the biological systems (Bradford et al. 2018). Traditional kinetic modeling, optimization, and state estimation of batch systems predict the product yield. However, accuracy of these models is low due to the reduction of vast number of metabolic reactions into a few parameters and small sets of differential equations. The several assumptions to simplify the complicated nonlinear differential equations have huge significant impacts on the model accuracy (Zhang et al. 2019). Therefore, data-driven models, also known as ML models, have been proven as a better alternative for modeling and optimization problems, since precision is high at the experimental conditions. ML is a discipline of artificial intelligence (AI) that uses massive datasets to uncover hidden patterns within them. It is accomplished by inventing and researching numerous statistical algorithms that can effectively generalize and achieve goals without explicit instructions. Therefore, ML can also be applied in an algal biotechnology. Mathematical models have investigated the basic mechanism, such as large-scale algal growth cultivations. However, the complex mathematical models are time consuming and complex. Therefore, ML models can generate reliable and accurate data that serve as a foundation for large-scale cultivation. Previous studies have included the application of AI (Peter et al. 2023) to monitor biomass growth in semi-batch cultivation of *Chlorella vulgaris* and the classification of microalgae for culture characterization (Chong et al. 2023).

The optimization of algal culture media composition can be regarded as a significant aspect in the production of sustainable goods for a variety of applications such as wastewater treatment, harvesting, nutrient recovery, and biofuel generation. The optimization process aids in the identification of more appropriate steps as well as the optimized composition for cost reduction. Several studies have found that optimizing algal cultivation techniques increases biomass and lipid yields (Breuer et al. 2013; Zavřel et al. 2015). Various optimization frameworks assist in effectively simulating and forecasting their growth and production dynamics and identifying optimal cultivation strategies. Microalgae are frequently cultivated for a variety of purposes, including food additives, colors, and antibiotics (Foley et al. 2011). However, the production cost remains too high for large volumes in a low-value market. For example, the cost of algae-based biodiesel is around $10 per gallon (Davis et al. 2011). By comprehending various optimization challenges, a system engineer can play a vital role in optimizing the economics of algal-based products and processes.

This chapter focuses on the application of optimization processes, AI, and ML-driven decision-making for various process optimization in algal cultivation. The models developed for the algal cultivation, nutrient recovery, algae wastewater treatment, and biofuel production from algal biomass are discussed along with the potentials and limitations of AI/ML modeling systems. In addition to that, state-of-the-art predictive analysis, ML-driven optimization, and decision-making processes are also discussed. Application of the ML techniques to algal cultivation monitors and optimizes nutrient levels, temperatures, and light conditions, which helps in understanding feasible ways to enhance the efficiency of the process. In similar way, the optimal time for algal harvesting based on growth patterns, ambient conditions, and feedstock characteristics will be analyzed. Real-time monitoring of the water quality and removal of efficient nutrient by algae by ML modeling also improves the productivity and sustainability of the waste water treatment and nutrient recovery. Nutrient recovery contributes to circular economy by recovering the nutrients to reuse and reduce the external fertilizer dependency. Lipid extraction and conversion, feedstock characteristics, and operating conditions of the process controls the bio-oil yield and cost-effectiveness. Applications of ML modeling optimizes the effect of the challenges encountered during cultivation strategies, which are also explained along with the mitigation strategies.

18.2 Algal Cultivation and ML-Driven Growth Optimization

Conventional modeling techniques have been widely explored, therefore these methods have advantage of better understanding over ML. One significant difference between ML and traditional approaches is that ML primarily improves prediction accuracy, whereas traditional models are concerned with inferring relationships between input variables (Azzolina et al. 2019). Traditional modeling has been applied in various optimization processes for algal cultivations. Wang et al. (2023) created and successfully replicated a complicated technique for *Spirulina platensis* biofilm production, from inoculation adhesion to growth to biofilm creation and water erosion. The results showed that surface furrow-patterned motor plates were the most engineerable supporting materials for this algal biofilm-attached culture (Wang et al. 2023). Apart from many useful benefits, traditional modeling has certain disadvantages, such as process simplification, data monitoring and lack of information, following certain sets of protocol, the higher cost compared to other systems, and many more. Therefore, to overcome these drawbacks, ML modeling is recommended for optimization processes in algal growth cultivation. Bello et al. (2017) created a dynamic mathematical modeling system of microalgae cultivation for the simulation of microalgae production in a high-rate algal wastewater pond (HRAP). The behavior of the HRAP system was investigated by solving mass balance equations for various variables such as light intensity and gas–liquid mass transfer. The impact of various factors such as kinetic and operating parameters such as CO_2 supply, maximum growth rate, pond depth, and dilution rates was also evaluated. High yield accuracies were obtained comparing to the previous literatures. This study concluded that the presented model can be used effectively for the simulation and design of HRAP in a variety of experimental and environmental conditions (Barros et al. 2015). Greene et al. (2021) developed and validated a thermal and biological model for forecasting algal production in large-scale culture systems. To estimate output with temporal and geographical precision, the developed model incorporated geospatial data, inputs from reactor geometry, and a maximum of six strain-specific characteristics. The experimental dataset produced the lowest average accuracies of 0.24% and 5.32% (Greene et al. 2021). Wang et al. (2023) developed a mathematical model that accurately predicted the complicated process of *S. platensis* biofilm growth, from inoculation adhesion to biofilm formation and water erosion (Wang et al. 2023). Using a compartmental approach, Nauha and Alopaeus (2015) created a mathematical model for outdoor algal cultivation simulation. The developed model could predict system hydrodynamics, mass transfer, and algae growth. The model results were also validated with experimental measurements, demonstrating high accuracy (Nauha and Alopaeus 2015). Figueroa-Torres et al. (2018) provided findings from a study on starch accumulation that used a kinetic model capable of predicting the nitrogen-limited dynamics of mixotrophic algae growth and production of starch and lipid. The findings highlighted the potential of kinetic modeling to be used as an efficient optimizing tool for establishing large-scale alga to fuel cultivation strategies (Figueroa-Torres et al. 2018). However kinetic models are difficult to construct as the size of reaction network increases, which complicates the fundamental understanding of the parameter relationship. ML models are widely applied and recommended to systems where multiple parameters such as operating conditions, feedstock properties, and reaction kinetics play major role on final output (Costello and Martin 2018; Ali et al. 2020; Saleh and Al-Hammadi 2021).

ML is a subclass of AI that uses statistical analysis approaches to predict system variables. In an enormous dataset, ML algorithms can readily find complicated relationships (Antonakoudis et al. 2020; Chaurasia and Mondal 2022; Sundui et al. 2021). ML has numerous applications, including decision-making, optimization, and many others. ML-based approaches have the potential to break down traditional barriers and aid in the development of sustainable growing operating conditions for algal cultivation methods. There are three types of ML modeling—supervised learning, unsupervised learning, and reinforcement learning—which are further subdivided into algorithms such as classifications, regression, clustering, decision-making, and many others, as shown in Fig. 18.2.

Classification is further divided into k-nearest neighbor, random forest (RF), Gaussian process, support vector machine (SVM), naïve Bayes classifier, and decision tree (DT). DT regression, linear regression, ridge regression, support vector regression, multivariate regression, and Lasso regression are subclasses in regression technique. K-means clustering, hierarchical clustering, Gaussian mixture, DBSCAN clustering, and mean shift clustering are the subclasses in the clustering technique. Decision-making of reinforcement learning is further classified into Q-learning, R-learning and TD-learning. These modeling algorithms can be used to predict and optimize growth conditions that are important in algal biotechnologies.

Various studies have been done for the prediction of different quantities such as yield and compositions and optimization of parameters with the help of ML, as presented in Table 18.1. Peter et al. (2023) sought to improve the recycling culture media strategy for semi-batch *Chlorella vulgaris* cultivation. Based on the trained model, the iteration model produced an interpretation with more than 92% accuracy for every dataset (Peter et al. 2023).

Sangué Djandja et al. (2023) developed models for predicting nitrogen content and bio-oil yield in algal biomass solvothermal liquefaction. RF and extreme gradient boosting models (XGB) were created for this purpose. The coefficient of determination (R^2) and root mean square error (RMSE) value for the XGB model was 81.61% and 8.08 wt% for product yield and 93.33% and 0.6 wt% for nitrogen content. The study also demonstrated the importance of the hydrogen-bonding donor of the solvent over the polarity and refractive index (Sangué Djandja et al. 2023). Lai et al. (2023) created an ML model for predicting algal biomass based on Moderate Resolution Imaging Spectrometer (MODIS) data, which was successfully applied

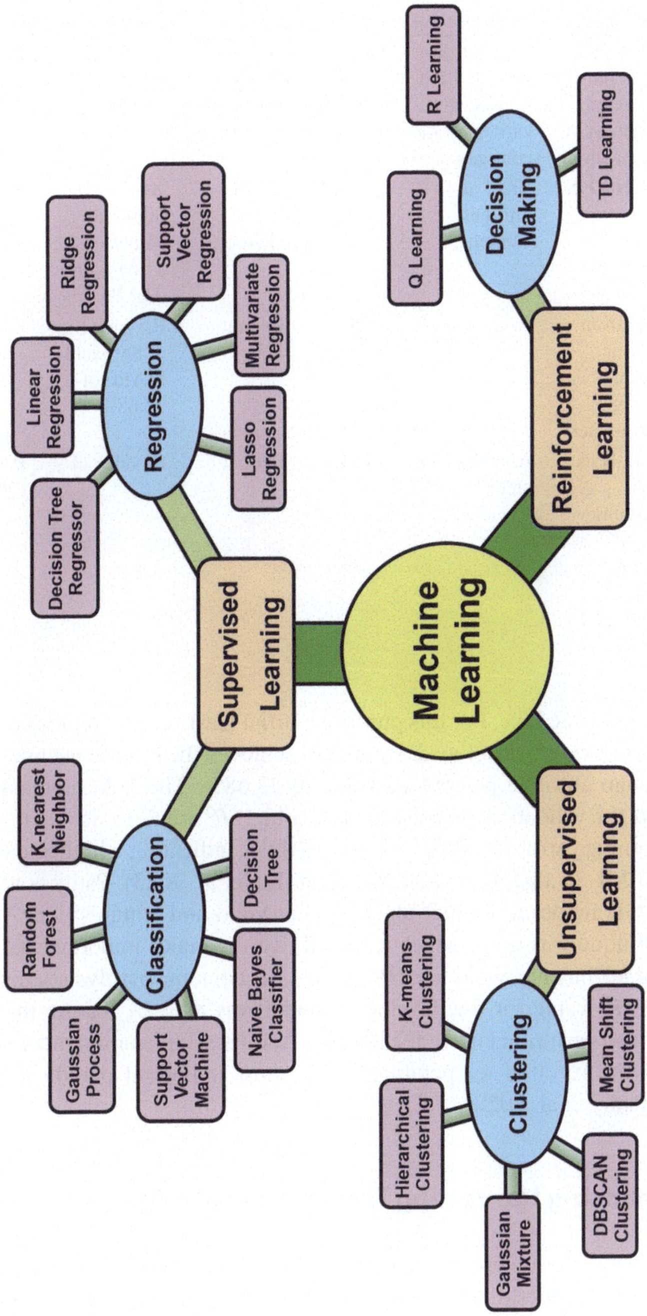

Fig. 18.2 Different types of ML algorithms

Table 18.1 Literature studies of ML in algal cultivations

Dataset quantity	Model input	Model output	Model type	Software used	Performance measures	References
3000 images	Culture ratio, mixing ratio of recycled medium to fresh medium	Microalgae growth, lipid, protein, and carbohydrate contents	One-way ANOVA	IBM SPSS Statistics version 26	92% of R^2 value	Peter et al. (2023)
290	C, H, N, O, S, ash, protein, lipid, carbohydrate	Yield and nitrogen content	RF and XGB	Scikit-learn (Python)	XGB gave higher R^2 values of 81.61% and 93.33%	Sangué Djandja et al. (2023)
156	Inputs from moderate resolution imaging spectrometer	Beu value	PLS, SVM, RF, and XGB	Python 3.7	RF gave R^2 of 85% and MAPE of 22.68%	Lai et al. (2023)
91	13 pyrolysis inputs containing H/C, N/C, ash content, pyrolysis temperature, and time	Algal biochar yield	XGB	Python	R^2 value of 84%	Pathy et al. (2020)

to a Chinese eutrophic lake. For this purpose, partial least squares regression, SVM, RF, and XGB trees were developed. RF model achieved the highest accuracy (85%) with a low mean absolute percentage value of 22.68%. The field biomass dataset was also used for validation, which revealed a high R^2 value of 86% with a mean absolute percentage error (MAPE) value of 18.42%, indicating a higher and further application of ML in algal biomass estimation (Lai et al. 2023). Pathy et al. (2020) created an XGB model to predict algal biochar yield and composition. A feature selection technique was used, which revealed that the major important parameters influencing algal biochar yield are H/C, N/C, ash content, pyrolysis temperature, and time. The R^2 value for the developed model was 84%. A feature importance analysis was also performed using the Shapley Additive Explanation method, which revealed that the reaction temperature is the most important parameter for this experiment (Pathy et al. 2020).

18.3 Enhancing Harvesting Techniques with AI

Although algal cultivation process has numerous advantages, yet few challenges are constraining the real-world applications such as, it is extremely difficult due to the small cell size of 1–20 m and the water suspension (Chen et al. 2011; Lam and Lee 2012). Despite being grown in a closed photobioreactor, the mass ratio of algal biomass to water is considered low (Suali and Sarbatly 2012). When it comes to offering several energy and environmental advantages, microalgae are among the most promising sources. Nevertheless, advancements in their intense uses are hampered by the poor productivity and high expense of algae growth. Low cell densities also make biomass collection one of the costliest methods (Roostaei et al. 2018). However, certain thin-layer cascade photobioreactors are capable of attaining biomass densities above 30 g/L (Schädler et al. 2020).

Aside from that, some factors influence algal cultivation parameters. Many microalgae, such as *phytoplankton*, are highly temperature sensitive, whereas macroalgae, such as *seaweed*, prefer warmth and can go dormant in cold or low temperatures. This restricts algaculture to warmer regions of the world or areas with limited indoor facilities. Most microalgae species prefer temperatures ranging from 65 to 90 °C, which is another bottleneck. One of the most important challenges is meeting consistent nutrient requirements. As a specific bioreactor will have a much higher concentration of algae cells than a typical open waterway, hence it is critical to supplement nutrients to stimulate the desired rapid growth. However, adding nutrient solutions to water for algae growth can reduce a venture's profits. Low-cost nutrient sources such as sludge, fertilizer runoff, and others can be used for this purpose. Since many algae stocks are used for animal and human consumption, developing the wrong species or introducing a contaminant from the water supply could cause serious problems. Harvesting and processing algae is a difficult task. Companies have reported that harvesting and dewatering accounted for 30% or more of the costs for algae grown specifically for oil production. Figure 18.3 depicts

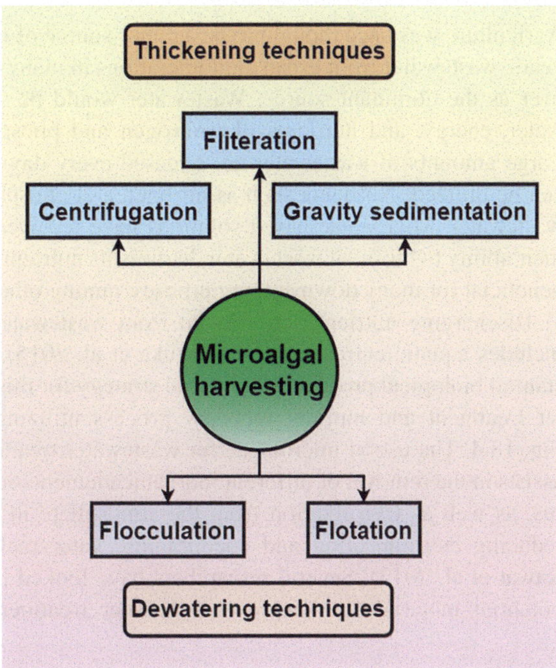

Fig. 18.3 Harvesting techniques for algal cultivation (Ananthi et al. 2021; Barros et al. 2015)

various harvesting techniques, including thickening and dewatering, which are further classified as filtration, centrifugation, gravity sedimentation, flocculation, and flotation.

Harvesting and dewatering are two of the most important stages in microalgae processing. The goal is to raise total solid matter to 10–25% of total dry matter or even to dry a product (Barros et al. 2015). These two operations can be completed in one or more sequential phases, depending on the equipment employed. When the algal cultivation reaches the final processing stage, the collected biomass is separated into several components, such as lipids and proteins. Based on many studies, the process of harvesting costs between 20% and 30% of the entire production expenses (Danquah et al. 2009; Rawat et al. 2013). It was also found that the cost of dewatering and harvesting equipment for open pond-farmed algal biomass might account for up to 90% of the overall cost (Amer et al. 2011). To commercialize algal-based biofuel, the cost of algal commodities must be reduced by ten times (Draaisma et al. 2013). Harvesting techniques attempt to remove growth medium from algal biomass to enable subsequent downstream processes such as extraction (Singh and Patidar 2018). Several harvesting methods for biomass collection include filtering, centrifugation, flocculation, and floatation (Tan et al. 2020). Dewatering primarily aims to isolate microbial biomass from a bulk liquid solution for the generation of dry biomass.

18.4 Wastewater Treatment and Nutrient Recovery Through ML

Agriculture was once thought to be a major source of nutrient loading, but in recent years, wastewater from urban agglomerations in many developing nations has taken over as the dominant source. Wastewater would be seen as a source of recycled water, energy, and nutrients like nitrogen and phosphorus (Robles et al. 2020). Large amounts of wastewater are released every day from numerous sources that can be utilized. Nutrients such as nitrogen and phosphorus can be reclaimed from wastewater. Microalgae-based solutions have received interest in this area due to their ability to bio-treat wastewater, harness its nutrients, and generate algal biomass beneficial for many downstream purposes, among other benefits.

Discharging nutrient-rich effluent from wastewater treatment plants primarily includes aquatic eutrophication (Renuka et al. 2015). Water treatment using substantial biological processes is a viable strategy for this recovery. The full wastewater treatment and nutrient recovery process utilizing microalgae is depicted in Fig. 18.4. The use of microalgae for wastewater treatment is advantageous since it assists in the removal of different inorganic elements such as nitrogen and phosphorus, as well as CO_2 fixation from the atmosphere in the presence of light, hence reducing eutrophication and encouraging water reclamation (Díaz et al. 2022; Rawat et al. 2013). Several researchers have looked into microalgae bacteria as a potential material for low-cost wastewater treatment, particularly for nitrogen

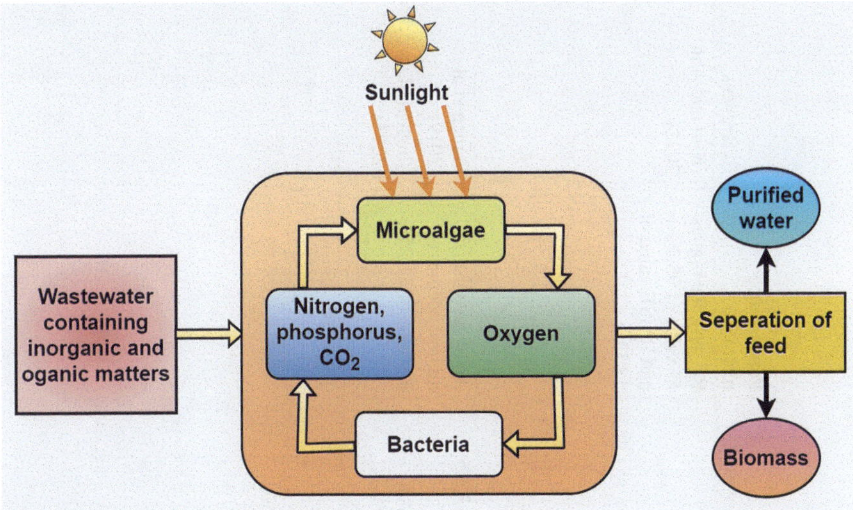

Fig. 18.4 Nutrient recovery and wastewater treatment from algal biomass (Acién Fernández et al. 2018)

removal, while also producing biomass for resource recovery (Fallahi et al. 2021; Jiang et al. 2021; Saravanan et al. 2021; Yang et al. 2020; Zhang et al. 2020, 2021).

Due to complex nature, employing algae for wastewater has several drawbacks. Microalgae development can be influenced by nutrient content in wastewater, inoculum level, light intensity, and a variety of other factors. As a result, optimizing these parameters is critical if the primary goal is to produce biomass from wastewater (Patel and Brahmbhatt 2016). For optimization and decision-making processes, many modeling methodologies can be used. Compared to traditional kinetic or transport models, ML algorithms and dynamic models for maximizing algal system efficiency have demonstrated intriguing results because they offer a better knowledge of biological processes concerning uncertainty (Iratni and Chang 2019; Patel and Brahmbhatt 2016). Additionally, these models can provide a thorough knowledge of the metabolism of algal bacteria, in which many factors such as temperature, light intensity, nitrogen limitation, and dissolved inorganic carbon play significant and unique roles (Blanken et al. 2014; Lee et al. 2015). Despite being more traditional and less precise, several kinetic models can also be used to forecast algal growth, biomass output, and the interaction between various process factors although they have certain drawbacks of less prediction accuracies.

Several ML studies have been conducted to enhance nutrient recovery from algae systems and to produce algal biomass that can be utilized for various applications, as provided in Table 18.2. Kim and Ahn (2022) used data from various monitoring stations in the Han River basin to construct a model for predicting *chlorophyll-a* concentrations as a proxy for the algal population.

The relationship between *chlorophyll-a* and the input factors was also investigated. Datasets containing meteorological and daily water quality from 2012 to

Table 18.2 Literature studies of ML in wastewater treatment and nutrient recovery

Dataset quantity	Model input	Model output	Model type	Software used	Performance measures	References
Water quality and meteorological data collected from 2012 to 2020	Total nitrogen, pH, water temperature, total phosphorus, electrical conductivity, dissolved oxygen, and minimum, mean, and maximum air temperature	*Chlorophyll-a* concentrations	RF, SVM and ANN	Python	RF gave the highest R^2 value of 95.2% and RMSE value of 3.152	(Kim and Ahn 2022)
550	Monoalgal and mixed algal cultures	Classification of cultures	ANN	Python 3.6.5	ANN gave the highest accuracy	Franco et al. (2019)
481	Light intensity, CO_2 content, and initial inoculum level	Biomass production	DT	MATLAB version 9.2; R2017b	R^2 value of 81.25%	Singh et al. (2023)

2020 were obtained from the Korean Meteorological Administration for model construction. Three models were created using R^2, RMSE, and mean absolute error as performance parameters: RF, SVM, and artificial neural network (ANN). When 78 trees were used, the RF model had the greatest R^2 score. Total organic carbon (27%), total nitrogen (19%), pH (13%), water temperature (8%), total phosphorus (8%), electrical conductivity (7%), dissolved oxygen (6%), minimum air temperature (AT; 4%), mean AT (3%), and maximum AT (3%) were found to be the most important features for predicting *chlorophyll-a* (Kim and Ahn 2022). Franco et al. (2019) created an ANN-based model for discriminating between monoalgal and mixed algal cultures. This was accomplished by visualizing the microalgae spectral fingerprints from light absorption measurements of several algal species. In this investigation, *Nostoc* sp., *Scenedesmus almeriensis*, *Spirulina platensis*, and *Chlorella vulgaris* were used. Both forms of algal absorption were used as input parameters for model building and validation. The outcomes showed that the ANN model was for recognizing microalgae species in monoalgal cultures, categorizing mixed algal cultures, and estimating the approximate composition of mixed algal cultures (Franco et al. 2019). Singh et al. (2023) created an ML-based optimization model for microalgae biomass production using wastewater. To predict several optimal combinations of descriptor variables that result in a high optimum growth rate and biomass production, the DT model was employed to evaluate microalgae datasets. With a general R^2 value of 81.25%, the DT model produced 18 descriptor variable combinations, resulting in improved biomass output. Seven combinations were validated experimentally using 2 freshly isolated bacteria, which projected a 10% inaccuracy following experimental validation. Compared to other factors, feature significance demonstrated that the medium's initial inoculum level and nitrogen/phosphorus ratio had the greatest effect on microalgae development (Singh et al. 2023).

18.5 Biofuel Production Optimization with AI

Burning fossil fuels (coal, natural gas, and oil) generates roughly 82% of the energy used in the present scenario, which raises concerns about the depletion of fossil fuel sources and the escalation of carbon emissions-related environmental pollution. As a result, shifting energy supply from fossil fuels to biofuels is critical. Currently, large proportions of biofuels produced globally are derived from various basic organic resources, which are categorized into three generations (Brennan and Owende 2010; Zhu et al. 2016). First-generation biofuels contain feedstocks that humans can consume, including crops (sugarcane, maize, pea, oil, and many more). The second generation is those derived from lignocellulose feedstock, which are nonedible agricultural wastes such as stems, leaves, and husks (Kalnes et al. 2017; Naik et al. 2010; Zhu 2015). Although these two generations can only meet a portion of the world's energy requirement, they rely heavily on the available cultivable area, which is one of their disadvantages. High cultivation space is required for their manufacturing, which is generated from the human consumption of agricultural products. Therefore, growing microalgae—unicellular photosynthetic microorganisms that can transform CO_2 and light into biomass and high-energy molecules, which are the building blocks of biofuels—has been suggested as a way to get over this drawback (Abomohra et al. 2017). Figure 18.5 depicts the process of producing biofuel from different algal resources. These feedstocks can produce methane, syngas, biohydrogen, biodiesel, bioethanol, and biobutanol (Liew et al. 2014).

Microalgae-based fuels offer several benefits over other feedstocks, such as quick algae growth, CO_2 consumption, and O_2 release while growing, or the ability to survive and grow in wastewater while cleaning it. The production of oil from microalgae basically involves three steps. Growing microalgae in open ponds or photobioreactors, collecting them, and then extracting the oil is the initial step. The two most expensive and energy-intensive steps in the manufacture of microalgal biofuels are cultivation and harvesting, which account for around 40% and 20–30% of the total, respectively (Molina Grima et al. 2004; Niyogi 1999). High lipid content also produces an essential effect in lowering the processing cost per unit of biomass products. Many factors influence the lipid content and composition of microalgae, including nitrogen, phosphorus, metals (because iron is involved in the basic enzymatic activities of photosynthesis, iron is one of the most effective trace metals), carbon effect, salinity, pH, temperature, and many others (Wan et al. 2014). A growing number of studies on lipid-rich microalgal culture for biodiesel generation have piqued the public's curiosity. Lipids are the microbial organisms' energy and carbon stores (Vitova et al. 2014). In addition to providing energy for biological processes related to microalgal cell multiplication, such as DNA (deoxyribonucleic acid) replication, nuclear division, *cytokinesis*, and the formation and release of

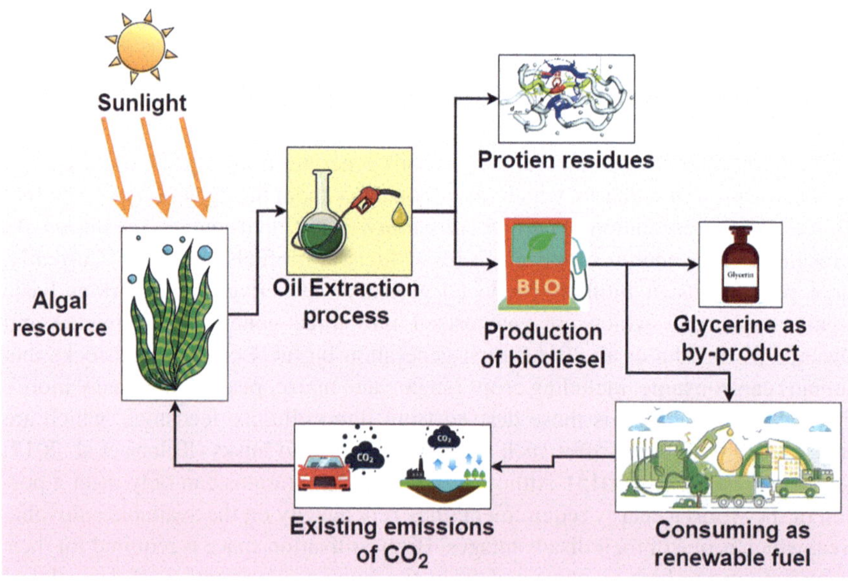

Fig. 18.5 Biofuel production from algal resources (Bajhaiya et al. 2012; Oves et al. 2022)

daughter cells, lipids also ensure the survival of microalgal cells at night and during periods of variable light intensity. The amount and ratio of saturated and unsaturated fatty acids is another crucial aspect in determining microalgae's potential as a biofuel feedstock. The majority of the oil generated by microalgae is made up of unsaturated fatty acids, including palmitoleic (16:1), oleic (18:1), linoleic (18:2), linolenic (18:3), and palmitic (16:0) saturated fatty acids. Stearic saturated fatty acids make up a very little amount of the oil (Rahman et al. 2019). It is well-known that microalgae's biological activities are strongly dependent on their growing conditions. Unfavorable circumstances can cause changes in the cellular composition of microalgae and their growth rate and lipid accumulation. According to studies, nitrogen and phosphorus concentration are the most effective factors in increasing lipid formation in microalgae by many times. So, for the optimization of parameters such as efficiency of biodiesel production, microalgal growth rate, and lipid accumulation, ML techniques could be very much useful, due to their ability to understand the highly complicated relationships, along with the reaction kinetics compared to the traditional modeling systems.

ML algorithms can be quite useful in identifying optimal settings, the most influential factors, and the best strain selection for improving lipid synthesis. They can also be used to identify and optimize various lipid extraction techniques. Various studies of ML applications in biofuel production are listed in Table 18.3. Coşgun et al. (2023) created a DT ML model to investigate essential aspects of algal biomass and lipid production in the generation of renewable fuels. A total of 4670 data points were retrieved from the experimental results presented in 102 published research publications for this purpose.

The input variables for this study were microalgae species, growing settings, light intensity, CO_2 quantity, nutrient concentrations, reactor type, stress conditions, cell disruption techniques, and lipid extraction solvents. The DT model produced 11 distinct input combinations that resulted in high biomass productivity and 13 combinations that resulted in high lipid content. It was determined that the ideal growing conditions for maximal lipid output in the production of biofuels can be identified using ML algorithms and that this can guide the development and implementation of additional experimental tasks (Coşgun et al. 2023). Tanhaemami et al. (2019) used flow cytometry and multistage ML to find label-free markers of algal lipid accumulation. A simple approach based on feature selection and linear regression analysis was introduced to combine statistical information from cell populations with and without labels and to find models for accurate label-free single-cell quantification. This was expected based on the lipid content of algal cells (*Picochlorum soloecismus*). Regression analysis, genetic algorithms, and model weighting techniques were shown to work together to develop a unique set of models and weights using statistical aspects of unlabeled cell population data. This approach can extract enough information to offer effective label-free measurement of lipid concentration in *Picochlorum soloecismus* across time under nitrogen deprivation (Tanhaemami et al. 2019). Aty et al. (2022) developed an ML platform for identifying experimental lipid phase behavior from small-angle X-ray scattering patterns by pre-training on synthetic data (Aty et al. 2022). Shafrai et al. (2023) used an ANN model to

Table 18.3 Literature studies of ML applications in biofuel production

Dataset quantity	Model input	Model output	Model type	Software used	Performance measures	References
4670	Microalgae species, cultivation conditions, light intensity, CO_2 amount, nutrient concentrations, reactor type, stress conditions, cell disruption methods, and lipid extraction solvents	Biomass and lipid production	DT	MATLAB version number 9.4; R2018a	R^2 value of 89.6%	Coşgun et al. (2023)
23	Label-free quantification signals	Lipid content in algal cells	Flow cytometry and multistage ML	–	Gave high accuracy and matched with validation set	Tanhaemami et al. (2019)
–	Raw, experimental SAXS spectra	Lipid phase behavior	Convolutional neural network	Python 3.8	R^2 value >99%	Aty et al. (2022)
147	Rendering temperature and duration, enzyme preparation, and concentration	Fat yield in lipid extraction	ANN	Python in Google Colaboratory	Least mean relative error of 5.16%	Shafrai et al. (2023)
231	Microalgae class, pre-cultivation stage deciding factors, and operating variables	Biomass productivity and wastewater treatment capability	DT	MATLAB version 9.2; R2017a	R^2 value of 80%	Singh and Mishra (2021)
19	Glucose, yeast extract, and $NaNO_3$	Lipid productivity, biomass concentration, and lipid yield	Response surface methodology and artificial neural network	Python	Gave high accuracies with the validated experiment	Mohamed et al. (2013)

simulate the biocatalytic technique of lipid extraction. Mean relative error was used to assess the accuracy of the 8 hidden layers, each including 64 neurons that made up the model. The resulted model had an error value of 5.16%. The generated models produced the best ideal values for applied concentration, temperature, and rendering duration. This value produced a fat yield for each enzyme preparation that was more than 98% (Shafrai et al. 2023). Singh and Mishra (2021) used a DT model to investigate the impact of various combinations of predictor factors for wastewater treatment using algae and biomass production. Microalgae class, pre-cultivation stage decision factors, and operational variables were among the predictor variables, while biomass production and wastewater treatment capabilities were chosen as the outcome variables. The DT model yielded 8 different combinations for boosting biomass output with an accuracy of more than 80% and 10 different combinations for excellent nitrogen removal efficiency. The study indicated that the DT model can be utilized to build high-throughput experimental designs that can aid in the development of large-scale, highly efficient wastewater treatment (Singh and Mishra 2021). For *Tetraselmis* sp. medium optimization, Mohamed et al. (2013) used ANN models and response surface techniques to maximize lipid production. The habitat in which FTC209 was formed was mixotrophic. Three medium variables affecting biomass concentration, Xmax, and lipid yield, Pmax/Xmax, were the input variables for this model. These were glucose (an organic C source), $NaNO_3$ (major N source), and yeast extract (supplemental N, amino acids, and vitamins) (Mohamed et al. 2013).

18.6 Data Integration and Decision Support Systems

Apart from the effective development of data-driven models for various applications of algae biotechnology, data heterogeneity (including the types and quality of datasets) could be a huge problem. Algal biotechnology encompasses a wide range of datasets, including metabolomics, genetic, and environmental data. As a result of the varied formats, sizes, and units, integrating this heterogeneous dataset is difficult. Similarly, data quality varies among different sources, and a dataset can sometimes contain trivial inaccuracies. As a result, analyzing this dataset can impede integration. Algal biotechnology is an area that draws on expertise from various disciplines, including chemistry, biology, engineering, energy management, environmental science, and others. Collaboration within these fields and data integration from numerous sources, techniques, and studies can be a substantial disadvantage. Several types of datasets derived from industrial applications can be quite sensitive, and data privacy and security might be difficult to integrate. There can be multiple legal requirements for data management and reporting in applications such as biofuel production; these rules further complicate the integration system. Environmental factors such as light, temperature, nutrient availability, and many more greatly affect microalgae development and algal farming. Therefore, incorporating this temporal and geographical variability into a dataset is extremely tough, however extremely beneficial to the optimization and prediction processes. Aside from data gathering, data storage, archiving, and retrieval are also critical processes.

Creating extensive datasets for genetics, biochemistry, and algal growth is also crucial. Gathering a comprehensive dataset, including data extraction and organizing the data, and managing and identifying important input and output variables more pertinent to algal cultivations and growth processes are the essential preliminary steps required in this study. This is followed by developing appropriate techniques and algorithms for data modeling. Making appropriate tables for algae strains, operational and experimental settings, and other pertinent data is also crucial. A variety of data analysis methods can be employed to comprehend the behavior and relationships of a dataset and eliminate undesirable input factors such as partial dependence plot, which helps in the identification of dependence of different input parameters on the output variables. It is necessary to perform procedures like data pre-processing for gathered data to eliminate outliers, treat missing data values, and either standardizing or normalizing input factors. This aids in improving the produced model's accuracy. Once effective models with great accuracy have been developed, the behavior and pattern of the entire process can be studied using various visualization tools such as SHAP plot, precision-recall curve, cumulative explained variance plot, KS plot and many more. This helps in the identification of the important and most influential input parameters out of all of them, which also helps to understand unknown parameter effects on product yield. To ensure the model is operating as intended, there are other options for validation besides using an experimental setup. However, any validation should be conducted objectively.

AI systems are widely used in many applications; thus, it is critical to comprehend how they decide to establish responsibility and foster confidence. To handle AI-driven judgments for various reasons, it is imperative to consider transparency and interpretability as critical elements. Building trust among diverse users and researchers is facilitated by transparent systems. It also facilitates the detection of prejudice and discrimination in AI algorithms. Understanding weight distributions among all input characteristics and decision-making processes is aided by this. As a result, all hidden patterns can be found, and subsequent mitigation techniques can be applied to improve performance. Transparent models facilitate identification of various mistakes that arise in the decision-making processes. More comprehensible faults make locating them simpler; identifying their underlying causes and implementing various fixes can contribute to their reduction. Due to this, researcher can more readily explain and comprehend the reaction chemistry behind these processes, and AI models are easier to interpret. Transparent AI systems can also significantly impact decision-making processes. Transparency in AI-driven models can also aid in identifying any possible risks that can be there. This aids in the creation of risk mitigation plans. The interpretability and transparency in AI-driven judgments are helpful in decision-making processes and the production of algae culture models.

18.7 Future Directions and Challenges

Algal biotechnology will keep expanding as demand for more affordable, easily accessible, and environmentally friendly energy sources rises. Thus, this will lead to the improvement of more optimized processes for the development of algae, the generation of biofuel, and the recovery of nutrients; all of these results will add to the importance of algae as a renewable energy source. Numerous studies based on the AI/ML techniques are developing in this discipline as technology progresses. Biotechnology also uses concepts such as precision cultivation, which assists in extremely accurate control of several factors. Apart from ML, deep learning methods, such as computational neural networks for image analysis in algae growth, are now becoming popular as optimization techniques for improving analytical technologies for the generated product. It also includes the discovery of stress variables that rely on visual signals, the detection of specific algae species, and the estimation of bioproducts. To predict phenotypic (individual) features in algae, ML can also be used to analyze different types of genomic data or information about the structure and function of an organism's genome. Reinforcement learning along with the supervised and unsupervised ML algorithms can be investigated to create various control schemes in algae biotechnologies. For this purpose, new developments in increasing transparency, such as the use of explainable AI-based models to improve system comprehension, are also required. In algal production, edge computing—a developing computer program that refers to a range of networks close to the users—is becoming popular since it makes real-time decision-making techniques possible. Because models are created right at the edge, dataset analysis can be completed more quickly and different algal parameters could be adjusted. In several applications of algal cultivation, ML models are also used to forecast the strains of the algae. This includes identifying strains with high lipid contents and other characteristics. In addition to these developments, techniques like collaborative AI platforms, digital systems for algae systems, automated monitoring and control systems, and resilience modeling are becoming more and more significant instruments.

The interpretability and transparency in AI-driven judgments are helpful in decision-making processes and the production of algae culture models. A system's effective development, operations, and deployment can be significantly impacted by addressing issues including data quality, model robustness, and scalability. Missing values in a dataset can lead to biased and incorrect model training and prediction accuracy. Therefore, a high-quality dataset free of missing values is necessary. Imputation techniques like mean imputations can be used to handle missing values to tackle this issue. The reliability of created ML models is affected by the possibility of errors or outliers in algae-based datasets. Robust pre-processing procedures, outlier detection techniques, data validation, and data analysis can also aid in the discovery of gathered data imperfections to alleviate this problem. One of the most crucial data pre-processing techniques is scalability, which is converting a dataset into a standard format to aid in developing ML algorithms designed to better comprehend data behavior and changes. Enhancing the resilience of ML models also necessitates domain shift, which is modifying the statistical distribution of data between several domains, such as training, testing, and validating for a model. Numerous transfer learning strategies and domain adaption on newly gathered datasets can assist in addressing domain shift concerns. One of the most prevalent issues with all ML models is overfitting, which has to be fixed. This occurs when noise is introduced into the models during training, preventing the models from producing accurate predictions when fresh testing data are fed in. Simpler models and explainable techniques like SHAP and LIME analyses, which promote interpretability and model-agnostic explanation, can tackle this issue. Real-time monitoring of developed models during the algal cultivation is difficult task that is needed to be monitored appropriately. Apart from that, deployment of developed ML models for a small-scale cultivation is technically and economically challenging, and it might also deliver inappropriate results. Algal cultivation itself is a very complicated procedure, so it becomes difficult to model the convolute biological and environmental conditions. ML has provided wide ranges of applications in the biodiesel production from algal biomass, as it helps in the optimization of different affecting parameters such as algal composition, reaction temperature, growth condition, lipid extraction methods, and nutrient supply. It also helps in the generation of cost-effective methods, which reduce resource utilization and waste materials. A vast amount of algal biomass dataset can help in the development of ML models with high accuracies; this will identify accurate biodiesel yield at process operating conditions. However, certain challenges of ML could result into many consequences such as development of complicated ML models. Improper dataset could result into overfitted models that will lead into generation of inaccurate product yields. Nutrient recovery and wastewater treatment might also suffer similar problems by ML modeling since initial data collection is significant. If the dataset used to train the model is inaccurate then it can result into biased ML models resulting into bad prediction, which can ultimately affect the wastewater treatment strategies. ML algorithms are highly advanced, which makes it difficult for the understanding of decision-making strategies, which could also lead to the improper design of photobioreactors. Inaccuracies could also lead to the improper chemical reaction and kinetics during the nutrient recovery processes. Apart from the product yields and efficiency of systems, generation of high amount of wastage and high energy consumption can be mitigated by this strategy. Therefore, it is very important for the suitable development of ML models that datasets are reliable and ML algorithms are well-trained and tested, to avoid above consequences.

18.8 Conclusion

The integration of machine learning and artificial intelligence in algae biotechnology holds great potential for revolutionizing and streamlining procedures within the industry, which are discussed in this chapter. Integrating data-driven AI with algae culture offers many applications that greatly enhance product efficiency and promote sustainability and development. One of the most important lessons discovered is that cultivating circumstances can be optimized by examining past data. This mostly entails adjusting factors like temperature, light intensity, and fertilizer levels, all of which improve the selectivity of ethanol and other goods. Real-time environmental condition monitoring, control systems, various automated cultivation parameter modifications, and ML-driven simulations provide a wealth of knowledge on the dynamic behavior of algal cultivations. These simulations reduce resource waste by enabling researchers to investigate novel approaches and procedures without requiring tests.

Collaborative AI facilities are also emerging to let the scientific community share and exchange knowledge about combined research investigations from several domains, including chemistry and biology. These platforms aid in accelerating the creation of AI model best practices, enhancing transparency, and encouraging a team-based and collaborative approach. Deep insight into the algae systems is consistently provided by ML-driven analysis, and there is always room for improvement. Researchers can optimize cultivation techniques and procedures by using patterns and correlations found in datasets. This can also reveal previously undiscovered patterns in the models and improve their robustness. In the end, combining AI and ML with algae cultivation can address several current problems and advance the field into new frontiers for improving effective and novel ideas. As these technologies advance, ML/AI-driven models will be essential to realizing the full potential of algae-based goods across a wide variety of applications.

Acknowledgments The authors would like to thank the Indian Institute of Petroleum and Energy for their continuous encouragement and support.

References

Abomohra AEF, El-Sheekh M, Hanelt D (2017) Screening of marine microalgae isolated from the hypersaline Bardawil lagoon for biodiesel feedstock. Renew Energy 101:1266–1272

Acién Fernández FG, Gómez-Serrano C, Fernández-Sevilla JM (2018) Recovery of nutrients from wastewaters using microalgae. Front Sustain Food Syst 2:1–13

Ali I, Al-Arfaj AA, Saleh TA (2020) Carbon nanofiber-doped zeolite as support for molybdenum based catalysts for enhanced hydrodesulfurization of dibenzothiophene. J Mol Liq 304:112376

Amer L, Adhikari B, Pellegrino J (2011) Technoeconomic analysis of five microalgae-to-biofuels processes of varying complexity. Bioresour Technol 102(20):9350–9359

Ananthi V, Balaji P, Sindhu R, Kim S-H, Pugazhendhi A, Arun A (2021) A critical review on different harvesting techniques for algal based biodiesel production. Sci Total Environ 780(146):467

Antonakoudis A, Barbosa R, Kotidis P, Kontoravdi C (2020) The era of big data: genome-scale modelling meets machine learning. Comput Struct Biotechnol J 18:3287–3300

Aty HA, Strutt R, McIntyre N, Allen M, Barlow NE, Páez-Pérez M, Seddon JM, Brooks N, Ces O, Gould IR (2022) Machine learning platform for determining experimental lipid phase behaviour from small angle X-ray scattering patterns by pre-training on synthetic data. Digit Discov 1(2):98–107

Azzolina D, Baldi I, Barbati G, Berchialla P, Bottigliengo D, Bucci A, Calza S, Dolce P, Edefonti V, Faragalli A, Fiorito G, Gandin I, Giudici F, Gregori D, Gregorio C, Ieva F, Lanera C, Lorenzoni G, Marchioni M et al (2019) Machine learning in clinical and epidemiological research: isn't it time for biostatisticians to work on it? Epidemiol Biostat Public Health 16(4):1–3

Bajhaiya AK, Suseela MR, Ramteke PW (2012) Approaches and prospectives for algal fuel. In: The science of algal fuels. Cellular origin, life in extreme habitats and astrobiology. Springer, Dordrecht, pp 43–62

Barros AI, Gonçalves AL, Simões M, Pires JCM (2015) Harvesting techniques applied to microalgae: a review. Renew Sust Energy Rev 41:1489–1500

Bello M, Ranganathan P, Brennan F (2017) Dynamic modelling of microalgae cultivation process in high rate algal wastewater pond. Algal Research 24:457–466

Blanken W, Janssen M, Cuaresma M, Libor Z, Bhaiji T, Wijffels RH (2014) Biofilm growth of Chlorella sorokiniana in a rotating biological contactor based photobioreactor. Biotechnol Bioeng 111:2436–2445

Borowitzka MA (2015) Algal biotechnology. In: The algae world cellular origin. Springer, Dordrecht, pp 319–338

Bradford E, Schweidtmann AM, Zhang D, Jing K, del Rio-Chanona EA (2018) Dynamic modeling and optimization of sustainable algal production with uncertainty using multivariate Gaussian processes. Comput Chem Eng 118:143–158

Brennan L, Owende P (2010) Biofuels from microalgae—a review of technologies for production, processing, and extractions of biofuels and co-products. Renew Sust Energy Rev 14(2):557–577

Breuer G, Lamers PP, Martens DE, Draaisma RB, Wijffels RH (2013) Effect of light intensity, pH, and temperature on triacylglycerol (TAG) accumulation induced by nitrogen starvation in Scenedesmus obliquus. Bioresour Technol 143:1–9

Chaurasia AK, Mondal P (2022) Enhancing biohydrogen production from sugar industry wastewater using Ni, Ni–Co and Ni–Co–P electrodeposits as cathodes in microbial electrolysis cells. Chemosphere 286:131728

Chen CY, Yeh KL, Aisyah R, Lee DJ, Chang JS (2011) Cultivation, photobioreactor design and harvesting of microalgae for biodiesel production: a critical review. Bioresour Technol 102(1):71–81

Chong JWR, Khoo KS, Chew KW, Ting HY, Show PL (2023) Trends in digital image processing of isolated microalgae by incorporating classification algorithm. Biotechnol Adv 63:108095

Coşgun, A., Günay, M. E., & Yıldırım, R. (2023). Machine learning for algal biofuels: a critical review and perspective for the future. In *Green Chemistry* 25, 9, 3354–3373. Royal Society of Chemistry. https://doi.org/10.1039/d3gc00389b

Costello Z, Martin HG (2018) A machine learning approach to predict metabolic pathway dynamics from time-series multiomics data. Npj Syst Biol Appl 4(1):19

Danquah MK, Gladman B, Moheimani N, Forde GM (2009) Microalgal growth characteristics and subsequent influence on dewatering efficiency. Chem Eng J 151(1–3):73–78

Davis R, Aden A, Pienkos PT (2011) Techno-economic analysis of autotrophic microalgae for fuel production. Appl Energy 88(10):3524–3531

Díaz V, Leyva-Díaz JC, Almécija MC, Poyatos JM, del Mar Muñío M, Martín-Pascual J (2022) Microalgae bioreactor for nutrient removal and resource recovery from wastewater in the paradigm of circular economy. Bioresour Technol 363:127968

Draaisma RB, Wijffels RH, Slegers PM, Brentner LB, Roy A, Barbosa MJ (2013) Food commodities from microalgae. Curr Opin Biotechnol 24(2):169–177

Fallahi A, Rezvani F, Asgharnejad H, Khorshidi E, Hajinajaf N, Higgins B (2021) Interactions of microalgae-bacteria consortia for nutrient removal from wastewater: a review. Chemosphere 272:129878

Figueroa-Torres GM, Pittman JK, Theodoropoulos C (2018) Model-based fed-batch algal cultivation strategy for enhanced starch production. Comput Aided Chem Eng 43:1595–1600

Foley PM, Beach ES, Zimmerman JB (2011) Algae as a source of renewable chemicals: opportunities and challenges. Green Chem 13(6):1399–1405

Franco BM, Navas LM, Gómez C, Sepúlveda C, Acién FG (2019) Monoalgal and mixed algal cultures discrimination by using an artificial neural network. Algal Res 38:101419

Greene JM, Quiroz D, Compton S, Lammers PJ, Quinn JC (2021) A validated thermal and biological model for predicting algal productivity in large scale outdoor cultivation systems. Algal Res 54:102224

Iratni A, Chang NB (2019) Advances in control technologies for wastewater treatment processes: status, challenges, and perspectives. In: IEEE/CAA journal of automatica sinica, vol 6, issue 2. Institute of Electrical and Electronics Engineers, Piscataway, pp 337–363

Jiang L, Li Y, Pei H (2021) Algal–bacterial consortia for bioproduct generation and wastewater treatment. Renew Sustain Energy Rev 149:111395

Kalnes T, Shonnard DR, Marker T (2017) Green diesel: a second generation biofuel green diesel: a second generation biofuel. Int J Chem Reactor Eng

Kim KM, Ahn JH (2022) Machine learning predictions of chlorophyll-a in the han river basin. Korea Journal of Environmental Management *318*. https://doi.org/10.1016/j.jenvman.2022.115636

Lai L, Zhang Y, Cao Z, Liu Z, Yang Q (2023) Algal biomass mapping of eutrophic lakes using a machine learning approach with MODIS images. Sci Total Environ 880:163357

Lam MK, Lee KT (2012) Microalgae biofuels: a critical review of issues, problems and the way forward. Biotechnol Adv 30(3):673–690

Lee E, Jalalizadeh M, Zhang Q (2015) Growth kinetic models for microalgae cultivation: a review. Algal Res 12:497–512

Leng L, Wei L, Xiong Q, Xu S, Li W, Lv S, Lu Q, Wan L, Wen Z, Zhou W (2020a) Use of microalgae based technology for the removal of antibiotics from wastewater: a review. Chemosphere 238:124680

Leng L, Xu S, Liu R, Yu T, Zhuo X, Leng S, Xiong Q, Huang H (2020b) Nitrogen containing functional groups of biochar: an overview. Bioresour Technol 298:122286

Liew WH, Hassim MH, Ng DKS (2014) Review of evolution, technology and sustainability assessments of biofuel production. J Clean Prod 71:11–29

Mohamed MS, Tan JS, Mohamad R, Mokhtar MN, Ariff AB (2013) Comparative analyses of response surface methodology and artificial neural network on medium optimization for Tetraselmis sp. FTC209 grown under mixotrophic condition. Sci World J 2013:948940

Molina Grima E, Belarbi E-H, Acién Fernández FG, Robles Medina A, Chisti Y (2004) Recovery of microalgal biomass and metabolites: process options and economics. Biotechnol Adv

Naik SN, Goud VV, Rout PK, Dalai AK (2010) Production of first and second generation biofuels: a comprehensive review. Renew Sust Energ Rev 14(2):578–597

Nauha EK, Alopaeus V (2015) Modeling outdoors algal cultivation with compartmental approach. Chem Eng J 259:945–960

Niyogi KK (1999) PHOTOPROTECTION REVISITED: genetic and molecular approaches. Annu Rev Plant Physiol Plant Mol Biol 50:333–359

Oves MA, Qari H, Ismail IM (2022) Biofuel formation from microalgae: a renewable energy source for eco-sustainability. Curr World Environ 17(1):4–19

Patel KA, Brahmbhatt PK (2016) A comparative study of the RSM and ANN models for predicting surface roughness in roller burnishing. Procedia Technol 23:391–397

Pathy A, Meher S, P, B. (2020) Predicting algal biochar yield using eXtreme Gradient Boosting (XGB) algorithm of machine learning methods. *Algal Res 50*:102006. https://doi.org/10.1016/j.algal.2020.102006

Peter AP, Chew KW, Pandey A, Lau SY, Rajendran S, Ting HY, Munawaroh HSH, Van Phuong N, Show PL (2023) Artificial intelligence model for monitoring biomass growth in semi-batch Chlorella vulgaris cultivation. *Fuel 333*. https://doi.org/10.1016/j.fuel.2022.126438

Rahman A, Prihantini NB, Nasruddin (2019) Fatty acid of microalgae as a potential feedstock for biodiesel production in Indonesia. AIP Conf Proc 2062:1789

Rawat I, Ranjith Kumar R, Mutanda T, Bux F (2013) Biodiesel from microalgae: a critical evaluation from laboratory to large scale production. Appl Energy 103:444–467

Renuka N, Sood A, Prasanna R, Ahluwalia AS (2015) Phycoremediation of wastewaters: a synergistic approach using microalgae for bioremediation and biomass generation. Int J Environ Sci Technol 12(4):1443–1460

Robles Á, Capson-Tojo G, Gales A, Viruela A, Sialve B, Seco A, Steyer JP, Ferrer J (2020) Performance of a membrane-coupled high-rate algal pond for urban wastewater treatment at demonstration scale. Bioresour Technol 301:122672

Roostaei J, Zhang Y, Gopalakrishnan K, Ochocki AJ (2018) Mixotrophic microalgae biofilm: a novel algae cultivation strategy for improved productivity and cost-efficiency of biofuel feedstock production. Sci Rep 8(1):12528

Ryu B, Kim S (2013) Isolation and biological activities of peptides from marine microalgae by fermentation. In: Marine proteins and peptides. Wiley, New York, pp 441–448

Sajjad N, Orfali R, Perveen S, Rehman S, Sultan A, Akhtar T, Nazir A, Muhammad G, Mehmood T, Ghaffar S, Al-Taweel A, Jilani MI, Iqbal M (2022) Biodiesel production from alkali-catalyzed transesterification of Tamarindus indica seed oil and optimization of process conditions. Molecules 27(10):3230

Saleh TA, Al-Hammadi SA (2021) A novel catalyst of nickel-loaded graphene decorated on molybdenum-alumina for the HDS of liquid fuels. Chem Eng J 406:125167

Sangué Djandja O, Shan YQ, Fan L, Wu Y, Akim Salami A, Lu X, Duan PG, Kang S (2023) Machine learning aids solvothermal liquefaction of algal biomass: prediction of nitrogen content and bio-oil yield. Fuel 353:129284

Saravanan A, Kumar PS, Varjani S, Jeevanantham S, Yaashikaa PR, Thamarai P, Abirami B, George CS (2021) A review on algal-bacterial symbiotic system for effective treatment of wastewater. Chemosphere 271:129540

Savage PE (2012) Algae under pressure and in hot water. Science 338(6110):1039–1040

Schädler T, Neumann-Cip AC, Wieland K, Glöckler D, Haisch C, Brück T, Weuster-Botz D (2020) High-density microalgae cultivation in open thin-layer cascade photobioreactors with water recycling. Appl Sci (Switzerland) 10(11):3883

Shafrai AV, Prosekov AY, Vechtomova EA (2023) Modeling the biocatalytic method of lipid extraction using artificial neural networks. Information (Switzerland) 14(8):452

Singh V, Mishra V (2021) Exploring the effects of different combinations of predictor variables for the treatment of wastewater by microalgae and biomass production. *Biochem Eng J 174*. https://doi.org/10.1016/j.bej.2021.108129

Singh G, Patidar SK (2018) Microalgae harvesting techniques: a review. J Environ Manag 217:499–508

Singh V, Verma M, Chivate MS, Mishra V (2023) Machine learning-based optimisation of microalgae biomass production by using wastewater. J Environ Chem Eng 11(6):111387

Suali E, Sarbatly R (2012) Conversion of microalgae to biofuel. Renew Sustain Energy Rev 16(6):4316–4342

Sundui B, Ramirez Calderon OA, Abdeldayem OM, Lázaro-Gil J, Rene ER, Sambuu U (2021) Applications of machine learning algorithms for biological wastewater treatment: updates and perspectives. Clean Technol Environ Policy 23(1):127–143

Tan JS, Lee SY, Chew KW, Lam MK, Lim JW, Ho SH, Show PL (2020) A review on microalgae cultivation and harvesting, and their biomass extraction processing using ionic liquids. Bioengineered 11(1):116–129

Tanhaemami M, Alizadeh E, Sanders CK, Marrone BL, Munsky B (2019) Using flow cytometry and multistage machine learning to discover label-free signatures of algal lipid accumulation. *Phys Biol 16*(5). https://doi.org/10.1088/1478-3975/ab2c60

Veerabadhran M, Gnanasekaran D, Wei J, Yang F (2021) Anaerobic digestion of microalgal biomass for bioenergy production, removal of nutrients and microcystin: current status. J Appl Microbiol 131(4):1639–1651

Vitova M, Bisova K, Kawano S, Zachleder V (2014) Accumulation of energy reserves in algae: from cell cycles to biotechnological applications. Biotechnol Adv 33(6):1204–1218

Wan M, Jin X, Xia J, Rosenberg JN, Yu G, Nie Z, Oyler GA, Betenbaugh MJ (2014) The effect of iron on growth, lipid accumulation, and gene expression profile of the freshwater microalga Chlorella sorokiniana. Appl Microbiol Biotechnol 98(22):9473–9481

Wang Y, Li L, Zhao D, Zhou W, Chen L, Su G, Zhang Z, Liu T (2023) Surface patterns of mortar plates influence Spirulina platensis biofilm attached cultivation: experiment and modeling. Algal Res 71:103079

Yang J, Shi W, Fang F, Guo J, Lu L, Xiao Y, Jiang X (2020) Exploring the feasibility of sewage treatment by algal–bacterial consortia. Crit Rev Biotechnol 40(2):169–179

Zavřel T, Sinetova MA, Búzová D, Literáková P, Červený J (2015) Characterization of a model cyanobacterium Synechocystis sp.: PCC 6803 autotrophic growth in a flat-panel photobioreactor. Eng Life Sci 15(1):122–132

Zhang D, Del Rio-Chanona EA, Petsagkourakis P, Wagner J (2019) Hybrid physics-based and data-driven modeling for bioprocess online simulation and optimization. Biotechnol Bioeng 116(11):2919–2930

Zhang B, Li W, Guo Y, Zhang Z, Shi W, Cui F, Lens PNL, Tay JH (2020) Microalgal-bacterial consortia: from interspecies interactions to biotechnological applications. Renew Sustain Energy Rev 118:109563

Zhang C, Ho SH, Li A, Fu L, Zhou D (2021) Co-culture of Chlorella and Scenedesmus could enhance total lipid production under bacteria quorum sensing molecule stress. J Water Process Eng 39:101739

Zhu L (2015) Microalgal culture strategies for biofuel production: a review. Biofuels Bioprod Biorefining 9(6):801–814

Zhu LD, Li ZH, Hiltunen E (2016) Strategies for lipid production improvement in microalgae as a biodiesel feedstock. BioMed Res Int 2016:8792548

Yield Enhancement in Algal Production and Genetic Engineering Tools for Microalgal Biorefinery

19

Ritu Kumari, Vikas Kumar Patel, Priya Kumari, and Sachin Kajla

Contents

Abstract

Addressing the ecological constraints that the globe is currently facing necessitates the development of food, feed, and bioproducts that are environmentally

---

R. Kumari · V. K. Patel · P. Kumari · S. Kajla (✉)
Environment Research Group, R&D and Scientific Services Department, Tata Steel Limited, Jamshedpur, Jharkhand, India
e-mail: sachin.kajla@tatasteel.com

S. K. Arya et al. (eds.), *Value Added Products From Bioalgae Based Biorefineries: Opportunities and Challenges*,
https://doi.org/10.1007/978-981-97-1662-3_19

conscientious. Algae, which include cyanobacteria, eukaryotic microalgae, and seaweeds, provide a workable foundation for achieving this. Algae are safe for both humans and animals to eat, require limited space and resources, and may be modified to generate a wide range of beneficial bio-based products. In the past, compared with different microbial systems like bacteria and yeast, the genetic toolkit available for tackling the primary barriers of getting microalgal cultivation techniques to an economically feasible scale was rather minimal. The invention of innovative microalgal bioproducts for industrial purposes will be accelerated by broadening the molecular techniques available for modifying the genome of microalgae, which will raise product yields and promote the shift toward sustainable and environmentally benign products. With the goal to enable microalgae to serve as a more effective biomanufacturing system, we highlight important recent developments in this chapter, including the creation of microalgal expression vectors, the identification of genetic regulatory elements (promoters and transcription factors), the optimization of transformation techniques, and the creation of novel strain improvement methods.

Keywords

Microalgae · Genetic engineering · Metabolic engineering · Gene editing

Abbreviations

ACS	Acetyl-CoA synthetase
BKT	Beta-carotene ketolase
bzip	Basic leucine zipper
CRISPR/Cas9	Clustered regularly interspaced short palindromic repeats CRISPR/CRISPR-associated protein 9
crRNA	cis-Repressed RNA
CrtR	Beta-carotene oxygengenase gene
CrtR-B	Carotenoid hydroxylase
CSO	Chlamy Sequence Optimizer
DCW	Dry cell weight
DHA	Docosahexaenoic acid
DOF	DNA-binding-with-one-finger
DSBs	Double-strand breaks
EMS	Ethyl methane sulphonate
EPA	Eicosapentaenoic acid
FAMEs	Fatty acid methyl esters
glgC	Glucose-1-phosphate adenylyltransferase
GM	Genetically modified
HR	Homologous recombination
MAT	Malonyl-CoA:ACP transacyclase
NHEJ	Non-homologous end joining

NobZIP1	Basic-region/leucine zipper transcription factor
PEPC	Phosphoenolpyruvate carboxylase
PUFAs	Polyunsaturated fatty acids
RNAi	RNA interference
SSNs	Sequence-specific nucleases
TAGs	Triacylglycerols
TALENs	Transcription activator-like effector nucleases
taRNA	Trans-activating RNA
UTRs	Untranslated regions
ZEP	Zeaxanthin epoxidase gene
ZNFs	Zinc-finger proteins

19.1 Introduction

The overabundance of fuels derived from petroleum is contributing to a scarcity of fossil fuels and an upsurge in emissions of greenhouse gases into the atmosphere. When coal is ignited, fossil fuels emit carbon dioxide (CO_2), which is a primary contributor to an array of ecological and environmental issues. Production expenses develop as a consequence of the enormous quantity of particles generated and the necessity to dispose them off. Research has indicated that an alarming rise in CO_2 content in the environment is the major cause of global temperature rise. Fossil fuels must be substituted by sustainable bioenergy alternatives to address these challenges. The microalgal biorefinery approach is a means of resolving these challenges (Goswami et al. 2020). The development of microalgae, encompassing both cyanobacteria and eukaryotic taxa, is aimed at establishing a long-term foundation for the biological production of many beneficial products such as medications, feeds, resources, energy sources, chemicals, and foods. Algal-derived chemicals display affordable synthesis due to rapid development, efficient allocation of resources, and inherent genetic diversity. Several species of microalgae have been approved by the U.S. Food and Drug Administration (FDA) to be safe for human intake in contrast to other microbes such as yeast and bacteria. This designation, known as GRAS (Generally Recognized as Safe), may lower the cost of purifying algal-derived medicines and enhance the potential for economically viable food and agricultural supplies (Torres-Tiji et al. 2020).

Value-adding co-products must be developed from the complete biomass of microalgae in order to make commodities generated from them economically feasible. Microalgae naturally possess a wide range of economically viable substances, such as vitamins, pigments, lipids, and the biomass itself, which is known to be extremely beneficial (Goswami et al. 2020).

Synthetic biology can additionally be utilized to generate high-value recombinant protein derivatives via these algae. Transforming production from laboratory settings to an industrial setting has emerged as the biggest constraint toward yielding more algal bioproducts on an industrial basis. To accomplish this, strain optimization and the establishment of genetic toolkits for strain manipulation that produces high quantities of desired outcomes will be critical. Methods involving genetic engineering including RNA interference (RNAi), zinc-finger proteins (ZNFs), transcription activator-like effector nucleases (TALENs), and clustered regularly interspaced short palindromic repeats CRISPR/CRISPR-associated protein 9 (CRISPR/Cas9) are employed to improve both the amount and the quality of a targeted biomolecule (Vidya and Arunkumar 2022). Metabolic engineering frequently involves altering cellular processes by overexpressing or silencing genes that promote the release of desired products. With regard to the commercialization of microalgae biorefineries, these modified microalgae assist in minimizing the techno-economic barriers. On the contrary, integrating recombinant proteins in microalgae entails carefully incorporating an external gene encoding a protein of interest into the algal genome. This enables the protein to be highly expressed and directed to a particular subcellular area of the algal cells. The genetic heterogeneity among species has limited the implementation of genetic tools from a single species to another, which has caused a difficulty in the utilization of molecular techniques in algae. However, as genome sequencing becomes more accessible, new microalgal species may be identified with ease, and the techniques can be tailored to fit their genetic profile (Naghshbandi et al. 2020). As a consequence, more broadly applicable molecular toolkits are being aimed at facilitating numerous kinds of microalgae as commercial production platforms, and research in this area is expanding. However, the hurdles of scaling up microalgal cultivation persist in operational expansion along with economically achievable execution. The reality seems quite straightforward, for an industrial procedure to be productive; an appropriate biological platform must be coupled with the correct engineering systems in a customized approach depending on specific surroundings, facilities, market demands, regulations, and expenditures. As a consequence, there is a paradox: whereas modifying genes delivers an approach toward profitability for a biorefinery facility, the developmental expenses present may constrain the possibilities exposed for how a specific microalgae procedure could be technologically advanced and thus interfere with exploitation initiatives.

A key ecological and safety concern is the adverse effects of escaping mass-cultured genetically modified (GM) algae-derived variants on natural ecosystems. Even with precautionary measures, the likelihood of accidentally releasing GM algae resulting in negative environmental impacts cannot be governed out. Considering these issues and the shortage of evidence necessary to directly evaluate the nature and scale of real risks associated with GM algae should be focused by the researchers.

19.2 Novel Machinery Development for Algal Genetic Modification and Yield Improvement

The creation and optimized utilization of new genetic components and transformation methods have resulted in current developments in the field of microalgae, putting them on par with various industrial microbes. A more comprehensive collection of molecular techniques will be necessary for the advancement of microalgae as a

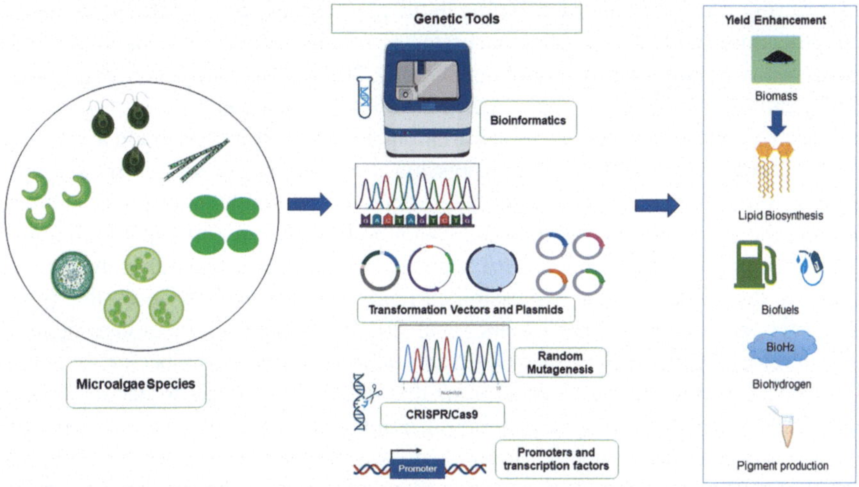

Fig. 19.1 Genetic strategies for developing microalgae as a foundation for industrial biomanufacturing

productive platform for constructed biochemical. The alteration of the algal genome to generate ideal amounts of the desired by-products will be possible with a broad range of validated, reliable resources (Gopal and Kumar 2013) (Fig. 19.1).

19.2.1 Expression Vectors

Building expression vectors containing particular genetic components for obtaining a specific phenotype or function inside a model organism is known as synthetic biology. For the vast majority of commercially essential microorganisms, Golden Gate Modular Cloning (MoClo) toolkits are now easily accessible, enabling it to be simple for researchers to generate such vectors by offering a library of genetic building blocks that serve as a manual for the process of designing and constructing them (Engler et al. 2014; Moore et al. 2016; Celińska et al. 2017; Martella et al. 2017). The initial MoClo toolkit was designed recently by Crozet et al. (2018) for the microalga *Chlamydomonas reinhardtii*, containing over 100 gene parts that represent 67 distinct genetic elements. Promoters, untranslated regions (UTRs), terminators, tags, reporters, antibiotic resistance genes, and introns are among the gene components. These components operate as a defined library of modular elements that can be used to create codon-optimized expression vectors in *C. reinhardtii* resulting in high gene expression (Crozet et al. 2018). Furthermore, a similar modular arrangement was utilized for developing a preliminary high-capacity gene stacking toolkit for *Nannochloropsis oceanica* CCMP1779. It included several adaptable promoters, resistance markers, and reporter genes into Gateway-compatible vectors that could be coupled in multiple manners to achieve multi-gene expression. When numerous enzymes act together in the production of desired products, pathway engineering is made easier by attaining multi-gene expression. Metabolic engineering procedures will become more accurate and reliable with the introduction of the novel gene stacking system (Poliner et al. 2019, 2020). Using sequence algorithmic optimization, computational approaches have also shown value for improving the efficacy of gene expression vectors. Variations in expression of as much as 65-fold was observed when nuclear gene expression in *C. reinhardtii* was correlated into various algorithms to develop numerous versions of a synthetic enzyme, demonstrating the significance of appropriate codon optimization for attaining high transgene expression. The *C. reinhardtii* sequence optimization technique that yields maximal heterologous gene expression currently offers a publicly accessible pipeline and code through the web (Weiner et al. 2018). The same group of investigators also developed Chlamy Sequence Optimizer (CSO), a chloroplast-specific sequence improvement program for *C. reinhardtii* that allows individuals to decide among three experimentally verified codon optimization algorithms. When combined with the recently introduced modular toolkits, these simple techniques for obtaining an expression vector of the ideal codon sequence streamline the engineering phase thereby opening up additional time for scientists to concentrate on discovering possible new algal products and scaling up strategies. It can additionally speed up screening processes and elevate the expression of recombinant proteins and natural metabolites if the collection of specified resources and molecular methods for all of the essential microalgal strains utilized in the biotechnological field are further developed (Weiner et al. 2020).

19.2.2 Recent Progress in Genome Editing and Regulation

Site-directed nucleases, which have been created to target almost any desired DNA sequence for manipulation, have applications in genome editing processes. They may be utilized to knock out genes or introduce donor DNA with existing DNA to insert transgenes. The process of inducing double-strand breaks (DSBs) in DNA stimulates the cell's DNA repair machinery. Several distinct species are being used to apply these methods, with a majority of them through DNA-binding domains from transcription activator-like nucleases (TALENs) and zinc-finger nucleases (ZFNs). However, over the last 10 years, a more successful approach consisting of RNA-guided DNA endonucleases and clustered regulatory interspaced short palindromic repeats (CRISPR)/Cas (CRISPR-associated) has developed (Gaj et al. 2013).

19.2.2.1 RNA Interference (RNAi)

Fire et al. (1998) made the initial discovery of the RNA interference (RNAi) mechanism in *Caenorhabditis elegans*. This method of analyzing genes and how they operate that targets mRNA molecules is known as reverse genetics. RNA interference (RNAi) can silence a specific gene's sequence by targeting complementary sequences on mRNA (Fire et al. 1998; Ketting 2011). In microalgae, alterations other than transcriptional inhibition and mRNA degradation were additionally discovered. For example, RNA interference (RNAi) was used to suppress translation during *Phaeodactylum tricornutum* research. Another study by Koblenz and Lechtreck (2005) used artificial induction of RNAi in *C. reinhardtii* by ammonium depletion, leading to the expression of inverted repeats regulated by the NIT1 promoter (Koblenz and Lechtreck 2005). According to Levitan et al. (2015), RNA interference (RNAi) was applied to examine how nitrogen stress impacts lipid production (Levitan et al. 2015). In *P. tricornutum*, nitrate reductase gene expression was downregulated, leading to a rise in lipid levels. Although it has major downsides, RNA interference (RNAi) is an efficient method for modifying the functioning of genes and silencing them. It may cause unintended irregular and partial repressions of the transformants and gene targets, respectively, off-target alterations, and silencing of desired phenotypes by influencing RNAi transgene (Boettcher and McManus 2015). This prompted the creation of further practical and successful methods that might be applied to focus on mutagenesis.

19.2.2.2 Zinc-Finger Nucleases (ZFNs)

To find methods for manipulating and optimizing the genome, experts designed sequence-specific nucleases (SSNs) that could be programmed. Among these are zinc-finger nucleases (ZFNs), which are protein-based nucleases made from the FokI endonuclease of *Flavobacterium okeanokoites*. The N- and C-terminals of the endonuclease, correspondingly, contain a non-specific DNA-binding domain and a DNA-binding domain (Kim et al. 1996). Zinc-finger proteins get combined with the C-terminal of FokI to form ZFNs. According to Durai et al. (2005), they detect and attach to the 3-bp site on the desired sequence. Two FokI domains are induced by dimerization and break DNA causing double-strand breaks (DSBs). As an outcome, the breaks are repaired by homologous recombination (HR) or non-homologous end joining (NHEJ) when the repair process in DNA starts functioning (Durai et al. 2005). Thus, target DNA is restored. The N- and C-terminals of the endonuclease, respectively, contain a non-specific DNA-binding domain and a DNA-binding domain. Zinc-finger proteins are synthesized with the C-terminal of FokI to form ZFNs (Urnov et al. 2010). ZFNs have been added to a mutant strain of *Chlamydomonas* that was developed especially with a cop3 gene insert into an aphVIII marker. The marker became dysfunctional due to the insertion. Eventually, the nuclease cleaved the gene following the introduction of ZFNs, resuming the functioning of aphVIII (Sizova et al. 2013). ZFNs need to be specifically programmed for every target sequence because designing for each target demands arrays of ZFNs with a high degree of specificity and affinity for the desired sequence. Furthermore, Ng et al. (2017) discovered that it is productive only when the FokI domains dimerize (Ng et al. 2017). Additional negative aspects of ZFNs include their pricey construction, off-target consequences, and complicated design (Fayyaz et al. 2020).

19.2.2.3 Transcription Activator-Like Effector Nucleases (TALENs)

TALENs were discovered while investigating the *Xanthomonas* genus. Through the secretion of effector proteins, these phytopathogenic bacteria adversely affected plants. According to Nemudryi et al. (2014), the potential to bind DNA mimicked eukaryotic transcription factors (TFs), initiating target genes. These proteins are composed of 12–27 tandem arrays of protein repeat domains with 34 amino acids (aa) in each of 3 domains, i.e., DNA binding, transcriptional activation, and nuclear localization signal (NLS) domains. Artificially designed TALE (dTALE) was introduced in a *C. reinhardtii* experiment to watch how efficiently they could stimulate genes (Jankele and Svoboda 2014). Hla3, a membrane protein, and Hsp70A/RbcS2, a promoter, were studied independently. Since the protein has been linked to the absorption of inorganic carbon (Ci), in the prior instance, Ci was accumulated. This demonstrated its significance in the transfer of HCO_3 in surroundings minimal in CO_2 (Weyman et al. 2015). In the following research, the promoter raised the concentration of mRNA and protein by enabling the ARS1 and ARS2 genes to be overexpressed. TALEN was utilized in *P. tricornutum* to inactivate the urease gene, thereby altering its function (Daboussi et al. 2014).

19.2.2.4 CRISPR/Cas9

Repeated sequences that include clustered, regularly interspaced, short palindromic repeats (CRISPR) and associated genes (Cas genes) contribute bacteria immunity against invaders. According to Gasiunas et al. (2012) and Jinek et al. (2012), this mechanism consists of the Cas9 protein with endonuclease activity and an RNA module called CRISPR RNA, which guides Cas9 to the target sequence and results in location-specific double-strand breaks (Gasiunas et al. 2012; Jinek et al. 2012).

CRISPR/Cas9 escalated genome editing and was recently implemented to enhance the microalgae engineering process. In 2013, it was first applied to plants by multiple researchers. It is an RNA-guided nuclease complex following basic complementary sequence base pairing regulations. CRISPR/Cas9 represents one of the most prominent methods for gene modification for its effective, manageable, and accurate outcomes (Feng et al. 2013; Nekrasov et al. 2013; Jeon et al. 2017). Jiang and colleagues attempted the first application of CRISPR/Cas9 editing in microalgae in 2014, but their approach was largely unsuccessful in producing numerous successful mutants because of Cas9 toxicity in constitutively expressed *C. reinhardtii* (Jiang et al. 2014). Even though Cas9 toxicity still appeared to be common, the same researchers demonstrated effective CRISPR-mediated nucleotide replacement of *C. reinhardtii* subsequently while employing co-transformation with Cas9/intron-sgRNA constructs and the arginosuccinate gene or acetolactate synthetase gene (Jiang and Weeks 2017). This team also used oligonucleotide-directed mutagenesis (ODM) and was able to edit the *C. reinhardtii genome*, although with relatively low mutagenic rates (Jiang et al. 2017). An alternative team achieved greater success by substituting the orthologous Cpf1 ribonucleoproteins for Cas9, leading to indel mutations with a rate of 0.02% and DNA replacement frequencies of as much as 10% in a strain lacking a cell wall (Ferenczi et al. 2017). Another technique for preventing the toxicity of Cas9 involves utilizing its variant, Cas12a (Naduthodi et al. 2019). Cas9/dCRISPRi was utilized in *C. reinhardtii* CC400 to inhibit the CrPEPC1 gene, which elevated the lipid levels. Research findings on the nblA gene of the *Synechocystis elongatus* UTEX 2973 have been successfully carried out recently and this approach was applied to repress multiple genes in *Synechocystis* sp. PCC 6803 (Wendt et al. 2016; Yao et al. 2016). By prohibiting competitive pathways that may have an impact on lipid production, microalgae may yield more lipids. Lipid degradation, starch synthesis, and β-oxidation are some of the associated paths. The dCas9 variant, which is formed by combining dCas9 with an activator molecule, is suitable to boost biomass yield for lipids. The variant commences the process of fatty acid synthesis, which generates the precursors essential to begin lipogenesis. In the cell-walled *C. reinhardtii* strain CC-125, Greiner et al. (2017) managed to effectively modify variables for photoreceptor gene knockouts through CRISPR/Cas9 and ZFNs (Greiner et al. 2017). Later, Baek et al. (2018) developed knockout variants of the zeaxanthin epoxidase gene that substantially enhanced zeaxanthin and lutein extraction for the industrial enrichment of hen diets (Baek et al. 2018). Furthermore, this approach was developed into CRISPR-interference (CRISPRi) that utilizes a dead Cas9 with a mutant cleavage domain resulting in control instead of alterations of the desired gene. It was first shown in microalgae by suppressing a *C. reinhardtii* gene responsible for carbon partitioning to divert higher carbon flux toward growing lipid product amounts (Kao and Ng 2017). By eliminating the glycogen anabolic gene, glucose-1-phosphate adenylyltransferase (glgC), the CRISPR/Cas9 approach also appears to be profitable for modifying the cyanobacterial genome to obtain succinate production (Li et al. 2016b). This ultimately leads to the shifting of carbon metabolic flux toward organic acid accumulation. It was recently discovered that certain multicellular cyanobacteria contain special CRISPR/Cas machinery that are capable of being used to insert DNA segments to specific locations inside the *Escherichia coli* genome to induce RNA-guided DNA transposition. Additionally, CRISPRi has been thoroughly used to restrict multiple genes within different cyanobacterial types, expanding their potential for photosynthetic activity and yielding fatty alcohols (Hou et al. 2019; Strecker et al. 2019). Additional RNA-based mechanisms involving base pairing with the mRNA-targeted sequence are being used to govern the expression of genes in cyanobacteria (Sakai et al. 2015). These structures, also referred to as riboregulators, are built up of two RNA fragments called trans-activating RNA (taRNA) and cis-repressed RNA (crRNA), which collaborate for regulating target mRNA's translation through blocking entry to the small ribosomal subunit by RNA:RNA interactions (Isaacs et al. 2004). In *Synechocystis* sp. PCC 6803, this pathway has been utilized currently for tuning the expression of the gene encoding the AbrB-like transcriptional regulator, cyabrB2, leading to a 50-fold dynamic range in protein expression (Ueno et al. 2017). The main advantage of manipulating microalgal genomes by performing site-directed gene editing tools, such as the CRISPR/Cas9 system, is the capability to fine-tune gene expression and regulate positional effects, such as transgene silencing, produced by random integration. The detection of "neutral sites" locations that can be transformed without changing phenotypes within cyanobacteria genomes has largely made the latter feasible. Furthermore, the production of gene knockout mutants instead of the incorporation of transgenes expressing proteins serves as the primary constraint of CRISPR-mediated gene editing in microalgae. Thus, even though metabolic engineering has proved to be an advantageous method to promote the synthesis of naturally produced algal compounds, it is yet to be applied for the synthesis of high-value recombinant protein products in microalgae. This could be related, in particular, to *C. reinhardtii*, the primary organism used to express recombinant proteins, having an inadequate nuclear homologous recombination (HR) machinery for executing homology-mediated editing. Non-homologous end joining (NHEJ), among the two cellular DNA repair processes, re-ligates both ends of DSBs, leading to indel mutations for the desired gene, although homologous recombination (HR) is a more reliable process of two identical sequences interacting and exchanging data; as such, HR must be performed for inserting foreign DNA into a specific sequence of the genome (Gaj et al. 2013). Since *C. reinhardtii* rarely encounters HR, NHEJ-mediated gene editing has proved to be substantially better so far. Future advances facilitating high-efficiency HR-mediated editing in *C. reinhardtii* could greatly progress microalgae protein engineering (Ng et al. 2015). The CRISPR/Cas9 system can be exploited to functionally characterize novel genes in addition to editing genes. Through metabolic engineering, microalgae may synthesize more lipids, pigments, biohydrogen, and other essential chemical and biological compounds (Baek et al. 2018).

19.2.2.5 Genetic Transformation

To minimize the time needed on post-transformational screening, the plasmid vector DNA must be introduced into the algal genome with high efficacy after designing and cloning a suitable expression vector. For microalgal transformations, various approaches have been effective, involving biolistic particle bombardment, glass bead agitation, electroporation, *Agrobacterium*-mediated method, silicon carbide whiskers, and nanoparticles (Jinkerson and Jonikas 2015). Electroporation serves as considerably among the most prevalent and efficient method, as it is less demanding of labor and especially efficient for *C. reinhardtii* nuclear transformations, yielding up to 100 better transformants and generating fewer false positives than conventional approaches. This method employs an electrical pulse to form micropores in the cell wall, which facilitate to incorporate exogenous DNA (Yamano et al. 2013; Mini et al. 2018). Muñoz et al. (2018) identified the best voltage settings, cell concentrations, light intensities, and fragment sizes for electroporation with four distinct microalgal species. When constructing transformation experimentation, these findings can be used to estimate suitable circumstances (Muñoz et al. 2018). Moreover, the initial successful chloroplast transformation of *Nannochloropsis* occurred when a nuclear-targeted construct was electroporated and was observed to be integrated into the plastid genome. Transformation of the *Nannochloropsis* chloroplast was formerly tough due to the species' relatively small cell size (2 M) along with unidentified cell wall components; meanwhile, this discovery broadens the way to chloroplast transformation becoming an additional standard tool for this critically necessary industrial species. Although electroporation is still the most common transformation process used by numerous microalgal species, additional techniques might prove useful in specific instances (Gan et al. 2018). For example, Velmurugan and Deka (2018) discovered that particle bombardment, polyethylene glycol-mediated, and agrobacterium-mediated methods have been widely employed for haptophyte transformation, with particle bombardment delivering the greatest advantages (Velmurugan and Deka 2018). Particle bombardment transformation has been shown to be successful in *P. tricornutum*, where biolistic delivery of CRISPR/Cas9 ribonucleoproteins was successful in giving rise to double-gene knockouts of a photoreceptor-encoding gene with 65–100% performance, establishing a DNA-free editing mechanism for this particular organism (Serif et al. 2018). Furthermore, the glass bead strategy, which consists of spinning algal cells in a tube with glass beads, has been appropriately utilized for inserting artificial microRNA into *C. reinhardtii* with the objective to modulate gene expression and raise fatty acid content (Wang et al. 2017).

19.2.2.6 Transcriptional Engineering

By manipulating transcriptional elements, such as transcription factors (TFs), that modulate gene expression by linking to particular motifs in DNA sequences and interacting with RNA polymerase either to up- or downregulate transcription of the gene, transcriptional engineering is focused on controlling multiple genes in metabolic processes simultaneously (Bajhaiya et al. 2017). The majority of TFs in microalgae are recognized by predictions, and their primary roles are not known. Still, transcriptional regulatory systems that have been suggested utilizing RNA-seq information have associated TFs with the control of various cellular metabolism processes and their binding sites (Romero-Campero et al. 2016). The objective is to employ this genetic engineering application to maximize the microalgae's cultivation of distinct metabolites. Basic leucine zipper (bzip) transcription factors (TFs) are highly editable since they control the processes of metabolism, which involve lipid production, and serve an integral role in stress responses. Because of their inherent abilities to accumulate increased lipid levels, *Nannochloropsis* species have sparked interest currently as potential production strains (Ma et al. 2014, 2016).

The percentage of lipids in the transformant lines appeared greater than in wild-type strains when endogenous bzip TFs were overexpressed in *Nannochloropsis salina* (NsbZIP1), which was initiated by the tubulin (TUB) promoter, whereas in case of *N. oceanica* (NobZIP1) it was triggered by the HSP20 promoter. Utilizing Nile Red fluorescence, the lipids in *N. salina* were initially estimated. Lipid content and classes were subsequently assessed using thin-layer chromatography and high-performance liquid chromatography (Kwon et al. 2018). To put things in a larger context, it has been discovered that *N. oceanica* might accumulate as much as 50% of intracellular lipids, although the wild-type *N. salina* can only accumulate about 30% (Ma et al. 2014). On the contrary, the *N. oceanica* mutants showed a notable rise in intracellular and secreted volume of lipids accomplishing up to 16 times higher than the original strain. Besides, it was discovered that by silencing the uridine diphosphate (UDP) glucose dehydrogenase protein through metabolic engineering and overexpressing NobZIP, carbon flux was redirected from the synthesis of cell wall lipids to the making of triacylglycerol (TAG) lipids, which elevated lipid secretion broadly. Optimizing secretion and lipid production levels might contribute to cutting the total expenditure of commercialization by eliminating the necessity for extracting lipids from algae cells (Li et al. 2019).

Gene editing strategies were additionally used to knock out particular TFs in order to modify the metabolic processes, as compared to overexpressing TFs. Recently, knockout mutant *Nannochloropsis gaditana* strains of 20 TFs, which were shown by RNA-Seq to be suppressed in nitrogen-depleted surroundings, were generated based on CRISPR-Cas9 reverse genetics. Higher lipids were reported upon knocking out a predicted Zn(II)2Cys6 (ZnCys) binuclear domain cluster TF, and an end strain with a threefold rise in overall lipids and carbon partitioning to lipids almost 200% exceeding the control strain was produced upon further attenuation of ZnCys. TFs perform a vital function with regard to controlling the activity of genes and the biosynthesis of products. The subsequent phase in the development of this genetic application will be to discover TFs associated with mechanisms of interest and their possible homologs among various microalgae species, as well as establish uniform methods for transcriptional engineering throughout species (Ajjawi et al. 2017).

19.3 Microalgae for Biorefinery Facilities Through Genetic Engineering

Genes can be changed through genetic engineering to produce beneficial products. The fundamental steps in algal genetic manipulation are performed in genetic engineering, which include: (1) Choice of host, (2) Gene of interest, (3) Transformation, and (4) Identifying and screening. Of these, transformation techniques and selection are the two most crucial steps. A diverse set of biological and synthetic chemicals, categorized as safe products, are produced by microalgae. These products include proteins, biofuels, pigments, biohydrogen, and nutraceuticals. They grow relatively quickly and possess an increased lipid production rate. Transcriptional and translational genes can be upregulated or downregulated to regulate metabolic processes. To attain the desired effects, genes are either knocked in or knocked out (Ng et al. 2017) (Fig. 19.2).

Studies that demonstrate targets related to genetic and metabolic engineering are presented in Table 19.1 (Baek et al. 2018; Li et al. 2018; Rengel et al. 2018; Salas-Montantes et al. 2018; Sarnaik et al. 2018; Kamravamanesh et al. 2019; Liu et al. 2019; Seo et al. 2020; Velmurugan and Incharoensakdi 2020; Jia et al. 2022).

19.3.1 Lipid Synthesis

Microalgal lipids are widely utilized in industry for many purposes, which comprise the synthesis of triacylglycerols (TAGs) for biofuels and polyunsaturated fatty acids (PUFAs) such as omega-3 fatty acids for dietary supplementation. Microalgal lipids additionally are of major importance in the advancement of third-generation biofuels. Acetyl-CoA synthetase (ACS) and malonyl-CoA:ACP (acyl carrier protein) transacyclase (MAT) are two enzymes that are crucial in lipid biosynthesis process. ACS is desired for lipid manufacturing because it is considered to be an inhibitory enzyme that induces the transformation of acetate to acetyl-CoA, a crucial molecule

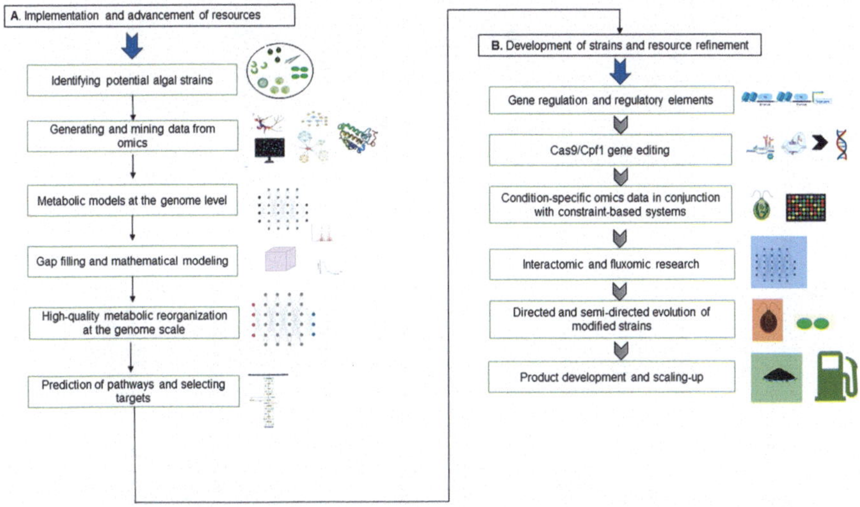

Fig. 19.2 The data and resource-driven strategy for microalgal bioengineering

in the fatty acid synthesis process (Rengel et al. 2018). MAT, on the contrary, is a type 2 fatty acid synthase (FAS) that functions throughout both fatty acid synthesis and polyketide formation in *Schizochytrium* sp. by synthesizing malonyl-ACP intermediates (Li et al. 2018; Misson et al. 2020). The increased expression of the endogenous chloroplast ACS gene in *C. reinhardtii* through constructs including the RBCS2 and HSP70A promoters contributed to a sixfold rise within ACS transcripts relative to wild-type, resulting in a 2.4-fold rise in TAG formation within the mutants under nitrogen diminished conditions (Anand et al. 2019). An additional 52% rise in TAG aggregation over the wild-type variant was reported when acetate was introduced to the medium. While glucose-fed batch fermentation was implemented, investigators obtained variants with a fivefold rise in MAT transcripts among the natural kind, with subsequent production of 47 g/L for docosahexaenoic acid (DHA) and 1.65 g/L for eicosapentaenoic acid (EPA) (Osanai et al. 2006). By hindering competitive routes, production of lipids can also be accelerated. One may immediately prohibit β-oxidation by eliminating genes, or indirectly by lowering acyl-CoA levels. Blocking the system that regulates phospholipid synthesis is an additional option. Phospholipids necessary for the creation of the cell wall were lacking as a consequence of the pathway's suppression. Studies have indicated that lipid synthesis rose when phosphoenolpyruvate carboxylase (PEPC) was blocked. When CRISPRi was applied for gene knockdown, *C. reinhardtii*'s lipid profile rose by 94%. *P. tricornutum* was used to investigate and create an alternative strategy. After the introduction of the genetically modified malic enzyme, the amount of lipid formed was raised by a factor of 2.5. Applying adverse conditions has a possibility to considerably raise lipid accumulation (Ludwig and Bryant 2012).

A lot of researchers have examined the variations in lipid generation among microalgae through genetic and metabolic alteration (Table 19.2) (Li et al. 2016b;

Table 19.1 Recent research on genetic and metabolic engineering targets

Targets	Origin organism	Host organism	Function	Product pathway	References
Basic-region/leucine zipper transcription factor (NobZIP1)	*Nannochloropsis oceanica*	*Nannochloropsis oceanica*	Regulates key genes involved in lipid and cell wall carbohydrate metabolism	Lipid biosynthesis	Zhipeng et al. (2018)
DNA-binding-with-one-finger transcription factor (Dof11)	*Chlamydomonas reinhardtii*	*Chlamydomonas reinhardtii*	Carbon and nitrogen substance metabolism	Lipid biosynthesis	Salas-Montantes et al. (2018)
Plastidic acetyl-CoA synthetase (ACS2)	*Chlamydomonas reinhardtii*	*Chlamydomonas reinhardtii*	Catalyzes conversion of acetate to acetyl-CoA	Lipid biosynthesis	Rengel et al. (2018)
Malony-CoA: ACP (acyl carrier protein) transacyclase (MAT)	*Schizochytrium* sp. *MYA1381*	*Schizochytrium* sp. *MYA1381*	Catalyzes formation of malonyl-ACP	Lipid biosynthesis	Zhipeng et al. (2018)
PTP plastidial pyruvate transporter (PtPTP)	*Phaeodactylum tricornutum*	*P. tricornutum*	Significant role in the pyruvate metabolism, plastids' de novo fatty acid metabolism	Fatty acid synthesis	Seo et al. (2020)
Beta-carotene oxygenase (CRTR)	*Synechococcus elongatus PCC7002*	*Synechococcus elongatus PCC7002*	Catalyzes conversion of beta-carotene to zeaxanthin	Carotenoid biosynthesis	Sarnaik et al. (2018)
Hexose-H+ symporter (GalP)	*Escherichia coli*	*Synechococcus elongatus* PCC 7942	Facilitates uptake of glucose across the cell wall	Carotenoid biosynthesis	Sarnaik et al. (2018)
Zeaxanthin epoxidase gene (ZEP)	*Chlamydomonas reinhardtii* (strain CC-4349)	*Chlamydomonas reinhardtii* (strain CC-4349)	Important role in the xanthophyll cycle and abscisic acid (ABA) biosynthesis	Carotenoid biosynthesis	Baek et al. (2018)
Carotenoid hydroxylase (CrtR-B)	*Haematococcus pluvialis*	*Synechocystis* sp. PCC6803	Catalyzes conversion of canthaxanthin to astaxanthin	Carotenoid biosynthesis	Liu et al. (2019)

Beta-carotene ketolase (BKT)	*Haematococcus pluvialis*	*Synechocystis* sp. PCC6803	Catalyzes conversion of echinenone to canthaxanthin	Carotenoid biosynthesis	Liu et al. (2019)
Malonyl-CoA: ACP transacylase (MAT)	*Schizochytrium* sp.	*Schizochytrium* sp.	Transfer of the malonyl 74 moiety to holo-ACP, forming malonyl-ACP intermediates that participate in fatty acid synthesis	Polyunsaturated fatty acids (PUFAs) and lipid synthesis	Li et al. (2018)
Glucose-1-phosphate adenylyltransferase (glgC)	*Synechococcus elongatus PCC 7942*	*Synechococcus elongatus PCC 7942*	Catalyzes alpha-D-glucose-1-phosphate diphosphate to ADP-glucose	Catalyzes alpha-D-glucose-1-phosphate diphosphate to ADP-glucose	Velmurugan et al. (2020)
Sedoheptulose-1,7-bisphosphatase (SBPase), citrate synthase (gltA), and succinate transporters (phosphoenolpyruvate carboxylase, PEPC)	*Synechococcus elongatus PCC 11801*	*Synechococcus elongatus PCC 11801*	Enzymes in the succinate pathway	Succinate synthesis	Sengupta et al. (2020)
Phosphate transport system permease protein (PstA)	*Synechocystis* sp. *PCC 6714*	*Synechocystis* sp. *PCC 6714*	Transports phosphate	Glycogen synthesis	Kamravamanesh et al. (2019)
DNA-binding-with-one-finger (DOF) transcription factor, long-chain acyl-CoA synthetase (LACS2), and citrate synthase (CIS1)	*Chlamydomonas reinhardtii*	*Chlamydomonas reinhardtii*	LACS2: activating fatty acids to fatty acyl-CoA thioesters CIS1: a rate-limiting enzyme in the tricarboxylic acid cycle	Lipid metabolism	Jia et al. (2022)

Table 19.2 Studies focused on enhancing lipid biosynthesis in various microalgae through genetic and metabolic engineering methods

Algal strain	Targeted genes	Strategy	Effect on lipid synthesis	References
Phaeodactylum tricornutum	3-Ketoacyl-coA synthase (MaKCS)	Heterologous expression	0.2% content of nervonic acid (C24:1) in overall fatty acid methyl esters (FAMEs)	Fan et al. (2018)
Schizochytrium sp.	Malonyl-CoA: ACP transacylase (MAT)	Overexpression	10.1% and 24.5% increased total fatty acid (TFA) and PUFA content, respectively; TFA, DHA, and EPA yields increased by 39.6%, 81.5%, and 172.5%, respectively	Li et al. (2018)
Nannochloropsis oceanica	Malonyl-CoA-acyl carrier protein transacylase (NoMCAT)	Overexpression	36% increased TFA and 31% increased neutral lipid content	Chen et al. (2017)
Nannochloropsis oceanica	Type 2 diacylglycerol acyltransferase (DGAT2)	Overexpression	Enhanced TAG biosynthesis and an increase in neutral lipid content by 69%	Li et al. (2016a)
Chlamydomonas reinhardtii	Transformed with acyl-ACP thioesterases (TE) from *Dunaliella tertiolecta* (DtTE)	Transformation	Lipid content increased by ~56%	Tan and Lee (2017)
Chlamydomonas reinhardtii	Phospholipase A2 (PLA2)	Gene knockout	Increased lipid production by up to 64.25%	Shin et al. (2019)
Nannochloropsis oceanica CCMP1779	Multiple fatty acid desaturases (FAD)	Construction and overexpression	Enhanced ω-3 long-chain polyunsaturated fatty acids (LC-PUFAs) and EPA production	Poliner et al. (2018)

P. tricornutum	Glycerol-3-phosphate acyltransferase 2 (GPAT2)	Overexpression	2.9-fold increase in TAG content; altered fatty acid profile in TAGs	Wang et al. (2020)
Neochloris oleoabundans	GPAT, LPAAT (lysophosphatidic acid acyltransferases), and DGAT	Heterologous expression	Single-gene expression resulted in 1.3- and 1.4-fold increased TFA and TAG content, respectively; multi-gene expression resulted in 1.2-fold increase in TFA and TAG	Muñoz et al. (2019)
N. oleoabundans	LPAAT1 and DGAT2	Overexpression; heterologous expression	Co-expression resulted in 1.6- and 2.1-fold increased TFA and TAG content, respectively; 1.9- and 2.1-fold increased TFA and TAG productivities, respectively	Chungjatupornchai and Fa-Aroonsawat (2020)
P. tricornutum	PtDGAT2B and OtElo5	Overexpression; heterologous expression	Co-expression resulted in 37-fold increased TAG (from N-replete to starved conditions) compared to 1.8-fold in wild-type, respectively; high DHA and docosapentaenoic acid (DPA) content in TAG	Haslam et al. (2020)
Nannochloropsis oceanica	NoAP2	Knockout	40% increased neutral lipids during growth phase	Südfeld et al. (2021)

(continued)

Table 19.2 (continued)

Chlorella protothecoides	Malic enzyme (ME gene)	Overexpression	Lipid content increased by 2.8-fold	Yan et al. (2019)
Chlorella vulgaris	Omega-3 desaturase (ω-3 FAD)	Overexpression	16–17% increment in oil content	Norashikin et al. (2018)
P. tricornutum	GPAT and DGAT	Co-overexpression	2.6-fold rise in total lipid content	Zou et al. (2018)
Coccomyxa sp.	FAT and DGAT	Co-overexpression of endogenous *FAT* and *DGAT2*	1.4-fold increase in lipid productivity	Kasai et al. (2018)
Chlorella sp. *HS2*	bZIP	Overexpression of endogenous HSbZIP1	113% increase in FAME yields	Lee et al. (2020)

Chen et al. 2017; Fan et al. 2018; Kasai et al. 2018; Li et al. 2018; Norashikin et al. 2018; Poliner et al. 2018; Zou et al. 2018; Muñoz et al. 2019; Shin et al. 2019; Yan et al. 2019; Haslam et al. 2020; Lee et al. 2020; Wang et al. 2020; Chungjatupornchai and Fa-Aroonsawat 2021; Südfeld et al. 2021).

19.3.2 Ethanol Biosynthesis

Yeast fermentation is capable of turning polysaccharides derived from microalgae, such as starch from eukaryotes and glycogen from cyanobacteria, to bioethanol. In the stages of fermentation, glycogen prevails over starch considering it is smaller in size and more readily soluble. The glycogen metabolism has been thoroughly researched using the model cyanobacterial isolates *Synechocystis* sp. PCC 6803, *Synechococcus elongatus* PCC 7942, *Synechococcus* sp. NKBG 15041c, and *Synechococcus* sp. PCC 7002, where production is influenced by nitrogen-limiting conditions. Prior transcriptional study revealed that in nitrogen-depleted *Synechococcus* sp. NKBG15041c, six out of eight glycogen synthesis genes were elevated; however, they were downregulated or unaltered in *Synechocystis* sp. PCC 6803 and *Synechococcus* sp. PCC 7002 (Badary et al. 2018; Kamravamanesh et al. 2019).

This resulted in the recent finding of a novel feature for glycogen synthesis regulated by genes under nitrogen restriction in *Synechococcus* sp. NKBG 15041c, which contributes to increased glycogen yields relative to other model strains (Badary et al. 2018). A point mutation in the phosphate transporter gene resulting from mutagenesis was additionally reported to improve glycogen productivity in *Synechocystis* sp. PCC 6714. The alteration of ethanol biosynthesis can also result in the production of other valuable by-products (Kamravamanesh et al. 2019). Recent research discovered that eliminating glgC in *S. elongatus* PCC 7942 while overexpressing two enzymes involved in the ethanol formation process raised the ethanol and exopolysaccharide production (Velmurugan and Incharoensakdi 2020). Another investigation utilized the identical ethanol synthesis genes that have been employed to boost succinic acid yields in *S. elongatus* PCC 7942 to evaluate that *S. elongatus* PCC 11801 has a better chance of producing succinate than *S. elongatus* PCC 7942 (Sengupta et al. 2020).

19.3.3 Pigments

Carotenoids, chlorophyll, zeaxanthin, phycobilin, lycopene, canthaxanthin, phycocyanin, and various other pigments are generated by microalgae. The objectives of many current investigations are aimed to promote the production of the two primary xanthophylls, astaxanthin and zeaxanthin, in several microalgae and cyanobacteria. Algal biopigments are a secure and potential sustainable resource of pigments that can be consumed as an alternative to harmful and unproductive synthetic pigments. Phytoene is transformed into lycopene in the carotenoid synthesis process, which is necessary in the formation of carotenoids (Yusuf et al. 2017). Carotenoid production may be increased by enhancing the gene expression accountable for its synthesis. Lycopene facilitates the development of valuable carotenoid compounds. *Synechococcus elongatus* strain PCC 7942, a cyanobacterium lacking zeaxanthin and lutein formation in its carotenoid pathway, is an ideal strain for the biosynthesis of these antioxidants. The exogenous beta-carotene oxygengenase gene (CrtR) from *S. elongatus* strain PCC 7002 was introduced by Sarnaik et al. (2018) in order to expand the carotenoid route for the synthesis of only zeaxanthin. The strain exhibited a zeaxanthin content of 9 mg/g dry cell weight (DCW) in autotrophic environments and 8 mg/g DCW in mixotrophic conditions, accompanied by the CrtR gene along with the incorporation of the galactose transporter from *E. coli* for improved biomass output (Sarnaik et al. 2018). Zeaxanthin was originally introduced into *Chlorella zofingiensis* through chemical mutagenesis, which was only capable to produce up to 7 mg/g DCW under photosynthetic conditions in addition to 4 mg/g DCW under mixotrophic conditions (Huang et al. 2018). The most widely utilized strain for astaxanthin production is *Haematococcus pluvialis*, which has been reported to aggregate between 9 and 36 mg/g DCW based upon growth conditions, but since it is readily impacted by environmental parameters, alternative cultivation isolates are currently being explored. Although the genes responsible for beta-carotene ketolase (BKT) and carotenoid hydroxylase (CrtR-B) cannot be naturally expressed in *Synechocystis* sp., both enzymes can be genetically constructed through exogenous DNA from *H. pluvialis* under controlled conditions of the cpc560 promoter (for BKT) and psbA2 promoter (for CrtR-B). In nitrogen-depleted limitations, this efficient gene expression enabled the transformation of the astaxanthin precursor molecules, echinenone and zeaxanthin, toward 4.8 mg/g DCW astaxanthin (Liu et al. 2019). Likewise, *C. reinhardtii* UVM4 strain that failed to synthesize high-value ketocarotenoids became capable in accumulating astaxanthin after the malfunctioning endogenous BKT gene was synthetically modified by codon optimization (Perozeni et al. 2020a). Furthermore, the endogenous phytoene desaturase coding sequence (pds) was expressed excessively in *Haematococcus pluvialis* chloroplast through optimizing codons. The most potent variants recorded astaxanthin production of about 34.3 mg/L whereas 18.0 mg/L was observed in natural strain after successfully inserting a nuclear encoded gene in the chloroplasts of *H. pluvialis* (Galarza et al. 2018).

19.4 Modern Applications of Genome Shuffling and Random Mutagenesis

Evolutionary engineering has been extensively employed as a genetic tool to promote the advancement of algae-producing strains. Employing non-recombinant methodologies having repetitious procedure initially provides genetic diversity to a population. Next, selective stresses and rigorous screening are applied to discover novel isolates that exhibit the phenotype of interest. Two of the most prevalent approaches for introducing genetic diversification are genome shuffling, which is the mechanism of generating genetic permutations across genome-scale recombination as an outcome of sexual recombination or via fusing protoplasts, and random mutagenesis, which utilizes chemical or physical mutagens to induce mutations all over the genome (Biot-Pelletier and Martin 2014).

19.4.1 Random Mutagenesis

Traditional methods of strain development for microalgae rely entirely on mutagenesis inducing genetic diversification. This can happen since random mutagenesis might be applied with minimal or no knowledge of the genetic foundations or life cycle, and thus can be more easily resolved than using synthetic biology or sexual breeding (Fu et al. 2016). On the contrary, the method can be laborious, since desired mutations are often rare and phenotypic modifications are marginal, demanding much iteration paired with vigorous screening. The most prevalent process for random mutagenesis is exposure to ultraviolet (UV) light, that is an inexpensive, safe, and accessible way to trigger gene mutations. This approach has been utilized effectively to engineer the *C. reinhardtii* strains UVM4 and UVM11, leading to promoted transgene expression with negligible positional outcomes mainly since the genes engaged in the transgene suppression process are assumed to be knocked out (Neupert et al. 2009). Different types of radiation are also used presently for stimulating genetic diversity, like 60Co-γ irradiation, which was found to increase lipid amounts of *Chlorella pyrenoidosa* by almost 20%. The resulting variants in this investigation had higher concentrations of saturated fatty acids and indicated robust growth that surpassed that of the parental strain with wild-type characteristics (Wang et al. 2018). A distinctive method known as atmospheric and room temperature plasma (ARTP) enhances the genetic variation of an entire population substantially more than an isolated mutagen through facilitating multiple mutagenic agents (such as UV radiation, charged particles, neutral reactive species, and free radicals) to work upon the cells concurrently and generate an array of variations at the same time. The ARTP approach was adopted to lower the chlorophyll levels of *C. reinhardtii* while increasing hydrogen production. Due to a surge in photosynthesis rate, the resultant mutant generated hydrogen gas at an average of three times the productivity of the original strain (Ban et al. 2019).

Chemical mutagens provide a substitute approach to physical mutagens. One widely utilized mutagen that results in point mutations through nucleotide substitutions is ethyl methane sulphonate (EMS). The lipid profile of microalgae has been transformed and biomass yields have grown using EMS. *Nannochloropsis oculata* and *Chlorella* sp. cultures displayed an increase in overall biomass formation of 15% and 5%, respectively (Al-Saedi et al. 2019). *Tetraselmis* sp. was subjected to different EMS concentrations at different times followed by the selection and screening of the most rapidly evolving mutant colonies (Kumar et al. 2018). As a consequence, the biomass density rose by 30%, and the lipid profile improved by 20–30%, majorly caused by an upsurge in the proportion of monounsaturated fatty acids. With the aim to elevate the secretion of violaxanthin, *Chlorella vulgaris* was mutagenized using EMS. This was accomplished by creating a single-point mutation that halted the lutein synthesis, altering the autofluorescence of cells. The violaxanthin pathway eventually observed an energy shift because of this mutation, which raised biomass output by 30% and the formation of pigment by threefold (Kim et al. 2020).

19.4.2 Genome Shuffling

Reorganizing the genomes is a reliable method to add genetic variation, and it operates well in conjunction with random mutagenesis to efficiently improve strains. This approach, though, is restricted to strains that exhibit a distinctive regulated sexual life cycle; in other words, the strain should be competent to generate protoplasts compatible with cell fusions. Though genome shuffling is being used to rapidly evolve strains of bacteria and yeasts with success, it is still not utilized to the same extent in algae (Biot-Pelletier and Martin 2014). Out of all the algae species that can undergo laboratory-induced sexual recombination, *C. reinhardtii* possesses a highly characterized genome, ensuring genome shuffling is easier to use. Another study found that by deploying sexual recombination for reorganizing the entire genome, the amount of recombinant protein expressed in *C. reinhardtii* was produced in just 17 weeks, enhancing the amount produced by 15-fold to over 2% overall soluble protein. Protoplast fusion is a possible method of genome shuffling for species lacking a well-defined sexual cycle. In order to create protoplasts artificially, the cell walls are removed by enzymes, and cells are subsequently induced chemically or electrically for greater membrane fluidity. This causes the protoplasts to fuse together, which triggers genetic recombination (Echeverri et al. 2019). Current research adopted this technique to enhance the nutritional composition of *C. vulgaris*, whereby the cells are stripped of their cell walls using lysozyme and then placed in media having polyethylene glycol to induce cell fusion. This led to the development of a novel strain of *C. vulgaris* with greater levels of unsaturated fatty acids, improved amino acid content, and a faster growth rate (Kusumaningrum and Zainuri 2018).

19.5 Challenges Associated with Large-Scale Cultivation of Microalgae

Multiple challenges arise when growing microalgae and cyanobacteria on a larger scale, involving the need to optimize both upstream and downstream operations. The main obstacle to the industrial application of microalgae biomass continues to be its expensive cost of production. The approximate expenses vary between 290 € and 570 € per kilogram of dry weight, contingent upon the particular manufacturing technique utilized (Oostlander et al. 2020). The most expensive components of microalgae have capital expenditures for manufacturing facilities and operational costs involving liquid movement, lighting, regulating the temperature, cleaning, and harvesting (Medeiros and Moreira 2022; Maroušek et al. 2023). For microalgae-growing businesses, financial viability usually comes from massive-scale operations. Massive microalgae cultivation necessitates an array of skills, including those of engineers, scientists, chemical professionals, and biological researchers, particularly phycologists. In contrast, just one individual can run a laboratory-scale photobioreactor (PBR). The algae sector is struggling with a lack of algal expertise; proficient phycologists are in great demand and getting harder to find. To satisfy rising demands, however, leading colleges have started offering relevant training and courses to develop an emerging cadre of professionals and researchers in the field of algae. Eliminating labor expenses and introducing automation will both be crucial for cutting cultivation expenditures. Social obstacles still exist despite all these advancements, especially when it comes to accepting algae as a food source (Novoveská et al. 2023).

Since microalgae are capable of expanding quickly under a range of ecological stresses, their introduction into the ecosystem may have a detrimental effect on native organisms due to competing with them for resources. This is a serious concern because it can result in algal blooms in surrounding water bodies. These algal blooms release toxins into the surroundings and compete with native species for vital nutrients. Algal blooms are harmful, but they can be minimized by increasing scientific research accessibility and employing specific algicides to eradicate surplus algae. Before employing algae for the large-scale production of fuel, it is essential to take its water requirements into account. Based on the specific species of microalgae, 3.5–3726 liters of water would be needed to produce 1 kg of biodiesel (Pugazhendhi et al. 2020).

Scalability is a further problem with algal bioenergy because, according to estimates, 12 million hectares would be required to be planted for the cultivation and extraction of algal oil to fulfil US oil demand (Hannon et al. 2010). To promote expenditures and utilization in huge-scale algal cultivation, which has the largest commercial sites determining no more than 5 ha, it is critical to tackle the ambiguities surrounding this practice. There is disagreement over how to distinguish between open and closed cultivation structures, particularly when it comes to the greater upfront expenses of closed systems versus the obstacles of open systems. Although open-pond systems generate 98% of the algal biomass, they are very vulnerable to outside complications like contaminants, and stressful environments (Patel et al. 2012).

It is significant to first consider how much algae biofuels cost in comparison to different fuel options. In October 2021, ethanol (E85) cost $2.73 per gallon whereas for gasoline, about $3.25 per gallon constituted the price throughout the United States. However, the expense of algal biofuel exceeds $8 per gallon. Algal biofuel costs approximately three times as much as ethanol whereas it costs more than twice as much as gasoline. Additionally, the cost of algae is much higher than that for various biofuel sources (Bourbon and Science 2021). According to Bourbon and Science (2021), the market price of biodiesel B20 was $3.21 per gallon in October 2021, while the rate of biodiesel B99/B100 was $3.80 per gallon (Price et al. 2022). The difficulties involved in producing algal biomass on an enormous scale are briefly discussed by Price et al. (2022).

A broad range of techniques exist for growing algae to be utilized as a biofuel, such as tubular photobioreactor systems, bubble column photobioreactor systems, and raceway systems. The energy consumption and cost expenditures of these approaches are evaluated by Dasan et al. Out of the three, tubular photobioreactor arrangements that have increased biomass productivity demand the greatest energy input, with an approximate annual consumption of 1777.70 MWh (Dasan et al. 2019; Price et al. 2022). By contrast, the annual demand for open-pond systems and bubble column photobioreactors is approximately 1140 MWh. A significant portion of this energy is used for reusing solvent by evaporating it, heat reactions, and evaporating water. It is advantageous for organizations to analyze the life cycle expenses of their algal biofuel operations because it enables them to increase earnings without spending money on things that will become obsolete. The species may be released as an outcome of large-scale production, which might result in a negative effect on the surrounding environment. Because of recent developments in metabolic engineering, which result in genetically altered algae, the threat is now more significant. To more accurately evaluate the risks to the surroundings, research must be done on the impacts of the potential release of genetically modified organisms (GMOs) or substances. The influence on overall biodiversity will rely on how extensively the variant is released and how suitable it is for capturing scarce resources (Dasan et al. 2021). This makes it feasible to thoroughly evaluate any random biological or biochemical generation from a bioprocess and reduce environmental hazards. The effects on ecosystems, the climate, and the quantity and quality of water are the potential hazards associated with massive operations of algae production. There are differences in risks between methods of closed and open cultivation. In terms of water consumption, large-scale structures utilize an immense amount of water, either fresh or salt water, depending on the particular species (Novoveská et al. 2023). Open systems are more likely to experience substantial evaporation than closed systems. According to Slade and Bauen (2013), open-pond systems are usually more affordable than photobioreactor structures (Slade and Bauen 2013). For each kilogram of algae generated, open-pond applications are projected to cost between $0.3 and $0.4. Open-pond systems require an excessive amount of water. Furthermore, the growing capacity of algae cultivated in open-pond systems may be adversely influenced by stresses including solar variability and inadequate water temperature (Trilokesh and Uppuluri 2021). Many strategies, such as the use of continuous cultures, greater biomass densities, biorefinery concepts, genetic alterations (direct and indirect), improved automation and computation, and bigger culture units, are being put forward to address these issues. The latter seeks to promote the professionalization of algae production while addressing the issue of economies of scale (Novoveská et al. 2023).

19.6 Economic Limitations or Market Barriers Involved for Worldwide Adoption of Genetically Modified Microalgae

The term "genetically modified" and the abbreviation "GM" are generally regarded negatively in the general population's view. This is mostly because people lack knowledge of the underlying technology. Likewise, there is a lot of misleading information out there that is spread by broad, exploitative businesses that forced the development and introduction of genetically modified crops solely for their financial gain (Schiano di Visconte et al. 2019).

There is a great deal of attraction, although not broadly applicable, in GM algae to enhance, in comparison to wild-type strains, the yield of feedstock products in high-density cultures, to ensure a commercially viable production of algae-derived biomass for biofuels and value-based products. Many non-GM genetic techniques exist to promote algae phenotypes, which could trigger environmental risks; however, most nations only monitor transgenic and recombinant microbes. After it was determined, the present U.S. Department of Agriculture Animal & Plant Health Inspection Service (USDA-APHIS) and Environmental Protection Agency (EPA) laws exempted their process. Algenol appears to have acquired permission from the State of Florida for scaling up cultivation of GM cyanobacteria in outdoor enclosed bioreactors to manufacture ethanol. Huge-scale examination of genetically modified strains has been requested, and at least a single approval has been accorded currently. The latter document indicates that the majority of genetically altered algae grown in "enclosed" systems may be exempt from present federal regulations. To date, there is, however, minimal information available on possible effects and no authorized releases of engineered algae in open-pond cultivation facilities (Henley et al. 2013). The high expenses of strain growth (strain transformation and choice, strain robustness and banking, toxicity evaluation), including generation and downstream manufacturing, regulatory issues, and biosafety concerns, for example the possible necessity of registration and inspection, are the key drawbacks when applying GM methods to microalgae. When a genetically engineered microalga is raised in huge quantities in open-pond systems (the preferred cultivation technique), it involves releasing the algae again into the natural environment. Microalgae cannot become sterile to facilitate industry-controlled distribution, unlike genetically altered plants, even when alternate growth regulation methods, like auxotrophy, are applied. Additionally, there is a genuine and significant financial risk related to protecting the intellectual property of strains that could grow airborne and naturally isolate themselves from nearby places. In contrast, compared to concerns about biodiversity loss or ecological damage, it is this economic risk to business entities that might trigger GM microalgae methods to be hindered, limited, or possibly stopped on an enormous scale. Closed-system photobioreactors functioning at smaller scales offer accurate control over a release that must be achieved, but they additionally possess fewer yield capacities, influencing the purpose of the biorefinery facilities. However, the main advantage of using GM techniques to generate or improve commercial microalgal isolates is the upgraded overall economics generated by the development of novel and valuable molecules, increased growth, raised bioprocess efficacy, or any combination of these. If the production rates are sufficiently high, phototrophic GM microalgae possess the potential to revolutionize economic processes. As previously stated, it is vital to cultivate them at a suitable scale and safeguard business and environmental objectives. The laws regulating genetically modified microalgae must be modified efficiently in this context. Currently, the European Union (EU) classifies organisms developed through gene editing, including those with transformed native genes, as GM. The only variations that are not classified as GM are breeding and mutagenesis (chemical or UV radiation). However, recent improvements in the CRISPR/Cas and RNAi technologies pose the potential of developing a regulatory gap to which the research community has not yet fully responded (Spicer and Molnar 2018, Schiano di Visconte et al. 2019).

19.7 Ethical Considerations Related to Genetically Modified Organisms (GMOs) and Public Acceptance of Algae-Based Products

Algae are ideal candidates for genetic manipulation since they demonstrate rapid growth rates, effortless genetic engineering and identification processes, and readily transferable cultivating and processing approaches from the biofuel sector to the by-product manufacturing procedure. The public's mistrust of genetically modified products and their lack of knowledge about products obtained from algae represent two of the main issues. Strangely, a large portion of the food that Americans eat today is already produced with components that have been genetically modified. According to statistics, about nine out of every ten acres of domestic corn, cotton, soybeans, sugar beets, and canola have been raised utilizing genetically manipulated seeds. Numerous algae strains have already been granted GRAS recognition by the FDA. "GRAS," or "Generally Recognized as Safe," is an abbreviation for a classification granted by the FDA to a particular substance that is safe for usage by humans. *Arthrospira platensis*, *C. reinhardtii*, *Auxenochlorella protothecoides*, *C. vulgaris*, *Dunaliella bardawil*, *Euglena gracilis*, *H. pluvialis*, and *Schizochytrium* are among the algae species that are presently granted GRAS status (Torres-Tiji et al. 2020; Diaz et al. 2023). This is an important breakthrough toward turning algae into a dependable source, and it should help to increase the acceptance of algae throughout society as an achievable future resource for people.

19.8 Future Directions

Genetic modification may not only mitigate the overall expense of microalgal biofuels by 15–20%, but additionally generate capital through the extraction of other value-added substances and bioproducts. To develop better-performing biorefineries, the possibilities of metabolic and genome modification as an appealing strategy for engineering newly developed strains of microalgae must be fully explored (Fajardo et al. 2020). The expenditure decreased by 85% when natural microalgal isolates are paired with transgenic strains. The Aquaculture Raceway Integrated Design (ARID), an open-pond cultivation system, was also highlighted in the report. Through consuming fewer resources and improving productivity, this approach lowers costs by 16% (Spicer and Molnar 2018). The opportunities for a genetic engineering approach on microalgae are infinite, and these methods are versatile enough to create metabolic networks. This can assist in the exploitation of various products derived from the transformed microorganisms, as well as in resolving of economic and environmental sustainability issues. A better understanding of the algal genome is vital in order to lessen reliance on fossil fuels while expanding the efficiency of algal biorefinery (Naghshbandi et al. 2020).

Algomics, or the utilization of genomic and post-genomic strategies on algal cells, is currently utilized for this cause. Algomics aids scientists in the wet laboratory by recommending experimental methodologies (Koussa et al. 2014). Algomics can be utilized for interpreting metabolic pathways, genome sequences, and the outcomes of modifying genomes, reaction mechanisms and enzymes, function of proteins, the investigation of existing processes, and their alterations. This strategy, subsequently, has certain drawbacks. Incomplete sequencing and knowledge of the algal genome is constrained by complicated algal cell wall and organelle structures, gene silencing techniques, and compartmentalized target loci (Dahlin and Guarnieri 2016). At present, knowledge of algal pathways and algal response at the molecular level is insufficient whereas, only certain algal species possess genetic toolboxes, contributing to lesser sequenced species and databases than other microorganisms (Park et al. 2019). The complete sequencing of the algal genome remains of significant interest. Some algal genomes have been completely sequenced, whereas others are still being worked on. This will not solely aid in understanding the structure of the algal genome; however, it will additionally expedite algal research and engineering products, leading to the tailoring of ideal strains and a decline in dependability for fossil fuels. Nonetheless, the genetic manipulation of algae has negative effects and risks, including the relationship of these engineered species with natural ones and the emergence of unidentified species. This necessitates caution by appropriate ecological evaluation with the goal to avoid synthetically produced algae from evolving to combat biological traits aimed at eliminating the wild algae (Fayyaz et al. 2020).

19.9 Conclusion

Microalgae have tremendous possibilities as a renewable and secure biological manufacturing system for economically beneficial molecules. Algal products must be commercially viable and harnessed to deal with diminishing reserves and rising populations. Genome editing and modification is a critical step toward establishing the algal marketplace while ensuring that it is environmentally and financially sustainable. Genetic modification may manipulate the algal genome to enhance lipid generation, carbon fixation, pigment biosynthesis, and other beneficial by-products. Algomics and genome sequencing initiatives will assist in accomplishing objectives and target aspects in biomass yields with more precise and effective methods. The integration of microalgal multi-omics datasets and sophisticated molecular methods provides an efficient and predictable strategy for strain improvement. While the advancement of genetic machinery in algae remains beneath that of yeast and bacteria, new findings are significantly broadening the molecular toolbox and minimizing the engineering steps, enabling accelerated research and inventions in the subject while addressing financial hurdles associated with scaling up.

References

Ajjawi I, Verruto J, Aqui M, Soriaga LB, Coppersmith J, Kwok K, Peach L, Orchard E, Kalb R, Xu W, Carlson TJ (2017) Lipid production in *Nannochloropsis gaditana* is doubled by decreasing expression of a single transcriptional regulator. Nat Biotechnol 35(7):647–652

Al-Saedi RKM, Najem AM, Abed İJ (2019) Impact of the chemical mutagenesis on biomass and productivity of some micro green algae. Eurasia Proc Sci Technol Eng Math 6:88–95

Anand V, Kashyap M, Samadhiya K, Kiran B (2019) Strategies to unlock lipid production improvement in algae. Int J Environ Sci Technol 16:1829–1838

Badary A, Takamatsu S, Nakajima M, Ferri S, Lindblad P, Sode K (2018) Glycogen production in marine cyanobacterial strain *Synechococcus* sp. NKBG 15041c. Mar Biotechnol 20:109–117

Baek K, Yu J, Jeong J, Sim SJ, Bae S, Jin E (2018) Photoautotrophic production of macular pigment in a *Chlamydomonas reinhardtii* strain generated by using DNA-free CRISPR-Cas9 RNP-mediated mutagenesis. Biotechnol Bioeng 115(3):719–728

Bajhaiya AK, Moreira JZ, Pittman JK (2017) Transcriptional engineering of microalgae: prospects for high-value chemicals. Trends Biotechnol 35(2):95–99

Ban S, Lin W, Luo Z, Luo J (2019) Improving hydrogen production of *Chlamydomonas reinhardtii* by reducing chlorophyll content via atmospheric and room temperature plasma. Bioresour Technol 275:425–429

Biot-Pelletier D, Martin VJ (2014) Evolutionary engineering by genome shuffling. Appl Microbiol Biotechnol 98:3877–3887

Boettcher M, McManus MT (2015) Choosing the right tool for the job: RNAi, TALEN, or CRISPR. Mol Cell 58(4):575–585

Bourbon E, Science A (2021) Clean cities alternative fuel price report. https://afdc.energy.gov/files/u/publication/alternative_fuel_price_report_october_2021.pdf. Accessed 20 Dec 2023

Celińska E, Ledesma-Amaro R, Larroude M, Rossignol T, Pauthenier C, Nicaud JM (2017) Golden Gate Assembly system dedicated to complex pathway manipulation in *Yarrowia lipolytica*. Microb Biotechnol 10(2):450–455

Chen JW, Liu WJ, Hu DX, Wang X, Balamurugan S, Alimujiang A, Yang WD, Liu JS, Li HY (2017) Identification of a malonyl CoA-acyl carrier protein transacylase and its regulatory role in fatty acid biosynthesis in oleaginous microalga *Nannochloropsis oceanica*. Biotechnol Appl Biochem 64(5):620–626

Chungjatupornchai W, Fa-Aroonsawat S (2021) Enhanced triacylglycerol production in oleaginous microalga *Neochloris oleoabundans* by co-overexpression of lipogenic genes: plastidial LPAAT1 and ER-located DGAT2. J Biosci Bioeng 131(2):124–130

Crozet P, Navarro FJ, Willmund F, Mehrshahi P, Bakowski K, Lauersen KJ, Pérez-Pérez ME, Auroy P, Gorchs Rovira A, Sauret-Gueto S, Niemeyer J (2018) Birth of a photosynthetic chassis: a MoClo toolkit enabling synthetic biology in the microalga *Chlamydomonas reinhardtii*. ACS Synth Biol 7(9):2074–2086

Daboussi F, Leduc S, Maréchal A, Dubois G, Guyot V, Perez-Michaut C, Amato A, Falciatore A, Juillerat A, Beurdeley M, Voytas DF (2014) Genome engineering empowers the diatom *Phaeodactylum tricornutum* for biotechnology. Nat Commun 5(1):3831

Dahlin LR, Guarnieri MT (2016) Recent advances in algal genetic tool development. Curr Biotechnol 5(3):192–197

Dasan YK, Lam MK, Yusup S, Lim JW, Lee KT (2019) Life cycle evaluation of microalgae biofuels production: effect of cultivation system on energy, carbon emission and cost balance analysis. Sci Total Environ 688:112–128

Dasan YK, Lam MK, Yusup S, Lim JW, Lee KT, Show PL, Tan IS, Foo HCY (2021) Cultivation of *Chlorella vulgaris* in sequential flow photobioreactor system: influence of recycled culture medium on growth, lipid and protein content. IOP Conf Ser Earth Environ Sci 721(1):012013

Diaz CJ, Douglas KJ, Kang K, Kolarik AL, Malinovski R, Torres-Tiji Y, Molino JV, Badary A, Mayfield SP (2023) Developing algae as a sustainable food source. Front Nutr 9:3147

Durai S, Mani M, Kandavelou K, Wu J, Porteus MH, Chandrasegaran S (2005) Zinc finger nucleases: custom-designed molecular scissors for genome engineering of plant and mammalian cells. Nucleic Acids Res 33(18):5978–5990

Echeverri D, Romo J, Giraldo N, Atehortúa L (2019) Microalgae protoplasts isolation and fusion for biotechnology research. Rev Colomb Biotecnol 21(1):101–112

Engler C, Youles M, Gruetzner R, Ehnert TM, Werner S, Jones JD, Patron NJ, Marillonnet S (2014) A golden gate modular cloning toolbox for plants. ACS Synth Biol 3(11):839–843

Fajardo C, De Donato M, Carrasco R, Martínez-Rodríguez G, Mancera JM, Fernández-Acero FJ (2020) Advances and challenges in genetic engineering of microalgae. Rev Aquac 12(1):365–381

Fan Y, Yuan C, Jin Y, Hu GR, Li FL (2018) Characterization of 3-ketoacyl-coA synthase in a nervonic acid producing oleaginous microalgae *Mychonastes afer*. Algal Res 31:225–231

Fayyaz M, Chew KW, Show PL, Ling TC, Ng IS, Chang JS (2020) Genetic engineering of microalgae for enhanced biorefinery capabilities. Biotechnol Adv 43:107554

Feng Z, Zhang B, Ding W, Liu X, Yang DL, Wei P, Cao F, Zhu S, Zhang F, Mao Y, Zhu JK (2013) Efficient genome editing in plants using a CRISPR/Cas system. Cell Res 23(10):1229–1232

Ferenczi A, Pyott DE, Xipnitou A, Molnar A (2017) Efficient targeted DNA editing and replacement in *Chlamydomonas reinhardtii* using Cpf1 ribonucleoproteins and single-stranded DNA. Proc Natl Acad Sci 114(51):13567–13572

Fire A, Xu S, Montgomery MK, Kostas SA, Driver SE, Mello CC (1998) Potent and specific genetic interference by double-stranded RNA in *Caenorhabditis elegans*. Nature 391(6669):806–811

Fu W, Chaiboonchoe A, Khraiwesh B, Nelson DR, Al-Khairy D, Mystikou A, Alzahmi A, Salehi-Ashtiani K (2016) Algal cell factories: approaches, applications, and potentials. Mar Drugs 14(12):225

Gaj T, Gersbach CA, Barbas CF (2013) ZFN, TALEN, and CRISPR/Cas-based methods for genome engineering. Trends Biotechnol 31(7):397–405

Galarza JI, Gimpel JA, Rojas V, Arredondo-Vega BO, Henríquez V (2018) Over-accumulation of astaxanthin in *Haematococcus pluvialis* through chloroplast genetic engineering. Algal Res 31:291–297

Gan Q, Jiang J, Han X, Wang S, Lu Y (2018) Engineering the chloroplast genome of oleaginous marine microalga *Nannochloropsis oceanica*. Front Plant Sci 9:439

Gasiunas G, Barrangou R, Horvath P, Siksnys V (2012) Cas9–crRNA ribonucleoprotein complex mediates specific DNA cleavage for adaptive immunity in bacteria. Proc Natl Acad Sci 109(39):E2579–E2586

Gopal GJ, Kumar A (2013) Strategies for the production of recombinant protein in *Escherichia coli*. Protein J 32:419–425

Goswami RK, Agrawal K, Mehariya S, Molino A, Musmarra D, Verma P (2020) Microalgae-based biorefinery for utilization of carbon dioxide for production of valuable bioproducts. In: Chemo-biological systems for CO2 utilization. CRC Press, Boca Raton, pp 203–228

Greiner A, Kelterborn S, Evers H, Kreimer G, Sizova I, Hegemann P (2017) Targeting of photoreceptor genes in *Chlamydomonas reinhardtii* via zinc-finger nucleases and CRISPR/Cas9. Plant Cell 29(10):2498–2518

Haslam RP, Hamilton ML, Economou CK, Smith R, Hassall KL, Napier JA, Sayanova O (2020) Overexpression of an endogenous type 2 diacylglycerol acyltransferase in the marine diatom *Phaeodactylum tricornutum* enhances lipid production and omega-3 long-chain polyunsaturated fatty acid content. Biotechnol Biofuels 13:1–17

Hannon M, Gimpel J, Tran M, Rasala B, Mayfield S (2010) Biofuels from algae: challenges and potential. Biofuels 1(5):763–784

Henley WJ, Litaker RW, Novoveská L, Duke CS, Quemada HD, Sayre RT (2013) Initial risk assessment of genetically modified (GM) microalgae for commodity-scale biofuel cultivation. Algal Res 2(1):66–77

Hou S, Brenes-Álvarez M, Reimann V, Alkhnbashi OS, Backofen R, Muro-Pastor AM, Hess WR (2019) CRISPR-Cas systems in multicellular cyanobacteria. RNA Biol 16(4):518–529

Huang W, Lin Y, He M, Gong Y, Huang J (2018) Induced high-yield production of zeaxanthin, lutein, and β-carotene by a mutant of *Chlorella zofingiensis*. J Agric Food Chem 66(4):891–897

Isaacs FJ, Dwyer DJ, Ding C, Pervouchine DD, Cantor CR, Collins JJ (2004) Engineered riboregulators enable post-transcriptional control of gene expression. Nat Biotechnol 22(7):841–847

Jankele R, Svoboda P (2014) TAL effectors: tools for DNA targeting. Brief Funct Genomics 13(5):409–419

Jeon S, Lim JM, Lee HG, Shin SE, Kang NK, Park YI, Oh HM, Jeong WJ, Jeong BR, Chang YK (2017) Current status and perspectives of genome editing technology for microalgae. Biotechnol Biofuels 10(1):1–18

Jia B, Yin J, Li X, Li Y, Yang X, Lan C, Huang Y (2022) Increased lipids in Chlamydomonas reinhardtii by multiple regulations of DOF, LACS2, and CIS1. Int J Mol Sci 23(17):10176

Jiang WZ, Weeks DP (2017) A gene-within-a-gene Cas9/sgRNA hybrid construct enables gene editing and gene replacement strategies in *Chlamydomonas reinhardtii*. Algal Res 26:474–480

Jiang W, Brueggeman AJ, Horken KM, Plucinak TM, Weeks DP (2014) Successful transient expression of Cas9 and single guide RNA genes in *Chlamydomonas reinhardtii*. Eukaryot Cell 13(11):1465–1469

Jiang WZ, Dumm S, Knuth ME, Sanders SL, Weeks DP (2017) Precise oligonucleotide-directed mutagenesis of the *Chlamydomonas reinhardtii* genome. Plant Cell Rep 36:1001–1004

Jinek M, Chylinski K, Fonfara I, Hauer M, Doudna JA, Charpentier E (2012) A programmable dual-RNA-guided DNA endonuclease in adaptive bacterial immunity. Science 337(6096):816–821

Jinkerson RE, Jonikas MC (2015) Molecular techniques to interrogate and edit the *Chlamydomonas* nuclear genome. Plant J 82(3):393–412

Kamravamanesh D, Slouka C, Limbeck A, Lackner M, Herwig C (2019) Increased carbohydrate production from carbon dioxide in randomly mutated cells of cyanobacterial strain *Synechocystis* sp. PCC 6714: bioprocess understanding and evaluation of productivities. Bioresour Technol 273:277–287

Kao PH, Ng IS (2017) CRISPRi mediated phosphoenolpyruvate carboxylase regulation to enhance the production of lipid in *Chlamydomonas reinhardtii*. Bioresour Technol 245:1527–1537

Kasai Y, Tsukahara T, Ikeda F, Ide Y, Harayama S (2018) Metabolic engineering using iterative self-cloning to improve lipid productivity in *Coccomyxa*. Sci Rep 8(1):11742

Ketting RF (2011) The many faces of RNAi. Dev Cell 20(2):148–161

Kim YG, Cha J, Chandrasegaran S (1996) Hybrid restriction enzymes: zinc finger fusions to Fok I cleavage domain. Proc Natl Acad Sci 93(3):1156–1160

Kim J, Kim M, Lee S, Jin E (2020) Development of a *Chlorella vulgaris* mutant by chemical mutagenesis as a producer for natural violaxanthin. Algal Res 46:101790

Koblenz B, Lechtreck KF (2005) The NIT1 promoter allows inducible and reversible silencing of centrin in *Chlamydomonas reinhardtii*. Eukaryot Cell 4(11):1959–1962

Koussa J, Chaiboonchoe A, Salehi-Ashtiani K (2014) Computational approaches for microalgal biofuel optimization: a review. Biomed Res Int 2014:649453

Kumar SD, Sojin K, Santhanam P, Dhanalakshmi B, Latha S, Park MS, Kim MK (2018) Triggering of fatty acids on *Tetraselmis* sp. by ethyl methanesulfonate mutagenic treatment. Bioresour Technol Rep 2:21–28

Kusumaningrum H, Zainuri M (2018) Improvement of nutrition production by protoplast fusion techniques in *Chlorella vulgaris*. J Food Process Technol 9(1):711. ISSN: 2157-7110

Kwon S, Kang NK, Koh HG, Shin SE, Lee B, Jeong BR, Chang YK (2018) Enhancement of biomass and lipid productivity by overexpression of a bZIP transcription factor in *Nannochloropsis salina*. Biotechnol Bioeng 115(2):331–340

Lee H, Shin WS, Kim YU, Jeon S, Kim M, Kang NK, Chang YK (2020) Enhancement of lipid production under heterotrophic conditions by overexpression of an endogenous bZIP transcription factor in *Chlorella* sp. HS2. J Microbiol Biotechnol 30(10):1597

Levitan O, Dinamarca J, Zelzion E, Lun DS, Guerra LT, Kim MK, Kim J, Van Mooy BA, Bhattacharya D, Falkowski PG (2015) Remodeling of intermediate metabolism in the diatom *Phaeodactylum tricornutum* under nitrogen stress. Proc Natl Acad Sci 112(2):412–417

Li DW, Cen SY, Liu YH, Balamurugan S, Zheng XY, Alimujiang A, Yang WD, Liu JS, Li HY (2016a) A type 2 diacylglycerol acyltransferase accelerates the triacylglycerol biosynthesis in heterokont oleaginous microalga *Nannochloropsis oceanica*. J Biotechnol 229:65–71

Li H, Shen CR, Huang CH, Sung LY, Wu MY, Hu YC (2016b) CRISPR-Cas9 for the genome engineering of cyanobacteria and succinate production. Metab Eng 38:293–302

Li Z, Meng T, Ling X, Li J, Zheng C, Shi Y, Chen Z, Li Z, Li Q, Lu Y, He N (2018) Overexpression of malonyl-CoA: ACP transacylase in *Schizochytrium* sp. to improve polyunsaturated fatty acid production. J Agric Food Chem 66(21):5382–5391

Li DW, Balamurugan S, Yang YF, Zheng JW, Huang D, Zou LG, Yang WD, Liu JS, Guan Y, Li HY (2019) Transcriptional regulation of microalgae for concurrent lipid overproduction and secretion. Sci Adv 5(1):eaau3795

Liu Y, Cui Y, Chen J, Qin S, Chen G (2019) Metabolic engineering of *Synechocystis* sp. PCC6803 to produce astaxanthin. Algal Res 44:101679

Ludwig M, Bryant DA (2012) Acclimation of the global transcriptome of the cyanobacterium *Synechococcus* sp. strain PCC 7002 to nutrient limitations and different nitrogen sources. Front Microbiol 3:145

Ma Y, Wang Z, Yu C, Yin Y, Zhou G (2014) Evaluation of the potential of 9 *Nannochloropsis* strains for biodiesel production. Bioresour Technol 167:503–509

Ma XN, Chen TP, Yang B, Liu J, Chen F (2016) Lipid production from *Nannochloropsis*. Mar Drugs 14(4):61

Maroušek J, Gavurová B, Strunecký O, Maroušková A, Sekar M, Marek V (2023) Techno-economic identification of production factors threatening the competitiveness of algae biodiesel. Fuel 344:128056

Martella A, Matjusaitis M, Auxillos J, Pollard SM, Cai Y (2017) EMMA: an extensible mammalian modular assembly toolkit for the rapid design and production of diverse expression vectors. ACS Synth Biol 6(7):1380–1392

Medeiros DL, Moreira ÍTA (2022) Microalgae biomass production from cultivation in availability and limitation of nutrients: the technical, environmental and economic performance. J Clean Prod 370:133538

Mini P, Demurtas OC, Valentini S, Pallara P, Aprea G, Ferrante P, Giuliano G (2018) Agrobacterium-mediated and electroporation-mediated transformation of *Chlamydomonas reinhardtii*: a comparative study. BMC Biotechnol 18(1):1–12

Misson LE, Mindrebo JT, Davis TD, Patel A, McCammon JA, Noel JP, Burkart MD (2020) Interfacial plasticity facilitates high reaction rate of *E. coli* FAS malonyl-CoA: ACP transacylase, FabD. Proc Natl Acad Sci 117(39):24224–24233

Moore SJ, Lai HE, Kelwick RJ, Chee SM, Bell DJ, Polizzi KM, Freemont PS (2016) EcoFlex: a multifunctional MoClo kit for E. coli synthetic biology. ACS Synth Biol 5(10):1059–1069

Muñoz CF, de Jaeger L, Sturme MH, Lip KY, Olijslager JW, Springer J, Wolbert EJ, Martens DE, Eggink G, Weusthuis RA, Wijffels RH (2018) Improved DNA/protein delivery in microalgae—a simple and reliable method for the prediction of optimal electroporation settings. Algal Res 33:448–455

Muñoz CF, Weusthuis RA, D'Adamo S, Wijffels RH (2019) Effect of single and combined expression of lysophosphatidic acid acyltransferase, glycerol-3-phosphate acyltransferase, and diacylglycerol acyltransferase on lipid accumulation and composition in *Neochloris oleoabundans*. Front Plant Sci 10:1573

Naduthodi MIS, Mohanraju P, Südfeld C, D'Adamo S, Barbosa MJ, Van Der Oost J (2019) CRISPR–Cas ribonucleoprotein mediated homology-directed repair for efficient targeted genome editing in microalgae *Nannochloropsis oceanica* IMET1. Biotechnol Biofuels 12:1–11

Naghshbandi MP, Tabatabaei M, Aghbashlo M, Aftab MN, Iqbal I (2020) Metabolic engineering of microalgae for biofuel production. In: Biofuels from algae: methods and protocols. Springer, Berlin, pp 153–172

Nekrasov V, Staskawicz B, Weigel D, Jones JD, Kamoun S (2013) Targeted mutagenesis in the model plant *Nicotiana benthamiana* using Cas9 RNA-guided endonuclease. Nat Biotechnol 31(8):691–693

Nemudryi AA, Valetdinova KR, Medvedev SP, Zakian, SÁ (2014) TALEN and CRISPR/Cas genome editing systems: tools of discovery. Acta Naturae (англоязычная версия) 6(3(22)):19–40

Neupert J, Karcher D, Bock R (2009) Generation of *Chlamydomonas* strains that efficiently express nuclear transgenes. Plant J 57(6):1140–1150

Ng AH, Berla BM, Pakrasi HB (2015) Fine-tuning of photoautotrophic protein production by combining promoters and neutral sites in the cyanobacterium *Synechocystis* sp. strain PCC 6803. Appl Environ Microbiol 81(19):6857–6863

Ng IS, Tan SI, Kao PH, Chang YK, Chang JS (2017) Recent developments on genetic engineering of microalgae for biofuels and bio-based chemicals. Biotechnol J 12(10):1600644

Norashikin MN, Loh SH, Aziz A, San Cha T (2018) Metabolic engineering of fatty acid biosynthesis in Chlorella vulgaris using an endogenous omega-3 fatty acid desaturase gene with its promoter. Algal Res 31:262–275

Novoveská L, Nielsen SL, Eroldoğan OT, Haznedaroglu BZ, Rinkevich B, Fazi S, Robbens J, Vasquez M, Einarsson H (2023) Overview and challenges of large-scale cultivation of photosynthetic microalgae and cyanobacteria. Mar Drugs 21(8):445

Oostlander PC, van Houcke J, Wijffels RH, Barbosa MJ (2020) Microalgae production cost in aquaculture hatcheries. Aquaculture 525:735310

Osanai T, Imamura S, Asayama M, Shirai M, Suzuki I, Murata N, Tanaka K (2006) Nitrogen induction of sugar catabolic gene expression in *Synechocystis* sp. PCC 6803. DNA Res 13(5):185–195

Park S, Nguyen THT, Jin E (2019) Improving lipid production by strain development in microalgae: strategies, challenges and perspectives. Bioresour Technol 292:121953

Patel B, Tamburic B, Zemichael FW, Dechatiwongse P, Hellgardt K (2012) Biofuels: a credible prospective? ISRN Renew Energy 2012:1. https://doi.org/10.5402/2012/631574

Perozeni F, Cazzaniga S, Baier T, Zanoni F, Zoccatelli G, Lauersen KJ, Wobbe L, Ballottari M (2020a) Turning a green alga red: engineering astaxanthin biosynthesis by intragenic pseudogene revival in *Chlamydomonas reinhardtii*. Plant Biotechnol J 18(10):2053–2067

Perozeni F, Cazzaniga S, Baier T, Zanoni F, Zoccatelli G, Lauersen KJ, Wobbe L, Ballottari M (2020b) A high-capacity gene stacking toolkit for the oleaginous microalga, *Nannochloropsis oceanica* CCMP1779. Algal Res 45:101664

Poliner E, Pulman JA, Zienkiewicz K, Childs K, Benning C, Farré EM (2018) A toolkit for *Nannochloropsis oceanica* CCMP 1779 enables gene stacking and genetic engineering of the eicosapentaenoic acid pathway for enhanced long-chain polyunsaturated fatty acid production. Plant Biotechnol J 16(1):298–309

Poliner E, Cummings C, Newton L, Farré EM (2019) Identification of circadian rhythms in *Nannochloropsis* species using bioluminescence reporter lines. Plant J 99(1):112–127

Poliner E, Clark E, Cummings C, Benning C, Farre EM (2020) A high-capacity gene stacking toolkit for the oleaginous microalga, Nannochloropsis oceanica CCMP1779. Algal Res 45:101664 https://doi.org/10.1016/j.algal.2019.101664

Price B, Ayala-Martinez A, Havill N, Zintel M, Flores A, Balzani E, Kopf S, Oyler J, Do T (2022) Barriers to accessibility of algal biofuels. Intersect 15(2)

Pugazhendhi A, Nagappan S, Bhosale RR, Tsai PC, Natarajan S, Devendran S, Al-Haj L, Ponnusamy VK, Kumar G (2020) Various potential techniques to reduce the water footprint of microalgal biomass production for biofuel—a review. Sci Total Environ 749:142218

Rengel R, Smith RT, Haslam RP, Sayanova O, Vila M, Leon R (2018) Overexpression of acetyl-CoA synthetase (ACS) enhances the biosynthesis of neutral lipids and starch in the green microalga *Chlamydomonas reinhardtii*. Algal Res 31:183–193

Romero-Campero FJ, Perez-Hurtado I, Lucas-Reina E, Romero JM, Valverde F (2016) ChlamyNET: a Chlamydomonas gene co-expression network reveals global properties of the transcriptome and the early setup of key co-expression patterns in the green lineage. BMC Genomics 17(1):1–28

Sakai Y, Abe K, Nakashima S, Ellinger JJ, Ferri S, Sode K, Ikebukuro K (2015) Scaffold-fused riboregulators for enhanced gene activation in *Synechocystis* sp. PCC 6803. Microbiologyopen 4(4):533–540

Salas-Montantes CJ, González-Ortega O, Ochoa-Alfaro AE, Camarena-Rangel R, Paz-Maldonado LMT, Rosales-Mendoza S, Rocha-Uribe A, Soria-Guerra RE (2018) Lipid accumulation during nitrogen and sulfur starvation in *Chlamydomonas reinhardtii* overexpressing a transcription factor. J Appl Phycol 30:1721–1733

Sarnaik A, Nambissan V, Pandit R, Lali A (2018) Recombinant *Synechococcus elongatus* PCC 7942 for improved zeaxanthin production under natural light conditions. Algal Res 36:139–151

Schiano di Visconte G, Spicer A, Chuck CJ, Allen MJ (2019) The microalgae biorefinery: a perspective on the current status and future opportunities using genetic modification. Appl Sci 9(22):4793

Sengupta S, Jaiswal D, Sengupta A, Shah S, Gadagkar S, Wangikar PP (2020) Metabolic engineering of a fast-growing cyanobacterium *Synechococcus elongatus* PCC 11801 for photoautotrophic production of succinic acid. Biotechnol Biofuels 13:1–18

Seo S, Kim J, Lee JW, Nam O, Chang KS, Jin E (2020) Enhanced pyruvate metabolism in plastids by overexpression of putative plastidial pyruvate transporter in *Phaeodactylum tricornutum*. Biotechnol Biofuels 13(1):1–11

Serif M, Dubois G, Finoux AL, Teste MA, Jallet D, Daboussi F (2018) One-step generation of multiple gene knock-outs in the diatom *Phaeodactylum tricornutum* by DNA-free genome editing. Nat Commun 9(1):3924

Shin YS, Jeong J, Nguyen THT, Kim JYH, Jin E, Sim SJ (2019) Targeted knockout of phospholipase A2 to increase lipid productivity in *Chlamydomonas reinhardtii* for biodiesel production. Bioresour Technol 271:368–374

Sizova I, Greiner A, Awasthi M, Kateriya S, Hegemann P (2013) Nuclear gene targeting in *Chlamydomonas* using engineered zinc-finger nucleases. Plant J 73(5):873–882

Slade R, Bauen A (2013) Micro-algae cultivation for biofuels: cost, energy balance, environmental impacts and future prospects. Biomass Bioenergy 53:29–38

Spicer A, Molnar A (2018) Gene editing of microalgae: scientific progress and regulatory challenges in Europe. Biology 7(1):21

Strecker J, Ladha A, Gardner Z, Schmid-Burgk JL, Makarova KS, Koonin EV, Zhang F (2019) RNA-guided DNA insertion with CRISPR-associated transposases. Science 365(6448):48–53

Südfeld C, Hubáček M, Figueiredo D, Naduthodi MI, Van Der Oost J, Wijffels RH, Barbosa MJ, D'Adamo S (2021) High-throughput insertional mutagenesis reveals novel targets for enhancing lipid accumulation in *Nannochloropsis oceanica*. Metab Eng 66:239–258

Tan KWM, Lee YK (2017) Expression of the heterologous Dunaliella tertiolecta fatty acyl-ACP thioesterase leads to increased lipid production in Chlamydomonas reinhardtii. J Biotechnol 247:60–67

Torres-Tiji Y, Fields FJ, Mayfield SP (2020) Microalgae as a future food source. Biotechnol Adv 41:107536
Trilokesh C, Uppuluri KB (2021) Biobutanol from lignocellulosic biomass and microalgae: scope, technology, and economics. In: Sustainable biofuels. Elsevier, Amsterdam, pp 163–223
Ueno K, Sakai Y, Shono C, Sakamoto I, Tsukakoshi K, Hihara Y, Sode K, Ikebukuro K (2017) Applying a riboregulator as a new chromosomal gene regulation tool for higher glycogen production in *Synechocystis* sp. PCC 6803. Appl Microbiol Biotechnol 101:8465–8474
Urnov FD, Rebar EJ, Holmes MC, Zhang HS, Gregory PD (2010) Genome editing with engineered zinc finger nucleases. Nat Rev Genet 11(9):636–646
Velmurugan N, Deka D (2018) Transformation techniques for metabolic engineering of diatoms and haptophytes: current state and prospects. Appl Microbiol Biotechnol 102:4255–4267
Velmurugan R, Incharoensakdi A (2020) Heterologous expression of ethanol synthesis pathway in glycogen deficient Synechococcus elongatus PCC 7942 resulted in enhanced production of ethanol and exopolysaccharides. Front Plant Sci 11:74
Vidya D, Arunkumar K (2022) Molecular engineering/metabolic engineering-based advanced biotechnological approach in microalgal biorefinery. In: Micro-algae: next-generation feedstock for biorefineries: contemporary technologies and future outlook. Springer, Berlin, pp 145–163
Wang C, Chen X, Li H, Wang J, Hu Z (2017) Artificial miRNA inhibition of phosphoenolpyruvate carboxylase increases fatty acid production in a green microalga *Chlamydomonas reinhardtii*. Biotechnol Biofuels 10(1):1–11
Wang W, Wei T, Fan J, Yi J, Li Y, Wan M, Wang J, Bai W (2018) Repeated mutagenic effects of 60Co-γ irradiation coupled with high-throughput screening improves lipid accumulation in mutant strains of the microalgae *Chlorella pyrenoidosa* as a feedstock for bioenergy. Algal Res 33:71–77
Wang X, Liu SF, Li RY, Yang WD, Liu JS, Lin CSK, Balamurugan S, Li HY (2020) TAG pathway engineering via GPAT2 concurrently potentiates abiotic stress tolerance and oleaginicity in *Phaeodactylum tricornutum*. Biotechnol Biofuels 13(1):1–14
Weiner I, Atar S, Schweitzer S, Eilenberg H, Feldman Y, Avitan M, Blau M, Danon A, Tuller T, Yacoby I (2018) Enhancing heterologous expression in *Chlamydomonas reinhardtii* by transcript sequence optimization. Plant J 94(1):22–31
Weiner I, Feldman Y, Shahar N, Yacoby I, Tuller T (2020) CSO–a sequence optimization software for engineering chloroplast expression in *Chlamydomonas reinhardtii*. Algal Res 46:101788
Wendt KE, Ungerer J, Cobb RE, Zhao H, Pakrasi HB (2016) CRISPR/Cas9 mediated targeted mutagenesis of the fast growing cyanobacterium *Synechococcus elongatus* UTEX 2973. Microb Cell Factories 15:1–8
Weyman PD, Beeri K, Lefebvre SC, Rivera J, McCarthy JK, Heuberger AL, Peers G, Allen AE, Dupont CL (2015) Inactivation of P haeodactylum tricornutum urease gene using transcription activator-like effector nuclease-based targeted mutagenesis. Plant Biotechnol J 13(4):460–470
Yamano T, Iguchi H, Fukuzawa H (2013) Rapid transformation of Chlamydomonas reinhardtii without cell-wall removal. J Biosci Bioeng 115(6):691–694
Yan J, Kuang Y, Gui X, Han X, Yan Y (2019) Engineering a malic enzyme to enhance lipid accumulation in Chlorella protothecoides and direct production of biodiesel from the microalgal biomass. Biomass Bioenergy 122:298–304
Yao L, Cengic I, Anfelt J, Hudson EP (2016) Multiple gene repression in cyanobacteria using CRISPRi. ACS Synth Biol 5(3):207–212
Yusuf M, Shabbir M, Mohammad F (2017) Natural colorants: historical, processing and sustainable prospects. Nat Prod Bioprospect 7:123–145
Zou LG, Chen JW, Zheng DL, Balamurugan S, Li DW, Yang WD, Liu JS, Li HY (2018) High-efficiency promoter-driven coordinated regulation of multiple metabolic nodes elevates lipid accumulation in the model microalga *Phaeodactylum tricornutum*. Microb Cell Factories 17:1–8

GPSR Compliance

The European Union's (EU) General Product Safety Regulation (GPSR) is a set of rules that requires consumer products to be safe and our obligations to ensure this.

If you have any concerns about our products, you can contact us on ProductSafety@springernature.com

In case Publisher is established outside the EU, the EU authorized representative is:

Springer Nature Customer Service Center GmbH
Europaplatz 3
69115 Heidelberg, Germany

The manufacturer's authorised representative in the EU is Springer Nature Customer Service Centre GmbH, Europaplatz 3, 69115 Heidelberg, Germany. If you have any concerns regarding our products, please contact ProductSafety@springernature.com

Printed and bound by CPI Group (UK) Ltd, Croydon, CR0 4YY
01/10/2026
02225863-0018